LYaPAS:

A PROGRAMMING LANGUAGE FOR LOGIC AND CODING ALGORITHMS

ACM MONOGRAPH SERIES

Published under the auspices of the Association for Computing Machinery Inc.

Edited by ROBERT L. ASHENHURST *The University of Chicago*

A. FINERMAN (Ed.) University Education in Computing Science, 1968

A. GINZBURG Algebraic Theory of Automata, 1968

E. F. CODD Cellular Automata, 1968

G. ERNST AND A. NEWELL GPS: A Case Study in Generality and Problem Solving, 1969

M. A. GAVRILOV AND A. D. ZAKREVSKII (Eds.) LYaPAS: A Programming Language for Logic and Coding Algorithms, 1969

Previously published and available from The Macmillan Company, New York City

G. SEBESTYN Decision Making Processes in Pattern Recognition, 1963

M. YOVITS (Ed.) Large Capacity Memory Techniques for Computing Systems, 1962

V. KRYLOV Approximate Calculation of Integrals (Translated by A. H. Stroud), 1962

LYaPAS: A PROGRAMMING LANGUAGE FOR LOGIC AND CODING ALGORITHMS

Edited by M. A. GAVRILOV *and* A. D. ZAKREVSKII

Academy of Sciences
Moscow, USSR

Translated by Morton Nadler

BULL-General Electric
Paris, France

1969

ACADEMIC PRESS New York and London

ACADEMIC PRESS, INC.
111 Fifth Avenue, New York, New York 10003

United Kingdom Edition published by
ACADEMIC PRESS, INC. (LONDON) LTD.
Berkeley Square House, London W1X6BA

LIBRARY OF CONGRESS CATALOG CARD NUMBER: 69-18353

PRINTED IN THE UNITED STATES OF AMERICA

First published in the Russian language under the title
Logicheskiy Iazyk dlia Predstavlienya Algoritmov Sinteza
Releinykh Ustroistv. Izdatelstro "Nauka," Moscow, 1966.

TRANSLATOR'S FOREWORD

For many years I have been looking for a language which would permit logical design and simulation problems to be compactly, simply, and rapidly programmed, and which at the same time would be associated with a system which would yield highly efficient object programs, to answer a need arising out of my own work in optical reading machine research and development at BULL-General Electric. Stopgap systems were developed by programmers working for me, but neither they nor people in programming research and development in the company ever had the leisure or the facilities to undertake a more ambitious solution to the problem. Nor did anything I ever came across in the literature suggest that the problem had been solved definitively elsewhere, although many were seeking a solution, and many suggestive candidate systems had been put forward. Perhaps the nearest hit has been Iverson's language, but this also presents a certain number of disabling disadvantages for what I had in mind.

Through my work for the English translation of the Soviet *Cybernetics Abstracts Journal* I came to be aware of the existence of LYaPAS and the potential interest that it would present. Through the kind assistance of Mr. P. P. Parkhomenko, who visited BULL-General Electric, I eventually obtained a copy of the book and immediately realized that this was almost the ideal answer to my needs. Considering that these needs were representative of problems in advanced development and design automation in the construction of logical machines, I proposed a translation of the book to Professor Ashenhurst for the ACM monograph series, with the result that you are today reading these lines.

A. D. Zakrevskii remarks that the language is largely inspired by Iverson

(p. 8). One great practical inconvenience of Iverson's notation has been removed, namely, the two-dimensional arrangement of symbols, which are replaced in LYaPAS by a linear arrangement, better suited for practical utilization with present-day keyboards. Likewise, having set themselves the much more limited goal of a "Language for the Representation of Synthesis Algorithms" rather than a universal language, LYaPAS is not diluted with a number of operators considered to be extraneous to the problem.

The starting point for the design of LYaPAS was set theory; its highly dense and compact symbolism surely presents no difficulties to the mathematician or other scientist versed in that or related branches of mathematics. It was also designed for machines with relatively small memories—the typical Soviet machine of the late 1960's has 4096 words, although more generously endowed with bits (at least 32) than the typical Western machines. It is also heavily influenced by the fact that Soviet computers are almost exclusively endowed with purely numeric inputs and outputs (and octal at that).

Under these conditions there was very little incentive or encouragement to design the language around the type of mnemonic syllables that we find in the most widely used Western languages, FORTRAN, ALGOL, and COBOL (the last two of which are used in the Soviet Union). In LYaPAS one symbol is given by three octal digits and represents one fundamental LYaPAS entity, either an operator, an operand, or a modifier thereof, or a delimiter. The result is, of course, an enormous density of expression. Note the Cherry–Vaswani algorithm (p. 199) for reduced search in the Boolean problems, which requires only 84 symbols in its LYaPAS representation, with one additional subroutine of 56 symbols, or a total of 137 symbols! In fact LYaPAS has permitted its users to compile and run problems which would tax known systems currently used in the West on modern machines (except, of course, for supermachines like the GE 600 series or the CDC 6000 series). It is even used with the URAL–1 (1024 word drum memory) for LYaPAS debugging runs, if not for useful work.

In view of these remarkable characteristics, the language immediately suggests itself as a candidate for international standardization, in association with the three general purpose languages mentioned above. It seems to me, however, that to become really convenient for the average engineer in the fields for which it is intended, its external aspect would have to be modified somewhat, by permitting the majority of symbols to have an alternative mnemotechnic syllabic symbol, as in the three existing international languages. I hope that the appropriate bodies in ACM, AFIPS, and IFIP, with the collaboration of the founders of the language, will be able to achieve this goal.

This project would be all the more attractive in that the present volume offers a large selection of synthesis and related algorithms in their "publication" language, LYaPAS.

* * *

In the translation I was faced with several difficult decisions, not ordinarily required of the translator, even of scientific and mathematical literature. One of these decisions had to do with the replacement of Cyrillic letter symbols by Latin, Greek, or others, commonly used in the Western literature. Precisely because I conceive of LYaPAS as a "publication" language, I have adopted the principle of retaining the Cyrillic letters that are LYaPAS symbols, whereas the Cyrillic letters in metalanguage (used to talk about LYaPAS) were transliterated. In this connection, the italicized acronyms representing the name of a subroutine appearing in a LYaPAS program were translated, since they are not, as the reader will easily determine, LYaPAS symbols. For the convenience of the reader, an index of these names is given, together with their Cyrillic equivalents and LYaPAS call numbers.

Although, with the help of Dr. Zakrevskii, practically all of the typographical errors remaining in the original Russian edition have been eliminated, it is inevitable that in a work of this typographical complexity new errors will have crept in. For this, I am alone responsible and I ask the reader's pardon in advance.

This book has the interesting feature of presenting all programs in LYaPAS, even those parts of the system, such as the translator, which are necessarily written in machine (or assembly) language in a concrete realization. The original did present, in addition, some twenty-odd pages of URAL–1 machine code for the translator (TRALU) and the debugger (OPLU); these have been omitted. They are available to the interested reader in the original, and as they are purely numeric in presentation, do not require the intervention of the (natural language) translator.

I would like to take this occasion to express my thanks to Mr. Parkhomenko and Dr. Zakrevskii for their assistance in the form of original documents and texts concerning LYaPAS, and in particular, Dr. Zakrevskii and the members of his team for reading and painstakingly correcting the translation manuscript; to Professor Ashenhurst, for including this volume in the ACM monograph series; and to BULL-General Electric for certain facilities for producing the manuscript, particularly the time-sharing program for sorting the various index entries. (This program was prepared by my close collaborator and good friend M. André Oisel and manipulated by Mme. A. Fruchart.)

MORTON NADLER

Paris
August 1969

PREFACE TO THE RUSSIAN EDITION

One of the basic divisions of modern automatic control theory is the theory of switching circuits and finite automata. Its origin and development are connected with the study of the properties of and methods of constructing discrete computer, control, and monitoring devices, which are finding wide application both in the automatic control of production and other processes and in the automation of mental work (automatic computing and theorem proving, optimization of schedules, recognition of situations, etc.).

The present state of the theory of switching circuits and finite automata can be characterized, on the one hand by the intrinsic complexity of the problems considered, and on the other by the ever increasing connection of these problems with problems arising in other fields of contemporary cybernetics (coding theory, reliability theory, theory of experiments, programming theory, etc.).

The fundamental problem of switching circuit theory is the synthesis problem, which can be broken down in each concrete case into some set of other, more restricted problems. As a rule, these problems are characterized by the need to inspect many variants, which cannot be avoided in the general case but can only be mitigated to some degree by the application of improved algorithms. Therefore for nontrivial initial data the cost of solution of the synthesis problem is measured by an enormous volume of simple computational operations whose manual realization is impossible. This volume can be reduced only by abandoning the search for *optimal* solutions, i.e., by the use of *approximate* algorithms. However even in this case the complexity of the problems which can be solved will be substantially limited.

The "ceiling" on synthesis algorithms can be raised by *automating* their realization.

There exists another reason for the development of automatic synthesis methods which is no less important. At the present time there exists a large gap between theory, which is the framework in which synthesis algorithms are developed, and the engineering practice of construction of real automata, in which the known methods of synthesis are by far not fully utilized. It can be assumed that with further development of theory synthesis algorithms will become more and more perfected and complicated, i.e., less accessible to "manual" application. This growing contradiction can only be resolved by the automation of synthesis.

Problems in the field of synthesis can be solved on general-purpose computers. However the direct programming of problems in machine language requires large expenditures of time for the writing and checking of programs. In the last analysis the programs obtained are suitable only for the machines for which they have been written and are practically inaccessible to human examination.

Programming productivity in the field of synthesis problems can be sharply increased by the development of a special *language* oriented toward the representation of synthesis algorithms and by the development of the corresponding programming system for passage from this language to machine language.

Recently a tendency has been noted to develop, in addition to universal languages of the ALGOL-60 type, languages which are oriented toward the solution of given classes of problems, for example technical, economic, linguistic, etc. Such specialized languages are more efficient and convenient of application. Problems of a logical character, which include synthesis problems, form a broad and interesting class with its own specific nature. This class also requires its own language.

A number of projects have been carried out in this direction at the Siberian Physicotechnical Institute [SPTI] (Tomsk), where the LYaPAS language (the Russian acronym for the expression "logical language for the representation of synthesis algorithms") has been proposed. An automatic programming system has been constructed on this basis, and the possibilities of applying the language to the solution of various types of problems, particularly synthesis problems, have been explored. LYaPAS has been used at the SPTI and the Institute of Automation and Remote Control of the Academy of Sciences of the USSR to solve concrete synthesis problems, to obtain irredundant and minimal disjunctive normal forms of Boolean expressions, functional decompositions of Boolean functions, the synthesis of threshold and, in particular, majority logic circuits, minimization of the number of states of an automaton, the solution of certain problems in coding theory, and others. The results obtained testify to the high efficiency of the language.

In the development of LYaPAS the goal has been the maximum exploitation of the possibilities offered by present-day computers. In particular the following problems were solved: (1) investigation of the specific character of the computation processes involved in the solution of synthesis problems; (2) segmentation of synthesis algorithms and their formulation in the form of abstract operators; (3) raising the efficiency of the programming system by completing it with subroutines for introducing new operators into the language; optimization of these programs according to certain criteria (mainly the speed); (4) the development of programming methods based on the given language.

The development of a language oriented toward facilitating the programming of synthesis problems went beyond the framework of switching theory proper, and set up the preconditions for the successful solution of many other logical problems. Thus LYaPAS has been used for the construction of its own automatic programming system (for example, the compiler and the block for finding syntax errors were completely written in this language). Algorithms for the solution of systems of logical equations, the analysis of the connectedness of graphs, testing of Boolean identities, etc., have all been programmed in LYaPAS.

This book presents the results achieved in this direction; LYaPAS and the corresponding programming system are described, as well as algorithms for the solution of various problems in synthesis theory and certain logical problems outside this framework. The majority of these algorithms have an original character, and they are all presented in LYaPAS.

M. A. Gavrilov
A. D. Zakrevskii

1966

CONTENTS

Section B. The Automatic Programming System

An Algorithm for the Synthesis of Majority-Element Logical Circuits

V. L. Pavlov

Section C: Investigation of Automata

Simulation of Switching Circuits

A. A. Utkin

Algorithms for State Minimization of a Discrete Automaton

Yu. V. Pottosin

PART I: LYaPAS AND ITS PROGRAMMING SYSTEM

Section A: LYaPAS

DESCRIPTION OF LYaPAS

A. D. Zakrevskii

1. GENERAL PRINCIPLES OF LYaPAS

A. PURPOSE OF LYaPAS

The complexity of modern devices for the automatic processing of discrete information has become so great in many cases that the practical design of such devices on the intuitive level leads to enormous expenditures of means, materials, and time in design, and to excessively large and far from perfect solutions. It is therefore necessary to replace intuitive design methods by a rigorous theory of the synthesis of discrete automata.

The theory of synthesis of discrete automata is being intensively developed at the present time, but there are two serious obstacles to its resulting in an engineering tool. The first is that the practical methods developed within the framework of the theory are too complex for the broad circle of engineering designers. The second resides in the laboriousness of many of the well-known algorithms for the solution of practical synthesis problems; in a number of cases the realization of these algorithms is connected with the need to execute millions and billions of elementary computation steps.

Therefore the need arises to automate synthesis processes, according to the following criteria:

(a) the practical application of the system of automated synthesis must be fairly simple;

(b) the range of problems covered by the system must be sufficiently broad;

(c) the use of the system must yield high work productivity;

(d) the use of the system must favor further development of the theory of synthesis;

(e) the system must be sufficiently flexible not to become obsolete as the theory of synthesis develops, and must have the capacity to be perfected on the basis of this development; and

(f) the development of the system in a fairly short time and at relatively low cost must be a realistic goal.

Considering the present development of information processing techniques, it can be stated that the best way to satisfy the above requirements is to orient the solution toward general-purpose digital computers.

Of course, it must be taken into account that present [Soviet] general-purpose computers are specialized in the sense that they are intended principally for operations on numbers. The programming of problems of a logical character, among which are to be found synthesis problems, presents appreciable difficulties and requires special analysis and ingenuity. It can be facilitated by the development of a special language for the representation of algorithms to be used for the solution of logical problems of the type occurring in the synthesis of discrete automata, and by the development on the basis of this language of an automatic programming system for the problems of interest to us.

In the present chapter we describe such a language, called LYaPAS (the Russian acronym for the expression "logical language for the representation of synthesis algorithms"). LYaPAS has been developed in parallel with its programming program. In its development the following fairly clear goals were followed. In particular (even if precise formulations were lacking):

(a) LYaPAS must permit algorithms to be expressed compactly and clearly, so that LYaPAS can serve directly as a publication language;

(b) LYaPAS must utilize to the maximum the facilities of present computers; at the same time it must be freed from the technical details of programming and the characteristics of concrete computers, in order to be a common language for a broad class of computers;

(c) the programming programs corresponding to the language for concrete machines must be fairly compact; they must be able to be completely stored in core, enabling rapid interpretation and compilation; and

(d) the possibility must be offered of realizing all contacts with the computer at the level of LYaPAS; in particular, the checking out process for programs must be reduced to the process of checking out algorithms through the checking out of the LYaPAS programs.

B. TYPES OF WORDS

Expressions in LYaPAS are composed of strings of elements, called *words*.

Each word is the symbol for a definite concept; e.g., a variable, a constant, a certain operation on a variable (represented by an adjacent word in the string), etc. It is precisely the presence of this semantic content in the language elements which in the present case justifies the term "word."

Depending on their semantics, words are divided into types: the basic mass of words are the symbols for *operands* and *operators*, while certain other words play an auxiliary role, expressing the connections among the words in LYaPAS sentences or determining the mode of interpretation of the succeeding symbols.

The basic LYaPAS operands are the *variables*, whose values are represented by 32-bit binary codes and, in the general case, are not interpreted as numbers but as subsets of a certain abstract set of 32 elements. Such variables are *standard*, and they are fairly simply manipulated in modern computers.

The possibility is also provided of operating with compound variables, whose values are $32 \cdot \omega$-bit binary codes, where ω is a certain natural number, and the so-called *complexes*, also playing the role of operands in the language, comprise sets whose elements are analogous to the variables. A special system of indexing the elements of complexes permits the manipulation of both the complexes as a whole or their individual elements. In particular, the indexes of elements of complexes can be prescribed by the values of special standard variables, called *indexes*.

Aside from the variables, LYaPAS operands may be *constants*, composed of 32-bit binary codes.

A special type of constant can be prescribed directly by LYaPAS words, which in this case do not play the role of symbols of some particular 32-bit constants, but define at the right constants (of seven bits) in natural form, the remaining positions being filled by zeros.

This manner of prescribing constants is particularly convenient if these constants are interpreted as natural numbers and used, in particular, to number the sentences in the program.

We shall consider only such strings of LYaPAS words as constitute algorithms for the solution of given problems, and we shall call them L-*programs*, or simply programs, for the solution of these problems.

C. TWO LEVELS OF LYaPAS

LYaPAS has two levels. The *first level* is closer to machine language proper, is simpler, and is intended for the representation of not too complex algorithms; to it there corresponds a more compact programming program. At this level it is possible to manipulate directly such variables as are given by 32-bit binary codes, i.e., simple variables, indexes, and individual elements of complexes.

The operations at this level are represented by a special set of standard *first-level operators*, or l-*operators*, many of which are close to the elementary operations of general-purpose machines, i.e., bitwise disjunction, inversion, shift, count of the number of 1's in the code, etc. A particular role is played by the operator that assigns to one variable the value of another, the operators that increment the values of variables, and others. A small group of operators is used to control the order of execution of the program on the basis of the current information. Procedures are provided for the input of the initial values of the operands, and the output of the results, which are carried out by indexing—the values of the variables are accompanied by their symbols. Provision is made also for automatic checkout of programs (in LYaPAS). This consists in the output of the sequences of values generated during the course of evolution of a program for those variables and indexes which are tagged by a special code, and in the output of the path of realization of the program. A special regulatory mechanism, built into the checkout block, ensures the practical uniformity of the control information supplied over all loops of the checked-out program, independently of the relative configurations of these loops.

At the *second level* of LYaPAS the l-operators are generalized to the case of their application to compound variables (in programming practice this generalization corresponds to the joining of a number of memory registers). A number of the first level operations are also generalized to the case of manipulation of complexes. New operations are introduced. These are primarily the operations of union, intersection, and subtraction of the sets represented by the given complexes; the operations of finding the upper and lower bounds of sets, of finding in a given set subsets of the elements having a certain preassigned property, of finding a minimal interval of a set containing a given set, etc. The second level also provides for a special mode of execution of the operations occurring in the first level. For example, in case of binary operations executed on complexes, they are executed on all pairs composed of one element from each complex; the resultant complex is formed by canceling similar elements in the set of results obtained thereby. A particular place is occupied among the operators of the second level by sifting operators, which are optimized with respect to various concrete situations.

As additional examples we can mention an operator for testing the connectedness of a graph, an operator for optimal coverage of a given set, an operator for minimization of a Boolean function, and an operator for the simplification of a system of logical equations.

The second-level operators, called L-*operators*, are realized by means of suitable subroutines, expressed in LYaPAS. These subroutines may contain the symbols of other L-operators, representing a certain hierarchy of operators, expressed in the last analysis in first-level LYaPAS. LYaPAS can contain a practically unlimited number of L-operators, which thereby enrich the language in the process of its exploitation.

D. THE AUTOMATIC PROGRAMMING SYSTEM

The core of the automatic programming system provided for LYaPAS is the *translator*, which translates the expressions of first-level LYaPAS to the language of a concrete machine. Since the first level of LYaPAS is fairly simple, the translator is completely contained in core, occupying about 1000 registers, and translating at the rate of about 100 elementary machine operations for one symbol of the translated expression.

For each type of computer the translator can take on a concrete form; however, it can also be described well in a more abstract form, by means of LYaPAS itself.

To some degree the methods of input and output can also be bound to concrete machines.

However, the basic principle adopted for the automatic programming system is *universality*, expressed by the fact that all blocks of the system, except the translator (compiler, checkout program, the block for compounding variables, the block for optimizing memory allocation, etc.), are expressed in first-level LYaPAS, and can be translated to machine language at any moment by passing them through the translator. The use of this principle has substantially reduced the work of developing the programming programs, and these programs themselves are written in easily read form. In fact, every programming program is a program for the solution of a typical nontrivial logical problem and is an excellent touchstone for a programming language for logical problems.

In the realization of complex programs the interpretation and compilation modes are combined. This latter mode is entered, for example, when translating to machine language inner and close to inner loops of a program. Its combination with the interpretive mode avoids overload of core.

In principle, LYaPAS can be represented in any machine whose core store is not too small. LYaPAS symbols are coded in 9-bit binary codes (or by three octal digits) and are packed fully into the memory registers which provides maximum compactness of program representation. The representation of LYaPAS is somewhat more convenient and natural in single-address machines (which are in general more suitable for automatic programming) and in fixed-point machines. The problem is that the bits reserved for exponents are not easily utilized in the programming of logical problems.

E. PERSPECTIVES AND SOME COMPARISONS

It is hoped that the development and application of LYaPAS will open up perspectives for the efficient solution, of not only practical problems in the theory of discrete automata, but also a broad range of problems which can be expressed by means of set-theoretical concepts, which are at the foundation of LYaPAS. The application and development of LYaPAS can stimulate the development of the theory of synthesis in directions which take into account the possibilities of modern computers; and the development (on the basis of LYaPAS) of an experimental foundation for the theory of synthesis can enable new approaches to be applied in that theory (for example, statistical-experimental or analytical-comparative methods).

Doubts may arise on the part of the reader concerning usefulness of developing this language when such universal languages as, for example, ALGOL-60 [1, 2] exist, together with their automatic programming systems. We remark that more specialized languages, such as the one developed in the present article, can be substantially more efficient, just because they are specialized. Thus, the compactness and speed of the first-level LYaPAS programming program exceed by a factor of ten those of the known ALGOL-60 translators.

Of those languages known to the author, the closest in purpose to LYaPAS is the language of Iverson [3], whose results were taken into account in the development of LYaPAS.

The first version of LYaPAS was described in [4]. After a number of corrections and substantial additions, LYaPAS has taken on the form described in the sections which follow.

2. OPERANDS

A. REFERENCE SET

It may be considered that, in the last analysis, the language of set theory is the most universal of mathematical languages, and the language of the theory of finite sets is the most convenient for the discussion of problems of a logical character and in the orientation toward discrete computers. In this connection it is useful to represent all of the information concerning the problems to be solved by certain sets, their subsets, etc., and to interpret the process of the solution of these problems itself, in the language of set theory, expressing it by a certain sequence of definite operations on finite sets, which here represent the operands.

Passing now to the consideration of certain methods of representing operands, we introduce the *reference set* A of certain abstract elements

$$A = \{a_0, a_1, \ldots, a_{\mu-1}\},$$

the set $\tilde{A}$ of all subsets of the set A

$$\tilde{A} \equiv \{A_0, A_1, \ldots, A_{p-1}\}, \quad \text{where} \quad p = 2^{\mu},$$

and the set $\tilde{\tilde{A}}$ of all subsets of the set $\tilde{A}$

$$\tilde{\tilde{A}} \equiv \{\tilde{A}_0, \tilde{A}_1, \ldots, \tilde{A}_{q-1}\}, \quad \text{where} \quad q = 2^{p}.$$

Let us consider the variables ξ_i, whose values are subsets of the set $\tilde{A}$ or elements of the set $\tilde{\tilde{A}}$. It is convenient to interpret these variables as p-dimensional vectors with binary components

$$\xi_i \equiv (\xi_i^0, \xi_i^1, \ldots, \xi_i^{p-1}),$$

or more simply,

$$\xi_i \equiv \xi_i^0, \xi_i^1, \ldots, \xi_i^{p-1},$$

and to represent their values by p-bit binary codes, assuming that

$$\xi_i^{\epsilon} = 1 \leftrightarrow A_{\epsilon} \in \tilde{A}_j$$

if the value of the variable ξ_i is the set $\tilde{A}_j \subseteq \tilde{A}\,(\epsilon \in \{0, 1, \ldots, p-1\})$. Thus the notation $\xi_i = 101101\cdots01$ is equivalent to the proposition that the value of the variable ξ_i is the set $\{A_0, A_2, A_3, A_5, \ldots, A_{p-1}\}$.

This method of prescribing the values of the variables presents obvious advantages for use in digital computers which manipulate multiposition binary codes, usually interpreted as the codes of numbers. By varying the interpretation, these binary codes can be used to represent the most varied objects: numbers, Boolean functions, graphs, etc.

Generally speaking, the cardinality μ of a set A can be taken fairly arbitrarily; however, in the attempt to obtain maximal efficiency in the exploitation of modern computers, the value of μ is usefully taken on the basis of the properties of these computers. It is convenient to take

$$2^{\mu} \leq k < 2^{\mu+1},$$

where k is the number of bits in a memory register reserved for the mantissas of numbers. (We shall count on the use of only these bits, since in the execution of the elementary machine operations on which we shall orient our applications these bits are subject to homogeneous manipulation.)

Taking this into account, we shall adopt five as the standard value of μ; in this case the values of the variables ξ_i will be 32-bit binary codes, which are conveniently stored in the memory registers of the majority of modern computers.

At the same time, to simplify our illustrations, we shall adopt the convention that, in the discussion of concrete examples to clarify the text, we shall henceforth use the value $\mu = 3$ ($p = 8$).

We shall also provide for the use of other, nonstandard values of p, denoting the corresponding variables by ξ_i^*, and considering that $p = 32 \cdot \omega$, where ω is a special dimensional parameter, taking on values in the set of natural numbers.

If $2^5 \leq k < 2^6$ then for convenient representation of the variables ξ_i^*, ω memory registers will be required.

It is also useful to introduce the variables ξ_i^{**}, equivalent to certain sets of variables of the type ξ_i or ξ_i^*; the values of the variables ξ_i^{**} are appropriate sets of 32-bit codes and they can be interpreted as subsets of $\tilde{\tilde{A}}$.

B. SIMPLE AND COMPLEX VARIABLES, COMPLEXES, AND INDEXES

We shall introduce into LYaPAS the symbols for the following concrete types of operands.

Simple variables (or simply *variables*), denoted by the bold-face lowercase letters **a, b, c, d, e, f, g, h, i, j, k, l, m, n, p, q, r, s, t, u, v, w, x, y, z, ж, л, ф, ш, э, ю, я**, which play the role of the basic operands in LYaPAS, are of the standard type ξ_i and their values are 32-bit codes.

The dimensions of the variables can be increased by preceding their symbols by primes (′); this will signify that the value of the "primed" variable is a $32 \cdot \omega$-valued binary code in a set of ω registers. By convention, we shall call such variables compound.

Complexes (sets whose elements are analogous to the above variables) are denoted by the boldface upper-case letters, **A, B, C, D, E, F, G, H, I, J, K, L, M, N, P, Q, R, S, T, U, V, W, X, Y, Z, Ж, Л, Ф, Ш, Э, Ю, Я**. Their cardinalities are equal to σ_a, σ_b, σ_c, ..., respectively, and are, in general, variable. Complexes are prescribed by listing their elements (for example, $\mathbf{A} \equiv \{\mathbf{a}_0, \mathbf{a}_1, \ldots, \mathbf{a}_{\sigma_a - 1}\}$), each of which can be either a simple or a compound variable (but the dimensions of all elements of one complex must be the same). It is possible to manipulate either complexes as a whole or their individual elements. The selection of a concrete element in a complex is realized by means of *indexes*, denoted by the lower-case italics, *a*, *b*, *c*, *d*, *e*, *f*, *g*, *h*, *i*, *j*, *k*, *l*, *m*, *n*, *p*, *q*, *r*, *s*, *t*, *u*, *v*, *w*, *x*, *y*, *z*, *ж*, *л*, *ф*, *ш*, *э*, *ю*, *я*, whose values are interpreted as the numbers of the elements. By convention, we shall denote by $\mathbf{a}_b$ the $[b]$th element of the complex **A**, where $[b]$ is the natural number whose binary code coincides with the value of the index b (the positional binary code used in modern machines is understood here). We admit the equivalent notation **A***b*, i.e., $\mathbf{a}_b \sim Ab$. The codes for these two expressions are identical. Indexes can be LYaPAS operands on a level with simple variables, but the former cannot be primed, and the latter cannot be used for indexing elements of complexes.

Of the variable here introduced, the variable **я** has a special significance—it serves as a *random number generator*, to which no value can be assigned by programmed means. It is assumed that each occurrence of the variable **я** is a random vector of zeros and ones, in which the probabilities of 0 and 1 are equal, and in which the individual bits are independent.

Aside from these sets of variables and indexes there are analogously defined *auxiliary sets* for use in the subroutines corresponding to L-operators. The elements of auxiliary sets are denoted by the same symbols **a**, **b**, . . ., *a*, *b*, . . ., but are coded differently.

The variables and indexes represented by the letters ж, л, ф, ш, э, ю, я, will have a special significance (mainly in the automatic programming system), and therefore their utilization in programs on the same level as the other variables and indexes is not authorized.

C. CONSTANTS

LYaPAS operands can also be standard and natural constants.

The set of standard constants consists of the following 32-bit codes, each of which has as its symbol an appropriate 9-bit code. Aside from this, the set is constructed in the form of four complexes, which permits the same manipulation of standard complexes as it does of the elements of complexes.

c_0	1000	0000	0000	0000	0000	0000	0000	0000
c_1	0100	0000	0000	0000	0000	0000	0000	0000
c_2	0010	0000	0000	0000	0000	0000	0000	0000
	. . .	. . .	. . .	. . .	. . .	. . .	. . .	. . .
c_{37}	0000	0000	0000	0000	0000	0000	0000	0001

d_0	1010	1010	1010	1010	1010	1010	1010	1010
d_1	1100	1100	1100	1100	1100	1100	1100	1100
d_2	1111	0000	1111	0000	1111	0000	1111	0000
d_3	1111	1111	0000	0000	1111	1111	0000	0000
d_4	1111	1111	1111	1111	0000	0000	0000	0000
e_0	0101	0101	0101	0101	0101	0101	0101	0101
e_1	0011	0011	0011	0011	0011	0011	0011	0011
e_2	0000	1111	0000	1111	0000	1111	0000	1111
e_3	0000	0000	1111	1111	0000	0000	1111	1111
e_4	0000	0000	0000	0000	1111	1111	1111	1111
f_0	0000	0000	0000	0000	0000	0001	1000	0000
f_1	0000	0000	0000	0000	0000	0001	1100	0000
f_2	0000	0000	0000	0000	0000	0001	1110	0000
f_3	0000	0000	0000	0000	0000	0001	1111	0000
f_4	0000	0000	0000	0000	0000	0001	1111	1000
f_5	0000	0000	0000	0000	0000	0001	1111	1100
f_6	0000	0000	0000	0000	0000	0001	1111	1110
f_7	0000	0000	0000	0000	0000	0001	1111	1111
f_{10}	1111	1111	1111	1111	1111	1111	1111	1111
f_{11}	1111	1111	1111	1111	1111	1111	1111	1111

The last of these constants, f_{11} has particular significance. It is represented in any machine by ones in all bits of the memory register, regardless of the number of bits, whereas all the other constants can have ones only in 32 bits, designated in a definite way. This constant is intended mainly for use in the automatic programming system.

We now introduce 128 natural constants prescribed directly by program codes. The symbols of these constants are the corresponding natural numbers, expressed in the octal system as, for example:

0	0000	0000	0000	0000	0000	0000	0000	0000
1	0000	0000	0000	0000	0000	0000	0000	0001
2	0000	0000	0000	0000	0000	0000	0000	0010
	...	...	...	...	...	...	...	...
7	0000	0000	0000	0000	0000	0000	0000	0111
10	0000	0000	0000	0000	0000	0000	0000	1000
	...	...	...	...	...	...	...	...
176	0000	0000	0000	0000	0000	0000	0111	1110
177	0000	0000	0000	0000	0000	0000	0111	1111

D. SPECIAL COMPLEXES

As mentioned above in the listing of the standard constants, they are divided into the four complexes C, D, E, and F, belonging to the special complexes $A, B, C, D, E, F, G, H, I, J, K, L, M, N, P, Q$. The special complexes have fixed meanings, which will be discussed for certain of them in the present paragraph; others will be discussed in the description of the automatic programming system. The cardinalities of the special complexes are fixed.

The *working complex* K is a complex such that the value of the ith element is the contents of the ith register in core storage. In this definition the memory allocation problem becomes the problem of establishing a definite correspondence between the working complex and the other complexes and individual variables representing the information to be processed.

The complex A is the *complex of initial addresses* of all complexes **A**, **B**, . . ., **Я** in the working complex; i.e., we arrange that the element a_0 represent the index of that element of the working complex which is identified with the element $\mathbf{a}_0$, the element a_1 is put into the same correspondence with the element $\mathbf{b}_0$, etc.

We call the complex B the *complex of cardinalities* of the complexes **A**, **B**, . . ., **Я**; the value of b_0 will be σ_a—the cardinality of the set **A**, $[b_1] = \sigma_b$, $[b_2] = \sigma_c$, etc.

We denote by G the *complex of variables*, putting $g_0 = \mathbf{a}$, $g_1 = \mathbf{b}$, $g_2 = \mathbf{c}$, etc. Similarly, we define the *complex of indexes* H and the complexes of *auxiliary variables* I and *indexes* J.

3. OPERATORS OF FIRST-LEVEL LYaPAS

As has been pointed out previously, at the first level of LYaPAS it is basically intended to operate directly only on variables ξ_i whose values can be given by 32-bit binary codes, i.e., on simple variables, indexes, individual elements of complexes, and various types of constants.

We introduce here the set of first-level standard operators and the corresponding notation.

A. THE BASIC COMPUTATIONAL OPERATORS

The Disjunction Operator $\vee$

$$\xi_i \vee \xi_j = \xi_k \sim \xi_i^t \vee \xi_j^t = \xi_k^t, \quad \text{where} \quad t \in \{0, 1, \ldots, p-1\}.$$

Example: 01000110 $\vee$ 10001100 = 11001110.

The Conjunction Operator $\wedge$

$$\xi_i \wedge \xi_j = \xi_k \sim \xi_i{}^t \wedge \xi_j{}^t = \xi_k{}^t.$$

Example: 01000110 $\wedge$ 10001100 = 00000100.

The Exclusive Disjunction Operator $\oplus$

$$\xi_i \oplus \xi_j = \xi_k \sim \xi_i{}^t \oplus \xi_j{}^t = \xi_k{}^t.$$

Example: 01000110 $\oplus$ 10001100 = 11001010.

The Addition Operator—Modulo 2^p $+$

$$\xi_i + \xi_j = \xi_k \sim [\xi_i] + [\xi_j] = [\xi_k] \bmod 2^p.$$

Example: 01000110 + 10001100 = 11010010.

The Subtraction Operator—Modulo 2^p $-$

$$\xi_i - \xi_j = \xi_k \sim [\xi_i] - [\xi_j] = [\xi_k] \bmod 2^p$$

Example: 01000110 − 10001100 = 10111010.

The Multiplication Operator—Modulo 2^p $\times$

$$\xi_i \times \xi_j = \xi_k \sim [\xi_i]\cdot[\xi_j] = [\xi_k] \bmod 2^p.$$

Example: 01000110 $\times$ 10001100 = 01001000.

The Multiplication Operator With Round-off $\overline{\times}$

$$\xi_i \,\overline{\times}\, \xi_j = \xi_k \sim [\xi_k] = ([\xi_i]\cdot[E_j] - [\xi_i]\cdot[\xi_j] \quad \bmod 2^p):2^p.$$

Example: 01000110 $\times$ 10001100 = 00100110.

The Division Operator :

The expression $\xi_i:\xi_j = \xi_k$ signifies that the value of the variable ξ_k is the remainder after division of $[\xi_i]$ by $[\xi_j]$. At the same time, in the realization of division operations the integer part of the quotient is always represented by the value of the index я. For example, 00001101:00000101 = 00000011, and я takes on the value 00000010.

In both of these operations the value $\xi_j = 00\cdots0$ is not permitted. This must be taken into account in the program.

The Componentwise Inversion Operator ⌉

$$\xi_i \urcorner = \xi_k \sim \bar{\xi}_i^{\,t} = \xi_k^{\,t}.$$

Example: 01000110 ⌉ = 10111001.

The Order Inversion Operator I

$$\xi_i \, \mathrm{I} = \xi_k \sim \xi_i^{\,t} = \xi_k^{p-t-1}.$$

Example: 01000110 I = 01100010.

Leftmost 1 Position-Locating Operator ⊢

$$\xi_i \vdash = \xi_k \sim [\xi_k] = t_{\min} \quad \text{with} \quad \xi_i^{\,t} = 1.$$

Example: 00010110 ⊢ = 00000011.

When using this operator, by convention we put

$$00000000 \vdash = 10000000 \vdash = 00000000.$$

The Normalization Operator |←

$$\xi_i \,|\!\leftarrow\; = \xi_k \sim \xi_k^{\,t} = \begin{cases} \xi_i^{t+q} & \text{if} \quad t + q < p, \\ 0 & \text{if} \quad t + q \geq p, \end{cases}$$

where $q = [\xi_i \vdash]$. The effect of this operator is expressed by the shift of the operand to the left until a 1 appears in the extreme left position, if there is any in the code.

Example: 00010110 |← = 10110000, 00000000 |← 00000000.

The Left-Shift Operator <

$$\xi_i < \xi_j = \xi_k \sim \xi_k^{\,t} = \begin{cases} \xi_i^{t+q} & \text{if} \quad t + q < p, \\ 0 & \text{if} \quad t + q \geq p, \end{cases}$$

where $q = [\xi_j]$.

Example: 01101010 < 00000100 = 10100000.

The Right-Shift Operator >

$$\xi_i > \xi_j = \xi_k \sim \xi_k{}^t = \begin{cases} \xi_i^{t-q} & \text{if } t - q \geq 0, \\ 0 & \text{if } t - q < 0, \end{cases}$$

where $q = \lfloor \xi_j \rfloor$.

Example: 01101010 > 00000101 = 00000011.

The Cyclic-Shift Operator ←

$$\xi_i \leftarrow \xi_j = \xi_k \sim \xi_k{}^t = \xi_i^{(t+q) \bmod p}.$$

where $q = [\xi_j]$.

Example: 01101010 ← 00100100 = 10100110.

The Weighting Operator ∇

$\xi_i \nabla = \xi_k \sim [\xi_k]$ is the number of ones in the code ξ_i.

Example: 01101010 ∇ = 00000100.

The Reflection Operator $\veebar$

$$\xi_i \veebar \xi_j = \xi_k \sim \xi_k{}^t = \begin{cases} \xi_i^{t+2n} & \text{if } t^n = 0, \\ \xi_i^{t-2n} & \text{if } t^n = 1, \end{cases}$$

where $n = [\xi_j] = 0, 1, 2, 3, 4$, and t^n is the nth bit to the right in the binary positional code of the number t (beginning the count from 0).

If we consider a 32-bit code—the value of the variable ξ_i—as the development of the 5-dimensional cube, the effect of the reflection operator can be represented as the mirror exchange over the 4-dimensional edges, corresponding to the distinct values of one of the five binary variables prescribed by the operand ξ_j.

Examples:*

01001101 $\veebar$ 00000000 = 10001110,

01001101 $\veebar$ 00000001 = 00010111,

01001101 $\veebar$ 00000010 = 11010100.

For $[\xi_j] > \log_2 p$ the result of the operation will be 00000000.

* Translator's note: In the first example the bits are exchanged between adjacent columns, in the second between columns one removed, and in the last between the first and fourth, second and fifth, etc.

B. ASSIGNMENT OPERATORS

While the above listed operators admit a "static" interpretation, i.e., it may be considered that they prescribe a certain functional relation holding among the different variables, the operators given below can be interpreted only dynamically, based on the concept of change in the values of the variables with time.

The Assignment Operator ⇒

$$\xi_i \Rightarrow \xi_j$$

signifies that the variable ξ_j takes on the value of the variable ξ_i;

$$\xi_i \vee \xi_j \Rightarrow \xi_k$$

signifies that the variable ξ_k takes on the value of the result of the operation $\xi_i \vee \xi_j$, etc.

We shall introduce the following generalization of the operator ⇒ to the case of exchange of information among groups of variables of the type ξ and among complexes.

We define

$$(\xi_1\xi_2 \cdots \xi_k) \Rightarrow \mathbf{A},$$

where ξ_i can be the symbol of a simple variable, an index, or a constant, and where the cardinality σ_a of the complex **A** at the given moment is equal to n, to be the result of increasing the value of σ_a to $n + k$ with addition to the complex **A** of the elements with indexes $n, n + 1, \ldots, n + k - 1$ taking on the values of the variables $\xi_1, \xi_2, \ldots, \xi_k$, respectively.

The expression

$$\mathbf{A} \Rightarrow (\xi_1\xi_2 \cdots \xi_k),$$

where ξ_i can be only the symbol of a simple variable or index has the inverse sense; that is, if the cardinality σ_a of the set **A** up to the given moment has been equal to n, then the variables $\xi_1, \xi_2, \ldots, \xi_k$ take on the values of the elements of the complex **A** with indexes $n - k, n - k + 1, \ldots, n - 1$, respectively, and the cardinality of the complex **A** is reduced to $n - k$.

The Interchange Operator $\Leftrightarrow$

$$\xi_i \Leftrightarrow \xi_j$$

signifies that the variables ξ_i and ξ_j interchange their values.

The Natural Assignment Operator $\Leftarrow$

The expression

$$\xi_i \Leftarrow \psi_1\psi_2\psi_3\psi_4,$$

where the ψ_j are arbitrary symbols of the language (constant vectors with nine binary components), is realized by assignment to the variable ξ_i of the values defined by the vectors ψ_j in the following way:

$$\xi_i^t = \begin{cases} \psi_1^{t+4} & \text{for} \quad 0 \leq t < 5, \\ \psi_2^{t-5} & \text{for} \quad 5 \leq t < 14, \\ \psi_3^{t-14} & \text{for} \quad 14 \leq t < 23, \\ \psi_4^{t-23} & \text{for} \quad 23 \leq t < 32. \end{cases}$$

We shall denote in boldface type those LYaPAS symbols which coincide with their codes (for example, the code 065 corresponds to the symbol **065**). Then to assign to the variable **a** the value

1011 0011 0011 0111 1001 0011 0101 0101

it is sufficient to realize the operation

a $\Leftarrow$ **026 315 711 525.**

The Positive Element-Incrementation Operator $\triangle$

$$\triangle\, \xi_j \sim \xi_j + 1 \Rightarrow \xi_j.$$

Example: If $\xi_j = 10011011$, then the result of the operator $\triangle$ on the variable ξ_j is that the latter takes on the value 10011100.

The Negative Element-Incrementation Operator $\bar{\triangle}$

$$\bar{\triangle}\, \xi_j \sim \xi_j - 1 \Rightarrow \xi_j.$$

Example: If $\xi_j = 10011011$, then the result of the operator $\bar{\triangle}$ on the variable ξ_j is that the variable ξ_j takes on the value 10011010.

The Null-Value-Assignment Operator $\circ$

The result of the operator $\circ$ on the variable ξ_j is always that the variable ξ_j takes on the value 00000000.

The Maximum-Value-Assignment Operator $\bar{\circ}$

$$\bar{\circ}\ \xi_j \sim 0\ \urcorner \Rightarrow \xi_j.$$

The result of the operator $\bar{\circ}$ on the variable ξ_j is always that the variable ξ_j takes on the value 11111111.

C. THE IMPLICIT VARIABLE τ

Let us pass now to the dynamic interpretation of the operators, as it occurs in LYaPAS. A string of symbols in this language, representing the L-program, reflects a certain computation process in time, in which the program operators are realized in succession. In the standard realization of the program the operators are realized in that order in which their symbols appear in the program, reading from left to right. The result of the realization of the current operator is the current value of the implicit variable τ, whose symbol never appears in the program. This variable plays the role of the left operand (denoted in the representations of the operators by ξ_i), which makes possible greater program compactness through the connectedness of the computation processes. Thus

$$\xi_j\ \urcorner + 1 \wedge \xi_j \Rightarrow \xi_j$$

represents the following sequence of actions:

$$\begin{aligned} \xi_j &\Rightarrow \tau\ 00110100 \\ \tau\ \urcorner &\Rightarrow \tau\ 11001011 \\ \tau + 1 &\Rightarrow \tau\ 11001100 \\ \tau \wedge \xi_j &\Rightarrow \tau\ 00000100 \\ \tau &\Rightarrow \xi_j\, 00000100. \end{aligned}$$

The significance of this expression is the extraction (in the code representing the value of the variable ξ_j) of the rightmost 1. The computation process realized is represented by the column of codes at the right, showing the sequence of values taken on by the implicit variable τ for the initial value $\xi_j = 00110100$.

The program consists of *sentences*, separated from each other by the symbol §, accompanied by the number of the corresponding sentence, which is represented by the corresponding natural constant. Each sentence may contain an arbitrary number of words and must contain only once the symbol §, which appears in the sentence as its leftmost symbol. For convenience, in written form we adopt the convention of starting each new sentence of the program on a new line.

Here we have examined the standard order of operator realization in time in which, immediately after the current operator is realized, the operator whose symbol in the L-program follows the symbol of the current operator is realized. The standard order of operator realization can be altered only when one of the following program control operators is encountered.

D. CONTROL-TRANSFER OPERATORS

The Unconditional Transfer Operator $\rightarrow$

The pair of symbols $\rightarrow \xi_j$ signifies that, immediately after the realization of the preceding portion of the program has been completed, there begins the realization of the sentence with number $[\xi_j]$.

The Conditional Transfer-on-Zero Operator $\circ\!\rightarrow$

This operator becomes equivalent to the operator $\rightarrow$ when

(a) the operator $\circ\!\rightarrow$ is directly preceded by the operator $+$ or $-$ where it is found that $[\xi_i] \pm [\xi_j] \neq [\xi_i \pm \xi_j]$, i.e., either $[\xi_i] + [\xi_j] \geq 2^p$ (for the operator $+$) or $[\xi_i] < [\xi_j]$ (for the operator $-$), where ξ_i and ξ_j are, respectively, the left and right operands, subject to the effect of the operators $+$ and $-$; or

(b) the operator $\circ\!\rightarrow$ is not directly preceded by an operator $+$ or $-$, and the next value of the variable τ is 00000000.

In all other cases the standard order of program realization is conserved.

The Conditional Transfer-on-One Operator $|\!\rightarrow$

The effect of this operator is inverse to the effect of the operator $\circ\!\rightarrow$.[1]

The following two operators are introduced mainly to simplify the manipulation of subroutines. We introduce the concept of depth of realization of a program, with the convention that the start of the realization of a program has zero depth, and that the depth can only change upon encountering the operators $\nrightarrow$ and !.

[1] Translator's note: The condition on τ becomes $\tau \neq 0$.

The Exit Operator ⇸

The expression ⇸ $\xi_i\alpha$, where α is a certain arbitrary expression in LYaPAS, signifies that the depth of realization is to be increased by one, the "coordinates" of the expression α must be put into correspondence with the new value of the depth and stored, and transfer is to be realized to the sentence of number $[\xi_j]$.

The Return Operator !

When the operator ! is encountered at a nonzero depth, transfer is carried out to that expression whose "coordinates" have been stored at the present depth. The depth of realization is now decreased by unity. The realization of any program begins at zero depth.

These transfer operators have meaning only if the value of the variable whose symbol follows the operator symbol agrees with the number of one of the L-program sentences. By convention, this symbol can only be that of a natural constant.

The Transfer-to-Machine-Language Operator ↓

The pair of symbols $\downarrow \xi_i$ indicates passage to the realization of a certain program in machine language, the first of whose instructions is the $[\xi_i]$th element of the working complex. To ensure return to the realization of the L-program beginning at the symbol following the symbols $\downarrow \xi_i$, the machine language program must terminate with an instruction for unconditional transfer to the core register which is reserved for the values of the variable **ю**. The symbol ξ_i can be either a variable or an index.

This operator can be useful in those situations where it is desirable to express some computation process directly in machine language. Its utilization presupposes knowledge of certain characteristics of the automatic programming system (memory allocation, etc.). By convention, in the realization of transfer by any of the above operators the value of the implicit variable τ does not change. The following two operators have a combined character.

The Enumeration-of-Ones Operator $\dot{\mathbf{X}}$[2]

$$\xi_i \; \dot{\mathbf{X}} \; \xi_j\xi_k \sim \xi_i \;\circ\!\!\rightarrow \xi_j\xi_i \vdash \Rightarrow \xi_k \; C \; \xi_k \oplus \xi_i \Rightarrow \xi_i.$$

Restrictions are placed on the operand ξ_k—it can be an index only. The realization of the operator $\dot{\mathbf{X}}$ is accompanied by the assignment of the new value of this index to the implicit variable τ.

[2] Translator's note: In original LYaPAS the letter **X** with a bar across it, somewhat like an asterisk, is used. For typographical reasons it is replaced by the $\dot{\mathbf{X}}$ here.

Example: Let the initial value of the variable ξ_i = 01011010, the value of the index ξ_k arbitrary. Then after the realization of the operation $\xi_i \; \dot{\mathbf{X}} \; \xi_j \xi_k$, the variable ξ_i takes on the value 00011010, the index ξ_k the value 00000001. If the initial value of the variable ξ_i is 00000000, the realization of the sentence $[\xi_j]$ begins.

The Random-Enumeration-of-Ones Operator $\ddot{\mathbf{X}}$ [3]

This operator differs from the preceding one in that the next 1 in the code of the variable ξ_i is chosen at random, with equal probabilities of selecting each 1.

For both operators $\dot{\mathbf{X}}$ and $\ddot{\mathbf{X}}$ the left operand must be directly prescribed by the symbol appearing before the operator. (For example, an expression of the type $\mathbf{a} + \mathbf{b} \; \dot{\mathbf{X}} \; 5c$, cannot be used, when it is desired to operate with the value of the sum $\mathbf{a} + \mathbf{b}$.)

Directly after the operators $\dot{\mathbf{X}}$ and $\ddot{\mathbf{X}}$, as after the previously discussed operators →, |→, ○→, ⇻ only the symbol of a natural constant can appear.

The operators and symbols presented below are auxiliary and are intended mainly for use in the automatic programming system, all of whose blocks except the translator must be expressed directly in LYaPAS.

E. OPERATORS FOR EXCHANGING INFORMATION WITH THE EXTERNAL MEDIUM

The following operators play an auxiliary role, providing principally for the exchange of information between the memory of the machine executing the algorithm and the external medium.

The Print-One-Element Operator $*$

The realization of the operator $*$ consists of printing the current value of the implicit variable τ. Considering the characteristics of current [Soviet] computers, we shall require that τ be printed in octal code.

The Punch-One-Element Operator $\overset{*}{*}$

This operator punches the current value of τ. It has certain advantages, for example, when the information produced during the realization of an algorithm is intended for reinput to the machine.

The values of all elements of a complex are printed or punched in succession if and only if the operator $*$ or $\overset{*}{*}$ follows directly after the symbol of that complex (for example, in the case $\mathbf{A}*$).

[3] Translator's note: In original LYaPAS the symbol for this operator was the letter **X** with two bars across it (see note p. 21).

The Input Operator ↑

The information to be processed by the machine is set up in the form of a string of constants, which is divided initially into portions. At the start of each portion, its dimension (as well as the number of a certain element in the working complex of the machine) is indicated. To realize the operator ↑ the next portion is input—the constants appearing in it become the values of the elements of that portion of the working complex which begins with the denoted element and which coincides in cardinality with the input portion.

Aside from the working complex, corresponding to the core store of the machine, we introduce into consideration the external complex, corresponding to the auxiliary, peripheral storage of the machine. Information exchange between these two complexes is carried out by means of the following two operators.

The Data-Transfer-to-External-Complex Operator ↗

The expression ↗ $\xi_i\xi_j\xi_k$ signifies that the values of $[\xi_j]$ elements in succession in the working complex, beginning with element number $[\xi_i]$ are transferred to $[\xi_j]$ elements in succession in the external complex, beginning with element number $[\xi_k]$.

The Data-Transfer-from-External-Complex Operator ↙

This operator represents transfer of information in the opposite direction: the values of $[\xi_j]$ elements of the external complex, beginning with element number $[\xi_k]$ are transferred to the corresponding series of elements in core, beginning with element number $[\xi_i]$.

F. OTHER OPERATORS

The End-of-Algorithm Operator .

The period symbol "." plays a dual role: on the one hand, it serves as the sign for the end of the algorithm (the realization of the algorithm is terminated when this symbol is encountered); on the other hand, it always closes the expression representing the algorithm.

To some degree the following operator is beyond the scope of our structure, and is introduced in order to make possible a fuller utilization of the machine memory, specifically the complete utilization of the memory registers when they contain more than 32 bits, and also full utilization of the input/output channels of the machine.

The Unlimited-Shift Operator $\leftleftarrows$

The expression $\xi_i \leftleftarrows \xi_j$ can be interpreted in the following way. The entire contents of the register in which are stored the values of the variable ξ_i are shifted by $q - [\xi_j] \bmod 2^6 - 2^5$ positions, where the left shift corresponds to positive q, right shift to negative, and the positions freed are filled in by zeros.

G. SYNTAX OF FIRST-LEVEL LYaPAS

An expression in a first-level LYaPAS program is a certain string composed of the above-considered symbols. However not every string of these symbols is an L-program—only those that satisfy a set of conditions comprising the syntax of first-level LYaPAS.

Of course, not even every such string can be considered to be a sufficiently reasonable program. For example, it may contain obvious redundancy, permitting some parts of the string to be eliminated without altering the sense of the program. Nevertheless, any string satisfying the defined syntax is called an L-program, and must be reproduced uniquely by the translator from first-level LYaPAS to machine language. Therefore it can be said that the given syntax defines the domain of definition of the translator functions.

We define

$$\pi_1 \in \{\mathbf{a}, \mathbf{b}, \ldots, \textbf{я}\};$$
$$\pi_2 \in \{a, b, \ldots, \text{я}\};$$
$$\pi_3 \in \{0, 1, \ldots, 177\};$$
$$\pi_4 \in \{c_0, c_1, \ldots, c_{37}; d_0, d_1, \ldots, d_4; e_0, e_1, \ldots, e_4; f_0, f_1, \ldots, f_{11}\};$$
$$\pi_5 \in \{\mathbf{A}, \mathbf{B}, \ldots, \textbf{Я}; A, B, \ldots, Q\};$$
$$\pi_6 \in \{\mathbf{A}, \mathbf{B}, \ldots, \textbf{Я}; A, B, G, H, \ldots, Q\};$$
$$\pi_7 \in \{\mathbf{A}, \mathbf{B}, \ldots, \textbf{Я}\};$$
$$\rho_1 \in \{\vdash, |\!\leftarrow, \urcorner, \mathrm{I}, \nabla\};$$
$$\rho_2 \in \{+, -, \vee, \wedge, \oplus, \leftleftarrows, \times, \overline{\times}, >, <, \leftarrow, \vee, :\};$$
$$\rho_3 \subset \{\rightarrow, \cup\!\!\rightarrow, |\!\!\rightarrow, \nrightarrow\};$$
$$\rho_4 \in \{\circ, \bar{\circ}, \triangle, \overline{\triangle}\}.$$

We shall consider that if

$$\eta \in \{\eta_1, \eta_2, \ldots, \eta_m\} \qquad \text{and} \qquad \zeta \in \{\zeta_1, \zeta_2, \cdots, \zeta_n\},$$

then

$$\eta\zeta \in \{\eta_1\zeta_1, \eta_1\zeta_2, \ldots, \eta_1\zeta_n, \eta_2\zeta_1, \eta_2\zeta_2, \ldots, \eta_2\zeta_n, \ldots, \eta_m\zeta_1, \eta_m\zeta_2, \ldots, \eta_m\zeta_n\}.$$

We define

$$\begin{aligned}
\pi^1 &\in \{\pi_1,\ \pi_2\},\\
\pi^2 &\in \{\pi_2,\ \pi_3\},\\
\pi^3 &\in \{\pi_1,\ \pi_2,\ \pi_3,\ \pi_4\},\\
\pi^4 &\in \{\pi_1,\ \pi_2,\ \pi_6\pi^2\},\\
\pi\ &\in \{\pi^3,\ \pi_5\pi^2\}.
\end{aligned}$$

We shall call an α-string an arbitrary finite sequence of composite symbols α:

$$\begin{aligned}
\alpha \in \{&\pi,\ \rho_1,\ \rho_2\pi,\ \rho_3\pi_3,\ \rho_4\pi^4,\ *,\ \overset{*}{*},\ \Rightarrow \pi^4,\ \downarrow\ \pi^1,\ \pi_5*,\ \pi_5\overset{*}{*},\ \uparrow,\\
&\nearrow\ \pi^1\pi^1\pi^1,\ \swarrow\ \pi^1\pi^1\pi^1,\ \pi_7 \Rightarrow (\pi^1\pi^1\ldots\pi^1),\ (\pi^3\pi^3\ldots\pi^3) \Rightarrow \pi_7,\\
&\pi^4\ \dot{\mathbf{X}}\ \pi_3\pi_2,\ \pi^4\ \ddot{\mathbf{X}}\ \pi_3\pi_2,\ \pi^4 \Leftarrow \psi\psi\psi\psi,\ \pi^4 \Leftrightarrow \pi^4,\ !,\ \S\ \pi_3\},
\end{aligned}$$

considering that ψ is an arbitrary elementary symbol (i.e., one encoded by a single 9-bit binary code).

The symbols $\S\ \pi_3$ divide an α-string into sentences whose numbers are the values of the symbol π_3 appearing at the start of each sentence in the label $\S\ \pi_3$.

An L-program is an α-string closed by the symbol "." and satisfying the following conditions.

The numbers of the sentences must not be repeated and must compose a set N such that for all combinations $\rho_3\pi_3$, $\dot{\mathbf{X}}\ \pi_3\pi_2$ and $\ddot{\mathbf{X}}\ \pi_3\pi_2$ encountered in the given α-string the condition $\pi_3 \in N$ is satisfied.

Let us construct an oriented graph whose vertices correspond to the sentences of the program and whose branches correspond to possible transfers by the operators ρ_3, $\dot{\mathbf{X}}$, and $\ddot{\mathbf{X}}$, as well as to transfers to the immediately succeeding sentence (when the preceding sentence is not terminated by the symbol $\rightarrow \pi_3$ or !). We mark the branches corresponding to the transfers by the operator $\nrightarrow$ and the nodes corresponding to the sentences terminating with the symbol !.

The marked graph must not contain cycles which include the marked paths. In the graph obtained by deleting the marked branches there must be no paths from nodes corresponding to the start of a sentence (such a node must not be marked) to any marked node.

Let us now clarify the rules for determining the current values of the implicit variable τ. We consider that τ retains its preceding value for the operations $*$, $\overset{*}{*}$, $\Rightarrow \pi^4$, $\circ\!\rightarrow \pi_3$, $|\!\rightarrow \pi_3$, and in passage to a new sentence labeled $\S\ \pi_3$. In the realization of operation $\pi^4 \Leftrightarrow \pi^4$, τ obtains the new value of the right operand, and in the realization of the operations $\pi^4 \Leftarrow \psi\psi\psi\psi$, the new value of π^4. In the execution of the operations $\pi_5 *$, $\pi_5 \overset{*}{*}$, $\uparrow$, $\nearrow \pi^1\pi^1\pi^1$, $\swarrow \pi^1\pi^1\pi^1$, $\pi_7 \Rightarrow (\pi^1\pi^1 \ldots \pi^1)$, $(\pi^3\pi^3 \ldots \pi^3) \Rightarrow \pi_7$, and also in the realization of conditional transfer by the operators $\dot{\mathbf{X}}$ and $\ddot{\mathbf{X}}$, τ takes on the value zero.

G. THE SYMBOL ?

To make it possible to trace the order in which the working complex is used during program execution, we introduce a special symbol ?. Its presence after a combination $\pi_7\pi_2$, where

$$\pi_7 \in \{\mathbf{A}, \mathbf{B}, \ldots, \mathbf{Я}\} \qquad \text{and} \qquad \pi_2 \in \{a, b, \ldots, я\},$$

serves as a signal for testing the complex π_7 for "creeping" of the working complex into other operands, and as a signal to take measures to correct this if it occurs. The combination $B\pi_8??$, where $\pi_8 \in \{0, 1, \ldots, 37\}$, is a signal to find a location in the working complex for the complex whose number is prescribed by the value of π_8.

These operations are realized in the corresponding block of the programming system.

To clarify this description of the syntax of first-level LYaPAS, we note that a string of symbols which satisfies it can be extended by substituting $\pi_7\pi_2?$ for $\pi_7\pi_2$ and $B\pi_8??$ for $B\pi_8$. However, repeated substitutions of this type are inadmissible. for example, to obtain combinations of the form $\pi_7\pi_2??$.

4. EXAMPLES OF FIRST-LEVEL L-PROGRAMS

A. THE APPLICATION OF BINARY-RELATION MATRICES TO DESCRIBE ALGORITHMS

In the statement of an arbitrary problem and its solution algorithm, we shall use the latin alphabet and the other LYaPAS symbols, but with a different significance. To match this description to the corresponding L-program we shall agree to use C matrices of binary relations between the elements of two sets $A \equiv \{a_0, a_1, \ldots, a_{n-1}\}$ and $B \equiv \{b_0, b_1, \ldots, b_{m-1}\}$, defining these matrices in the following way.

$$C = \| A \prec B \|$$

signifies that $c_i^j = 1$ if $a_i \prec b_j$, where the symbol $\prec$ is the symbol of some binary relation, and $c_i^j = 0$ otherwise. In a number of cases it is more convenient to begin the numbering of the elements in the sets under consideration from one, i.e., to put $A \equiv \{a_1, a_2, \ldots, a_n\}$ and $B \equiv \{b_1, b_2, \ldots, b_m\}$.

We shall denote by $\| \{A\} \prec B \|$ a row matrix of the binary relation $\prec$ between the single element A of the set $\{A\}$ and the elements of the set B.

We define the expression

$$A::\| A \prec B \|$$

to mean that the complex **A** represents a matrix $\| A \prec B \|$, i.e., the rows of this matrix are the values of the elements of the complex **A**. Similarly,

$$\mathbf{a}::\| \{A\} \prec B \|$$

signifies that the value of the variable **a** is the row matrix $\| \{A\} \prec B \|$.

Let us give several examples of programs expressed in LYaPAS, in order of increasing complexity.

B. OBTAINING THE COORDINATES OF UNITY IN A CODE a

$$\S 0 \quad \mathbf{a}\ \dot{\mathbf{X}}\ 1b * \rightarrow 0$$
$$\S 1 \quad \cdots *$$

The result of the execution of this program is a sequence of codes giving the locations of unity in the code **a**. For example, if the initial value of **a** is 00110101, the resultant sequence will have the form:

00000010,
00000011,
00000101,
00000111.

C. OBTAINING THE LOWER BOUND OF A SET

Let $B \subseteq \tilde{A}$, where A is a certain set. The lower bound of the set B is its subset consisting of those elements $b_i \in B$ for which it is not possible to find $b_j \in B$ such that $b_j \subset b_i$.

We propose one possible form of a program for obtaining the lower bound of a set B, assuming $\mathbf{B}::\| B \ni A \|$.

$$
\begin{array}{ll}
\S 0 & \bar{\circ}\ b \\
\S 1 & \triangle\ b \oplus b_1 \circ \rightarrow 4\ \bar{\circ}\ a \\
\S 2 & \triangle\ a \oplus b_1 \circ\!\rightarrow 3a \oplus b \circ\!\rightarrow 2\mathbf{b}_b \urcorner \wedge \mathbf{b}_a \circ\!\rightarrow 1 \rightarrow 2 \\
\S 3 & \mathbf{b}_b * \rightarrow 1 \\
\S 4 & .
\end{array}
$$

For example, as a result of applying this program to the complex

10110010, 01000010,
01101101, 01100001,
10111011, 10001100,
11110101, 00001000,

we obtain the complex

10110010,
01000010,
01100001,
00001000.

D. QUINE SIMPLIFICATION

By Quine simplication we refer to the algorithm due to Quine for determining a reduced set $B \subseteq \tilde{A}$, proposed as one of the stages of minimization of Boolean functions in the class of disjunctive normal forms. Quine simplification consists in obtaining a subset of the set B whose elements are not absorbed by a kernel C of B (i.e., by the set of essential elements of the set B) or else enter into this kernel. An element $b_i \in B$ is called an essential element of B if it contains a certain element a_j of A not contained in any other element of B. It is also considered that the element $b_i \in B$ is absorbed by a set $B_j \subseteq B$ if $b_i \subseteq B_j^*$, where B_j^* is the union of all elements of B_j.

We present here a program for Quine simplification, in which

$$\mathbf{B}::\| B \ni A \|,$$

and the values of the variables **b**, **c**, **d**, and **e** are defined in the process of realization of the program in the following way.

$$\mathbf{b}::\| \{B^*\} \ni A \|,$$
$$\mathbf{c}::\| \{B^{**}\} \ni A \|,$$

where B^{**} is the set of elements of A such that each of them enters into at least two elements of B,

$$\mathbf{d}::\|\ \{B^*\backslash B^{**}\}\ \ni A\ \|,$$
$$\mathbf{e}::\|\ \{c^*\}\ \ni A\ \|.$$

$\S0\quad \bar{\circ}\ a\ \circ\ \mathbf{b}\ \circ\ \mathbf{c}$

$\S1\quad \triangle\ a \oplus b_1\ \circ\!\!\rightarrow 2\mathbf{b}\ \wedge\ \mathbf{b}_a \vee\ \mathbf{c} \Rightarrow \mathbf{cb}_a \vee\ \mathbf{b} \Rightarrow \mathbf{b} \rightarrow 1$

$\S2\quad \mathbf{c}\ \urcorner\ \wedge\ \mathbf{b} \Rightarrow \mathbf{d}\ \bar{\circ}\ a\ \circ\ e$

$\S3\quad \triangle\ a \oplus b_1\ \circ\!\!\rightarrow 4\mathbf{b}_a \wedge\ \mathbf{d}\ \circ\!\!\rightarrow 3b_a \vee\ \mathbf{e} \Rightarrow \mathbf{e} \rightarrow 3$

$\S4\quad \mathbf{d}\ \urcorner\ \wedge\ \mathbf{e} \Rightarrow \mathbf{f}\ \bar{\circ}\ a$

$\S5\quad \triangle\ a \oplus b_1\ \circ\!\!\rightarrow 6\mathbf{f}\ \urcorner\ \wedge\ \mathbf{b}_a\ \circ\!\!\rightarrow 5\mathbf{b}_a{*} \rightarrow 5$

$\S6\quad .$

5. OPERATIONS ON COMPOUND VARIABLES

Let us introduce four independent parameters of dimensionality ω_1, ω_2, ω_3, and ω_4, whose values, given by the indexes ж, л, ф, and ш, can be natural numbers. We shall mark the variables and the elements of the complexes, whose dimensions will be defined by these parameters, by the symbols ′, ″, ″′, and ″ ″ (preceding the corresponding symbols) respectively. We shall assume that the dimensions of all the elements in the same complex are identical. We shall admit the possibility of changing the values of the indexes ж, л, ф, and ш within reasonable limits during the course of realization of L-programs.

We define operations on variables of extended dimensions, or compound variables, as the generalization of the already considered operations on variables of the type ξ_i with the substitution of $p = 32$ by $p = 32 \cdot \omega$, where $\omega \in \{\omega_1, \omega_2, \omega_3, \omega_4\}$. Among the operators to which this generalization is extended we include the unary operators $\urcorner$, ∇, $|\!\!\leftarrow$, $\vdash$, $\triangle$, $\bar{\triangle}$, $\circ$, $\bar{\circ}$, $\mathbf{I}$ and the binary operators $+$, $-$, $\vee$, $\wedge$, $\oplus$, $<$, $>$, $\leftarrow$, $\underline{\vee}$.

We assume that the right operands of the operators $<$, $>$, $\leftarrow$, and $\underline{\vee}$ can only be of the type ξ_i (for which $p = 32$) and that the left operand of the operator $\underline{\vee}$ can be a compound variable only if the dimensional parameter ω takes on a value 2^n, where n is a natural number. The interpretation of the right operand of the operator $\underline{\vee}$ is broadened: $[\xi_j] = 0, 1, 2, 3, 4, 5, 6, \ldots$. We admit that the operators $+$, $-$, $\vee$, $\wedge$, and $\oplus$ can act on operands of different dimensions, i.e., we admit, for example, the operations $\mathbf{a} + {}'\mathbf{b}$, ${}'\mathbf{c}_i \oplus {}''\mathbf{d}$, etc. Before such operations are realized the dimensions of the operands are completed by adding to the right of the operand of the smaller dimensions a suitable number of zeros.

With the aim of simplifying the automatic programming system, we adopt the convention that the operators $\times$, $\overline{\times}$, $:$, and $\Leftarrow$ are not extended to complex variables.

We shall consider that the dimensions of the implicit variable τ are defined by the dimensions of the operands processed in the course of realization of the L-program. However, in order not to complicate the programming program, we assume that, under transfer operations, only values of τ which are of standard dimensions can be transmitted, as in sentence progression.

We shall assume that in the realization of the operations $'\mathbf{a} \Rightarrow \mathbf{b}$, $\mathbf{b}$ takes on the value defined by the left components of the complex variable $\mathbf{a}$.

We admit the possibility of utilizing expressions of the type $(\mathbf{a}\ ''\mathbf{b}\ '\mathbf{c}\ d) \Rightarrow \mathbf{C}$, in which "packing" (or "unpacking") of operands of various dimensions occurs in the complex $\mathbf{C}$.

A compound variable can be the left operand of an operator $\dot{\mathbf{X}}$ or $\ddot{\mathbf{X}}$ (but never a right operand).

The transfer conditions for the operators $\circ\!\!\rightarrow$, $\mapsto$, $\dot{\mathbf{X}}$, and $\ddot{\mathbf{X}}$ and their action on compound variable are defined analogously to the case of variables of the type ξ_i.

By convention we extend the complex of constants C, assuming that if

$$32 \leq [i] < 32 \cdot \omega,$$

then c_i is a compound variable with a single 1 in the ith position. Analogously we extend the complexes D and E.

Finally we consider that expressions of the type $'\mathbf{a}*$ or $''\mathbf{b}^*_*$ represent the output of the values of the corresponding compound variables for printing or punching.

Let us recall that the symbols $'$, $''$, $'''$, and $''''$ can be placed only before the symbols of variables $\mathbf{a}$, $\mathbf{b}$, . . . or symbols of elements of complexes, but not before indexes, which cannot be made compound.

6. HIERARCHICAL PROGRAMMING

A. GENERAL IDEA

Complex computational processes can usually be divided into "pieces" which have independent meaning and which can be associated with certain operators. The introduction of these operators into the language and the formation of the corresponding programs permit the effectiveness of the language (enriched thereby) to be progressively improved, since an ever greater part of the programming work is realized automatically. The representation of the computational processes also improves, the programs become more compact and readable, their assembly is speeded up, and checking out is facilitated.

We shall realize this general idea in LYaPAS.

B. THE PROBLEM OF FINDING A QUINE COVERAGE

Let us consider the concrete example of the coverage of a Quine table. This problem can be formulated in general form in the following way. A Boolean matrix U is given, where $u_i^j \in \{0, 1\}$, a certain subset b of its rows u_i and a subset c of its columns u^j are chosen, and it is required to find a subset $d \subseteq b$ of its rows such that together they cover all the columns in c, i.e., such that for each $u^j \in c$ there be found such $u_i \in d$ that $u_i^j = 1$. In this it is desirable that the coverage be "inexpensive," for example, that it contain the smallest possible number of elements.

We shall consider here a very simple approximate algorithm for the solution of this problem, in order to use it as an illustration of the method of segmenting algorithms.

We denote by Uxy the minor of the matrix U formed by the elements at the intersections of the rows in the set x and the columns in the set y. A minimal row (or column) of a minor is that row (or column) which contains the minimal number of ones. Analogously, we define the maximal rows and columns. If there are several such rows or columns any one of them can be taken, for example the first. The algorithm consists in the selection of a minimal column e_1 of the minor Ubc and in finding a maximal row f_1 in the minor $Ub'c$, where b' is the set of rows having unit elements in the intersections with the column e_1. After this we substitute the minor Ub_1c_1 for the minor Ubc, where $b_1 = b\backslash\{f_1\}$, and $c_1 = c\backslash c'$, and where c' is the set of columns having unit elements at the intersections with the row f_1. Then a minimal column e_2 is sought in Ub_1c_1, etc., until the set of columns c_i at some ith step is found to be empty. The set of rows found represents the solution

$$d = \{f_1, f_2, \ldots, f_{i-1}\}.$$

C. L-OPERATORS

It is obvious that the fundamental part of the above algorithm can be represented in the form of two operators having independent significance. We shall call such operators L-*operators* (in contradistinction to first-level operators, called l-*operators*), and the programs in which they are encountered *external programs* (in contradistinction to *internal programs* or *subroutines*, necessary for the realization of the L-operators themselves). We shall assign individual names to the L-operators as they are introduced, fairly arbitrary abbreviations of which, in italics, will serve as the symbols for these operators and will be used in the external program. After the symbol of an L-operator there must appear the list of its operands, closed by the symbol //.

For example, the above operation of finding the minimal column of a minor can be represented by the expression

$$\textit{mincol} \quad \mathbf{Ubcd} \ //,$$

where *mincol* is the symbol for the operator used to find a minimal column of the minor Ubc, and **U**, **b**, **c**, and **d** are the operands subject to its operation. The complex **U** represents the matrix U (i.e., the elements of the complex are the rows of the matrix), and **b** and **c** are the variables defining the sets b and c rows and columns of the matrix U, forming the minor Ubc:

$$\mathbf{b}::\|\ \{b\} \ni B\ \|, \qquad \mathbf{c}::\|\ \{c\} \ni C\ \|,$$

where B and C are the complete sets of rows and columns of U. The minimal column found in the minor is the value of the variable **d**:

$$\mathbf{d}::\|\ \{b'\} \ni B\ \|,$$

where b' is the set of rows of the minor having unit elements at the intersections with the column found.

By analogy, we define the operator *maxrow*, finding a maximal row of the minor, where the expression

$$\textit{maxrow} \quad \mathbf{Ubc}d \ //$$

signifies the operation of applying the given operator to the minor Ubc, represented as in the above example. In the present case the value of the index d will be the number of the maximal row found in the minor Ubc.

Using these two operators, we can write compactly the above algorithm for obtaining the coverage of a Quine table.

§0 $\circ\ \mathbf{n}$
§1 $\textit{mincol}\ \mathbf{Ubcd}\ //\ \textit{maxrow}\ \mathbf{Udc}b\ //$
$c_b \vee \mathbf{n} \Rightarrow \mathbf{n}\ \urcorner \wedge \mathbf{b} \Rightarrow \mathbf{bu}_b\ \urcorner \wedge \mathbf{c} \Rightarrow \mathbf{c}\ |\!\!\rightarrow 1.$

The result of applying this algorithm, a list of the rows of the minor forming the required coverage, will be given by the final value of the variable **n**.

D. SUBROUTINES AND THEIR FORMATION

In their turn, the L-operators *mincol* and *maxrow* must be represented by certain subroutines. We shall consider the rules for their formation.

We shall assume that the variables and indexes required by the subroutine are divided into internal and external. *External* operands of a subroutine are, as a rule, the operands of the subroutine of an L-operator. The remaining variables and indexes required for the realization of the computation process are *internal* to the given program.

For example, the operation *maxrow* **Udc**b can be realized by the following subroutine:

$$\begin{array}{ll} \S 0 & \circ\, b \circ c \\ \S 1 & \mathbf{d}\ \dot{\mathbf{X}}\ 2a\mathbf{u}_a \wedge \mathbf{c}\ \nabla - c \circ\!\rightarrow 1 + c \Rightarrow ca \Rightarrow b \rightarrow 1 \\ \S 2 & . \end{array}$$

In this program the external operands are **U**, **d**, **c**, and b, and the internal operands are a and c.

We shall adopt the following conventions in the formation of subroutines.

(a) The sentences are numbered in succession, beginning with 1. The initial sentence, if there are no transfers to it in the given program, is not numbered.

(b) Everywhere in the subroutine, the symbols α, β, γ, δ, ϵ, ζ, η, ϑ, κ, λ, μ, ν, ξ, π, ρ, σ substitute for the symbols of the external operands in strict correspondence with the positions of the operands in the list accompanying the symbol of the L-operator (for example, the symbol γ in the subroutine substitutes for the symbol of the external operator occupying the third position in the list).

(c) By convention, the complexes used in the subroutine (except for special complexes) always relate to external operands (this convention permits the automatic programming system to be simplified).

In the present example the subroutine will have the following form:

$$\begin{array}{ll} & \circ\, \delta \circ c \\ \S 1 & \beta\ \dot{\mathbf{X}}\ 2a\alpha_a \wedge \gamma\ \nabla - c \circ\!\rightarrow 1 + c \Rightarrow ca \Rightarrow \delta \rightarrow 1 \\ \S 2 & . \end{array}$$

E. AUXILIARY SETS OF VARIABLES AND INDEXES

To facilitate second-level programming, we introduce *auxiliary sets of variables* and *indexes*, employing the same symbols to denote them as above. It must be recalled that the sense of the symbols is defined by their positions—identical symbols will be differently encoded or interpreted, depending on whether they appear in a subroutine or in the external program. We agree to take all the variables and indexes which are internal operands of a subroutine only from these sets (except for the variable **я**). The danger of confusion of variables and indexes in subroutines and the external program is thereby avoided. In writing the external program the programmer need concern himself only with the external operands of those subroutines which correspond to the L-operators needed, without raising the question of their internal operands.

F. EXTENDED CONCEPT OF EXTERNAL OPERAND

Considering the characteristics of the chosen method of compilation, we declare that the elements of the list of operands of an L-operator can be, in particular, the symbols of variables, indexes, constants, complexes, and individual elements of complexes. Variables and complexes may be primed, by placing the prime symbol before the corresponding symbol.

The formal substitution of elements of the list in place of the symbols α, β, . . ., realized during compilation, permits the concept of external operand of a subroutine to be extended. For example, the external operands of a subroutine can be l-operators and even entire LYaPAS expressions. If a list element consists of more than a single symbol, it is called *composite*, and is enclosed in parentheses, except for primed variables and complexes. For example, expressions of the following type are admitted:

$$\textit{operator} \qquad 5'\mathbf{A} + (\mathbf{b}_q)''\mathbf{a}'\mathbf{m}(a + b \Rightarrow c)//.$$

It must be understood that in utilizing composite elements, the enclosing parentheses are removed when they are substituted for the symbols α, β, ...

More detailed information on the possibilities of formal substitution of list elements can be found in the compiler description [5].

G. "PRIMED" VARIABLES

Some discussion of "primed" variables is necessary. It is obvious that the dimensions of internal operands depend on the dimensions of the external ones, and that the "primed" external operands must be associated with "priming" of definite internal variables. We agree to represent the information necessary for executing this priming by listing the couplings among the corresponding external and internal operands. For example, (α **bdm**) signifies that if the external operand α is primed, then the internal variables **b**, **d**, and **m** must be primed in the same way and must, therefore, all have the same dimensions.

H. SUBROUTINE HEADER

Translation of second-level LYaPAS expressions to first-level language must be carried out by means of a special program called a compiler. To facilitate the operation of the compiler, the subroutines of L-operators are given *headers* containing certain characteristics of the subroutines.

The first symbol of a header is the number (code) of the L-operator; the second is its volume, expressed by the total number of symbols in the subroutine, and the third is the number of numbered sentences. These three numbers are set off by bold print. This signifies that they coincide with their codes. Then there follows in the above-described form the information necessary to realize priming, followed by the subroutine proper.

For example, the header

53 116 5 (β**a**) (γ **cd**)

belongs to the subroutine of operator number 53. This subroutine contains 116 symbols and has 5 sentences, numbered 1, 2, 3, 4, 5. When the external operand β is primed, the compiler must realize similar priming of the variable **a**, and when γ is primed, **c** and **d** are primed. For example, if in the third position of the list of operands in the given L-operator there appears ″**m**, then, everywhere in the subroutine generated by the compiler, the symbol ″ must be placed before the symbols **c** and **d**.

I. MULTITERMINAL L-OPERATORS

The L-operator *maxrow* is a two-terminal operator, and has one entry and one exit terminal. It will be useful to consider the more general case of multiterminal L-operators. We shall adopt the following conventions.

We shall assume that the end of the subroutine corresponds to the *basic exit terminal* of the L-operator. By convention *additional* exit terminals can be represented in the corresponding subroutine by the transfer symbols ($\dot{\mathbf{X}}$, $\ddot{\mathbf{X}}$, $\rightarrow$, $\not\leftrightarrow$, $\circ\!\!\rightarrow$, $|\!\!\rightarrow$), after which is placed the symbol of the external operand (α, β, ...), having in the present case a particular meaning.

In composing the external program it is sufficient to substitute in place of the indicated symbol accompanying the symbol of the L-operator the number of any sentence of the external program, in order to connect to this sentence the corresponding terminal of the L-operator.

The start of the subroutine is the *basic entry terminal*. If the L-operator has *additional* entry terminals, they must necessarily correspond to the starts of some given sentences of the subroutine, and connections to them are represented in the external program by means of the same transfer operators, followed by the symbol of the L-operator and the number of the corresponding sentence of the subroutine, represented by a natural constant.

J. SUBROUTINE CHARACTERISTICS

Complex subroutines can have a multilevel hierarchical structure composed of L-operators, many of which can again be expressed in terms of L-operators. The rules for forming the subroutines of such L-operators have their specific features.

Since the internal variables and indexes of such subroutines are taken from the same set as the variables and indexes of the subroutines of the subordinate L-operators, a careful informational matching is necessary. The following rule is established for this purpose.

In forming each subroutine it is necessary to employ its internal variables and indexes economically, using them in alphabetical order, and noting the last one. In composing subroutines containing L-operators it is necessary to take into account which of the variables and indexes have already been used in the subroutines corresponding to these L-operators.

The information necessary for the programmer is incorporated into the characteristic of the subroutine in the following order:

(a) the symbol of the L-operator realized by the given subroutine;

(b) the list of numbers of the L-operators utilized in the subroutine, in order of appearance in the subroutine;

(c) the list of external operands with declaration of the type of each operand (п—a variable, и—an index, к—a complex, ч—a number, о—an operator, в—an expression), tagging with the symbol + those operands whose values do not change during the execution of the subroutine, and with the symbols ? and ?? those complex operands whose symbols are encountered in the subroutine accompanied by ? or ??;

(d) the last (alphabetically) of the variables and indexes used in the subroutine;

(e) the list of subsets of external operands whose elements can be identified (for example, the presence in this list of the expression $\alpha\beta$ signifies that, at each appearance in the subroutine of the symbols α and β, a single symbol of a concrete operand can be substituted);

(f) the list of subsets of external operands, including those whose dimensions in the given subroutine must always coincide.

These lists are separated in the given subroutine by the symbol /, the elements of a list are separated by commas, the elements of lists (b) and (f) are enclosed in parentheses.

For example, the characteristic of a subroutine realizing a certain abstract L-operator *oper*,

oper (5, 7, 3)/αч βо, γк +, δп, ϵк?, ζп/**d**, *e*/$\delta\zeta$/($\gamma\epsilon$)($\delta\zeta$)

signifies, in particular, that in the subroutine the L-operators with numbers 5, 7, and 3 are used, in that order, that the operand α is a certain natural number given by a natural constant, that the value of the complex given by the operand γ does not change during the execution of the subroutine, that the symbols of the complex designated by the operand ϵ are encountered in the program, accompanied by the symbol ? which reflects the growth of this complex during the evolution of the program, that in place of the symbols δ and ξ in the subroutine there can be placed the symbol of a single operand, that the dimensions of the operands γ and ϵ must agree, etc.

If any of the external operands is marked by the sign +, and at the same time it is indicated that it can be identified with some other operand, then it must be considered that in this identification the sign + loses its significance.

K. CERTAIN RECOMMENDATIONS FOR SECOND-LEVEL PROGRAMS

Let us consider certain characteristics of automatic compilation which must be taken into account in second-level programming.

As we have already seen, two forms of reference are admitted in the utilization of L-operators. The first, corresponding to the basic entry terminal of the subroutine and containing the list of operands, is called the *direct form.* The second, corresponding to the use of the additional terminals of the subroutine and containing only the subroutine code and the number of the required terminal, is called *indirect.*

The essence of compilation, which occurs during the inspection of the external program from left to right, consists in the following. Each time the direct form of reference to an L-operator is encountered, the subroutine of this L-operator is inserted into the first-level program being synthesized, after "adjustment" according to the instructions in the operand list. When the indirect form is encountered, the action of the compiler is limited to programming the transfer to one of the additional terminals of the subroutine in the external program, where the set of operands is not altered.

If, in the external program, several direct forms of reference to a single L-operator are required, the synthesized first-level program will have the same number of inclusions of the subroutine for this operator, differing from each other only in that they will be "adjusted" to different sets of external operands.

Therefore, if the volume of the subroutine of a certain L-operator is fairly great, it is recommended to limit it to a single utilization of the direct form, carrying out the "adjustment" of its operands in the external program and forming an additional sentence at the start of which appears the direct form.

This approach should also be used when the L-operator has several entry terminals. The point is that, after analysis of the indirect form of reference to some given L-operator, the compiler programs transfer to the terminal indicated in the indirect form in that subroutine of the L-operator which appears first in the synthesized first-level program. Therefore, if some subroutine with several entrances appears several times in the external program, all transfers to additional entry terminals of this subroutine will be connected to the first occurrence of the subroutine.

L. INCREASING THE EFFICIENCY OF EXPLOITING CONCRETE COMPUTERS

Certain computers are fitted with special devices (parameter registers, loop counters, etc.) which permit complex computational processes to be realized at the level of elementary machine operations. Taking these concrete properties into account permits the efficiency of utilization of the machine to be increased. On the other hand, the inclusion in the developed language of a set of operators permitting these individual features to be exploited is undesirable, since it leads to a more cumbersome language, and would require knowledge of the specific features of the concrete machine and, in the last analysis, programming for it.

Therefore LYaPAS is given an abstract form, in order to be applicable to practically any general-purpose computer. The problem of raising the efficiency of exploitation of a concrete machine is solved in the following way.

Direct contact with the machine language is realized exclusively by the translator, which must take into account the properties of the given machine in the composition of the working program. To facilitate the solution of this problem by the translator, provision is made for a certain extension of LYaPAS through the introduction of additional operators which are special to the given machine and the given translator. The system of "machine plus translator" can be characterized as a whole by a set of costs of the operators of the extended language—the cost of an operator can be, for example, the number of machine cycles necessary for its execution. The problem of equivalence transformations of L-programs then arises, directed toward reducing the total cost of realization. This problem can be solved by a special *optimizer* which processes an L-program expressed in the above-described abstract language to a form optimal for various concrete machines. The optimization of subroutines, described in abstract form, is obviously particularly useful at the time they are entered into the subroutine library stored in the memory of a concrete machine.

Since subroutines are optimized only once, it can be carried out with particular care, while the subsequent use of already optimized subroutines does not require any additional outlay of time.

The optimizer is expressed in abstract LYaPAS, and can therefore be put into operation on any machine, like any block of the automatic programming system except the translator.

If some program or subroutine has not passed through the optimization stage of a concrete machine and concrete translator, it must be reproduced unconditionally by the other blocks of the automatic programming system. The difference will be only that the realization will be connected (in the general case) with a larger expenditure of machine time.

Certain of the functions of the optimizer can be realized by the programmer if he is interested in increasing the speed of realization of the program where an automatic optimizer is not provided. For example, if in some machine there appears a single index register, then, in the composition of the working program, the translator connects this register with the index a. Therefore the use of the index a in internal loops in the program is recommended.

7. SIMPLE OPERATIONS ON COMPLEXES

As has been noted previously, the further development of LYaPAS proceeds along the lines of extending the set of second-level operators, or

L-operators. We shall examine a series of fairly simple L-operators, and the corresponding subroutines, which enable a number of first-level operators to be extended in a certain manner to complexes. The subroutines brought to attention can serve as simple examples of the programming of concrete problems.

A. REDUCTION OF SIMILAR TERMS

The complex α may contain several equal elements. The operator *redsim* "reduces similar terms," i.e., it leaves only a single element of each type (elements which are not equal in this case are of different type) and forms from them the complex β.

$$redsim \quad \alpha\kappa+,\ \beta\kappa/-,\ c/\alpha\beta/(\alpha\beta)$$

40 65 4

$\bar{\circ}\ b \circ c$

§1 $\triangle\, b \oplus b_\alpha \circ\!\!\rightarrow 4\ \bar{\circ}\ a$

§2 $\triangle\, a \oplus c \circ\!\!\rightarrow 3\alpha_a \oplus \alpha_b \circ\!\!\rightarrow 1 \rightarrow 2$

§3 $\alpha_b \Rightarrow \beta_c \triangle\, c \rightarrow 1$

§4 $c \Rightarrow b_\beta$.

B. FINDING LOWER LIMIT

We shall say that the variable **a** absorbs the variable **b** if for any $i \in \{0, 1, \ldots, p-1\}$ there holds $a^i \leftarrow b^i$, where $\leftarrow$ is the implication sign, and if $\mathbf{a} \neq \mathbf{b}$. The lower limit of a complex is the set of those of its elements which do not absorb any of the other elements of this complex. The lower limit of a complex α (if this complex does not contain equal elements) is realized by the operator *lowlim*, which executes this limit in the complex β.

$$lowlim \quad \alpha\kappa+,\ \beta\kappa/-,\ c/\alpha\beta/(\alpha\beta)$$

41 74 4

$\bar{\circ}\ b \circ c$

§1 $\triangle\, b \oplus b_\alpha \circ\!\!\rightarrow 4\ \bar{\circ}\ a$

§2 $\triangle\, a \oplus b_\alpha \circ\!\!\rightarrow 3\alpha_b \urcorner \wedge \alpha_a \mid\!\rightarrow 2a \oplus b \circ\!\!\rightarrow 2 \rightarrow 1$

§3 $\alpha_b \Rightarrow \beta_c \triangle\, c \rightarrow 1$

§4 $c \Rightarrow b_\beta$.

C. FINDING UPPER LIMIT

The operator *uplim* finds the upper limit of a complex α, if this complex does not contain equal elements, defined as the set of those elements of the complex which are not absorbed by any other elements of the same complex.

The result is represented by the complex β.

$$\textit{uplim} \quad \alpha\text{к}+, \beta\text{к}/-, c/\alpha\beta/(\alpha\beta)$$

42 74 4

$$\bar{\circ}\ b \circ c$$

§1 $\triangle\, b \oplus b_\alpha \circ\!\!\rightarrow 4\ \bar{\circ}\ a$

§2 $\triangle\, a \oplus b_\alpha \circ\!\!\rightarrow 3\alpha_a\ \urcorner \wedge \alpha_b \mid\!\rightarrow 2a \oplus b \circ\!\!\rightarrow 2 \rightarrow 1$

§3 $\alpha_b \Rightarrow \beta_c\ \triangle\ c \rightarrow 1$

§4 $c \Rightarrow b_\beta.$

D. COMPRESSION OF A COMPLEX

In the realization of the operator *compress* **AbC**//, a complex **C** is formed of those elements of the complex **A** which are tagged by ones in the code of the variable **b**.

$$\textit{compress} \quad \alpha\text{к}+, \beta\text{п}, \gamma\text{к}/-, b/\alpha\gamma/(\alpha\gamma)$$

43 33 2

$$\circ\ b$$

§1 $\beta\ \dot{\mathbf{X}}\ 2a\ \alpha_a \Rightarrow \gamma_b\ \triangle\ b \rightarrow 1$

§2 $b \Rightarrow b_\gamma.$

E. THE UNION OPERATOR

This operator adds to the complex α those elements of the complex β which are not present in α.

$$\textit{union} \quad \alpha\text{к}, \beta\text{к} + / -, c/(\alpha\beta)$$

44 70 4

$$\bar{\circ}\ b\ b_\alpha \Rightarrow c$$

§1 $\triangle\, b \oplus b_\beta \circ\!\!\rightarrow 4\ \bar{\circ}\ a$

§2 $\triangle\, a \oplus b_\alpha \circ\!\!\rightarrow 3\alpha_a \oplus \beta_b \circ\!\!\rightarrow 1 \rightarrow 2$

§3 $\beta_b \Rightarrow \alpha_c\ \triangle\ c \rightarrow 1$

§4 $c \Rightarrow b_\alpha.$

F. THE INTERSECTION OPERATOR

This operator forms the complex γ of those elements which are common to the complexes α and β (more exactly, of those elements of the complex α which appear in the complex β).

$$\textit{inters}\quad \alpha\text{к}+,\ \beta\text{к}+,\ \gamma\text{к}/-,\ c/\alpha\gamma/(\alpha\beta\gamma)$$
$$\mathbf{45\ 62\ 3}$$
$$\bar{\circ}\ b \circ c$$
$$\S 1\quad \triangle\ b \oplus b_\alpha \circ\!\!\rightarrow 3\ \bar{\circ}\ a$$
$$\S 2\quad \triangle\ a \oplus b_\beta \circ\!\!\rightarrow 1\alpha_b \oplus \beta_a \mid\!\!\rightarrow 2\alpha_b \Rightarrow \gamma_c \triangle\ c \rightarrow 1$$
$$\S 3\quad c \Rightarrow b_\gamma.$$

G. THE DIFFERENCE OPERATOR

The difference operator forms the complex γ from those elements of the complex α which are not present in the complex β.

$$\textit{differ}\quad \alpha\text{к}+,\ \beta\text{к}+,\ \gamma\text{к}/-,\ c/\alpha\gamma/(\alpha\beta\gamma)$$
$$\mathbf{46\ 66\ 4}$$
$$\bar{\circ}\ b \circ c$$
$$\S 1\quad \triangle\ b \oplus b_\alpha \circ\!\!\rightarrow 4\ \bar{\circ}\ a$$
$$\S 2\quad \triangle\ a \oplus b_\beta \circ\!\!\rightarrow 3\alpha_b \oplus \beta_a \circ\!\!\rightarrow 1 \rightarrow 2$$
$$\S 3\quad \alpha_b \Rightarrow \gamma_c \triangle\ c \rightarrow 1$$
$$\S 4\quad c \Rightarrow b_\gamma.$$

H. CONVOLUTION OF A COMPLEX

The expression *convol* $\odot$ **Ab**// is equivalent to the expression

$$\mathbf{a}_0 \odot \mathbf{a}_1 \odot \cdots \odot \mathbf{a}_{\sigma a-1} \Rightarrow \mathbf{b},$$

where $\odot$ is a binary operator.

$$\textit{convol}\quad \alpha\text{о},\ \beta\text{к}+,\ \gamma\text{п}/-,\ a/(\beta\gamma)$$
$$\mathbf{47\ 35\ 2}$$
$$\circ\ a\ \beta_a \Rightarrow \gamma$$
$$\S 1\quad \triangle\ a \oplus b_\beta \circ\!\!\rightarrow 2\gamma\alpha\beta_a \Rightarrow \gamma \rightarrow 1$$
$$\S 2\quad .$$

I. CONVOLUTION WITH COMPRESSION

The operation *concom* ⊙ **Acb** // is distinguished from the operation *convol* ⊙ **Ab** // by the participation of only those elements of **A** in the formation of the value of the variable **b** that are marked by ones in the code of the value of the variable γ.

$$
\begin{array}{ll}
\textit{concom} & \alpha\text{о}, \beta\text{к}+, \gamma\text{п}, \delta\text{п}/-, a/(\beta\delta) \\
 & \mathbf{50\ 34\ 2} \\
 & \gamma\ \dot{\mathbf{X}}\ 2a\beta_a \Rightarrow \delta \\
\S 1 & \gamma\ \dot{\mathbf{X}}\ 2a\delta\alpha\beta_a \Rightarrow \delta \rightarrow 1 \\
\S 2 & .
\end{array}
$$

J. SCALAR PRODUCT

We define the scalar product of complexes **A** and **B** with respect to the binary operator ⊙ as a complex **C** for which $\sigma_c = \min(\sigma_a, \sigma_b)$ and where $i \in \{0, 1, \ldots, \sigma_c - 1\}$ $\mathbf{c}_i = \mathbf{a}_i \odot \mathbf{b}_i$.

The operation of obtaining this complex is realized by the operator *proscal*, for which in the given case $\odot = \alpha$, $\mathbf{A} = \beta$, $\mathbf{B} = \gamma$, and $\mathbf{C} = \delta$.

$$
\begin{array}{ll}
\textit{proscal} & \alpha\text{о}, \beta\text{к}+, \gamma\text{к}+, \delta\text{к}/-, a/\beta\delta, \gamma\delta/(\beta\gamma\delta) \\
 & \mathbf{51\ 60\ 3} \\
 & \bar{\circ}\ ab_\beta - b_\gamma\ \circ\!\rightarrow 1b_\gamma \Rightarrow b_\delta \rightarrow 2 \\
\S 1 & b_\beta \Rightarrow b_\delta \\
\S 2 & \triangle\ a \oplus b_\delta\ \circ\!\rightarrow 3\beta_a\alpha\gamma_a \Rightarrow \delta_a \rightarrow 2 \\
\S 3 & .
\end{array}
$$

K. CARTESIAN PRODUCT

The cartesian product of the complexes **A** and **B** with respect to the binary operator ⊙ is the complex **C** for which $\mathbf{c}_i = \mathbf{a}_j \odot \mathbf{b}_k$, where $i = j + k \cdot \sigma_b$ and $i \in \{0, 1, \ldots, \sigma_c - 1\}$, $j \in \{0, 1, \ldots, \sigma_a - 1\}$, $k \in \{0, 1, \ldots, \sigma_b - 1\}$. The operation of obtaining a complex with these properties is realized by the operator *cartes*, where $\odot = \alpha$, $\mathbf{A} = \beta$, $\mathbf{B} = \gamma$ and $\mathbf{C} = \delta$.

$$
\begin{array}{ll}
\textit{cartes} & \alpha\text{о}, \beta\text{к}+, \gamma\text{к}+, \delta\text{к}??/-, c/(\beta\gamma\delta) \\
 & \mathbf{52\ 64\ 3} \\
 & b_\beta \times b_\gamma \Rightarrow b_\delta??\ \bar{\circ}\ b\ \circ\ c \\
\S 1 & \triangle\ b \oplus b_\gamma\ \circ\!\rightarrow 3\ \bar{\circ}\ a \\
\S 2 & \triangle\ a \oplus b_\beta\ \circ\!\rightarrow 1\beta_a\alpha\gamma_b \Rightarrow \delta_c\ \triangle\ c \rightarrow 2 \\
\S 3 & .
\end{array}
$$

L. CARTESIAN PRODUCT WITH REDUCTION OF SIMILAR TERMS

The operation *cartred* combines the functions of the operators *cartes* and *redsim*, and is of interest in that the reduction is carried out during the calculation of the cartesian product, thereby enabling economical utilization of the working complex. For the operator *cartred*, α is the binary operator with respect to which the cartesian product of the complexes β and γ is constructed, with assignment of the result to the complex δ.

cartred αo, βк+, γк+, δк?/**a**, d/($\beta\gamma\delta$)
53 112 5 (β**a**)
$\bar{\text{o}}$ b o c
§1 $\triangle b \oplus b_\gamma$ o→ 5 $\bar{\text{o}}$ a
§2 $\triangle a \oplus b_\beta$ o→ $1\beta_a\alpha\gamma_b \Rightarrow$ **a** $\bar{\text{o}}$ d
§3 $\triangle d \oplus c$ o→ 4**a** $\oplus\ \delta_d$ o→ 2 → 3
§4 **a** $\Rightarrow \delta_c$? $\triangle c \to 2$
§5 $c \Rightarrow b_\delta$.

M. LOWER LIMIT OF A CARTESIAN PRODUCT

The operator *carlowlim* is equivalent to the effect of the sequential application of the operators *cartes* and *lowlim* and, as in the case of the operator *cartred*, provides economical exploitation of the working complex. The operands α, β, γ, and δ are defined in the same way as for the operator *cartred*.

carlowlim αo, βк+, γк+, δк?/**a**, d/($\beta\gamma\delta$)
54 133 5 (β**a**)
$\bar{\text{o}}$ b o c
§1 $\triangle b \oplus b_\gamma$ o→ 5 $\bar{\text{o}}$ a
§2 $\triangle a \oplus b_\beta$ o→ $1\beta_a\alpha\gamma_b \Rightarrow$ **a** $\bar{\text{o}}$ d
§3 $\triangle d \oplus c$ o→ 4**a** ⌉ $\wedge\ \delta_d$ o→ $2\delta_d$ ⌉ $\wedge$ **a** |→ 3
$\bar{\triangle}\ c\delta_o \Rightarrow \delta_d\ \bar{\triangle}\ d \to 3$
§4 **a** $\Rightarrow \delta_c$? $\triangle c \to 2$
§5 $c \Rightarrow b_\delta$.

N. UPPER LIMIT OF A CARTESIAN PRODUCT

The operator *caruplim* is defined analogously to the operator *carlowlim*, except that in place of the lower, the upper limit of the cartesian product of two complexes is found.

caruplim αo, βк+, γк+, δк?/**a**, d/($\beta\gamma\delta$)
55 133 5 (β**a**)
$\bar{\circ}\ b \circ c$
§1 $\triangle\ b \oplus b_\gamma \circ\!\!\rightarrow 5\ \bar{\circ}\ a$
§2 $\triangle\ a \oplus b_\beta \circ\!\!\rightarrow 1\beta_a\alpha\gamma_b \Rightarrow \mathbf{a}\ \bar{\circ}\ d$
§3 $\triangle\ d \oplus c \circ\!\!\rightarrow 4\delta_d\ \urcorner \wedge \mathbf{a} \circ\!\!\rightarrow 2\mathbf{a}\ \urcorner \wedge \delta_d \mid\!\!\rightarrow 3$
$\bar{\triangle}\ c\delta_c \Rightarrow \delta_d\ \bar{\triangle}\ d \rightarrow 3$
§4 $\mathbf{a} \Rightarrow \delta_c?\ \triangle\ c \rightarrow 2$
§5 $c \Rightarrow b_\delta.$

O. THE CARTESIAN PRODUCT OF A COMPLEX BY A VARIABLE

The operator *carcomva* differs from the operator *cartes* only in that the second of the multiplied complexes contains but a single element, represented by the variable γ.

carcomva αo, βк+, γп, δк/−, $a/\beta\delta$/($\beta\gamma\delta$)
56 37 2
$\bar{\circ}\ ab_\beta \Rightarrow b_\delta$
§1 $\triangle\ a \oplus b_\beta \circ\!\!\rightarrow 2\beta_a\alpha\gamma \Rightarrow \delta_a \rightarrow 1$
§2 .

P. THE CARTESIAN PRODUCT OF A VARIABLE BY A COMPLEX

The operator *carvacom* differs from the operator *cartes* only in that the first of the multiplied complexes contains but a single element, represented by the variable β.

carvacom αo, βп, γк+, δк/−, $a/\gamma\delta$/($\beta\gamma\delta$)
57 37 2
$\bar{\circ}\ ab_\gamma \Rightarrow b_\delta$
§1 $\triangle\ a \oplus b_\gamma \circ\!\!\rightarrow 2\beta\alpha\gamma_a \Rightarrow \delta_a \rightarrow 1$
§2 .

Q. INFORMATION EXCHANGE

The following very simple operators can be useful in programming the exchange of information between complexes.

The operator *transfer* assigns to the complex β the value of the complex α.

$$\textit{transfer}\quad \alpha\text{K}+,\ \beta\text{K}/-,\ a/(\alpha\beta)$$
$$\mathbf{60\ 35\ 2}$$
$$\bar{\text{o}}\ a$$
$$\S 1\quad \triangle\ a \oplus b_\alpha\ \circ\!\rightarrow 2\alpha_a \Rightarrow \beta_a \rightarrow 1$$
$$\S 2\quad b_\alpha \Rightarrow b_\beta.$$

If it is not required to preserve the previous value of the complex α, it is possible to limit the operation to the transfer to the complex β of the characteristics of the complex $\alpha(a_\alpha$ and $b_\alpha)$, using the expression

$$a_\alpha \Rightarrow a_\beta b_\alpha \Rightarrow b_\beta,$$

which is so simple that its substitution by a special L-operator is hardly justified.

In certain cases it may be necessary to obtain the interchange of information between two complexes, which is also programmed simply:

$$a_\alpha \Leftrightarrow a_\beta b_\alpha \Leftrightarrow b_\beta.$$

The operator *cleanup* assigns the value zero to all elements of a complex α.

$$\textit{cleanup}\quad \alpha\text{K}/-,\ a$$
$$\mathbf{61\ 26\ 2}$$
$$\bar{\text{o}}\ a$$
$$\S 1\quad \triangle\ a \oplus b_\alpha\ \circ\!\rightarrow 2\ \circ\ \alpha_a \rightarrow 1$$
$$\S 2\quad .$$

REFERENCES

1. Bottenbruch, G., "The Structure of ALGOL-60 and Its Application" [Russian translation]. Moscow, IL, 1963.
2. Languages for Aiding Compiler Writing, *in* "Symbolic Languages Data Processing," pp. 184–204. New York, London, 1962.
3. Iverson, K. E., "A Programming Language." New York, London, 1963.
4. Zakrevskii, A. D., LYaPAS—a Logical Language for the Representation of Synthesis Algorithms [in Russian] (Materials of the scientific seminar on the theoretical and applied problems of cybernetics). Kiev, 1964.
5. Zakrevskii, A. D., The LYaPAS Compiler. This volume.

REPRESENTATION OF INPUT INFORMATION TO THE LYaPAS COMPILER

M. Ya. Tovshtein

1. INTRODUCTION TO PS-LYaPAS

Algorithms written in LYaPAS are translated to machine language by means of the programming system PS-LYaPAS. As we know, LYaPAS has two levels. A program written in the second level is sequentially transformed to a first-level program. These transformations are carried out by the blocks of the programming system. The machine program (MP) is composed by the translator, which receives first-level LYaPAS.

Figure 1 shows the general scheme of PS-LYaPAS. Let us briefly acquaint ourselves with the purpose of each individual block.

If in the course of checkout of an algorithm the need arises for some given correction of the L-program, it is necessary to indicate those points in the program that are to be changed. On the basis of these indications the *corrector* block composes the corrected variant of the L-program.

It can happen that in the composition of L-programs or in their preparation for input to the machine (coding, punching) an error may be committed, resulting in an expression that makes no sense in LYaPAS (for example, $\S\, a \wedge b$). The block *synter* warns about the presence of such situations, searching for syntactic errors and putting out information about them.

In the process of checking out L-programs it is frequently desirable to know the order in which the sentences are executed (the so-called trace, or

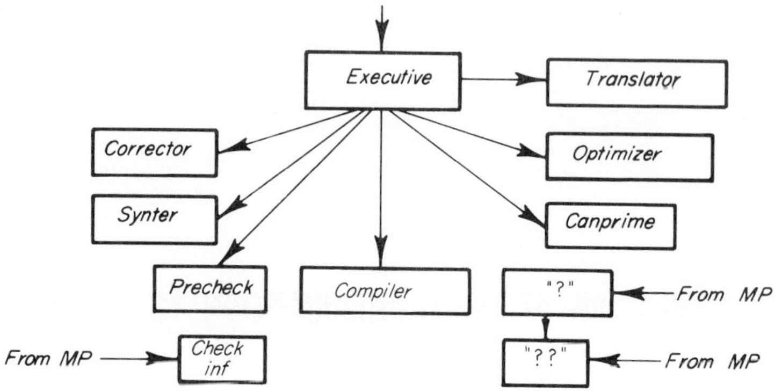

FIG. 1. Block diagram of programming system PS-LYaPAS.

trajectory of the algorithm) and the values of certain external operands. From the indications furnished by the programmer, the block for preparation of checkout (*precheck*) forms operators in the L-program for calling up the block *checkinf*, which produces a listing of checkout information. The block *checkinf* operates during the execution of MP, providing for the output of information by printing or punching.

An important role in PS-LYaPAS is played by the *compiler*. By its aid the L-operators encountered in the algorithm are selected from a pre-existing library, and the corresponding subroutines are included in the L-program. After operation of the compiler, the program is represented basically by expressions of first-level LYaPAS. However, operations on complex variables can still be encountered in it.

In this case the block *canprime* operates, which "cancels primes." In place of operations on complex variables it substitutes operators for calling up programs that carry out these operations. These programs, written in first-level LYaPAS, are switched into the program obtained as a result of the compiler action. After *canprime* operation, the L-program is ready for processing by the translator, which translates it directly to machine program.

The programming system provides for the following possibilities:

(1) the solution of problems with changing sets of input values of the operands without repeated construction of the machine program;

(2) the use of the final values of the operands of one problem as input data for subsequent programs; and

(3) the solution of a problem independently of the preceding one together with its initial data.

Introduced into the machine, PS-LYaPAS is able to process and execute several problems, successively inputting the required information. The programmer informs PS-LYaPAS about the number of problems to be translated by the value of an element k_1 in the working complex K.

2. CODING L-PROGRAMS

Before being input to the machine, LYaPAS expressions must be coded. This means that each symbol of the language is represented by a definite three-position octal code in a system of codes associated one-to-one with the set of LYaPAS symbols. A sequence of codes is obtained that represents the words that must be read by the machine. (In further discussion we shall identify this sequence of *word codes* with the sequence of *words*, and also call it an L-program.) The L-program is compactly represented; several word codes are "packed" into one register of the computer. The chosen system of codes is shown in Table I.

To find the code of a word of the language in the table, it is necessary to add the number of the column in which the symbol appears to the code found at the left of the row in which the symbol is found.

An element of a complex is denoted by two symbols. The first (complex) indicates the name of the complex to which the given element belongs; the second (of a natural constant or index) indicates the ordinal number or index of the element in the complex.

The coding is also composed according to this symbolism; i.e., first the code of the complex is written and then the code of the index.

Examples:

the variable $\mathbf{m}$—254
index p—316
complex $\mathbf{P}$—116
$\mathbf{p}_i$—the ith element of $\mathbf{P}$—116 310
auxiliary variable $\mathbf{m}$—214
auxiliary index p—356

An index can play the same role as a simple variable. Frequently this can be conveniently exploited. The encoded L-program is transferred to punched tape or cards, by means of which it is introduced into the machine's memory.[1]

[1] Translator's note: It will be seen from the above that even in 1966 Soviet programmers could not count on internal processing of alphanumeric codes, and the entire process of transforming an external program to digital code had to be done by hand.

TABLE I

Type of word	Code	0	1	2	3	4	5	6	7	10	11	12	13	14	15	16	17
Operators	000	/	§	(	)	!	↓	.	⇐	△	$\bar{\triangle}$	○	$\bar{\circ}$	⇒	⇔	↗	↙
	040	?		$\dot{\mathbf{X}}$	$\ddot{\mathbf{X}}$	→	⇻	○→	↦	▽	$\mathbf{I}$	↤	⊢	┐	↑	*	$\overset{*}{*}$
Complexes	100	**A**	**B**	**C**	**D**	**E**	**F**	**G**	**H**	**I**	**J**	**K**	**L**	**M**	**N**	**P**	**Q**
	140	*A*	*B*	*C*	*D*	*E*	*F*	*G*	*H*	*I*	*J*	*K*	*L*	*M*	*N*	*P*	*Q*
Variables	200																
	240	**a**	**b**	**c**	**d**	**e**	**f**	**g**	**h**	**i**	**j**	**k**	**l**	**m**	**n**	**p**	**q**
Indexes	300	*a*	*b*	*c*	*d*	*e*	*f*	*g*	*h*	*i*	*j*	*k*	*l*	*m*	*n*	*p*	*q*
	340																
Natural constants	400	0	1	2	3	4	5	6	7	10	11	12	13	14	15	16	17
	440	40	41	42	43	44	45	46	47	50	51	52	53	54	55	56	57
	500	100	301	102	103	104	105	106	107	110	131	112	113	114	115	116	117
	540	140	141	142	143	144	145	146	147	150	151	152	153	154	155	156	157
Standard constants	600	c_0	c_1	c_2	c_3	c_4	c_5	c_6	c_7	c_{10}	c_{11}	c_{12}	c_{13}	c_{14}	c_{15}	c_{16}	c_{17}
	640	d_0	d_1	d_2	d_3	d_4	e_0	e_1	e_2	e_3	e_4	f_0	f_1	f_2	f_3	f_4	f_5

[a] Note: the symbol ⧺, coded 034, does not appear in first-level LYaPAS. It is a prefix used when coding the names of subroutines. For example, the operator *maxrow*, p. 191, has the number **6**, as is stated in the header of this operator: **6 42 2**, i.e., number

3. INITIAL DATA FOR L-PROGRAMS

Since machines differ from each other in memory capacity and in their input/output devices, the programming system must be accompanied by instructions that indicate the specific representation of the input information for the concrete machine and the memory allocation during the operation of PS-LYaPAS. We shall describe the basic rules that must be respected in preparing the initial information for the L-program.

We shall agree that all operands will be represented at the input and output of the machine by the contents of 32 positions of a register, counted from the right. For example, in the 39-bit code of BESM-2 the operands will occupy the positions marked by X:

0 0 0 0 0 0 0 XXXXXXXXXXXXXXXXXXXXXXXXXXXXXXXX

Assigning the simple variables and indexes, the programmer gives definite values to the elements of the special complexes *G* and *H*, respectively. The positions of these complexes in memory are fixed, and must be

LYaPAS CODING SYSTEM

Type of word	Code	20	21	22	23	24	25	26	27	30	31	32	33	34	35	36	37
Operators	000	′	″	‴	⁗									⧺	//	$\odot$	$\otimes$
	040	$\leftleftarrows$	$\langle$	$\rangle$	$\leftarrow$	$\underline{\vee}$	:	$\times$	$\overline{\times}$	+	−	$\wedge$	$\vee$	$\oplus$			
Complexes	100	**R**	**S**	**T**	**U**	**V**	**W**	**X**	**Y**	**Z**	**Ж**	**Л**	**Ф**	**Ш**	**Э**	**Ю**	**Я**
	140	α	β	γ	δ	ϵ	ζ	η	ϑ	κ	λ	μ	ν	$\dot{\xi}$	π	ρ	σ
Variables	200																
	440	**r**	**s**	**t**	**u**	**v**	**w**	**x**	**y**	**z**	**ж**	**л**	**ф**	**ш**	**э**	**ю**	**я**
Indexes	300	*r*	*s*	*t*	*u*	*v*	*w*	*x*	*y*	*z*	ж	л	ф	ш	з	ю	я
	340																
Natural constants	400	20	21	22	23	24	25	26	27	30	31	32	33	34	35	36	37
	440	60	61	62	63	64	65	66	67	70	71	72	73	74	75	76	77
	500	120	121	122	123	124	125	126	127	130	131	132	133	134	135	136	137
	540	160	361	162	163	164	165	366	167	170	171	172	173	174	175	176	177
Standard constants	600	c_{20}	c_{21}	c_{22}	c_{23}	c_{24}	c_{25}	c_{26}	c_{27}	c_{30}	c_{31}	c_{32}	c_{33}	c_{34}	c_{35}	c_{36}	c_{37}
	640	f_6	f_7	f_{10}	f_{11}												

of operator, number of symbols in operator subroutine, number of sentences in operator. This operator is cited in *approshco*, p. 191, and must be coded there as ⧺ 006 α **c** α $c//$.

known to the programmer from the instruction manual. The manual also indicates the dimensions and addresses in the field F of working store assigned to the representations of the values of complexes and compound variables.

The set $\Omega = \{\mathbf{A}, \mathbf{B}, \ldots, \mathbf{Я}\}$ of complexes is divided into two subsets. For each complex in the first subset the programmer must state in advance how many registers of F are required to represent the value of this complex, so that when its cardinality changes it will not "creep" into other complexes. We call the complexes of this subset *fixed*. The programmer himself allocates the field F for them, assigning definite values to the corresponding elements of the complex A (starts of complexes).

Values are assigned to the elements of complex B (cardinalities of complexes) equal to the numbers of elements in the complexes if they are initial for the L-program. If a fixed complex is not initial, its cardinality is found by the program and is assigned to the corresponding element of B in the course of solution of the problem.

Complexes belonging to the other subset of Ω are called *floating*. They differ from fixed complexes in that it is not possible to indicate exactly their cardinalities before the start of solution. If such complexes are initial, the programmer assigns to them sections of F, noting their starts by the values of the corresponding elements of A. The dimensions of these sections should be established with the use of orientational values of the cardinalities of the floating complexes. If the floating complexes are not initial, but are used in the course of solution of the problem (we call them working complexes), the values of the corresponding elements of complex A should be zero. The cardinalities of the floating complexes are found by program.

For floating complexes PS-LYaPAS organizes checking of the "threat of intrusion" by one complex on the "territory" of another and (when necessary) the search for a place for them in a part of F not occupied by fixed complexes. It is clear that much machine time is expended in operation with elements of floating complexes. Therefore, if it is possible to assign them to the group of fixed complexes, this should be done.

Compound variables are represented similarly to fixed complexes. The distinguishing characteristic of the compound variable (which must be taken into account in programming!) is that its "start" is fixed by the corresponding simple variable. If the programmer knows the "start" and dimensions of a compound variable, he gives it the values of definite variables and indexes. (For example, for **″a** the "start" is indicated by the variable **a**, and the dimensions by the index л). Otherwise their calculation must be provided for in the L-program.

In allocating in F the complexes and compound variables it is necessary to observe the following order: first are located (one after the other) the fixed complexes, the compound variables, and then the floating complexes. The number of the first register in the section of F reserved for the floating complexes is indicated by the programmer by the value of the element k_3 of the complex K. In addition, the value of the element $k_4 \in K$ is given by the subset Ω_1 of the fixed complexes of the set Ω:

$$k_4 :: || \{\Omega_1\} \ni \Omega ||,$$

i.e., if the ith element of the set Ω belongs to the set Ω_1, then the ith component of the element k_4 has unit value. The values of the elements of the complexes A and B and the element k_4 must be in core before the translator starts to operate. The values of the initial operands are input to the machine after the translator has prepared the machine program.

4. CHECKOUT MODE

Let $\Psi = \Psi_1 \cup \Psi_2$ be the set of external operands of the L-program, whose values it is desired to list during the execution of the program in the checkout mode. The subsets Ψ_1 of variables and Ψ_2 of indexes are designated by the programmer by means of codes k_5 and k_6, respectively, in the following way:

$$k_5 :: \| \{\Psi_1\} \ni \Psi \|; \quad k_6 :: \| \{\Psi_2\} \ni \Psi \|.$$

Listing is accompanied by indexing, which indicates the code of the operand and the number of the sentence in the L-program where this operand takes on its current value (the variables are indexed by the octal numbers from 00 to 37, the indexes from 40 to 77). For example, for a 45-bit computer the checkout information has the following form (in octal notation):

32	02	340	0001	2531
33	41	000	0000	0012

This signifies that in the 32nd sentence the variable **c** and in the 33rd the index *b* have taken on the following values:

c—1110	0000	0000	0000	0001	0101	0101	1001
b—0000	0000	0000	0000	0000	0000	0000	1010

The number of times each operand in Ψ is listed is counted. An operand is excluded from the set Ψ if this number exceeds a certain limit. The limit is assigned by the programmer, and is given by the value of k_7, where the limit for the variables is represented by the components 14–22 and for the indexes by 23–31.

It can happen that some part of the checked algorithm will repeat a very large number of times because of an error not foreseen by the programmer. To prevent loss of machine time, the programmer must indicate some value of k_{10} not less than the total number of entries to the L-program up to the start of each sentence. It is recommended to choose simple examples in checkout, in order that the course of solution of the problem be simple to follow.

The values of the operands in the listed machine program of the operators $*$ and $\overset{*}{*}$ are also indexed; the value of a complex is listed (punched) with the code of the complex in prefix and the value of a compound variable with the code of the corresponding variable. Simple variables and indexes are listed in batches (about 20 codes in a batch), and labeled as in the checkout mode.

Before each listing the number of the problem solved is printed (punched).

The checkout process includes the search for syntax errors in the L-program, carried out by the block *synter*. We shall label the location of the symbol in the program—its position—by a pair of octal numbers: the ith symbol ($i = 1, 2, \ldots$) of the jth sentence ($j = 0, 1, \ldots, 177$) has the position j_i, where the symbol § has the position j_0, and the number of the sentence has the position j_1.

The error information is printed by the machine in the following order: the position of the symbol ξ_i whose presence in the given location of the L-program gives rise to the inadmissible combination of symbols, and the codes ξ_{i-2}, ξ_{i-1}, ξ_i.

5. CORRECTION

When it becomes necessary to introduce corrections into the L-program, the correction mode may be used.

We shall call the symbol with which an unaltered part of the program begins the left boundary, and the symbol with which this part of the program terminates the right boundary. The information for the correction program (*corrector*) consists of the positions of the boundaries of the unaltered parts of the L-program and the number and codes of the symbols that are to be introduced into the program. The indication of the unaltered part of the program has the form "rewrite from . . . to . . . ." The word *rewrite* (the symbol $\otimes$) is encoded by $\langle\otimes\rangle = 037$. The words *from* and *to* are represented by the positions of the left and right boundaries, respectively. The first number j of the position j_i is represented by the code $(400 + j)$ of a natural constant, the second number i by its three-digit octal positional code.

Information for the *corrector* will be called INCOR for short. The sign of the end of INCOR will be the code $\langle\odot\rangle = 036$. The symbols that must be inserted in the program are coded like LYaPAS words and indicated before the symbols $\otimes$ or $\odot$. In memory INCOR is located ahead of the L-program. The machine, under the control of INCOR, inserts the changes into the L-program. The corrected variant of the program is then processed. It can be listed or punched for further use.

Example: We denote all symbols of a program except the sentence symbols by lower-case latin letters. Then a given program has the form

$$\S\,0\,abcd \quad \S\,1\,efgh \quad \S\,2\,ijk.$$

Suppose it is necessary to remove the underlined symbols $\underline{\S}$, $\underline{0}$, $\underline{a}$, $\underline{b}$, $\underline{e}$, and insert the symbols indicated by the capital letters: A, B, C, D; i.e.,

$$\overset{ABC}{}\;\underline{\S}\,\underline{0}\,\underline{ab}cd \quad \S\,1\,\underline{e}fgh \;\overset{D}{}\; \S\,2\,ijk.$$

INCOR is written as follows:

$$3ABC \otimes \mathbf{0}_4\mathbf{1}_1 \otimes \mathbf{1}_3\mathbf{1}_5 1D \otimes \mathbf{2}_0\mathbf{2}_4.$$

After the operation of *corrector* the following program is obtained:[2]

$$ABC\ cd\ \S\ 1\ fgh\ D\ \S\ 2\ ijk.$$

6. WARNING

From the discussions of the *compiler* and *canprime* it can be seen that the volume of the L-program can grow substantially (the number of sentences can increase) during the operation of PS-LYaPAS. The number of symbols § is also increased by the checkout mode; the block *checkinf*, itself an L-program, is attached to the initial L-program. If during the processing of some L-program the moment arrives when the number of sentences in it exceeds 177_8, the constant

$$f_{10} = 1111 \quad 1111 \quad 1111 \quad 1111 \quad 1111 \quad 1111 \quad 1111 \quad 1111$$

is printed.

A different situation is also possible, in which, during the formation of the machine program, the translator detects that the field reserved for the program in the machine memory is already filled, and the L-program is not yet completely processed. The machine then prints the constant

$$e_4 = 0000 \quad 0000 \quad 0000 \quad 0000 \quad 1111 \quad 1111 \quad 1111 \quad 1111.$$

In these situations the programmer must analyze the problem algorithm for the purpose of representing it in the form of individual (at least two) parts, which can be processed sequentially.

Before solving a problem it is recommended that the programmer inspect the card file of L-operators available in the library. If the required L-operators are not found, the corresponding subroutines are entered into the library under definite numbers. During operation of the programming system the subroutines needed to carry out the algorithm are transferred to core, composing the "working library." Thus each solved problem has its working library. It is not necessary to use the external library, but if it

[2] Translator's note: See also the discussion of the URAL-1 corrector (p. 137), which may be clearer. See also *corrector* (p. 62).

is not used, the programmer himself must compose the working library from subroutines arranged in the strict order necessary for the operation of the *compiler*. If the *compiler* detects a discordance between the working-library structure and the established requirements, it signals this by printing the standard constant

$$d_4 = 1111\quad 1111\quad 1111\quad 1111\quad 0000\quad 0000\quad 0000\quad 0000.$$

Output of the value f_{10}, e_4, or d_4 signifies that the given problem cannot be run. In this case the machine automatically passes to the solution of the next problem, or stops if the number of processed L-programs is equal to the value of k_1.

TABLE II

Bit position (octal)	Operating mode of PS, prescribed by unity in this position
0	Carries out correction of the L-program
1	Searches for syntactic errors
2	Switches in *precheck*
3	Switches in *compiler* (the L-program is written in second-level LYaPAS)
4	Introduces the working library
5	Optimizes the L-program
6	The resultant values of the operators of the given problem are retained
7	Restores the information on the resultant values of the operands of the preceding problem
10	Introduces new initial data in the composition of MP
11–17	Not utilized
20	Print } the corrected L-program
21	Punch } the corrected L-program
22	Print } the compiled program
23	Punch } the compiled program
24	Print } the optimized L-program
25	Punch } the optimized L-program
26	Print } the machine program
27	Punch } the machine program
30–37	Not utilized

7. CALLUP OF THE SYSTEM BLOCKS

The programming system (in particular, its control block—the *executive*) needs to know whether the L-program requires the *corrector*, syntactic errors are to be sought in it, the checkout mode is used, the external library is required, and also at which level of LYaPAS the L-program is written. This information is supplied by the programmer in the form of the values of a definite code whose bit positions are uniquely assigned to the blocks of PS-LYaPAS. The presence of unity in any of the positions signifies the callup of the corresponding block of the programming system, the presence of zero its omission.

The machine can put out the result of the operation of *corrector*, *compiler*, *canprime*, or *translator* by printing or punching. The desired form of output is also indicated by a definite value of a certain code. The position of the code in which unity appears calls up the required output mode. For these purposes we shall use a single 32-bit code $k_2 \in K$. The values of its first 16 positions will indicate the callup (omission) of given blocks of the programming system, and the values of the following bits the output modes. The role of each bit in the code k_2 is shown in Table II.

8. CONCLUSION

The present chapter has given the information necessary for the operation of the programming system PS-LYaPAS. The programmer will obtain practical assistance from the instruction manual appended to PS-LYaPAS for each concrete machine.

PART 1: LYaPAS AND ITS PROGRAMMING SYSTEM

Section B. THE AUTOMATIC PROGRAMMING SYSTEM

THE PROGRAMMING SYSTEM PS-LYaPAS

A. D. Zakrevskii, M. Ya. Tovshtein, and N. R. Toropov

1. THE COMPOSITION OF THE SYSTEM

The purpose of the automatic programming system developed on the basis of LYaPAS is to form a machine program from a given L-program, either in the checkout mode or not, with output of information in terms of the input language.

The programming system PS-LYaPAS contains several blocks, whose purposes are clear from their designations: the block for correction of L-programs (*corrector*), the block for detecting syntactic errors (*synter*), the block for preparing checkout (*precheck*), the block for listing checkout information (*checkinf*), the *compiler*, the block for increasing the dimensions of variables—in other words, "the cancelling of primes"—(*canprime*), the *optimizer*, the *translator*, and two blocks for memory allocation, denoted by "?" and "??."

It is not necessary for all blocks to participate in the operation of the programming system on any given problem. A special control block (*executive*) in the system, guided by the value of a special code [1] supplied by the programmer, successively switches in only those blocks whose operation is foreseen. The machine program is formed directly by the *translator*. The blocks *checkinf*, "?," and "??" operate during the execution of the MP, if needed.

In later chapters, detailed descriptions of the blocks *compiler*, *translator*, and *synter* are given, and the procedure for checking out L-programs on a concrete machine "URAL-1" is discussed. Here we describe the working algorithms of other blocks of PS-LYaPAS, using the information supplied in [1].

2. CORRECTOR

In the course of checking out an L-program the need can arise for various corrections involving the removal, substitution, or addition of certain symbols to the L-program. It is not necessary to recode the program and repunch it on tape or cards in order to make such corrections; this procedure can lead to the appearance of new errors. It is sufficient to furnish the previous program with the corresponding information, according to which *corrector* composes the corrected version. The "new" program thereby obtained can be listed or punched before being subjected to subsequent processing.

Let us give the working algorithm for *corrector*. We shall assume that the L-program is a complex **A**, its corrected version a complex **B**, and INCOR is a complex **C**, where the value of each element of a complex is five LYaPAS words; the indexes *c*, *k*, and *g* serve as word counters in the complexes **A**, **B**, and **C**, respectively.

corrector

	$\bar{\circ}\ g\ \bar{\circ}\ k$	Transfer to **B** the codes in INCOR up to the operator $\langle \otimes \rangle$
§1	$⇻ 21 \oplus \langle \odot \rangle \circ\!\!\to 24b \oplus \langle \otimes \rangle \circ\!\!\to 3$ $c_{27} \oplus b + g \Rightarrow f$	
§2	$g \oplus f \circ\!\!\to 1 ⇻ 21 ⇻ 22 \to 2$	
§3	$\bar{\circ}\ c ⇻ 21 \Rightarrow u$	
§4	$⇻ 23 \oplus \langle \Leftarrow \rangle \mid\!\!\to 5c + 4 \Rightarrow c \to 4$	Inspection of expressions of the type $\xi \Leftarrow \psi_1\psi_2\psi_3\psi_4$, or *oper* ....
§5	$d \oplus \langle ⧺ \rangle \mid\!\!\to 7\ \triangle c \to 4$	
§7	$d \oplus \langle \S \rangle \mid\!\!\to 4 ⇻ 23 \oplus u \mid\!\!\to 4 ⇻ 21c$ $\Rightarrow v + b - 2 \Rightarrow c ⇻ 21 \oplus u$ $\circ\!\!\to 17 \Rightarrow h$	Testing of the position of the unaltered part of the L-program within a single sentence
§10	$⇻ 23h \circ\!\!\to 16d \oplus \langle \Leftarrow \rangle \mid\!\!\to 12d ⇻ 22$ $\bar{\circ}\ z$	Transfer codes relating to the operators $\Leftarrow$ and ⧺
§11	$\triangle z \oplus 4 \circ\!\!\to 10 ⇻ 23 ⇻ 22 \to 11$	
§12	$d \oplus \langle ⧺ \rangle \mid\!\!\to 14$	
§13	$d ⇻ 22 ⇻ 23 ⇻ 22 \to 10$	

§14	$d \oplus \langle § \rangle$	Transfer of codes from the "old" to the "new" program up to the sentence symbol of the right boundary
§15	$\Rightarrow hd \nrightarrow 22 \rightarrow 10$	
§16	$d \oplus b \mid\!\rightarrow 15c \Rightarrow v - 1 \Rightarrow c$	
§17	$\nrightarrow 21 + v \Rightarrow v$	Transfer the following codes up to the right boundary
§20	$c + 1 \oplus v \circ\!\!\rightarrow 1 \nrightarrow 23 \nrightarrow 22 \rightarrow 20$	
§21	$\Delta\, g{:}5 \times 11 \Rightarrow p\mathbf{c}_{\text{я}} \leftleftarrows p \wedge f_7 \Rightarrow b!$	Selection of code in INCOR
§22	$\Rightarrow e\ \Delta\, k{:}5 \times 11 \Rightarrow q100 - q \Rightarrow re \leftleftarrows r \Rightarrow qf_7 \leftleftarrows r \oplus f_{11} \wedge \mathbf{b}_{\text{я}} \vee q \Rightarrow \mathbf{b}_{\text{я}}!$	Transfer code to "new" L-program
§23	$\Delta\, c{:}5 \times 11 \Rightarrow p\mathbf{a}_{\text{я}} \leftleftarrows p \wedge f_7 \Rightarrow d!$	Selection of code in L-program
§24	.	

3. CHECKOUT MODE BLOCKS

The block *precheck* inserts into the L-program the symbols for calling up the block *checkinf*, which is prefixed to the start of the L-program, forming with it one unit. We shall agree that the zero sentence of the block *checkinf* "processes" the indexes of the L-program, the first sentence checks the variables, and the second sentence "tracks" loops. The required correction of the sentence numbers is carried out by the block *precheck*. If a variable, say **d**, or an index e appears in a set Ψ of operands listed during checkout, the following symbols are added at those points of the L-program where **d** and e take on new values: $45 \Rightarrow$ **я** $\nrightarrow 0$ for the index e, and $04 \Rightarrow$ **я** $\nrightarrow 1$ for the variable **d**.

After the symbols § δ (δ being the sentence number) the symbols $\Rightarrow$ **э**$\delta \Rightarrow$ **ю** $\nrightarrow 2$ are added. The variables **э**, **ю**, and **я** are taken from the auxiliary set.

In the algorithm presented below it is assumed that the L-program is represented by a complex **A** and the program obtained as a result of the operation by the complex **B**. For concreteness it is assumed that the value of each element of the complexes **A** and **B** is represented by five successive code words of LYaPAS.

precheck

§1	$\Delta\, j{:}5 \times 11 \Rightarrow p\mathbf{a}_{\text{я}} \leftleftarrows p \wedge f_7 \Rightarrow c$	Extraction of next code in L-program. Formation of complex **B**
§2	$\Delta\, k{:}5 \times 11 \Rightarrow p100 - p \Rightarrow qc \leftleftarrows q \Rightarrow p\ f_7 \leftleftarrows q \oplus f_{11} \wedge \mathbf{b}_{\text{я}} \vee p \Rightarrow \mathbf{b}_{\text{я}c} \Rightarrow b!$	
§3	$\bar{\circ}\, j\ \bar{\circ}\, k$	

§4	$\nrightarrow 1 \oplus \langle . \rangle \circ\!\!\rightarrow 23b \oplus \langle \S \rangle \circ\!\!\rightarrow 17b \oplus \langle \uparrow \rangle \circ\!\!\rightarrow 22b \oplus \langle ', '', ''', '''' \rangle \mapsto 5 \nrightarrow 1 \rightarrow 22$	Exit on the code $\langle \cdot \rangle$
§5	$b \oplus \langle \# \rangle \circ\!\!\rightarrow 16b \oplus \langle \Leftarrow \rangle \mapsto 6 \nrightarrow 1 \nrightarrow 1 \nrightarrow 1 \nrightarrow 1 \rightarrow 22$	Recognition of expressions in the L-program after which no "insertions" are made
§6	$b \oplus \langle (\rangle \circ\!\!\rightarrow 15b \oplus \langle \triangle, \bar{\triangle}, \circ, \bar{\circ} \rangle \circ\!\!\rightarrow 11 \; b \oplus \langle \nearrow, \swarrow \rangle \mapsto 7 \nrightarrow 1 \nrightarrow 1 \nrightarrow 1 \rightarrow 22$	
§7	$b \wedge f_1 \mapsto 22b \oplus \langle \Rightarrow, \Leftrightarrow \rangle \mapsto 10f \circ\!\!\rightarrow 11 \rightarrow 21$	Recognition of symbols of operands and operators
§10	$b \oplus \langle \mathbf{X}, \ddot{\mathbf{X}} \rangle \mapsto 20f \circ\!\!\rightarrow 14 \rightarrow 21$	
§11	$\bar{\circ} f \nrightarrow 1 \oplus \langle (\rangle \circ\!\!\rightarrow 15b \wedge f_0 \circ\!\!\rightarrow 22b \wedge 40 \circ\!\!\rightarrow 12 \langle 1 \rangle \Rightarrow e \rightarrow 13$	Recognition of situations of the type $\mathbf{A} \Rightarrow (\cdots)$
§12	$\langle 0 \rangle \Rightarrow e$	Cases where the operand is an element of a complex
§13	$b \Rightarrow gb \wedge 77 \vee c_{27} \Rightarrow c \nrightarrow 2 \langle \Rightarrow \rangle \Rightarrow c \nrightarrow 2 \langle \text{я} \rangle \Rightarrow c \nrightarrow 2 \langle \nrightarrow \rangle \Rightarrow c \nrightarrow 2e \Rightarrow c \nrightarrow 2 \rightarrow 4$	Formation of "insertions" after operands
§14	$\nrightarrow 1 \rightarrow 11$	
§15	$\nrightarrow 1 \oplus \langle) \rangle \mapsto 15 \rightarrow 22$	Transfer of symbols
§16	$\nrightarrow 1 \oplus \langle // \rangle \mapsto 16 \rightarrow 22$	
§17	$\nrightarrow 1 + \alpha \Rightarrow h \langle \Rightarrow \rangle \Rightarrow c \nrightarrow 2 \langle \text{э} \rangle \Rightarrow c \nrightarrow 2 \; h \Rightarrow c \nrightarrow 2 \langle \Rightarrow \rangle \Rightarrow c \nrightarrow 2 \langle \text{ю} \rangle \Rightarrow c \nrightarrow 2 \langle \nrightarrow \rangle \Rightarrow c \nrightarrow 2 \langle 2 \rangle \Rightarrow c \nrightarrow 2 \langle \text{э} \rangle \Rightarrow g \; \bar{\circ} f \rightarrow 4$	Formation of "insertions" after the symbols § δ with augmentation of the sentence numbers in the L-program
§20	$f \circ\!\!\rightarrow 4$	
§21	$c \Rightarrow s \bar{\triangle} kg \Rightarrow c \nrightarrow 2s \Rightarrow c \nrightarrow 2c \oplus \langle \Rightarrow, \Leftrightarrow \rangle \circ\!\!\rightarrow 11c \oplus \langle \dot{\mathbf{X}}, \ddot{\mathbf{X}} \rangle \circ\!\!\rightarrow 14$	Preservation of the operator code
§22	$\circ f \rightarrow 4$	
§23	.	

All the operands of this program are taken from the auxiliary set. The symbol α signifies the natural constant indicating the number of sentences in the *checking* program (which is not presented here).

The block *checkinf* puts out the variables and indexes of the L-program according to the list supplied by the programmer in the form of ones in the corresponding positions of the codes k_5 and k_6.

The working algorithm of the block *checkinf* can be grasped from the discussion given in [1]. During its execution the values of the "auxiliary" variables я and ю are utilized for indexing and finding the elements of the working complex identical to the corresponding operands; one special complex is reserved to accumulate the values of the operands for listing; the elements of the other complex are the counters of the number of times these values have been printed.

4. THE BLOCK FOR CANCELING PRIMES

If the L-program obtained by compilation contains operations on compound variables, the block *canprime* operates. The principles on which its algorithm is based are the following. The set of so-called prime operators is prepared in advance, consisting of the operations on compound variables represented in the form of complexes. The block *canprime* analyzes in turn the symbols of each operator of an L-program and the operands dependent on them. This group of symbols enables *canprime* to recognize whether an operation on compound variables is concerned, and if so to determine which prime operator can perform it. The call-up operators for the required prime operators are inserted into the L-program with the required information. Certain tables are organized to permit the "addition" of definite prime operators to the L-program and to bring the sentence numbers into correspondence.

5. THE BLOCKS "?" AND "??"

The block "?" serves to watch over the "creeping" of one complex into another during the execution of MP. This is effected by comparing the address of the register containing the value of the last element of the growing complex with the values of the elements of complex A, the starts of all complexes. If a situation arises where "creeping" can occur, the block "??" is switched in to search for an available location in the machine memory for storing the complex.

To represent complex operands (our designation for complexes and "primed" variables) 50% of the working store and a part of external memory (for example, a magnetic drum) is reserved, not smaller in dimensions than the core store.

The block "??" excludes from consideration the part of core that represents fixed complexes. Using the values of those elements of complex B that represent the cardinalities of the floating complex, the block "??" allocates all floating complexes in core and external storage, trying to satisfy two conditions—to maximize the speed of execution of the machine program and to minimize the number of entries into the block "??" itself (the second condition is subordinate to the first). Statistics of access to floating complexes are taken into account, with some extrapolation to the future.

6. OPERATION OF *executive*

Let us consider PS-LYaPAS in the dynamic mode. All blocks of the programming system except the translator are represented in the form of L-programs. The translator has a small volume (about 1000 machine words). Thus PS-LYaPAS is fairly compact. In this form it is stored in external memory, together with the associated subroutine library.

The system is initially started by several machine instructions that bring PS-LYaPAS into core and transfer control to the *translator* (Fig. 1). The

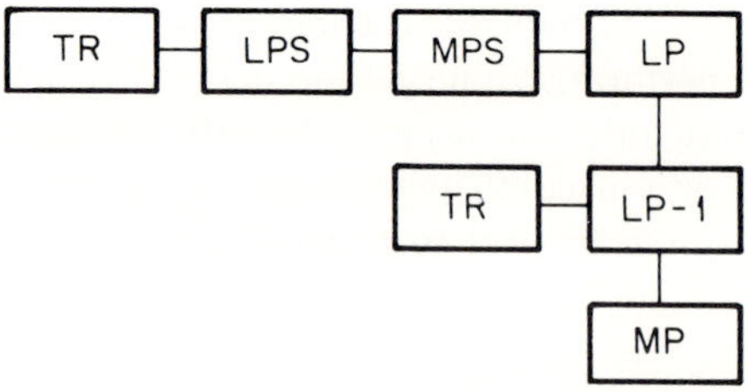

FIG. 1. Stages in the operation of the programming system.

translator (TR) translates the L-programs of the blocks of the programming system (LPS) to machine language. After the machine programming system (MPS) has been composed, one of its blocks (*executive*) begins to operate, sending MPS to external storage. Thus it is henceforth possible to use the machine representation of the system.

Subsequently, *executive* brings in the L-program (LP) and the accompanying information from the input reading device of the machine, making it possible to determine which of the system blocks must be brought into operation. Before *translator* begins to process the L-program reduced to first-level LYaPAS (LP-1), *executive* stores at definite points of core the blocks "?" and "??" (if needed) and the subroutine exercising subsequent control of the system (input of next L-program, exchange of information among various stages of solution of a problem, etc.).

This part of PS-LYaPAS, in contradistinction to the rest, is not erased during the formation (by *translator*) and execution of the machine program

(MP). Thus, during the execution of MP, core is divided into three parts in approximately the following proportions: 25% is occupied by the machine program, 50% by complexes and "primed" variables, and 25% by part of PS-LYaPAS, by special complexes necessary for the solution of the problem, the working registers of the machine program, and the programming system.

7. SUBROUTINE LIBRARY

Together with PS-LYaPAS a library of subroutines is stored in external memory in LYaPAS, fairly compactly. Each subroutine has a number, coinciding with the number of the L-operator realized by it. Before switching in the *compiler* the *executive* prepares the operation by inspecting the L-program for the numbers of the L-operators needed. Then, by a special table of subordination of L-operators and a table of addresses of their subroutines, the *executive* finds the subroutines necessary for the *compiler*, transfers them to core, and arranges them in the required order.

Also provided is the possibility of operating with an external library, composed, for example, of punched cards. In this case the programmer carries out the preliminary arrangement of the subroutines needed for compilation.

REFERENCES

1. Tovshtein, M. Ya., The Representation of Input Information to PS-LYaPAS. This volume.

TRANSLATOR FOR HIGH-SPEED COMPUTERS

M. Ya. Tovshtein

1. STRUCTURE OF THE TRANSLATOR

A. MACHINE-PROGRAM COMPILING

The translator is the core of the LYaPAS programming system. It accepts L-programs written in first-level LYaPAS and constructs machine programs (MP) from them. In addition to first-level LYaPAS, it can take into account expressions appearing in the L-program after operation of the optimizer [1].

The translator is a single-pass compiler [2], since it carries out direct translation from input language to machine language. The code of a LYaPAS operator can be considered to be the name of a certain machine program (consisting of one or several instructions) that is able to execute the given operator. Analyzing the l-operator, the translator selects from a preestablished set of such programs one that bears the given name. As a result of processing the entire L program, a sequence of programs is set up that executes the same functions and in the same order as the l-operators. This sequence of programs is the machine program corresponding to the given L-program.

This is the general scheme of construction of MP. To speak concretely, the process of forming a program that executes an l-operator consists of two steps. In the first step (*analysis*) all operands subject to an operator are recognized and the memory-register addresses in which the values of these operands are stored are found. These data are stored. In the second step (*synthesis*) the translator uses the set of so-called *macroinstructions*. A macroinstruction is a group of instructions (or a single instruction) with indeterminate addresses. The substitution in a certain order of the addresses found in the analysis phase and the adjunction of the program thereby obtained to the already constructed part of MP is the essence of the synthesis algorithm for machine operators.

B. SOME REMARKS ON NOTATION

In the construction of the translator account was taken of the operating experience with PP-LYaPAS-1 [3]. Thus, for example, the addressing principle of MP instructions was modified; tables of constants are formed with the aid of standard constants; the composed MP is easily adjusted for operation in any part of core; likewise, the translator can take up an L-program from any part of core. These changes have made it possible to increase the speed of the translator and to improve the quality of the composed machine program substantially.

The version of the translator presented below is in general oriented toward a three-address machine. However, the description is given in a form neglecting the concrete structure of the instructions. Therefore the basic principles of the algorithms can be utilized for constructing a translator for any machine.

The operation of the translator is described in LYaPAS. The blocks of the translator are represented in the form of L-operators, but without the use of Greek letters to denote the external operands, and without formation of characteristics and heads. The use of LYaPAS in this case makes it possible without great expense of time and means, to find optimal variants for certain problems arising in the construction of the translator, basically check out the algorithm without the use of the machine, and to write the translator in the language of a concrete machine. We shall use the symbols δ, ξ, η, ζ, and $\boldsymbol{\omega}_\eta$ to denote the operands of the translated L-program, where $\boldsymbol{\delta} \in \{0, 1, \ldots, 177\}$; $\boldsymbol{\xi} \in \{\mathbf{a}, \mathbf{b}, \ldots, \mathbf{я}\}$; $\eta \in \{a, b, \ldots, я\}$; $\boldsymbol{\omega}_\eta$ is an arbitrary element of some complex, $\boldsymbol{\Omega}$, $\boldsymbol{\Omega} \in \{\mathbf{A}, \mathbf{B}, \ldots, \mathbf{Я}, A, B, \ldots, Q\}$; $\boldsymbol{\zeta}$ is any of the above operands. If it is necessary to cite several operands of a single type, we shall use primes; for example, $\xi' \wedge \xi'' \Rightarrow \xi'''$. The constructed machine-language instruction (or group of instructions) will be represented in LYaPAS enclosed in braces. The machine program is represented by the complex **P**. The notation $\{\cdots\} \Rightarrow \mathbf{P}$ signifies adjunction of the instructions that carry out the expression in the braces to the machine program. The value of the so-called instruction counter then increases by a definite number. The notation $\{\cdots\} \subset \mathbf{P}$ signifies that the instructions "are written" at the same place as the instructions appended to MP directly before this; i.e., the value of the counter of formed instructions is decreased by the required amount, after which $\{\cdots\} \Rightarrow \mathbf{P}$ is performed.

We denote by $\llcorner \psi \lrcorner$ the number of the identity operand ψ of an element in the working complex K. For example, $\mid \mathbf{b}_0 \mid = [a_1]$, $a_1 \in A$, since $k_{[a_1]} = \mathbf{b}_0$. Since K is a complex, the value of whose ith element is represented in the ith register of core, we shall consider it possible to use the word "register" in place of "element of complex K" and "address" in place of "number of the element of complex K." We shall say that the register "stores," "contains," "fixes," or "represents" one or another value. By $\langle\varphi\rangle$ we denote the code of the symbol φ. We introduce the operation $\varphi \oplus \langle\alpha, \beta, \ldots, \gamma\rangle$, whose result is that the value of the implicit variable τ is equal to 0 if the value of the operand φ coincides with the code of one of the LYaPAS symbols $\alpha, \beta, \ldots, \gamma$ and different from zero otherwise. Other notation will be discussed as it arises in the text.

C. PHRASES OF AN L-PROGRAM

It is convenient to consider that the L-program is divided into phrases. A phrase is composed of an operator with the operands subject to it, if such exist. Very frequently a phrase is itself the operand for an operator that forms the next phrase, by virtue of the implicit variable τ. For example, the expression $a \urcorner \wedge b \Rightarrow c \Rightarrow$ **da** $\dot{\mathbf{X}}\, 2a \circ\!\!\rightarrow 1!$ contains the following phrases: **a** $\urcorner \Rightarrow \tau$, $\tau \wedge b \Rightarrow \tau$, $\tau \Rightarrow c$, $c \Rightarrow$ **d**, **a** $\dot{\mathbf{X}}\, 2a$, $a \circ\!\!\rightarrow 1$, !.

Sometimes, although a phrase appears in the place of an operand for the next operator, it does not appear as the value of τ. In this case we agree to consider the operand of such an operator to be the variable τ itself. For example, the expression § 1 $\urcorner \circ\!\!\rightarrow 2 \Rightarrow f$ consists of the phrases § 1, $\tau \urcorner \Rightarrow \tau$, $\tau \circ\!\!\rightarrow 2$, $\tau \Rightarrow f$. We shall agree that if the operand appearing in front of §, !, $\nrightarrow$, $\rightarrow$ is the variable τ, it forms the left operand of the phrase $\ldots \Rightarrow \tau$. For example, the expression $c \nrightarrow 1 \Rightarrow d \cdots b$ § $1 \Rightarrow a!$ consists of the phrases $c \Rightarrow \tau$, $\nrightarrow 1$, $\tau \Rightarrow d, \ldots, b \Rightarrow \tau$, § 1, $\tau \Rightarrow a$, $a \Rightarrow \tau$, !. During the analysis the translator, in particular, elucidates which of these situations holds.

From the viewpoint of construction it is convenient to divide phrases into several groups. Phrases formed by the operators $\triangle$, $\bar{\triangle}$, $\circ$, $\bar{\circ}$, $\Rightarrow$, $\Leftrightarrow$ are called *singular*. Phrases formed by the operators ∇, $\mathbf{I}$, $\mid\!\leftarrow$, $\vdash$, $\urcorner$, $\leftleftarrows$, $<$, $>$, $\vee$, :, $\times$, $\overline{\times}$, $+$, $-$, $\wedge$, $\vee$, $\oplus$, $-$ are called *nonsingular*. The codes of all these operators have definite distinguishing characteristics, facilitating the analysis of phrases. A separate group of phrases is formed by the operators $\dot{\mathbf{X}}$, $\ddot{\mathbf{X}}$, $\uparrow$, $*$, $\overset{*}{*}$, §, (,), !, ., $\Leftarrow$, $\nearrow$, $\swarrow$. The phrases formed by these operators will be called *nonstandard*, in contradistinction to the *standard* phrases of the first two groups. The analysis of these phrases differs somewhat from the analysis of standard phrases.

The group of phrases formed by the operators $\nrightarrow$, $\rightarrow$, $\circ\!\rightarrow$, $|\!\rightarrow$, $\downarrow$ will be called *transfer* phrases.

D. ROLES OF THE TRANSLATOR BLOCKS

The structure of the translator is evident from Fig. 1. Let us clarify the roles of its basic blocks.

The operation of the translator begins with the block *const.* It examines the L-program and composes a table of constants (TC), which contains the various constants needed for the operation of MP. In parallel with TC is constructed a table in which it is noted whether each sentence of the L-program is "good" or "bad." Sentences are "bad" when § δ appears in place of the left operand in phrases formed by the operators $\Rightarrow$, $*$, $\overset{*}{*}$, in transfer phrases, and in nonsingular phrases. The remaining sentences of the L-program are "good."

The separation of the set of sentences of the L-program into two groups is useful in order to determine whether it is necessary to construct the instructions for assigning the variable τ the value of the left operand in transfer operators.

The machine program is located directly after the TC. Therefore, the length of TC, fixed by some index w, defines the initial value of the instruction counter of the machine program (the role of this counter is played by

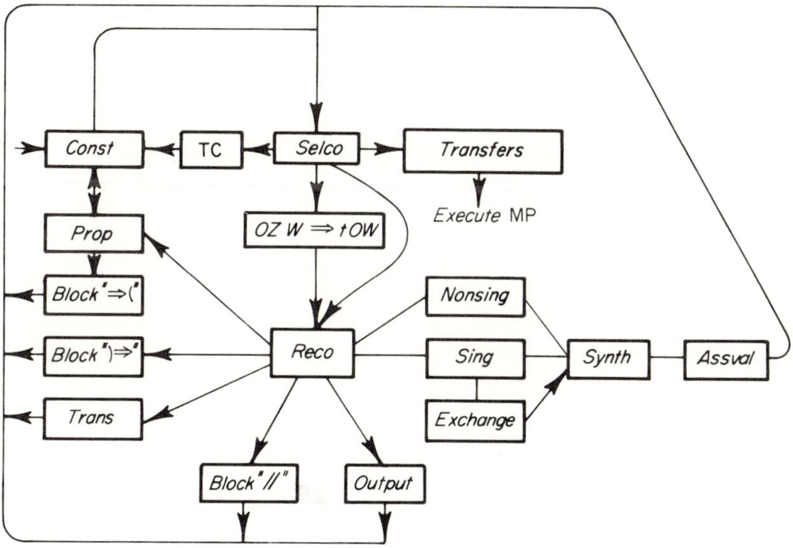

FIG. 1. Translator flow chart.

the index t). After operation of the block *const* the index w takes on its initial value (subsequently it serves to find the position of given constants in TC). The translator returns to the start of the L-program.

Then the block for recognizing codes (*reco*) is brought into operation; it analyzes the code of an operator to determine which block is to carry out the synthesis of the corresponding machine instructions, and transfers control to that block. In addition, *reco* finds or preserves the information on the left operand for each operator.

The block *prop* (processing of operands) finds (or forms) instructions needed to find the address of the register fixing the value of the given operand. If the operand is a constant, then the capability of the block *const* to find the address of the register with the value of this constant is used.

The blocks *sing* and *nonsing* collect information about the composition of singular and nonsingular phrases, respectively. The basic synthesis of instructions is carried out by the block *synth*. An individual block (*assval*) forms that part of *synth* which forms instructions that assign a value to any given operand as a result of an operation. Most frequently *assval* forms the (third) address in the instruction, carrying out the assignment of the result.

The block *trans* analyzes transfer phrases, synthesizes the codes for transfer instructions, and registers the positions of these instructions in the machine program together with the number of the sentence to which transfer is to be effected. After all the phrases of the L-program have been processed, a "rough" machine program is obtained with indeterminate addresses in parts of the transfer instructions. The purpose of the block *transfers* is to put these instructions into their working form. In subsequent paragraphs we shall discuss the working algorithms of the translator blocks.

2. SELECTION OF NEXT CODE IN THE L-PROGRAM

An L-program is given in compact form if several codes of LYaPAS words are densely "packed" into a single memory register of the computer. Then the L-program can be represented in the form of a matrix whose elements will be the codes of the words in the language (assuming that the bit positions not taking part in the representation of codes are at the extreme left; here they will be neglected):

$$S = \left\| \begin{matrix} s_0^0 & s_0^1 & \cdots & s_0^{h-1} \\ s_1^0 & s_1^1 & \cdots & s_1^{h-1} \\ \cdot & \cdot & \cdot & \cdot \\ \cdot & \cdot & \cdot & \cdot \\ \cdot & \cdot & \cdot & \cdot \\ s_0^{r-1} & s_{r-1}^1 & \cdots & s_{r-1}^{h-1} \end{matrix} \right\|,$$

where h is the maximal number of codes that can be represented in a memory register, and

$$r = \begin{cases} N/h & \text{if the number of program codes } N \text{ is proportional to the number } h; \\ [N/h] + 1 & \text{otherwise (here } [\cdots] \text{ denotes the integral part).} \end{cases}$$

In the process of phrase analysis the translator "inspects" the matrix S from left to right and from top to bottom. This examination is carried out by means of the block for selection of the next code (*selco*).

The address of the register representing the first row of the matrix S before the start of operation of the translator must be given by the value of some index, say y. It permits the position of the code $s_0{}^0$ to be found. In further operation the input information for *selco* consists of the coordinates of the code (the number of the row in S and the number of the code in the row) preceding the one to be selected. The coordinates of the selected code are "memorized" by the index z, and the code itself is fixed by the variable **a**.

The algorithm *selco* is very simple. The rows of the matrix S are shifted by the number of positions corresponding to the number (in the row) of the selected code, and the code is "cut out" by means of a constant f_7. To express the algorithm in LYaPAS we associate the matrix S with a complex **S** and construct a so-called scale of shifts—the complex $M = \{m_0, m_1, \ldots, m_{h-1}\}$, where the value m_j indicates the number of positions by which it is necessary to shift the value of the element from **S** in order to extract the jth LYaPAS code.

selco

§1 $\mathbf{s}_y \Leftarrow m_z \Rightarrow \mathbf{a} \triangle z \oplus h \mid\rightarrow 2 \triangle y \circ z$ When the code 000 is encountered,
§2 $\mathbf{a} \wedge f_7 \Rightarrow \mathbf{a} \circ\rightarrow 1.$ the next code is extracted

In the subsequent discussion we will assume that the L-program is represented as a linear sequence of LYaPAS codes (symbols) rather than as a matrix, and that the number of the code in it is given by the index z. We shall let the notation $\triangle z$ signify entry into *selco* and assume that after this operation the variable **a** takes on the value of the next code in the L-program.

3. TAKING ACCOUNT OF CONSTANTS AND "BAD" SENTENCES

As we have already mentioned, constants and "bad" sentences are taken account of by the block *const*, which forms the corresponding tables. The table of constants is constructed by the following algorithm. A constant that occurs otherwise than after § or transfer operators is compared with the standard constants. If there is none such, the TC constructed thus far is examined. If no such constant is found, it is added to TC and the cardinality of the complex **F** representing the table is increased by unity. The address of the register containing the constant is fixed by the variable **a**.

On detecting the combination § δ, the block *const* selects the next symbol and from it decides whether the sentence with number δ is a "bad" one. A table is filled out, represented by the complex $\mathbf{G} = \{\mathbf{g}_0, \mathbf{g}_1, \ldots, \mathbf{g}_{177}\}$; if the ith sentence is "good" $[\mathbf{g}_i] = 0$, otherwise $\mathbf{g}_i = c_0$. In general, it is uneconomical to use the complex **G** only for this purpose. Further, aside from this information $\mathbf{g}_i$ fixes the address of the instructions that begin the ith sentence. We shall call the complex **G** the table of starts of sentences (TSS). TSS is conveniently stored in core in such manner that $\llcorner \mathbf{g}_0 \lrcorner = 400$. Then the code of the sentence number directly indicates the position of the row of TSS in the computer memory.

The block *const* has four entries. The first entry—the start of the block—corresponds to that part of the algorithm where the set of sentences in the L-program is divided into two sets, and the dimensions of TC are defined. (In this stage the construction of TC has a fictitious character.) The second entry (§ 10) is used to process natural constants, the third (§ 11), when needed, to find the address of a register storing the required constant, and the fourth (§ 17) when constants are registered by the natural-assignment operator.

const

	$\circ\, y \circ z$	Setting to start of L-program
§1	$\triangle\, z$	
§2	$\mathbf{a} \oplus \langle . \rangle \circ\!\!\rightarrow 7 \circ \mathbf{da} \oplus \langle (\rangle \circ\!\!\rightarrow 15$ $\mathbf{a} \oplus \langle \Leftarrow \rangle \mid\!\!\rightarrow 3 \nrightarrow 17 \rightarrow 1$	
§3	$\mathbf{a} \wedge f_4 \oplus 40 \mid\!\!\rightarrow 4\mathbf{a} \oplus \langle ? \rangle \circ\!\!\rightarrow 1$ $\triangle\, z \rightarrow 1$	Recognition of constant appearing after a transfer operator
§4	$\mathbf{a} \oplus \langle \S \rangle \mid\!\!\rightarrow 6 \triangle z\mathbf{a} \wedge 177 \Rightarrow i \circ \mathbf{g}_i$ $\mathbf{a} \oplus \langle \Rightarrow \rangle \circ\!\!\rightarrow 5\mathbf{a} \wedge f_2 \oplus 40 \mid\!\!\rightarrow 6$	Registration of "bad" sentence
§5	$c_0 \Rightarrow \mathbf{g}_i \rightarrow 1$	
§6	$\mathbf{a} \oplus \langle /, \S, (,), \Leftarrow, !, \downarrow \rangle \circ\!\!\rightarrow 2\mathbf{a} \wedge f_0$ $\oplus c_{27} \mid\!\!\rightarrow 1 \nrightarrow 10 \rightarrow 1$	Recognition of natural constant

§7 $w \Rightarrow t \circ z \circ y \circ w \rightarrow reco\ 1$ §10 $\mathbf{a} \wedge 177 \Rightarrow \mathbf{z}$ §11 $\circ\ a \llcorner c_0 \lrcorner \Rightarrow \mathbf{a}$ §12 $\mathbf{z} \oplus c_a \circ\!\!\rightarrow 14 \triangle a \oplus 40 \mid\!\rightarrow 12$	Search for constant among standard constants
$\circ\ a \llcorner \mathrm{f}_0 \lrcorner \Rightarrow \mathbf{a}$ §13 $z \oplus \mathrm{f}_a \circ\!\!\rightarrow 14 \triangle a \oplus w \mid\!\rightarrow 13$ $\mathbf{z} \Rightarrow \mathrm{f}_w \triangle w \Rightarrow a$	Search for constant in TC (the index w points to the current cardinality of the complex **F**)
§14 $a + \mathbf{a} \Rightarrow \mathbf{a}!$	Determination of the register address storing the given constant
§15 $\triangle z\mathbf{a} \oplus \langle()\rangle \circ\!\!\rightarrow 16 \triangle \mathbf{da} \wedge f_0 \oplus c_{27}$ $\mid\!\rightarrow 15 \nleftrightarrow 10 \rightarrow 15$ §16 $\mathbf{d} \Rightarrow \mathbf{a} \nleftrightarrow 11 \rightarrow 1$	Registration of constants in an expression $(\cdots) \Rightarrow \Omega$ and the number of operands appearing between the parentheses
§17 $\circ\ d$ §20 $\triangle z\mathbf{a} \leftleftarrows n_d \Rightarrow \mathbf{a} \vee \mathbf{d} \Rightarrow \mathbf{d} \triangle d \oplus 4$ $\mid\!\rightarrow 20\mathbf{d} \Rightarrow \mathbf{z} \nleftrightarrow 11!$	Formation and registration in TC of the constant of the operator $\Leftarrow$

4. PROCESSING OPERANDS

A. ELEMENTS OF COMPLEXES

The role of *prop* was discussed in 1.D. Before describing the working algorithm of this block, let us note the following. In construction of the translator, memory was allocated so that the addresses of the registers fixing the values of LYaPAS variables and indexes can be found by the addition of the code of these operands to some preassigned fixed number ψ. In other words, the elements $k_j \in K$ for $200 + \psi \leq j \leq 277 + \psi$ are identified with the corresponding variables, and for $300 + \psi \leq j \leq 377 + \psi$ with the indexes. If $\psi = 0$, the address of the register is given directly by the code of the operand. The description of the algorithm *prop* directly follows this case. For $\psi \neq 0$ certain obvious operations are added to the algorithm.

In dealing with an element ω_η of a floating complex, *prop* forms the instructions to find $\llcorner \omega_\eta \lrcorner = \llcorner \omega_0 \lrcorner + \eta$, i.e., the address of the register fixing the value of this element. Two variables in K are used to store the value of ω_η: τ and φ, which are called implicit variables. We shall illustrate the execution of operations on the elements of complexes by an example. Let ω_η' and ω_η'' be arbitrary elements in the complex Ω, and let $\square$ be some binary operator. The expression $\omega_\eta' \square \omega_\eta''$ in MP is represented by instructions that do the following:

$$\llcorner \omega_0' \lrcorner + \eta \Rightarrow \tau k_\tau \Rightarrow \tau \llcorner \omega_0'' \lrcorner + \eta \Rightarrow \varphi k_\varphi \Rightarrow \varphi\tau \square \varphi,$$

or, more briefly, $\omega_\eta' \Rightarrow \tau\omega_\eta'' \Rightarrow \varphi\tau \square \varphi$. From this example it is evident that *prop* must know the position of the operand in the phrase with respect to the operators.

If Ω is a fixed complex and the number of its elements is given by the natural constant δ, then $\llcorner \omega_\delta \lrcorner = \llcorner \omega_0 \lrcorner + \delta$ is calculated by *prop* itself. If this number is a value of an index a with which a special index register of the machine is associated, *prop* finds only $\llcorner \omega_0 \lrcorner$ and notes it by a special tag, necessary in the synthesis of instructions for forming address-modification tags. Thus, operations on elements of a fixed complex require fewer machine instructions and less time for their execution.

B. ALGORITHM FOR PROCESSING OPERANDS

The block *prop* has two entry terminals. When the first one (§ 1) is used, the next code is extracted and is then investigated. In entry to the second terminal (§ 2) the next code is not extracted, but the code fixed by the variable **a** is immediately processed. The input information also contains a variable **f**. The values of its three last components indicate the place of the given operand in the phrase and that variable by means of which *prop* furnishes information about it to other blocks:

Number of component	30	31	32
Position of the operand indicated by unit value of the component	$\cdots \Rightarrow \xi$	$\cdots \square \xi \cdots$	$\xi \square \cdots$
Variable fixing the information about the operand	**e**	**d**	**e**

The block *prop* notes whether the value of the result of the operation is assigned to an element of the complex ω_η or to a different operand. In the former case, the variable **e** is assigned the value $\llcorner \tau \lrcorner$, tagged by a definite label, and the value of the variable **g** is $\llcorner \omega_0 \lrcorner$ and $\llcorner \eta \lrcorner$; in the latter case, $[\mathbf{g}] = 0$ and $[\mathbf{e}]$ is equal to the number of the register fixing the value of the operand. The value of **g** is used in the operation of the block *assval.*

prop

§1	$\triangle z$	
§2	$\circ \mathbf{ga} \wedge f_0 \circ\!\!\rightarrow 3\mathbf{a} \Rightarrow \mathbf{e} \wedge c_{30} \mid\!\!\rightarrow 14$ $\not\mapsto$ *const* $10\mathbf{a} \vee c_0 \Rightarrow \mathbf{e} \rightarrow 14$	Recognition of an element of a complex, processing of the constant
§3	$\mathbf{a} \Rightarrow c \llcorner \omega_0 \lrcorner \Rightarrow b \triangle z\mathbf{a} \wedge f_1 \mid\!\!\rightarrow 4$ $\mathbf{a} \oplus \langle \Rightarrow \rangle \mid\!\!\rightarrow$ *output* $\triangle z \rightarrow$ *block* "$\Rightarrow$ ("	Recognition of the expression $\boldsymbol{\Omega} \Rightarrow (\ldots)$, $\boldsymbol{\Omega}*$, $\boldsymbol{\Omega}^*_*$
§4	$b < 24 \Rightarrow dc \wedge 77 \Rightarrow c \wedge 40 \mid\!\!\rightarrow 12$ $c_c \wedge k_4 \mid\!\!\rightarrow 12\mathbf{a} \oplus \langle a \rangle \circ\!\!\rightarrow 10$ $\mathbf{a} \wedge c_{30} \mid\!\!\rightarrow 5 \not\mapsto$ *const* 10	Test: is the complex fixed?[a] The case ω_δ
§5	$\mathbf{a} < 10 \vee d \Rightarrow d\mathbf{f} \wedge 3 \circ\!\!\rightarrow 7$	Fixation of the values $\llcorner \omega_0 \lrcorner$ and $\llcorner \eta \lrcorner$
§6	$\{\omega_\eta \Rightarrow \tau\} \Rightarrow \mathbf{P} \llcorner \tau \lrcorner \vee c_1 \Rightarrow \mathbf{e}$ $\mathbf{f} \wedge 2 \circ\!\!\rightarrow 7 \{\omega_\eta \Rightarrow \varphi\} \subset \mathbf{P} \llcorner \varphi \lrcorner \Rightarrow \mathbf{d}$	Processing of left and right operands in the phrase
§7	$\mathbf{f} \wedge 4 \circ\!\!\rightarrow 15 d \Rightarrow \mathbf{g} \llcorner \tau \lrcorner \vee c_1 \Rightarrow \mathbf{e} \rightarrow 15$	
§10	$\mathbf{a} < 10 \vee d \Rightarrow d\mathbf{f} \wedge 3 \circ\!\!\rightarrow 7$	
§11	$\{\omega_a \Rightarrow \tau\} \Rightarrow \mathbf{P} \llcorner \tau \lrcorner \vee c_1 \Rightarrow \mathbf{ef} \wedge 2 \circ\!\!\rightarrow 7$ $\{\omega_a \Rightarrow \varphi\} \subset \mathbf{P} \llcorner \varphi \lrcorner \Rightarrow \mathbf{d} \rightarrow 7$	
§12	$\mathbf{a} \wedge c_{30} \mid\!\!\rightarrow 13\mathbf{a} \wedge 177 + \llcorner \omega_0 \lrcorner \Rightarrow \mathbf{e} \rightarrow 16$	Calculation of $\llcorner \omega_\delta \lrcorner = \llcorner \omega_0 \lrcorner + \delta$
§13	$\mathbf{a} \oplus \langle a \rangle \mid\!\!\rightarrow 5b \vee c_3 \Rightarrow \mathbf{e} \rightarrow 16$	Processing of ω_a; $\boldsymbol{\Omega}$ is a fixed complex
§14	$\mathbf{f} \wedge 2 \circ\!\!\rightarrow 15\mathbf{e} \Rightarrow \mathbf{d}$	Information about the right operand
§15	!	Tag of the case $\ldots \Rightarrow B_\psi$, $\psi \in \{0, 1, \ldots, 37)$
§16	$c \oplus \langle B \rangle \mid\!\!\rightarrow 14\mathbf{e} \vee c_2 \Rightarrow \mathbf{e} \rightarrow 14$	

[a] $k_4 \in K$ numbers in standard manner fixed complexes.

5. PHRASE ANALYSIS

A. CODE RECOGNITION

The block for code recognition (*reco*) is a nodal element of the translator; construction of the machine instructions that correspond to an L-program phase begins with the operation of this block. Before proceeding to the analysis of phrases and the subsequent synthesis of instructions, *reco* tests whether a free position remains in the field reserved in machine memory for MP. This is achieved by comparison of the value of the constructed instructions counter (the index t) with its known limiting value. (This is not reflected in the algorithm presented below.)

The block *reco* obtains information about the left operand of the phrase and fixes it by means of the variable **c**. The operator code is "memorized" in the variable **b**. On encountering the operator "§" *reco* registers in $\mathbf{g}_i \in \mathbf{G}$ "the start" of the ith sentence.

reco

§1	$\circ\, \mathbf{a}$	(See Section 6D concerning the role of h.)
§2	$1 \Rightarrow \mathbf{fa} \mid\!\rightarrow 3 \;\triangle\; z \;\circ\; h$	
§3	$\mathbf{a} \wedge f_1 \circ\!\rightarrow 4 \;\#\!\!\rightarrow prop\; 2 \rightarrow 1$	Recognition and processing of operand code. Conservation of information about the left operand of the phrase
§4	$\mathbf{e} \Rightarrow \mathbf{c}$	
	$\mathbf{a} \wedge 36 \mid\!\rightarrow 6\mathbf{a} \oplus \langle \S \rangle \mid\!\rightarrow 1 \;\triangle\; z\mathbf{a} \wedge 177 \Rightarrow i \llcorner \tau \lrcorner \Rightarrow \mathbf{eg}_i \wedge c_0 \circ\!\rightarrow 5$	Test of membership of the ith sentence in the class "good"
	$\mathbf{c} \wedge e_4 \oplus \mathbf{e} \circ\!\rightarrow 5$	Determination of the case where in the $i-1$ sentence the result is assigned to τ (for example, $a + b \;\S\; \delta, \ldots \Rightarrow c_d \;\S\; \delta$)
	$\langle \Rightarrow \rangle \Rightarrow \mathbf{b} \circ \mathbf{d} \;\#\!\!\rightarrow synth\; 1$	Synthesis of $\{\xi \Rightarrow \tau\}$ in cases of the type $\cdots \Rightarrow \mathbf{d} \;\S\; \delta + \cdots$
§5	$t \vee \mathbf{g}_i \Rightarrow \mathbf{g}_i \rightarrow 1$	Fixation of the "start" of the ith sentence

§6	**a** ∧ 30 ○→ 10**a** ⇒ **b** ∧ 40 ○→ 7	Recognition of one of the operators (,), ⇐, **Ẋ**, **Ẍ**, ⇸, →, ○→, |→. "Memorization" of the operator code
	a ⊕ 54 ∧ 54 |→ *nonsing* 1**a** ⊕ ⟨¬, ⊕⟩ ○→ *nonsing* 1	Recognition of operator codes and passage to the corresponding blocks for further processing
§7	**a** ⊕ ⟨↗, ↙⟩ ○→ *block* "↗↙" → *sing* 1	
§10	**a** ⊕ ⟨**Ẋ**, **Ẍ**, →, ⇸, ○→, |→⟩ ○→ *trans* 1	
	a ⊕ ⟨(,)⟩ ○→ *block* ")⇒"	
	a ⊕ ⟨!⟩ ○→ *trans* 11**a** ⊕ ⟨↓⟩ ○→ *trans* 14	The operator "." is compared with the instruction for passage to the register corresponding to the variable **ю**. The contents of this register define subsequent operation of PS-LYaPAS
	a ⊕ ⟨.⟩ ○→ *block* "⇐" {→ **ю**} ⇒ **P** → *transfers* 1	
§11	**a** ⊕ ⟨*, ⁑⟩ ○→ *output* {↑} ⇒ **P** → 1	

B. ANALYSIS OF SINGULAR PHRASES

In singular phrases an operand always follows the operator (and this permits singular phrases to be isolated into a separate group). For machines with index registers it is useful to distinguish the case where such an operand will be the index a, and the expressions $\triangle a$, $\bar{\triangle} a$, $\circ a$, $\bar{\circ} a$, $\zeta \Rightarrow a$, $\zeta \Leftrightarrow a$ are interpreted as instructions for changing the value of the index register. The phrases $\triangle \zeta$, $\bar{\triangle} \zeta$, $\circ \zeta$, $\bar{\circ} \zeta$ are replaced by $\zeta + 1 \Rightarrow \zeta$, $\zeta - 1 \Rightarrow \zeta$, $0 \Rightarrow \zeta$, and $f_{10} \Rightarrow \zeta$, respectively. The purpose of the block *sing* is to carry out these changes and to synthesize instructions with the aid of the block *synth.* The instructions for changing the values of the index register are formed directly by the block *sing.* The synthesis of instructions that execute the exchange operator is carried out separately by the block *exch.*

sing

§1	**b** ⊕ ⟨⇔⟩ ○→ *exch* 1 4 ⇒ **fb** ⊕ ⟨△, △̄⟩ |→ 2 △ **f**	Processing of the operand

§2 $\nrightarrow$ *prop* 1 $\circ$ **ae** $\oplus \langle a \rangle \mapsto$ 4**b** $\oplus \langle \Rightarrow \rangle \circ\!\rightarrow 6$
b $\oplus \langle \circ, \bar{\circ} \rangle \mapsto 3\{\bar{\circ}\ a\} \Rightarrow$ **Pb** $\oplus \langle \circ \rangle$
$\mapsto 7\{\bar{\circ}\ a\} \subset$ **P** $\rightarrow 7$
§3 $\{\triangle a\} \Rightarrow$ **Pb** $\oplus \langle \triangle \rangle \circ\!\rightarrow 7\{\bar{\triangle} a\} \subset$ **P** $\rightarrow 7$

Recognition of an expression $\xi \Rightarrow a$. Synthesis of an instruction for changing the value of the index register

§4 $\llcorner c_{37} \lrcorner \Rightarrow$ **db** $\oplus \langle \triangle, \bar{\triangle} \rangle \mapsto$ 5**b** $\vee$ 70
$\Rightarrow$ **be** $\Rightarrow$ **c** $\rightarrow 6$
§5 **b** $\oplus \langle \circ, \bar{\circ} \rangle \mapsto 6\langle \Rightarrow \rangle \Rightarrow$ **b** $\llcorner f_{10} \lrcorner \Rightarrow$ **c**
b $\oplus \langle \circ \rangle \mapsto 6 \circ$ **c**

Substitution of phrases by their equivalents

§6 $\nrightarrow$ *synth* 1 $\rightarrow$ *reco* 2
§7 $\rightarrow$ *reco* 1

C. ANALYSIS OF NONSINGULAR PHRASES

As is easily seen, a nonsingular phrase is the operand in a subsequent phrase. Exceptions are phrases of the type $\xi' : \xi''$, when the value of the remainder after division is not required.

nonsing

§1 **b** $\wedge$ 20 $\circ\!\rightarrow 2$
$2 \Rightarrow$ **f**
$\nrightarrow$ *prop* 1

In the analysis of phrases the block *nonsing* takes into account whether the following phrase is formed by the operator $\Rightarrow$

§2 $\triangle$ **za** $\oplus \langle \Rightarrow \rangle \mapsto 3$
$4 \Rightarrow$ **f**
$\nrightarrow$ *prop* 1 $\circ$ **a** $\rightarrow 4$
§3 $\llcorner \tau \lrcorner \Rightarrow$ **e**

As soon as this occurs, the operand after $\Rightarrow$ is examined. Otherwise the conditions are set up for forming the instruction that will assign the value of the result of the operation to the variable τ

§4 **b** $\oplus \langle \urcorner \rangle \mapsto$ *sing* 6
b $\vee$ 70 $\Rightarrow$ **b** $\llcorner f_{10} \lrcorner \Rightarrow$ **d**
$\rightarrow$ *sing* 6

The phrase $\xi \urcorner$ is "converted" to the phrase $\xi \oplus f_{10}$

In Section 4B we showed that if $\cdots \Rightarrow \boldsymbol{\omega}_\eta$, *prop* assigns to the variable **e** the value $\llcorner \tau \lrcorner$, with a special label. If now we turn to § 3 of *nonsing*, the role of this label for the block *assval* can be comprehended.

6. THE SYNTHESIS OF MACHINE INSTRUCTIONS

A. CHARACTERISTICS OF THE BLOCK *synth*

The input data for the block *synth* are the values of the variables **b**, **c**, **d**, **e**. The operator code represented by **b** indicates the position in core of the required macroinstruction and the procedure for completing the undetermined addresses, using **c**, **d**, **e**. For machines with index registers the block *synth*, constructing instructions which account for the dependence of the elements of a fixed complex on the index a, forms the modification tags for the corresponding instruction addresses. The presence of the variable **я** in the phrase is a signal to include in the program the mechanism for obtaining a random value of this variable.

In the construction of the block *synth* it should be taken into account that the synthesis of certain machine operations can be carried out, up to a certain stage, by a single algorithm. For example, instructions realizing the operators $+$, $-$, $\wedge$, $\vee$, $\oplus$ can be synthesized by a single algorithm; the same algorithm can be used for the synthesis of assignment operations. There is also much in common in the construction of instructions that perform the operations $\Leftarrow$ and $<$, $\times$ and $\overline{\times}$, $\nearrow$ and $\swarrow$, $\mathbf{\Omega} \Rightarrow (\cdots)$ and $(\cdots) \Rightarrow \mathbf{\Omega}$.

In the development of a translator for machines with index registers the correspondence between the register and the index a must be taken into account. For this purpose there is the block *regis*. The specific features of index registers leave their stamp on the construction of the block, but in general it must provide for: (1) formation of instructions fixing the values of the register in core; (2) formation of instructions assigning to the register the contents of some register in core; and (3) the formation of an instruction that changes the value of the register by a natural number, for example, $a \pm \delta \Rightarrow a$, $\delta \Rightarrow a$. In these cases the tag (1 in the 0th position) formed during processing of the constant by the block *prop* is taken into account.

Exit from the block *regis* can be to *synth* (case 1), to *reco* (cases 2, 3), or to *assval* (case 2).

B. REALIZATION OF l-OPERATORS IN LYaPAS

It is obvious that the dimensions of the translator, in particular, depend on the number of undetermined addresses in the macroinstruction instructions. Therefore in the construction of the macroinstructions the number of addresses determined during the operation of *synth* must be reduced as much as possible. For this it is sometimes convenient to utilize macroinstructions in the auxiliary (working) registers.

Let us consider algorithms for the realization of certain LYaPAS operators, denoting by the letters ж, л, ф the elements of the working complex K, which plays the role of the auxiliary variable (τ is the implicit variable; α, β, and γ are external operands)

weight

	$1 \Rightarrow$ ж $\circ$ л	Realization of the expression
§1	$\alpha \wedge$ ж $\circ\!\!\rightarrow 2 \,\triangle$ л	$\alpha \,\nabla \Rightarrow \beta$
§2	ж $< 1 \Rightarrow$ ж $\mid\!\!\rightarrow 1$л $\Rightarrow \beta$.	

invert

	$c_0 \Rightarrow$ ф$1 \Rightarrow$ ж $\circ$ л	Realization of the expression
§1	$\alpha \wedge$ ж $\circ\!\!\rightarrow 2$л $\vee$ ф $\Rightarrow$ л	$\alpha \, \mathrm{I} \Rightarrow \beta$
§2	ф $> 1 \Rightarrow$ фж $< 1 \Rightarrow$ ж $\mid\!\!\rightarrow 1$л $\Rightarrow \beta$	

cycshift

$\alpha \Rightarrow \tau\beta \wedge 37 \Rightarrow$ жτ $<$ ж $\Rightarrow$ л$40 -$ ж	Realization of the expression
$\Rightarrow$ ж	$\alpha \leftarrow \beta \Rightarrow \gamma$
$\tau >$ ж $\vee$ л $\wedge f_{10} \Rightarrow \gamma$	

reflect

	$\alpha \Rightarrow \tau a \Rightarrow$ ф$\beta \Rightarrow a - 5 \circ\!\!\rightarrow 1 \circ$ ж $\rightarrow 2$	Realization of the expression
§1	$1 < a \Rightarrow$ жτ $<$ ж $\wedge d_a \Rightarrow$ л	$\alpha \,\underline{\vee}\, \beta \Rightarrow \gamma$
	$\tau >$ ж $\wedge e_a \vee$ л $\Rightarrow$ ж	
§2	ф $\Rightarrow a$ж $\Rightarrow \gamma$.	

In macroinstructions relating to the operators $|\!\!\leftarrow$, $\vdash$, $:$, $\times$, $\overline{\times}$, instructions are included that provide for the representation of the operands as floating-point numbers, with exponent and mantissa [4]. Let us emphasize certain details in the realizations of these operators.

The Operator $|\!\!\leftarrow$**.** In normalization, the mantissa of an unnormalized number is shifted so that its first significant bit appears in the first bit of the register, and from the exponent of the number is subtracted a number equal to the number of positions by which the mantissa has been shifted. Since the values of the LYaPAS operands in the machine register are represented by 32-bit codes, an "exponent" is adjoined to this code to perform the normalization operations; the value of the operand is represented as a certain unnormalized number. After normalization the exponent must be cancelled; for this, the operation of logical multiplication by the constant f_{10} is applied to the result of the normalization.

The Operator $\vdash$**.** If from the order of the unnormalized number is subtracted the exponent of this number, written in normalized form, the difference indicates the number of zeros before the leftmost 1. This can be used to find the position of the leftmost 1. When this difference is obtained, the entire code in the register must be shifted so that this difference expresses the value of the LYaPAS operand.

The Operator : **.** To realize this operator the values of the operands are first represented as normalized numbers. Then division is carried out with blocked roundoff. The result, represented in the form of a 32-bit code, is assigned to the operand я. The remainder after division is calculated according to the description of this operator [5].

The Operators $\times$ **and** $\overline{\times}$**.** The values of the operands, represented as normalized numbers, are multiplied with blocked roundoff and normalization. The result of this multiplication, after certain transformations (shift, conjunction with f_{10}), gives the result of the operation $\overline{\times}$. For the operator $\times$ the operation of extraction of the lowest-order bits of the product is also carried out.

The random value of the variable **я** is obtained by the use of pseudorandom number generators [6]. Only the code representing the mantissa of an unnormalized number is used.

C. ALGORITHM FOR SYNTHESIS OF MACHINE INSTRUCTIONS

Below we present the algorithm for the operation of the block *synth*. Considering a three-address machine, it can be observed that the synthesis of instructions corresponding to the operators $+$, $-$, $\wedge$, $\vee$, $\oplus$, $\Rightarrow$ requires fewer machine operations than the synthesis of other instructions. The pattern for each of these operators is a single instruction with fixed operator code and three undetermined addresses, whose formation is the same for each of the operators. It is convenient to store these patterns in registers whose addresses coincide with the codes of the corresponding operators (or differ from them by a constant).

synth

§1	*regis* $\mathbf{bcdeP}//$	Test for the presence in the phrase of the index a and formation of instructions using the address register
	$\mathbf{c} \oplus \langle \text{я} \rangle \circ\!\!\rightarrow 2\mathbf{d} \oplus \langle \text{я} \rangle \mapsto 3$	Synthesis of a program for obtaining a random value of the variable **я**
§2	$\{\text{я}\} \Rightarrow \mathbf{P}$	
§3	$\mathbf{b} \oplus \langle +, -\ \wedge, \vee, \oplus \Rightarrow \rangle \mapsto 4$	Synthesis of instructions realizing the operators $\square \in \{+, -, \wedge, \vee, \oplus, \Rightarrow\}$ (for the operator "$\Rightarrow$"$k_{[d]} = 0$)
	$\{k_{[c]} \square k_{[d]}\} \Rightarrow \mathbf{P} \rightarrow 12$	
§4	$\mathbf{b} \oplus \langle \nabla, \text{I}, \mid\!\leftarrow, \mid\!- \rangle \mapsto 6$	Instructions that execute the operators *weight* and *invert* "adapt themselves" to the MP field. The value of the MP instruction counter t is used
	$\mathbf{b} \oplus \langle \mid\!\leftarrow, \mid\!- \rangle \mapsto 5$	
	$\{k_{[c]} \mid\!\leftarrow\} \Rightarrow \mathbf{Pb} \wedge 1 \circ\!\!\rightarrow 12$	
	$\{k_{[c]} \mid\!-\} \subset \mathbf{P} \rightarrow 12$	
§5	$\{k_{[c]} \nabla\} \Rightarrow \mathbf{Pb} \wedge 1 \circ\!\!\rightarrow 12$	
	$\{k_{[c]} \text{I}\} \subset \mathbf{P} \rightarrow 12$	
§6	$\mathbf{b} \oplus \langle \Lleftarrow, <, >, \leftarrow \rangle \mapsto 10$	Formation of programs that execute nonsingular phrases
	$\mathbf{b} \oplus \langle \leftarrow \rangle \mapsto 7 \{k_{[c]} \leftarrow k_{[d]}\} \Rightarrow \mathbf{P} \rightarrow 12$	
§7	$\{k_{[c]} \Lleftarrow k_{[d]}\} \Rightarrow \mathbf{Pb} \oplus \langle > \rangle \mapsto 12$	
	$\{k_{[c]} > k_{[d]} \wedge f_{10}\} \subset \mathbf{P} \rightarrow 12$	
§10	$\mathbf{b} \oplus \langle \veebar, : \rangle \circ\!\!\rightarrow 11 \{k_{[c]} \veebar k_{[d]}\} \Rightarrow \mathbf{P}$	
	$\mathbf{b} \wedge 1 \circ\!\!\rightarrow 12 \{k_{[c]} : k_{[d]}\} \subset \mathbf{P} \rightarrow 12$	
§11	$\{k_{[c]} \times k_{[d]}\} \Rightarrow \mathbf{Pb} \wedge 1 \circ\!\!\rightarrow 12$	
	$\{k_{[c]} \overline{\times} k_{[d]}\} \subset \mathbf{P}$	
§12	*assval* $\mathbf{egtP}//.$	

D. THE BLOCK *assval*

As has been noted, the block *assval* forms addresses in instructions realizing the assignment of the result of some operand. Thus, the synthesis of instructions for the majority of LYaPAS operators terminates with the execution of *assval*. The input information for *assval* consists of the values of the variables **e**, **g** and the index t, indicating the position of the instruction whose address must be determined by the block. After the block has operated the value of t is increased by unity; thus the block *reco* knows the address of the next instruction to be formed.

Let us consider the operation $\cdots \Rightarrow \omega_\eta$. The value of the element of the complex is assigned in two steps. In the first step this value is assigned to the implicit variable τ, in the second $\tau \Rightarrow \omega_\eta$. In this way the implicit variable retains the value of the result of the operation, and if it is used immediately afterwards the instruction $\{\omega_\eta \Rightarrow \tau\}$ is avoided. For example, the expression $\mathbf{a}_i \vee b \Rightarrow \mathbf{c}_j \oplus \mathbf{d}$ is represented in MP by instructions that execute $\mathbf{a}_i \Rightarrow \tau\, \tau \vee b \Rightarrow \tau\, \tau \Rightarrow \mathbf{c}_j\, \tau \oplus \mathbf{d}$. The instruction $\{\mathbf{c}_j \Rightarrow \tau\}$ is not constructed for the phrase $\mathbf{c}_j \oplus d$.

When an element of a floating complex is assigned by the block *assval*, it is necessary in general to test for the presence of the operator "?," as in the case of $\cdots \Rightarrow B_\psi$, for the presence of "??" ($\psi = 0, 1, \ldots, 37$). But sometimes it is unnecessary to make this test (for example, at certain stages of the synthesis of instructions for expressions of the type $\omega_\eta' \Leftrightarrow \omega_\eta''$ or $b_2 \Leftrightarrow b_3$). Therefore some operand is defined, say h, whose value will indicate whether (say $[h] = 0$) or not ($[h] \neq 0$) to inspect for the presence of "?" and "??." If one of these operators is present in the phrase to be processed, *assval* constructs the instructions for calling up the corresponding block.

assval

	$\mathbf{e} \oplus \langle a \rangle \circ\!\!\rightarrow$ *synth* 1 $\{\Rightarrow k_{[e]}\} \Rightarrow \mathbf{P}$	If $\cdots \Rightarrow a$ the block *regis* is called up, which forms the required instructions and executes transfer to the block *reco*
	$\mathbf{e} \wedge c_2 \mid\!\!\rightarrow 1$ $\mathbf{e} \wedge c_1 \circ\!\!\rightarrow 3$	Test for cases of the type $\cdots \Rightarrow B_\psi$ or $\cdots \Rightarrow \omega_\eta$
	$\{\tau \Rightarrow \omega_\eta\} \Rightarrow \mathbf{P}$	Here information about ω_η is used, storing the value of $\mathbf{g}$
§1	$h \mid\!\!\rightarrow 3 \triangle z\mathbf{a} \oplus \langle ? \rangle \mid\!\!\rightarrow 3$ $\triangle z\mathbf{a} \oplus \langle ? \rangle \circ\!\!\rightarrow 2$ $\{\nRightarrow ?\} \Rightarrow \mathbf{P} \rightarrow 3$	Synthesis of instructions for calling up the block "?"
§2	$\{\nRightarrow ??\} \Rightarrow \mathbf{P}$	. . . and the block "??"
§3	!.	

E. THE BLOCK *exch*

In the synthesis of instructions that carry out the operation of exchange of values of two operands, the basic block employed is *synth*. The block *exch* only analyzes the phrase formed by the operator $\Leftrightarrow$ and assigns various values to the operands **c**, **d**, and **e**. By means of these values *synth* constructs the corresponding assignment operations. During phrase analysis $\zeta' \Leftrightarrow \zeta''$ are denoted as such cases.

(1) ζ' and ζ'' are elements of complexes. Then to realize a phrase of the type $\omega_i \Leftrightarrow \omega_j$ instructions for executing $\omega_i \Rightarrow \tau\omega_j \Rightarrow \varphi\tau \Rightarrow \omega_j\varphi \Rightarrow \omega_i$ are constructed, where τ and $\varphi \in K$ are implicit variables. The instructions $\{\omega_i \Rightarrow \tau\}$ and $\{\omega_j \Rightarrow \varphi\}$ are also constructed during operation of *prop*. The instructions corresponding to the remaining two phrases are formed by the block *synth*.

(2) $\zeta' \equiv \omega_\eta$, $\zeta'' \not\equiv \omega_\eta$, i.e., a phrase of the type $\omega_i \Leftrightarrow \xi$ exists. The instructions realizing such a phrase execute $\omega_i \Rightarrow \tau\xi \Rightarrow \varphi\tau \Rightarrow \xi\varphi \rightarrow \omega_i$;

(3) $\zeta' \not\equiv \omega_\eta$, $\zeta'' \equiv \omega_\eta$, i.e., there exists a phrase of the type $\xi \Leftrightarrow \omega_i$. It is put into correspondence with the expression $\omega_i \Rightarrow \varphi\xi \Rightarrow \tau\varphi \Rightarrow \xi\tau \Rightarrow \omega_i$;

(4) ζ' and ζ'' are variables or indexes. The instructions $\{\zeta' \Rightarrow \tau\ \zeta'' \Rightarrow \zeta'\ \tau \Rightarrow \zeta'\}$ correspond to this case.

When one of the operands is the index a, the capability of the block *regis* to form a tag for conserving the value of the index register in a memory register and to assign the value represented by the register to the index register is used.

exch

§1	$\langle\Rightarrow\rangle \Rightarrow$ **bc** $\oplus \langle a\rangle \circ\!\!\rightarrow 11$	Recognition of the expression $a \Leftrightarrow \xi$
	$d \Rightarrow e6 \Rightarrow$ **f** $\nrightarrow$ *prop* **1d** $\Rightarrow f \circ$ **a**	Conservation of information about the right operand and processing of the left
	e $\wedge\ c_1 \circ\!\!\rightarrow 6$**c** $\wedge\ c_1 \circ\!\!\rightarrow 4$ $\nrightarrow$ *synth* $1e \Rightarrow$ **g**	Synthesis of instructions for phrases of the type $\omega_i \Leftrightarrow \omega_j$
§2	$\bar{\circ}\ h \llcorner \varphi \lrcorner \Rightarrow$ **c**	
§3	$\langle\Rightarrow\rangle \Rightarrow$ **b** $\nrightarrow$ *synth* $1f \Rightarrow$ **e** $\rightarrow$ *reco* 2	
§4	**e** $\wedge$ **e**$_4 \Rightarrow$ **e** $\nrightarrow$ *synth* $1\ \bar{\circ}\ h$ $c \Rightarrow$ **e**$\llcorner \varphi \lrcorner \Rightarrow$ **c**$\langle\Rightarrow\rangle \Rightarrow$ **b** $\nrightarrow$ *synth* $1 \llcorner \tau \lrcorner \vee c_1 \Rightarrow$ **e**	Processing of instructions of the type **a** $\Leftrightarrow \omega_i$
§5	$\circ\ h \llcorner \tau \lrcorner \Rightarrow$ **c** $\rightarrow 3$	
§6	$e \Rightarrow$ **ge** $\oplus \langle a\rangle \circ\!\!\rightarrow 10$**c** $\wedge\ c_1 \circ\!\!\rightarrow 7$	Detection of phrases of the type $\omega_i \Leftrightarrow a$ and **a** $\Leftrightarrow$ **b**
	e $\Rightarrow$ **c**$\llcorner \varphi \lrcorner \Rightarrow$ **e** $\nrightarrow$ *synth* 1 **c** $\Rightarrow$ **e**$\llcorner \tau \lrcorner \Rightarrow$ **c**$\langle\Rightarrow\rangle \Rightarrow$ **b** $\nrightarrow$ *synth* $1 \llcorner \tau \lrcorner \vee c_1 \Rightarrow$ **e** $\rightarrow 2$	Processing of expressions of the type $\omega_i \Leftrightarrow$ **a**

§7	$\llcorner \tau \lrcorner \Rightarrow \mathbf{e} \nrightarrow \mathit{synth}\ 1\ \bar{\circ}\ h\mathbf{c} \Rightarrow \mathbf{e}$ $f \Rightarrow \mathbf{c}\langle\Rightarrow\rangle \Rightarrow \mathbf{b} \nrightarrow \mathit{synth}\ 1\mathbf{c} \Rightarrow \mathbf{e} \rightarrow 5$	Processing of expressions of the type $\mathbf{a} \Leftrightarrow \mathbf{b}$
§10	$\mathbf{c} \wedge c_1 \mapsto 13\langle\Rightarrow\rangle \Rightarrow \mathbf{b} \rightarrow 12$	Detection of phrases of the type $\boldsymbol{\omega}_i \Leftrightarrow a,\ b \Leftrightarrow a$
§11	$5 \Rightarrow \mathbf{f} \nrightarrow \mathit{prop}\ 1\mathbf{e} \Rightarrow f \circ \mathbf{a}$ $\mathbf{e} \wedge c_1 \mapsto 13\mathbf{e} \Rightarrow \mathbf{c}\langle a\rangle \Rightarrow \mathbf{e}$	Analysis of expressions of the type $a \Leftrightarrow \boldsymbol{\omega}_i,\ a \Leftrightarrow \mathbf{b}$
§12	$\{a \Leftrightarrow \zeta\} \Rightarrow \mathbf{P}f \Rightarrow \mathbf{e} \rightarrow \mathit{reco}\ 2$	
§13	$\langle a\rangle \Rightarrow \mathbf{c}\llcorner \varphi \lrcorner \Rightarrow \mathbf{e} \nrightarrow \mathit{synth}\ 1\mathbf{c} \Rightarrow \mathbf{e}$ $\llcorner \tau \lrcorner \Rightarrow \mathbf{c}\langle\Rightarrow\rangle \Rightarrow \mathbf{b} \nrightarrow \mathit{synth}\ 1$ $\llcorner \tau \lrcorner \vee c_1 \Rightarrow \mathbf{e}\llcorner \varphi \lrcorner \Rightarrow \mathbf{c} \rightarrow 3$	Processing of expressions of the type $a \Leftrightarrow \boldsymbol{\omega}_i$

7. THE SYNTHESIS OF TRANSFER OPERATIONS

A. PRELIMINARY INSCRIPTION OF INSTRUCTIONS

In the synthesis of transfer operations it is frequently necessary to verify the left operand "for zero" and assign to it the value of the variable τ. For this, an instruction that executes the phrase $\zeta \wedge f_{10} \Rightarrow \tau$ is placed in front of the transfer instruction in MP, where ζ is the left operand in the transfer phrase (possibly $\zeta \equiv \tau$). As a result of this operation $\tau = 0$ if $\zeta = 0$ and $\tau = \zeta$ if $\zeta \neq 0$. In many digital computers in the execution of bitwise comparison ($\oplus$), conjunction ($\wedge$), and disjunction ($\vee$), as well as addition and subtraction ($+$ and $-$) of the mantissas of numbers, a control signal that can be used in the transfer instructions is generated. This is taken into account when a phrase $\zeta' \square \zeta''$ ($\square \in \{+, -, \vee, \wedge, \oplus\}$) is the left operand for a transfer operator $\circ\!\rightarrow$ or $\mapsto$; the transfer instruction is formed without preliminary synthesis of the instruction $\{\tau \wedge f_{10} \Rightarrow \tau\}$. In other cases, $\{\zeta \wedge f_{10} \Rightarrow \tau\}$ is always constructed before the instruction that executes the operator $\circ\!\rightarrow$ or $\mapsto$. Before the instructions corresponding to $\nrightarrow$ and $\rightarrow$ a command $\{\zeta \Rightarrow \tau\}$ is formed if (1) the left operand is not τ; (2) the sentence to which transfer is to be carried out is "bad"; (3) directly before these operators there does not occur $\nrightarrow$. In case (3) this is done because in the execution of the operator "!" there usually occurs assignment to τ of the value of the left operand.

At first the transfer instructions do not have an address part. It is substituted into these instructions after processing of all the phrases in the L-program. In order to be able to do this, a special table of transfer instructions (TTI) is constructed in the form of a complex $\mathbf{H}$. The element $\mathbf{h}_i \in \mathbf{H}$—a row of TTI— fixes in its 12 last components the value of the machine instruction counter at the instant the transfer instruction is formed, and in its 12 first components the number of the sentence appearing after the transfer symbol.

For a phrase $\downarrow \xi$ it is necessary to form instructions ensuring "memorization" of the address in MP of the instruction following the instruction "exit to machine language," composition of the instruction for return to the register with this address, and composition of the tag for transfer to the register whose address is given by $[\xi]$. Exit to machine language is conveniently realized by means of an instruction storing the number of the return register in the form of the address part of a transfer instruction.

B. TRANSFER INSTRUCTIONS WITH RETURN

Instructions that execute the operator $\nrightarrow$ must "memorize" the depth from which transfer is realized and the address of the instruction to which return is carried out. For this, k registers $R = \{r_0, r_1, \ldots, r_{k-1}\}$ are assigned, corresponding to the k depths. Each register contains an instruction of unconditional transfer, to which the address part is added when the instructions $\{\nrightarrow\}$ are executed in MP. To determine to just which of these registers the address is to be "attached," one further register is reserved, called the depth counter. Initially this counter indicates the address of register r_1 (first depth). After the corresponding address is inscribed in register $r_j \in R$, the value of the depth counter is increased by one.

At zero depth the operator "!" can be omitted; the address of the instruction following "!" is stored in the register r_0. In executing the operator "!" the machine "examines" the value of the depth counter. If it corresponds to the zero depth, the transfer instruction stored in r_0 is executed. Otherwise the value of this counter is reduced by unity, and the instruction stored in the register whose address is indicated by the counter is executed.

Instructions realizing the operators $\nrightarrow$ and ! can be constructed in the form of subroutines, located at the start of MP. Then when the symbols $\nrightarrow$ and ! are encountered in the L-program, only the instructions for calling up the corresponding subroutines are synthesized. We designate the synthesis of these instructions by $\{\nrightarrow\} \Rightarrow \mathbf{P}$ and $\{!\} \Rightarrow \mathbf{P}$.

C. OPERATORS FOR ENUMERATION OF 1's IN THE CODE

For the synthesis of instructions for the phrase $\alpha \dot{\mathbf{X}} \beta\gamma$ the following macroinstruction is used (with undetermined instruction addresses corresponding to the phrases with Greek letters):

$$a \Rightarrow \text{л}\alpha \Rightarrow \text{ж} \wedge f_{10} \circ\!\!\rightarrow \beta\text{ж} \;\vdash\; \Rightarrow ac_a \oplus \text{ж} \Rightarrow \text{ж}a \Rightarrow \gamma\text{л} \Rightarrow a\text{ж} \Rightarrow \alpha.$$

When this pattern is substituted in the machine program the instruction ж ∧ f_{10} ○→ β should be entered into TTI. Further the synthesis is conducted as for the operator ▽, Ⅰ, etc.

The algorithm that performs random equiprobable selection of 1's in the code with subsequent "erasure" and indication of their positions (the operator $\ddot{\mathbf{X}}$) is based on the following idea.

Let the code ζ contain n 1's. We divide $[0, 1]$ into n segments

$$\varphi_0 = \left[0, \frac{1}{n}\right], \quad \varphi_1 = \left[\frac{1}{n}, \frac{2}{n}\right], \quad \ldots, \quad \varphi_{n-1} = \left[\frac{n-1}{n}, \frac{n}{n}\right].$$

We take an arbitrary number l from a sequence of random numbers, uniformly distributed along $[0, 1]$ and find a $k \in [0, n-1]$ for which $l \in \varphi_k$. The position of a 1 that is to be "erased" from the code ζ is found from the value of k. This method ensures equiprobable "selection of the 1's" in the code ζ, and the randomness of the choice depends on how arbitrarily the number l is taken. To find l it is possible to use a generator of pseudorandom numbers.

Thus, three elements are essential to the execution of the operator $\ddot{\mathbf{X}}$: calculation of the number n of ones in the code, generation of the random number $l \in [0, 1]$, and determination of a k such that $k < l \cdot n \leq k + 1$. If $n = 1$, the number of the unique 1 in the code is found directly. The following expression corresponds to the phrase $\alpha \dot{\mathbf{X}} \beta\gamma$:

$\alpha \Rightarrow$ э $\Rightarrow$ ж ∧ f_{10} ○→ β ○ ла $\Rightarrow$ шж ▽ $\Rightarrow$ ф ⊕ 1 ○→ 1{l} ⊠ ф $\Rightarrow$ ф
§1   ж $\dot{\mathbf{X}}$ /a △ л − ф ○→ 1c_a ⊕ э $\Rightarrow$ αa $\Rightarrow$ γш $\Rightarrow$ a.

Here {l} is a generator for obtaining the number l, ⊠ is the operator for multiplication of normalized numbers; in the operator $\dot{\mathbf{X}}$ the second operand plays no role.

D. ALGORITHM FOR THE BLOCK *trans*

The basic explanations for the block constructing transfer instructions have already been given. Let us give the LYaPAS algorithm for the functioning of this block. This also includes those parts of the translator where the instructions { ↓ ξ}, {!}, {$\dot{\mathbf{X}}$}, and {$\ddot{\mathbf{X}}$} are synthesized.

trans

§1   **a** $\Rightarrow d$ △ zd ⊕ ⟨$\dot{\mathbf{X}}, \ddot{\mathbf{X}}$⟩ ○→ 12

   d ⊕ ⟨↦, →⟩ |→ 2**a** ∧ 177 $\Rightarrow i$ Test for the case →δ, member-
   g$_i$ ∧ c_0 ○→ 5 ↦ 2 → 5   ship of the sentence δ in the
                                                  group of "good"

§2	$\mathbf{c} \wedge e_4 \oplus \llcorner \tau \lrcorner \,|\!\!\rightarrow 3!c_3 \vee c_1 \wedge \mathbf{c}\,|\!\!\rightarrow 3$	Test: is left operand τ or an element of a complex?
	$\mathbf{b} \oplus \langle +, -, \wedge, \vee, \oplus \rangle \circ\!\!\rightarrow 5$	
	$\nrightarrow 3 \rightarrow 5$	
§3	$\llcorner f_{10} \lrcorner \Rightarrow \mathbf{d} \llcorner \tau \lrcorner \Rightarrow \mathbf{e}\langle \wedge \rangle \Rightarrow \mathbf{b}$	Synthesis of instructions
	$d \oplus \langle \nrightarrow, \rightarrow, !, \dot{\mathbf{X}}, \ddot{\mathbf{X}} \rangle \,|\!\!\rightarrow 4\langle \Rightarrow \rangle \Rightarrow \mathbf{b}$	$\{\zeta \wedge f_{10} \Rightarrow \tau\}$ or $\{\zeta \Rightarrow \tau\}$
§4	$\nrightarrow$ *synth* 1!	
§5	$d \oplus \langle \nrightarrow \rangle \circ\!\!\rightarrow 7\{\mathit{trans}\} \Rightarrow \mathbf{P}$	Synthesis and registration in TTI of transfer instructions
	$\nrightarrow 6 \rightarrow 10$	
§6	$\mathbf{a} < 10 \vee t \Rightarrow \mathbf{h}_s \triangle s!$	Completion of the rows of TTI (Index s is the row counter of TTI; initial value, $s = 0$)
§7	$\{\nrightarrow\} \Rightarrow \mathbf{P}$	
§10	$\llcorner \tau \lrcorner \Rightarrow \mathbf{e}d \Rightarrow \mathbf{b} \rightarrow$ *reco* 1	Exit from *trans*
§11	$\nrightarrow 2\{!\} \Rightarrow \mathbf{P} \rightarrow 10$	
§12	$d \wedge 1 \,|\!\!\rightarrow 13\{\dot{\mathbf{X}}\} \Rightarrow \mathbf{P} \rightarrow 10$	Synthesis of instructions for
§13	$\{\ddot{\mathbf{X}}\} \Rightarrow \mathbf{P} \rightarrow 10$	enumeration operators
§14	$\{\downarrow \xi\} \Rightarrow \mathbf{P} \rightarrow 10.$	Synthesis of instructions for exit to machine language

E. "WORKING" FORM OF TRANSFER INSTRUCTIONS

After composition of the "rough" machine program, the block *transfers* brings the transfer instructions of MP to their final form. It utilizes two tables: TTI and TSS. The rows of TTI indicate which instruction in MP is to be processed and communicates the number of the row in TSS in which is found the address of the instruction to which control is passed during the operation of MP. Examining successively all the rows of TTI, the block *transfers* forms the address parts of all transfer instructions.

8. SYNTHESIS OF INSTRUCTIONS FOR NONSTANDARD PHRASES

A. THE NATURAL ASSIGNMENT OPERATOR

The algorithm for the synthesis of instructions corresponding to the operator $\Leftarrow$ consists of finding in TC the row $\mathbf{f}_w$ containing the constant formed by the block *const* from the four codes following the operator $\Leftarrow$, and then forming the instruction $\{\mathbf{f}_w \Rightarrow \boldsymbol{\zeta}\}$, where $\boldsymbol{\zeta}$ is the left operand of $\Leftarrow$.

The case $\zeta \equiv \boldsymbol{\omega}_\eta$ is treated separately. In the analysis of a phrase $\boldsymbol{\omega}_\eta \Leftarrow \cdots$ the block *reco*, on encountering the code $\boldsymbol{\omega}_\eta$, constructs with the aid of *prop* the instructions for $\boldsymbol{\omega}_\eta \Rightarrow \tau$. It is obvious that they are not necessary to execute $\Leftarrow$. But *reco* only recognizes that just this operator appears after $\boldsymbol{\omega}_\eta$ on the next step, when the code $\langle \Leftarrow \rangle$ is recognized, and control is given to the block "$\Leftarrow$" for the synthesis. This block constructs the instruction $\{\mathbf{f}_w \Rightarrow \tau\}$ and by means of *assval* forms the instruction $\{\tau \Rightarrow \boldsymbol{\omega}_\eta\}$ in place of the instruction $\{\boldsymbol{\omega}_\eta \Rightarrow \tau\}$, which is erased. For this, before the synthesis, block "$\Leftarrow$" reduces the value of the index t by the number of instructions realizing $\boldsymbol{\omega}_\eta \Rightarrow \tau$.

Below we give the algorithm for synthesis of the instructions corresponding to the operator $\Leftarrow$. To correct the counter t we assume that for a machine with an index register the number of instructions corresponding to the phrase $\boldsymbol{\omega}_a \Rightarrow \tau$ is equal to χ; for the phrase $\boldsymbol{\omega}_\eta \Rightarrow \tau$ this number is equal to ϑ.

§1	$\mathbf{c} \Rightarrow \mathbf{e} \wedge c_1 \circ\!\!\rightarrow 3$	Test for situation $\boldsymbol{\omega}_\eta \Leftarrow \cdots$
	$\mathbf{d} \Rightarrow \mathbf{g}$	Conservation of information about $\boldsymbol{\omega}_\eta$
	$\vartheta \Rightarrow d\mathbf{g} > 10 \wedge f_7 \oplus \langle a \rangle \mid\!\!\rightarrow 2\chi \Rightarrow d$	Correction of instruction counter
§2	$t - d \Rightarrow t$	
§3	$\nrightarrow$ *const* $17\mathbf{a} \Rightarrow \mathbf{c}\langle \Rightarrow \rangle \Rightarrow \mathbf{b} \circ \mathbf{d}$	Preparation of the instruction $\{k_{[c]} \Rightarrow k_{[e]}\}$ for synthesis
	$\nrightarrow$ *synth* $1 \rightarrow$ *reco* 1	Synthesis of instructions. Passage to analysis of next phrase

B. PRINTING (PUNCHING) INSTRUCTIONS

The operator $*$ ($\overset{*}{*}$) is represented in MP by instructions that cause output from core of the values of the operands and complexes to be printed (punched). The translator block that forms these instructions has two entry terminals: the first is for the organization of the output of the values of complexes, the second for the formation of the instructions for output of the values of elements of a complex, indexes, variables, and constants. The first entry follows call-up from the block *prop*, the second, from the block *reco*. In the synthesis of output instructions for the value of a complex, indexing of the output code array must be assured. For this it is possible to use the code of the complex which is printed (punched) ahead of the array to be extracted.

To form the instructions for output of a fixed complex it is possible to take into account its "start" and cardinality, indicated by elements in A and B. These instructions conserve a constant form in the execution of MP. In the program for output of a floating complex these "executive" instructions must be constructed by other, "preparatory" instructions, using the values of the elements obtained from A and B and certain constants. Since such a program can become fairly long, it is useful to locate it at the head of MP.[1] The block *output* then has only to form the required call-up instructions.

C. THE BLOCK "$\nearrow\swarrow$"

On encountering the symbol $\nearrow$ or $\swarrow$ in the L-program the block *reco* transfers control to the block "$\nearrow\swarrow$." This block extracts from the L-program the operand codes following the operator code. Since it is known that these can only be variables or indexes, it is possible to find the addresses of the registers fixing the values of these operands without the aid of *prop*, and to use them for "completing" the corresponding patterns.

Basically the synthesis of instructions for the operators $\nearrow$ and $\swarrow$ follows a single algorithm. For the operator $\nearrow$ checking must be provided for the inscription of information in external memory (say, on a drum). This checking is performed by instructions that form a check sum for the codes written on the drum and compare it with the check sum of the same codes obtained during "readback" from the drum; if these sums do not agree, the recording on the drum must be repeated.

D. THE BLOCKS "$\Rightarrow$ (" AND ") $\Rightarrow$"

After it has been found that the operator $\Rightarrow$ appears after the symbol of a complex $\mathbf{\Omega}$, the block "$\Rightarrow$ (" operates, forming a program for the execution of the phrase $\mathbf{\Omega} \Rightarrow (\zeta_1\zeta_2 \cdots \zeta_k)$, where the ζ_i are the symbols of indexes or variables. In the first place block "$\Rightarrow$ (" constructs from known

[1] This is done immediately after operation of the block *const*, if the need is found during the inspection of the L-program. It is also convenient to use such a procedure in cases where the number of instructions corresponding to an operator exceeds a certain threshold (say five instructions). In the algorithm presented for the operation of the block *const* we have omitted this fact for simplicity.

code $\langle \mathbf{\Omega} \rangle$ the instruction for finding the number of the element in the complex K equivalent to the last element of the complex $\mathbf{\Omega}$: $\llcorner \omega_0 \lrcorner + \sigma_\Omega$. Further, in parallel with the selection from the L-program of the symbols of the operands, the corresponding assignment instructions are constructed and the number of operands selected is counted. As soon as the code $\langle) \rangle$ is encountered in the L-program, the number of the register containing such a number is found by means of *const,* and the instruction for correcting the cardinality of the complex is constructed, using this number. Each synthesis step is accompanied by the corresponding increment to t.

The block ") ⇒" is put into operation directly after *reco* has recognized a code $\langle (\rangle$. The structure of the macroinstruction on which the program is based is the same as for the macroinstruction corresponding to the phrase $\mathbf{\Omega} \Rightarrow (\zeta_1\zeta_2 \cdots \zeta_k)$. However, the synthesis of this program is somewhat more complicated. Thus, the instruction for finding the position of the last element of the complex $\mathbf{\Omega}$, which obviously must begin the subroutine, cannot be formed at first, since there is still no information about the complex. Therefore it is necessary to store the number of this instruction in MP. Successively examining the codes of the operands appearing in the parentheses, the block ") ⇒" first forms instructions for the assignment of the values of these operands to the elements of the complex. Just as in the block "⇒ (," the number of operands is counted and the number of the register with the required constant is found. The instructions for finding the last element of the complex, those for correcting the cardinality of the complex, and the assignment instructions are brought to final form from the code $\langle \mathbf{\Omega} \rangle$.

In the execution of the above programs, it is convenient to use the index register, when it exists in the machine, in MP for the execution of the assignment instructions.

9. "FLOATING" MACHINE PROGRAM

In the synthesis algorithms described, the value of the index t was used to indicate the position in core memory where the next-formed MP instruction was to be "placed." The macroinstructions were also "completed" by means of the index t (the reflection of this fact was somewhat suppressed in the notation $\{\cdots\} \Rightarrow \mathbf{P}$). Thus, the instructions obtained in synthesis were adjusted to the field in which MP was allocated. In using the translator in the programming system PS-LYaPAS it is necessary (1) that the place in core reserved for MP not be rigidly fixed, (2) that the machine program constructed be easily adjusted for operation in any part of memory, and (3) that the translator be able to read an L-program from any point in memory.

To satisfy these requirements, we do the following. Let the role of the machine instruction counter be played by two registers—in other words, two operands in the complex K: t and u, where t gives the location in which the constructed instruction is "written," and where the unknown addresses in this instruction (macroinstruction) are determined by means of u. Aside from this, the values of the index u are used for the construction of TSS, i.e., sentence five of the block *reco* takes the form §5 $u \vee \mathbf{g}_i \Rightarrow \mathbf{g}_i \rightarrow 1$. During instruction synthesis the values of t and u, as required, *simultaneously* obtain the same increment. The initial values of the indexes t and u are fixed at the start of operation of the translator, by the indexes э and ю, respectively, taken from the auxiliary set. During the operation of PS-LYaPAS the values of э and ю may evolve. The position of the working program and the place where it is formed evolve according to this.

We distinguish in the same set the index ш, adopting the convention that its value defines the initial address of the block of registers representing the L-program, and the index я, whose value after the termination of the translator operation is equal to the "length" of MP (together with the table of constants).

Thus the translator processes the L-program, whose first symbol is easily located by means of [ш], and constructs the machine program in absolute addresses, beginning with the address [ю + w]; the program is temporarily stored in the array of registers beginning with the register at address [$t + w$], where [w] is the number of rows in TC. (In the particular case, [ю] = [t].) After "insertion" of the absolute addresses in the transfer instructions of MP, the machine program is automatically transferred to the registers with addresses from [ю] to [ю + я]. But this is carried out by a different block of PS-LYaPAS, the *executive*.

10. CONCLUSION

This chapter has given the basic principles of translator construction for the input language, which is first-level LYaPAS. For the description we have used the logical language for representing algorithms, which permits the translator to be composed for any machine without particular outlays. The possibility exists of varying the structure of the translator and its individual blocks according to the features of concrete computers.

REFERENCES

1. Zakrevskii, A. D., Tovshtein, M. Ya., and Toropov, N. R., The Programming System PS-LYaPAS. This volume.
2. Carr, J., "Lectures on Programming" [Russian translation]. IL, Moscow, 1963.
3. Tovshtein, M. Ya., PP-LYaPAS-1—a Translator for First-Level LYaPAS. *Tr. SFTI*, No. 48 (1966).
4. Kitov, A. I., and Krinitskii, N. A., "Electronic Computers and Programming" [in Russian]. Fizmatgiz, Moscow, 1961.
5. Zakrevskii, A. D., Description of LYaPAS. This volume.
6. Buslenko, N. P., Golenko, D. I., *et al.* "The Method of Statistical Experiments [Monte-Carlo]" [in Russian]. Fizmatgiz, Moscow, 1962.

THE LYaPAS COMPILER

A. D. Zakrevskii

1. THE COMPILER FUNCTIONS

The compiler is that block of the automatic programming system which carries out the transition from a program written in second-level LYaPAS, which contains the symbols of L-operators, to a program that is functionally equivalent, but no longer contains the symbols of L-operators [1]. These latter must be replaced by the corresponding subroutines, "adjusted" to the prescribed sets of external operators. The expression obtained at the output of the compiler may differ from a first-level L-program only in that certain of the operands can be "primed," i.e., their values must be represented by consolidated registers of the machine.

Let us describe in greater detail the method of compilation adopted for the development of this compiler.

A. SUBSTITUTION OF OPERANDS

When a subroutine is substituted for the symbol of an L-operator in the processed program, the external operands must be defined in the subroutine. This is done by the substitution of the elements of the list of operands of the L-operator at the positions of the corresponding symbols $\alpha, \beta, \ldots, \sigma$ occurring in the subroutine. In the substitution of a composite element of the list the parentheses enclosing it are removed.

If a symbol $\psi \in \{\alpha, \beta, \ldots, \sigma\}$ follows a symbol A or B (these are the symbols of the complexes of starts and cardinalities of complexes) in the substitution, and if the element replacing it is the symbol of a certain complex, then in place of this element the symbol of a natural constant is substituted, representing the index of the given complex. For example, in the substitution of C for β, a_β is replaced by a_2.

B. PRIMED OPERANDS

If necessary during the compilation, the variables used by the subroutines are primed. The information needed for such priming is contained in the list of operands of the L-operator, where certain elements are primed, and in the header of the corresponding subroutine, where definite relations are reflected between the external and the internal operands of the subroutine.

Since it has been defined that complexes can only serve as external operands of subroutines (with the exception of special complexes, which are never primed), it is not necessary to provide a special mechanism for their priming; the procedure for substitution of a composite element of the list of operands is sufficient.

C. PROCESSING OF THE SYMBOLS ? AND ??

The subroutines processed by the compiler may contain the symbols ? or ??, which can be the signals to shift the complexes tagged by these symbols in the working complex. In the last analysis the presence of these symbols in a first-level program is connected with a definite loss of machine time in the execution of the program. This time is expended in supervising the tagged complexes, which are suspected of possibly interfering with each other. If the programmer can guarantee that certain complexes will not emerge from the place assigned to them in the working complex, these complexes can be included among the fixed ones (in contradistinction to floating ones) and their supervision is abandoned, with removal from the program of the symbols ? and ??, if they accompany the symbols of fixed complexes.

In these conditions the symbols ? and ?? are removed from the subroutines used in compilation by the compiler. The compiler utilizes the list of fixed complexes assigned by the value of the variable **э**, whose 32 binary components are associated in the standard correspondence with the complexes **А**, **Б**, . . . , **Я** (the special complexes are always fixed).

D. ASSEMBLY DURING COMPILATION

During compilation a new program is assembled, in which the direct citations of the L-operators are replaced by their subroutines (and the external program "pushed back"). If any L-operators figure in the substituted subroutine, the corresponding subroutines are inserted in it in turn, etc.

The sentences of the new program are numbered in a unified system, so that the sentences of the subroutines are renumbered. The new numbering is based on the linear ordering of the L-operators utilized, where the operator L_i is considered "to the left" of the operator L_j if one of the following conditions is satisfied:

(1) the operator L_j is subordinate to the operator L_i, i.e., appears in its subroutine;

(2) L_i and L_j enter directly into the same subroutine (or external program), where the direct form of citation of L_i is encountered earlier than the direct form of citation of L_j;

(3) the operator L_i is subordinate to the operator L_k, which is considered to be "to the left" of L_j on the basis of the above conditions.

The sentence numbers of the initial external program are preserved, and the next unused elements of the natural series are utilized for numbering the sentences of the subroutines in the order established by the linear ordering relation described above.

The renumbering of the sentences is accompanied by the necessary corrections of transfers within and between subroutines.

E. PARTICULARITIES OF THE OPERATOR $\Leftarrow$

Since four 9-bit codes follow the operator $\Leftarrow$, which can take on arbitrary values, they need not be analyzed during compilation, but are simply transcribed.

2. DESCRIPTION OF THE COMPILER

Two modes alternate during compilation. The first, called *preparatory mode,* sets in upon the encounter of a direct citation of an L-operator in the processed program, and carries out the necessary preparation of the information for the second mode. In the second, called *assembly mode,* the subroutine corresponding to the L-operator encountered is adjusted and assembled in the synthesized program. As soon as any L-operator is encountered in turn in the assembled subroutine (in direct form), the assembly mode is again replaced by the preparatory mode, and the information needed to complete adjustment and assembly of the subroutine is stored. The basic mass of processed information is represented by complexes. The initial information, i.e., the external program, and all the subroutines needed for the compilation together with their headers, in the order established by the linear ordering relation, form a sequence A' of 9-bit codes.

We shall assume that this information is stored in "packed" form in the complex **A**. To be concrete, we shall take the following method of packing: five 9-bit LYaPAS symbols are stored from left to right in a 45-bit machine register. At the same time, the standard 32-bit variables utilize only that part of the register obtained by dropping the 9 left and 4 right bits.

The unpacking operations, i.e., search and extraction from the complex **A** of an arbitrary element of the sequence A' (prescribed by number), are performed by a special unpacking block, using the operator $\Leftarrow$.

In an analogous condensed form the result of compilation is represented in the complex **B**—the resultant first-level L-program, in which primes may appear. The unpacking and packing of the complex **B** is carried out analogously.

As noted in the description of LYaPAS, a header appears before the subroutine, in which are listed in succession the number of the subroutine, its volume, the number of sentences, and information necessary for priming. We shall agree that the first symbol of the header (the number of the L-operator) always occupies the extreme left of the five positions in the machine register, and the others follow directly behind it. Then the volume of the subroutine (including the header) is defined by the number of occupied registers, each of which contains one element of the complex **A**.

If the total number of symbols in the subroutine is not a multiple of five, the last positions in the last element of the complex **A** occupied by the subroutine remain empty. The depth of the next subroutine is noted in the rightmost position of the next element of **A**; this next subroutine is stored in **A** beginning from the next element after that in which its depth is stored. We shall assume that the external program is realized at depth 1 and that if some subroutine is realized at depth i, a subroutine directly subordinate to it is realized at the depth $i + 1$. The depth of operation of the compiler is defined in agreement with this, and is initially denoted by the value of the index d.

When the next L-operator is encountered, its list of operands is analyzed, and tables, represented in the complexes **D**, **E**, and **C** are prepared for convenience of the subsequent "adjustment" of the subroutine.

The complex **C** prescribes the general list of all operands of those L-operators whose processing at the current moment has not been terminated and which are processed at various depths beginning with 2 and terminating with $[d]$. This list has a continuous dense numbering, and since certain of the operands can be represented by more than one symbol, the complex **E** defines the corresponding division of the complex **C**—if $j = \mathbf{e}_i$, then $\mathbf{c}_j$ is the first symbol representing the [i]th operand of the general list. The complex **D** defines in turn the order of dividing **E** into groups, corresponding to the different L-operators: [$\mathbf{d}_d$] is the number of the first operand of an L-operator that has been processed at the depth [d] and whose processing has not been terminated at the current moment.

After the set of external operands of the current subroutine has been prepared, the list of primed variables is composed. The set of these lists for the various depths forms the complex **H**, in which five adjacent elements correspond to each depth. The first of these is the general list of all variables subject to priming at the given depth, and the next four define the type of priming, giving the list of internal variables of the subroutines before whose symbols it is necessary to insert the symbol ′, ″, ‴, or ⁗.

Then the required subroutine is assembled, for which the details have already been described.

Transfers are corrected with the use of information contained in the complex **F**; [$\mathbf{f}_d$] represents the increment that must be given to all sentence numbers at the [d]th depth.

When the symbol "." is encountered, the depth is reduced and transfer is carried out to the preceding subroutine whose processing has not been terminated. The set of "return coordinates" for the various depths is given by the complex **G**.

The starts of all subroutines processed at various depths (at each depth there is a maximum of one subroutine whose processing has not terminated) is given by the complex **I**.

The operation of the compiler terminates with return to the zero level, upon encountering the symbol "." in the external program. Interruption of the work is also provided for certain types of programming errors (or machine errors): if the maximum number of sentences (177) has been exceeded, or if the numbers of the L-operators and the subroutines found do not coincide.

On termination of compilation, information is also given about the length of the synthesized program (b_1) and its total number of sentences (g).

3. THE L-PROGRAM OF THE COMPILER

Below we give the compiler program, written in first-level LYaPAS. All of the indexes figuring in it are taken from the auxiliary set. The program contains about 700 symbols and can be translated easily to machine language. The compiler is matched to the adopted system for coding LYaPAS symbols [2].

For convenience of reading the program we denote by $\langle\psi\rangle$ the symbol of an operand whose value is the code of the symbol ψ, and by

$$a \oplus \langle \psi_1, \psi_2, \ldots, \psi_n \rangle$$

the operation after whose execution the variable τ takes on the value 0 if the value of the index a coincides with one of the symbols $\psi_1, \psi_2, \ldots, \psi_n$. Output of the symbol $\overset{+}{\circ}$ is the error signal.

§0	$\circ b \circ c \circ v \bar{\circ} z \circ \mathbf{d}_2 \circ \mathbf{f}_1 \circ \mathbf{e}_0 5 \Rightarrow t\, 1 \Rightarrow d \Rightarrow h$ $\Rightarrow f \nrightarrow 2 \Rightarrow g\ \circ r2 \Rightarrow \mathbf{i}_1 \circ z$	Initialization
§1	$\nrightarrow 2 \oplus \langle \alpha, \beta, \ldots \rangle \circ\!\!\rightarrow 14x \oplus \langle \mathbf{a}, \mathbf{b}, \ldots \rangle$ $\circ\!\!\rightarrow 11x \oplus \langle // \rangle \circ\!\!\rightarrow 31x \oplus \langle \mathbf{A}, \mathbf{B}, \ldots \rangle$ $\circ\!\!\rightarrow 30x \oplus c_{36} - c_{36} \circ\!\!\rightarrow 47c \mapsto 13$ $x \oplus \langle \Leftarrow \rangle \circ\!\!\rightarrow 10x \oplus \langle \S \rangle \circ\!\!\rightarrow 7x \oplus \langle . \rangle$ $\circ\!\!\rightarrow 27x \oplus \langle ? \rangle \circ\!\!\rightarrow 20x \oplus \langle \rightarrow, \mapsto,$ $\circ\!\!\rightarrow, \nrightarrow \dot{\mathbf{X}}, \ddot{\mathbf{X}} \rangle \circ\!\!\rightarrow 22x \oplus \langle \# \rangle \mapsto 13$ $x \nrightarrow 6 \nrightarrow 2 \nrightarrow 6\ \bar{\circ}\ cz \rightarrow 1$	Commutator
§2	$\triangle f$	
§3	$f{:}5 \times 11 \Rightarrow i\mathbf{a}_{\text{Я}} \leftleftarrows i \wedge f_7 \circ\!\!\rightarrow 4 \Rightarrow x!$	Unpacking **A**
§4	$z \circ\!\!\rightarrow 2 \circ x!$	
§5	$u{:}5 \times 11 \Rightarrow i\mathbf{b}_{\text{Я}} \leftleftarrows i \wedge f_7 \Rightarrow y!$	Unpacking **B**
§6	$\Rightarrow jb{:}5 \times 11 \Rightarrow i100 - i \Rightarrow if_7 \leftleftarrows i \oplus f_{11}$ $\wedge \mathbf{b}_{\text{Я}} \Rightarrow \mathbf{b}_{\text{Я}}\, j \leftleftarrows i \vee \mathbf{b}_{\text{Я}} \Rightarrow \mathbf{b}_{\text{Я}} \triangle b!$	Packing **B**
§7	$x \nrightarrow 6 \nrightarrow 2 + v \nrightarrow 6 \rightarrow 1$	Sentence renumbering
§10	$\bar{\circ}\ zx \nrightarrow 6 \nrightarrow 2 \nrightarrow 6 \nrightarrow 2 \nrightarrow 6 \nrightarrow 2 \nrightarrow 6 \nrightarrow 2$ $\nrightarrow 6 \circ z \rightarrow 1$	Transcription after $\Leftarrow$
§11	$x \wedge 37 \Rightarrow ic_i \wedge \mathbf{h}_t \circ\!\!\rightarrow 13t \Rightarrow a + 5 \Rightarrow s$	Priming variables
§12	$\triangle a \oplus s \circ\!\!\rightarrow 44c_i \wedge \mathbf{h}_a \circ\!\!\rightarrow 12a - t + 17$ $\nrightarrow 6$	

§13	$x \nrightarrow 6 \rightarrow 1$	Transcription
§14	$\bar{\circ} zx \wedge 17 + d_d \Rightarrow a\mathbf{e}_a \Rightarrow k \triangle a \bar{\triangle} f \nrightarrow 3 \triangle f \bar{\circ} zx \oplus \langle A, B \rangle \circ\!\!\rightarrow 16$ $c \circ\!\!\rightarrow 15\ 2 \nrightarrow 6$	Operand substitution
§15	$\mathbf{c}_k \nrightarrow 6 \triangle k \oplus \mathbf{e}_a \mapsto 15c \circ\!\!\rightarrow 1$ $3 \nrightarrow 6 \rightarrow 1$	
§16	$\mathbf{c}_k \oplus \langle ', '', ''', '''' \rangle \mapsto 17 \triangle k$	Processing situations a_α, b_β
§17	$\mathbf{c}_k \oplus \langle \alpha, \beta, \cdots \rangle \mapsto 15\mathbf{c}_k \wedge 77 \vee c_{27}$ $\nrightarrow 6 \rightarrow 1$	
§20	$\nrightarrow 2 \oplus 40 \mapsto 21b - 1 \Rightarrow u \nrightarrow 5 \wedge 37$ $\Rightarrow ac_a \wedge \text{э} \mapsto 1\ 40 \nrightarrow 6\ 40 \nrightarrow 6 \rightarrow 1$	Processing symbols ? and ??
§21	$\bar{\triangle} fb - 2 \Rightarrow u \nrightarrow 5 \wedge 37 \Rightarrow a\mathbf{c}_a \wedge \text{э}$ $\mapsto 1\ 40 \nrightarrow 6 \rightarrow 1$	
§22	$x \nrightarrow 6 \nrightarrow 2 \oplus \langle 0, 1, 2, \ldots \rangle \mapsto 23x + v$ $\nrightarrow 6 \rightarrow 1$	Correction for transfers
§23	$x \oplus \langle \# \rangle \circ\!\!\rightarrow 24x \wedge 17 + \mathbf{d}_d \Rightarrow a$ $\mathbf{e}_a \Rightarrow a\mathbf{c}_a + w \nrightarrow 6 \rightarrow 1$	
§24	$\nrightarrow 2 \Rightarrow kf \Rightarrow l\mathbf{i}_d \Rightarrow f \bar{\circ} z \nrightarrow 3 + v$ $\Rightarrow m \bar{\triangle} f$	
§25	$\circ z \nrightarrow 3 + f - 1 \Rightarrow f \nrightarrow 3f \Rightarrow j \bar{\triangle} x$ $\oplus d \mapsto 26 \nrightarrow 2 \oplus k \mapsto 26l \Rightarrow f \nrightarrow 2$ $+ m \nrightarrow 6 \rightarrow 1$	Search for additional entry terminal
§26	$\bar{\circ} zj + 3 \Rightarrow f \nrightarrow 3 + m \Rightarrow m \bar{\triangle} f \rightarrow 25$	
§27	$\bar{\triangle} d \circ\!\!\rightarrow 45\mathbf{g}_d \Rightarrow f\mathbf{f}_d \Rightarrow v \bar{\triangle} d\mathbf{f}_d \Rightarrow w \triangle d$ $t - 5 \Rightarrow t \rightarrow 1$	Reduction of depth
§30	$x \wedge 37 \Rightarrow ic_i \wedge \mathbf{h}_0 \circ\!\!\rightarrow 13 \circ a$	Priming complexes
§46	$\triangle a \oplus 5 \circ\!\!\rightarrow 44c_i \wedge \mathbf{h}_a \circ\!\!\rightarrow 46a + 17$ $\nrightarrow 6 \rightarrow 13$	
§31	$f \Rightarrow \mathbf{g}_d b \Rightarrow u \triangle d \circ zt + 5 \Rightarrow t \circ c$	
§32	$\bar{\triangle} u \Rightarrow p \nrightarrow 5 \oplus \langle \# \rangle \mapsto 32h \Rightarrow f \nrightarrow 3$ $+ h - 1 \Rightarrow f \nrightarrow 3 \nrightarrow 2 \triangle u \nrightarrow 5$ $\oplus x \mapsto 44 \triangle f \Rightarrow h \bar{\circ} z \nrightarrow 2 \circ z$ $v \Rightarrow wg \Rightarrow v \Rightarrow \mathbf{f}_d + x \Rightarrow g - 200$ $\mapsto 44f \Rightarrow \mathbf{i}_d\mathbf{d}_d \Rightarrow e\mathbf{e}_e \Rightarrow a$	Back up in **B** and search for new subroutine

§33	$\triangle\ u \oplus b \circ\!\!\rightarrow 36 \nrightarrow 5 \oplus \langle (\rangle \circ\!\!\rightarrow 34$ $y \Rightarrow \mathbf{c}_a \triangle\ ay \oplus \langle ', '', ''', '''' \rangle \circ\!\!\rightarrow 33 \rightarrow 35$	Preparation of set of operands for new subroutine
§34	$\triangle\ u \nrightarrow 5 \oplus \langle) \rangle \circ\!\!\rightarrow 35 y \Rightarrow \mathbf{c}_a \triangle\ a \rightarrow 34$	
§35	$\triangle\ ea \Rightarrow \mathbf{e}_e \rightarrow 33$	
§36	$\triangle\ de \Rightarrow \mathbf{d}_d p \Rightarrow bt \Rightarrow a + 5 \Rightarrow s \triangle\ d$	
§37	$\circ\ \mathbf{h}_a \triangle\ a \oplus s \mid\!\!\rightarrow 37$	
§40	$\nrightarrow 2\ \overline{\triangle}\ fx \oplus \langle (\rangle \mid\!\!\rightarrow 1 \triangle f \nrightarrow 2 \wedge 17 + \mathbf{d}_d$ $\Rightarrow a\mathbf{e}_a \Rightarrow a\mathbf{c}_a \oplus \langle ', '', ''', '''' \rangle \circ\!\!\rightarrow 42$	
§41	$\nrightarrow 2 \oplus \langle) \rangle \mid\!\!\rightarrow 41 \rightarrow 40$	
§42	$\mathbf{c}_a - 17 + t \Rightarrow p$	
§43	$\nrightarrow 2 \oplus \langle) \rangle \circ\!\!\rightarrow 40 x \wedge 37 \Rightarrow ac_a \vee \mathbf{h}_t$ $\Rightarrow \mathbf{h}_t c_a \vee \mathbf{h}_p \Rightarrow \mathbf{h}_p \rightarrow 43$	
§47	$c \Leftrightarrow r \rightarrow 13$	
§44	$\langle ♁ \rangle *$	Error signal
§45	$x \nrightarrow 6.$	End

REFERENCES

1. Zakrevskii, A. D., Description of LYaPAS. This volume.
2. Tovshtein, M. Ya., The Representation of Input Information to PS-LYaPAS. This volume.

DETECTION OF SYNTACTIC ERRORS IN L-PROGRAMS

N. A. Usacheva

The programming system PS-LYaPAS processes expressions constructed in strict agreement with the description of LYaPAS [1]. At the same time, it is possible that combinations of symbols that contradict the rules will appear in the L-programs. Therefore, before the translator begins to operate, the L-program is analyzed for the purpose of detecting such combinations. This analysis is performed by the block *synter*—the block of search for syntactic errors.

Let us consider the more general problem of constructing a search algorithm for syntactic errors in an arbitrary language U. Then, basing ourselves on this algorithm, we shall construct the search algorithm for syntactic errors in LYaPAS.

1. SEARCH ALGORITHM FOR SYNTACTIC ERRORS IN AN ARBITRARY LANGUAGE *U*

Let $X = \{x_1, x_2, \ldots, x_n\}$ be a finite set of arbitrary symbols (words), called a dictionary.

Let us construct the set $U = \{u_1, u_2, \ldots\}$ of sequences of dictionary words. For the set X let us prescribe a syntax—a set of rules for constructing sequences of words. With the prescription of the syntax the language U is divided into subsets of syntactically correctly constructed sequences (correct) and sequences violating some syntactic rule (incorrect). Before a sequence in some construction can be used it is necessary to establish its correctness. To this end the syntax is given to the computer and a program is composed that permits the correctness of any sequence in U to be established.

To prescribe a syntax we represent it in the form of a finite automaton A as given in Chomsky's work [2].

Let the set of internal states $\{a_1, a_2, \ldots, a_k\}$ of the automaton coincide with the set of syntactic rules; a_0 is the initial state. The set of input symbols $\{\rho_1, \rho_2, \ldots, \rho_n\}$ coincides with the dictionary X. The transition function of the automaton $\delta(a_i, \rho_j)$ is prescribed by a flow table. If the word x_j does not satisfy the rule a_i, then $\delta(a_i, \rho_j) = a_0$.

If several equivalent rows exist in the flow table, only one is retained. After this reduction of the flow table, several dictionary words will correspond to each input symbol. We shall call these words syntactically equivalent. The flow table of the automaton representing the syntax of the language will be called the syntax table.

We shall construct the algorithm for testing the correctness of an arbitrary sequence from the prescribed table in the following manner.

Let $u_i = x_i^1, x_i^2, \ldots, x_i^\lambda \in U$ be an arbitrary sequence of words, and let λ be the length of the sequence.

An $x_i^1 \in u_i$ is chosen, and the value $\delta(a_0, \rho_k^1)$ in the syntax table is found, where ρ_k^1 is the input symbol corresponding to the word x_i^1. (Below ρ_k^m is the input symbol corresponding to x_i^m.) If $\delta(a_0, \rho_k^1) = a_j^1$, then $x_i^2 \in u_i$ is chosen and $\delta(a_j^1, \rho_k^2)$ is found, etc. If $(a_j^1, \rho_k^m) = a_0$, the word $x_i^m \in u_i$ and the words preceding it form an incorrect sequence. We shall say that an error has been committed at the word x_i^m, and the word x_i^m is termed erroneous.

In this case the position of the error is stored (the number of the erroneous word in the sequence) with the "neighborhood" of the error (several words preceding the erroneous one). Next the word x_i^{m+1} is chosen, and $\delta(a_0, \rho_k^{m+1})$ is found, etc., until the number of tested words coincides with the value of λ or the next tested word is a word signifying the end of the sequence. The last stage of the algorithm is the output of the accumulated error information.

The search algorithm for LYaPAS syntactic errors is based on this algorithm. We first construct the LYaPAS syntax table.

2. THE LYaPAS SYNTAX TABLE

In the composition of the LYaPAS syntax table the groups of syntactically equivalent words are sharply distinguished. These are the group of complexes, the group of indexes, the group of variables, the groups of natural and standard constants, the group of operators $\triangle$, $\bar{\triangle}$, $\circ$, $\bar{\circ}$, etc.

To each state we assign a certain number—the number of the column of the syntax table corresponding to this state. Each is a 7-bit binary number. The numbering begins from zero. The input symbols are numbered analogously.

The LYaPAS syntax is given in the table T [see Table I]. In each row of the table at the left, besides the input symbol number, are given the words to which this input symbol corresponds. At the top of each column, besides the number of states, are indicated the expressions whose construction is prescribed by the rule corresponding to this state. We shall adopt the following notation for certain expressions. Let

$$\xi \in \{\mathbf{a}, \mathbf{b}, \ldots, \text{я}\}, \quad \eta \in \{a, b, \ldots, \text{я}\},$$
$$\delta \in \{0, 1, \ldots, 177\} \epsilon \in \{c_0, \ldots, c_{37}, d_0, \ldots, d_4, e_0, \ldots, e_4, f_0, \ldots, f_{11}\},$$
$$\mathbf{\Omega} \in \{\mathbf{A}, \mathbf{B}, \ldots, \text{Я}, A, \ldots, Q\}, \quad \boldsymbol{\omega}_\eta \in \mathbf{\Omega}, \ \diamond_1 \in \{\vdash, |\!\leftarrow, \urcorner, I\},$$
$$\diamond_2 \in \{+, -\}, \qquad \diamond_3 \in \{\vee, \wedge, \oplus\},$$
$$\diamond_4 \in \{\leftleftarrows, \times, \bar{\times}\}, \qquad \diamond_5 \in \{\langle, \rangle, \leftarrow, \underline{\vee}\},$$
$$\diamond_6 \in \{\triangle, \bar{\triangle}, \circ, \bar{\circ},\}, \qquad \diamond_7 \in \{', '', ''', ''''\}.$$

We shall denote by the symbol τ the result of execution of the preceding operation not assigned to a concrete operand. We shall tag the operand symbol by a prime at the left if its value is represented by a complex variable, for example $'\xi$, $'\boldsymbol{\omega}_\eta$, $'\tau$.

For compact inscription in memory the syntax table is condensed; instead of repeating the state number, this number is written with the number of its repetitions, for which 7 binary positions are used. For example, after condensation of the table T, the 24th row takes the form (in octal notation)

072 025 000 075,

i.e., the state with number 72 is repeated 25 times, and the state with number 000 is repeated 75 times. The condensed form of the table T is represented in core by a sequence of 106 registers. Six codes of the table are "packed" into each register (45-bit registers are assumed). The leftmost bits of the register remain free. Each row of the table T is represented by a series of registers in the sequence. The start of each series in the sequence is indicated in the table P of "starts of rows." For example, the 24th row of the table T is represented by a series of two registers, beginning with the 80th register of the sequence.

TABLE I

THE TABLE T

			$\Uparrow \xi, \Uparrow \eta$	$\boldsymbol{\Omega} \Uparrow (\ldots)$	$\tau \not\!\not\rightarrow \delta$	$\dot{\mathbf{X}}\, \delta_\eta, \Diamond_6\, \xi$	$\Uparrow \omega_\eta$	$\Diamond_6\, {}'\xi \not\!\not\rightarrow \delta$	$\Uparrow \omega_\eta?$	ϵ^*, δ^*	$\S\, \delta$	$\bigcirc \rightarrow \delta$	$'\tau \bigcirc \rightarrow \delta$	$\Diamond_6\, \xi, {}'\xi^*$
		000	001	002	003	004	005	006	007	010	011	012	013	014
δ	000	030	030	030	030	030	030	030	030	030	030	030	030	030
η	001	031	031	031	031	031	031	301	031	031	031	031	031	031
ξ	002	031	031	031	031	031	031	031	031	031	031	031	031	031
$\Longrightarrow$	003	112	112	112	112	112	112	112	112	112	112	112	112	112
$\Leftrightarrow$	004	112	000	000	000	000	000	000	000	000	000	000	000	000
$\boldsymbol{\Omega}$	005	077	077	077	077	077	077	077	077	077	077	077	077	077
$\longrightarrow$	006	065	065	065	065	065	065	065	065	065	000	060	060	060
$\Diamond_1$	007	026	026	026	026	026	026	026	026	026	026	026	033	033
$\Diamond_2$	010	101	101	101	101	101	101	101	101	101	101	101	102	102
$\Diamond_3$	011	041	041	041	041	041	041	041	041	041	041	041	045	045
!	012	004	004	004	004	004	004	004	004	004	004	004	004	004
§	013	055	055	004	055	055	055	055	055	055	000	055	051	051
$\overset{*}{*}, *$	014	004	004	000	004	004	044	054	004	010	004	004	014	014
$\bigcirc \rightarrow\mid\rightarrow$	015	062	062	055	062	062	062	062	062	000	062	000	000	061
$\not\!\not\rightarrow$	016	062	062	062	062	062	062	062	062	062	062	062	063	063
$\Diamond_7$	017	114	114	114	114	114	114	114	114	114	114	114	114	114
ϵ	020	030	030	030	030	030	030	030	030	030	030	030	030	030
(	021	105	105	105	105	105	105	105	105	105	105	105	105	105
$\ddagger$	022	124	124	124	124	124	124	124	124	124	124	124	124	124
$\Diamond_4$	023	040	040	040	040	040	040	040	040	040	040	040	000	000
$\Diamond_6$	024	072	072	072	072	072	072	072	072	072	072	072	072	072
$\Diamond_5$	025	040	040	040	040	040	040	040	040	040	040	040	044	044
)	026	121	000	000	000	000	000	000	000	000	000	000	000	000
$\dot{\mathbf{X}}, \ddot{\mathbf{X}}$	027	057	057	000	000	000	057	000	000	000	000	000	000	000
$\Leftarrow$	030	123	000	000	000	000	000	000	000	000	000	000	000	000
:	031	071	071	070	071	071	071	071	071	071	071	071	000	000
?	032	007	000	000	000	000	007	000	004	000	000	000	000	000
$\nearrow, \swarrow$	033	107	107	107	107	107	107	107	107	107	107	107	107	107
$\underline{\downarrow}$	034	111	111	111	111	111	111	111	111	111	111	111	111	111
$\overline{\uparrow}$	035	021	021	021	021	021	021	021	021	021	021	021	021	021
$\cdot$	036	122	122	122	122	122	122	122	122	122	122	122	122	122

TABLE I (continued)

		$\Rightarrow {}'\omega_\eta$	$\Rightarrow {}'\xi$	$\Diamond_2 {}'\xi \circ\!\rightarrow \delta$	$\Rightarrow {}'\omega_\eta ?$	$\downarrow \xi, \uparrow$	$\Diamond^{6'} \xi \S \delta$	$\mathbf{\Omega}^*$	$(\ldots) \Rightarrow \mathbf{\Omega}$	$\tau \S \delta$	$\Diamond_3 \xi$	$\Diamond_2 \xi$	ϵ, δ
		015	016	017	020	021	022	023	024	025	026	027	030
δ	000	030	030	030	030	030	030	030	030	000	000	000	030
η	001	031	031	031	031	031	031	031	031	027	027	027	031
ξ	002	031	031	031	031	031	031	031	031	000	000	000	031
$\Rightarrow$	003	112	112	115	112	000	000	000	000	112	112	112	112
$\Leftrightarrow$	004	000	000	000	000	000	000	000	000	000	000	000	000
$\mathbf{\Omega}$	005	077	077	077	077	077	077	077	077	000	000	000	000
$\rightarrow$	006	060	060	060	060	060	000	065	065	000	047	047	047
$\Diamond_1$	007	033	033	033	033	000	000	000	000	026	026	026	026
$\Diamond_2$	010	102	102	102	102	000	000	000	000	101	101	101	101
$\Diamond_3$	011	045	045	045	045	000	000	000	000	041	041	041	041
!	012	004	004	004	004	004	004	004	004	004	004	004	004
§	013	051	041	051	051	051	051	051	051	054	054	054	054
$\overset{*}{*}, *$	014	014	014	014	014	000	000	000	024	004	004	004	004
$\circ\!\rightarrow\,\mid\!\rightarrow$	015	061	061	061	000	000	000	000	000	062	062	066	000
$\nrightarrow$	016	063	063	063	063	063	063	063	063	050	050	050	050
$\Diamond_7$	017	114	114	114	114	114	114	114	114	000	000	000	000
ϵ	020	030	030	030	030	030	030	030	030	000	000	000	000
(	021	105	105	105	105	105	105	105	105	000	000	000	000
$\nplus$	022	124	124	124	124	124	124	124	124	000	000	000	000
$\Diamond_4$	023	000	000	000	000	000	000	000	000	040	040	040	040
$\Diamond_6$	024	072	072	072	072	072	072	072	072	000	000	000	000
$\Diamond_5$	025	044	054	044	044	000	000	000	000	040	040	040	040
)	026	000	000	000	000	000	000	000	000	000	000	000	000
$\dot{\mathbf{X}}, \ddot{\mathbf{X}}$	027	057	057	000	000	000	000	000	000	000	000	000	000
$\Leftarrow$	030	000	000	000	000	000	000	000	000	000	001	000	000
:	031	000	000	000	000	000	000	000	000	071	071	071	071
?	032	020	000	000	014	000	000	000	000	000	000	000	000
$\nearrow, \swarrow$	033	107	107	107	107	107	107	107	107	000	000	000	000
$\downarrow$	034	111	111	111	111	111	111	111	111	000	000	000	000
$\uparrow$	035	021	021	021	021	021	021	021	021	000	000	000	000
.	036	122	122	122	122	122	122	121	122	000	000	000	000

TABLE I (continued)

		ξ	$'\xi$	$'\tau$	$\diamondsuit_2{}'\ \xi$	$\rightarrow\delta$	$\tau\rightarrow\delta$	$\diamondsuit_6{}'\ \xi\rightarrow\delta$	$\diamondsuit_4,\ \diamondsuit_5$	$\xi\ \diamondsuit_3$	$\diamondsuit_3\ \mathbf{\Omega}$	$'\xi\ \diamondsuit_5\ \mathbf{\Omega}$	$'\xi\ \diamondsuit_5$
		031	032	033	034	035	036	037	040	041	042	043	044
δ	000	030	000	000	000	000	000	000	026	026	026	033	033
η	001	031	027	031	031	031	000	000	026	026	026	033	033
ξ	002	031	000	000	000	000	000	000	026	026	000	000	033
$\Rightarrow$	003	112	112	112	112	000	000	000	000	000	000	000	000
$\Leftrightarrow$	004	112	112	000	000	000	000	000	000	000	000	000	000
$\mathbf{\Omega}$	005	077	000	000	000	000	000	000	042	042	000	000	043
$\rightarrow$	006	047	000	000	000	000	000	000	000	000	000	000	000
$\diamondsuit_1$	007	026	033	033	033	000	000	000	000	000	000	000	000
$\diamondsuit_2$	010	101	102	102	102	045	045	045	000	000	000	000	000
$\diamondsuit_3$	011	041	045	045	045	055	055	055	000	000	000	000	000
!	012	004	000	000	000	000	000	000	000	000	000	000	000
§	013	054	000	000	000	000	000	000	000	000	000	000	000
$\underset{*}{*}$, *	014	004	014	014	014	000	000	000	000	000	000	000	000
$\circ\rightarrow\mid\rightarrow$	015	062	061	061	064	000	000	000	000	046	000	000	000
$\not\mapsto$	016	050	000	000	000	000	000	000	026	026	000	000	000
$\diamondsuit_7$	017	000	000	000	000	000	000	000	000	000	000	000	000
ϵ	020	000	000	000	000	000	000	000	000	000	000	000	033
(	021	000	000	000	000	000	000	000	000	000	000	000	000
$\#$	022	000	000	000	000	000	000	000	000	000	000	000	000
$\diamondsuit_4$	023	040	000	000	000	000	000	000	000	000	000	000	000
$\diamondsuit_6$	024	000	000	000	000	000	000	000	000	000	000	000	000
$\diamondsuit_5$	025	040	044	044	044	000	000	000	000	000	000	000	000
)	026	000	000	000	000	000	000	000	000	000	000	000	000
$\dot{\mathbf{X}}$, $\ddot{\mathbf{X}}$	027	057	057	000	000	000	000	000	000	000	000	000	000
$\Leftarrow$	030	023	000	000	000	000	000	000	000	000	000	000	000
:	031	071	000	000	000	000	000	000	000	000	000	000	000
?	032	000	000	000	000	000	000	000	000	000	000	000	000
$\nearrow$, $\swarrow$	033	000	000	000	000	000	000	000	000	000	000	000	000
$\downarrow$	034	000	000	000	000	000	000	000	000	000	000	000	000
$\uparrow$	035	000	000	000	000	000	000	000	000	000	000	000	000
·	036	000	000	000	000	000	000	000	000	000	000	000	000

TABLE I (continued)

		′ξ ◇₃	′ξ ◇₃ ◇₇	τ →	τ ⇸	◇₆′ ξ ℘	⇸ ǂ	Ẋ ǂ	τ ℘	℘	Ẋ ǂ δ	Ẋ, Ẍ	◇₆′ ξ →
		045	046	047	050	051	052	053	054	055	056	057	060
δ	000	033	000	036	003	022	067	056	025	011	073	073	037
η	001	033	000	000	000	000	000	000	000	000	000	000	000
ξ	002	033	000	000	000	000	000	000	000	000	000	000	000
⇒	003	000	000	000	000	000	000	000	000	000	000	000	000
⇔	004	000	000	000	000	000	000	000	000	000	000	000	000
Ω	005	043	000	000	000	000	000	000	000	000	000	000	000
→	006	000	000	000	000	000	000	000	000	000	000	000	000
◇₁	007	000	000	000	000	000	000	000	000	000	000	000	000
◇₂	010	000	000	000	000	000	000	000	000	000	000	000	000
◇₃	011	000	000	000	000	000	000	000	000	000	000	000	000
!	012	000	000	000	000	000	000	000	000	000	000	000	000
§	013	000	000	000	000	000	000	000	000	000	000	000	000
$\overset{*}{*}$, *	014	000	000	000	000	000	000	000	000	000	000	000	000
○→|→	015	000	000	000	000	000	000	000	000	000	000	000	000
⇸	016	000	000	000	000	000	000	000	000	000	000	000	000
◇₇	017	046	000	000	000	000	000	000	000	000	000	000	000
ϵ	020	033	000	000	000	000	000	000	000	000	000	000	000
(	021	000	000	000	000	000	000	000	000	000	000	000	000
ǂ	022	000	000	000	000	000	000	000	000	000	000	053	052
◇₄	023	000	000	000	000	000	000	000	000	000	000	000	000
◇₆	024	000	000	000	000	000	000	000	000	000	000	000	000
◇₅	025	000	000	000	000	000	000	000	000	000	000	000	000
)	026	000	000	000	000	000	000	000	000	000	000	000	000
Ẋ, Ẍ	027	000	000	000	000	000	000	000	000	000	000	000	000
⇐	030	000	000	000	000	000	000	000	000	000	000	000	000
:	031	000	000	000	000	000	000	000	000	000	000	000	000
?	032	000	000	000	000	000	000	000	000	000	000	000	000
↗, ↙	033	000	000	000	000	000	000	000	000	000	000	000	000
⤓	034	000	000	000	000	000	000	000	000	000	000	000	000
⤒	035	000	000	000	000	000	000	000	000	000	000	000	000
•	036	000	000	000	000	000	000	000	000	000	000	000	000

TABLE I (continued)

		′τ ○→	○→, ⇻	◇₆ ′ξ ⇻	◇₂′ ξ ○→	→	◇₂ ξ ○→	τ ○→ ⧺ δ	:Ω, ◇₆ Ω	..	◇₆	Ẋ δ, Ẍ ⧺ δδ	⇒ Ω
		061	062	063	064	065	066	067	070	071	072	073	074
δ	000	013	012	006	017	035	004	004	004	004	000	000	005
η	001	000	000	000	000	000	000	000	000	004	004	005	005
ξ	002	000	000	000	000	000	000	000	000	004	004	000	000
⇒	003	000	000	000	000	000	000	000	000	000	000	000	000
⇔	004	000	000	000	000	000	000	000	000	000	000	000	000
Ω	005	000	000	000	000	000	000	000	000	070	070	000	000
→	006	000	000	000	000	000	000	000	000	000	000	000	000
◇₁	007	000	000	000	000	000	000	000	000	000	000	000	000
◇₂	010	000	000	000	000	000	000	000	000	000	000	000	000
◇₃	011	000	000	000	000	000	000	000	000	000	000	000	000
!	012	000	000	000	000	000	000	000	000	000	000	000	000
§	013	000	000	000	000	000	000	000	000	000	000	000	000
⁑, *	014	000	000	000	000	000	000	000	000	000	000	000	000
○→|→	015	000	000	000	000	000	000	000	000	000	000	000	000
⇻	016	000	000	000	000	000	000	000	000	000	000	000	000
◇₇	017	000	000	000	000	000	000	000	000	000	115	000	000
ε	020	000	000	000	000	000	000	000	000	004	000	000	000
(	021	000	000	000	000	000	000	000	000	000	000	000	000
⧺	022	052	052	052	052	052	052	000	000	000	000	000	000
◇₄	023	000	000	000	000	000	000	000	000	000	000	000	000
◇₆	024	000	000	000	000	000	000	000	000	000	000	000	000
◇₅	025	000	000	000	000	000	000	000	000	000	000	000	000
)	026	000	000	000	000	000	000	000	000	000	000	000	000
Ẋ, **Ẍ**	027	000	000	000	000	000	000	000	000	000	000	000	000
⇐	030	000	000	000	000	000	000	000	000	000	000	000	000
:	031	000	000	000	000	000	000	000	000	000	000	000	000
?	032	000	000	000	000	000	000	000	000	000	000	000	000
↗, ↙	033	000	000	000	000	000	000	000	000	000	000	000	000
↓	034	000	000	000	000	000	000	000	000	000	000	000	000
↑	035	000	000	000	000	000	000	000	000	000	000	000	000
•	036	000	000	000	000	000	000	000	000	000	000	000	000

TABLE I (continued)

		⇑, Ω	, Ω	Ω	ξ ◇₂ Ω	ξ ◇₂	ξ, ◇₂	ξ, ◇₂ Ω	◇₆, Ω	(	Ω ⇑)	↙, ↗	↙ ξ, ↗ ξ
		075	076	077	100	101	102	103	104	105	106	107	110
δ	000	015	033	031	027	027	034	034	014	105	000	000	000
η	001	015	033	031	027	027	034	034	014	105	106	110	111
ξ	002	000	000	000	000	027	034	000	000	105	106	110	111
⇒	003	000	000	120	000	000	000	000	000	000	000	000	000
⇔	004	000	000	000	000	000	000	000	000	000	000	000	000
Ω	005	000	000	000	000	000	103	000	000	000	000	000	000
→	006	000	000	000	000	000	000	000	000	000	000	000	000
$\diamond_1$	007	000	000	000	000	000	000	000	000	000	000	000	000
$\diamond_2$	010	000	000	000	000	000	000	000	000	000	000	000	000
$\diamond_3$	011	000	000	000	000	000	000	000	000	000	000	000	000
!	012	000	000	000	000	000	000	000	000	000	000	000	000
§	013	000	000	000	000	000	000	000	000	000	000	000	000
$\overset{*}{*}$, *	014	000	000	023	000	000	000	000	000	000	000	000	000
○→|→	015	000	000	000	000	000	000	000	000	000	000	000	000
⇸	016	000	000	000	000	000	000	000	000	000	000	000	000
$\diamond_7$	017	000	000	000	000	116	116	000	000	000	000	000	000
ε	020	000	000	000	000	027	034	000	000	000	000	000	000
(	021	000	000	000	000	000	000	000	000	000	000	000	000
ǂ	022	000	000	000	000	000	000	000	000	000	000	000	000
$\diamond_4$	023	000	000	000	000	000	000	000	000	000	000	000	000
$\diamond_6$	024	000	000	000	000	000	000	000	000	000	000	000	000
$\diamond_5$	025	000	000	000	000	000	000	000	000	000	000	000	000
)	026	000	000	000	000	000	000	000	121	002	000	000	000
$\dot{\mathbf{X}}$, $\ddot{\mathbf{X}}$	027	000	000	000	000	000	000	000	000	000	000	000	000
⇐	030	000	000	000	000	000	000	000	000	000	000	000	000
:	031	000	000	000	000	000	000	000	000	000	000	000	000
?	032	000	000	000	000	000	000	000	000	000	000	000	000
↗, ↙	033	000	000	000	000	000	000	000	000	000	000	000	000
↓	034	000	000	000	000	000	000	000	000	000	000	000	000
↑	035	000	000	000	000	000	000	000	000	000	000	000	000
·	036	000	000	000	000	000	000	000	000	000	000	000	000

TABLE I (continued)

		↙ ξξ	⇒	⇒ ◇7	◇7	◇6 ◇7	ξ ◇2 ◇7	(. . .) ⇒	**Ω** ⇒	(. . .)	•	⇐	⫲
		111	112	113	114	115	116	117	120	121	122	123	124
δ	000	000	000	000	000	000	000	000	000	000	000	000	000
η	001	021	001	000	000	000	000	000	000	000	000	000	000
ξ	002	021	001	016	032	014	034	000	000	000	000	000	000
⇒	003	000	000	000	000	000	000	000	000	117	000	000	000
⇔	004	000	000	000	000	000	000	000	000	000	000	000	000
Ω	005	000	075	076	076	104	103	024	000	000	000	000	000
→	006	000	000	000	000	000	000	000	000	000	000	000	000
$\diamond_1$	007	000	000	000	000	000	000	000	000	000	000	000	000
$\diamond_2$	010	000	000	000	000	000	000	000	000	000	000	000	000
$\diamond_3$	011	000	000	000	000	000	000	000	000	000	000	000	000
!	012	000	000	000	000	000	000	000	000	000	000	000	000
§	013	000	000	000	000	000	000	000	000	000	000	000	000
$\overset{*}{*}$, *	014	000	000	000	000	000	000	000	000	000	000	000	000
○→|→	015	000	000	000	000	000	000	000	000	000	000	000	000
⇸	016	000	000	000	000	000	000	000	000	000	000	000	000
$\diamond_7$	017	000	113	000	000	000	000	000	000	000	000	000	000
ϵ	020	000	000	000	000	000	000	000	000	000	000	000	000
(	021	000	000	000	000	000	000	000	106	000	000	000	000
⫲	022	000	000	000	000	000	000	000	000	000	000	000	000
$\diamond_4$	023	000	000	000	000	000	000	000	000	000	000	000	000
$\diamond_6$	024	000	000	000	000	000	000	000	000	000	000	000	000
$\diamond_5$	025	000	000	000	000	000	000	000	000	000	000	000	000
)	026	000	000	000	000	000	000	000	000	000	000	000	000
$\dot{\mathbf{X}}$, $\ddot{\mathbf{X}}$	027	000	000	000	000	000	000	000	000	000	000	000	000
⇐	030	000	000	000	000	000	000	000	000	000	000	000	000
:	031	000	000	000	000	000	000	000	000	000	000	000	000
?	032	000	000	000	000	000	000	000	000	000	000	000	000
↗, ↙	033	000	000	000	000	000	000	000	000	000	000	000	000
↓	034	000	000	000	000	000	000	000	000	000	000	000	000
↑	035	000	000	000	000	000	000	000	000	000	000	000	000
•	036	000	000	000	000	000	000	000	000	000	000	000	000

In order to find the input symbol corresponding to a LYaPAS word in a simple way, we construct a table P as follows. The values of the elements P_i $(0 \leq i \leq 77)$ are the starts of rows corresponding to the operators of the language, and the P_j $(100 \leq j \leq 105)$ are the starts of rows corresponding to the operands of the language. Some of the elements of the table P can coincide, since a single row of the table T corresponds to several LYaPAS words. For example, $[P_{10}] = [P_{11}] = [P_{12}] = [P_{13}] = 120$, since 120 is the number of the register in the sequence with which the inscription of the 24th row begins, corresponding to the operators $\triangle$, $\bar{\triangle}$, $\circ$, $\bar{\circ}$. The table P is "packed" in core into 12 registers.

3. SEARCH ALGORITHM FOR SYNTACTIC ERRORS IN L-PROGRAMS

The syntactic-error search algorithm described in Section 1 will be called the basic algorithm. For concrete languages it is convenient to test the correctness of certain expressions outside the basic algorithm. For LYaPAS such expressions are of the form

(1) $\mathbf{a} \Leftarrow \psi_1\psi_2\psi_3\psi_4$;
(2) *oper* $\alpha\beta\gamma\delta//$;
(3) $\mathbf{a} - b \rightarrow 1 \ldots$;
§1 $\mathbf{c} \Rightarrow d \ldots.$

The search for errors in L-programs is performed by the block *synter*. The operating algorithm is written in LYaPAS. Let us clarify the meaning of the fundamental operands.

We associate the complexes **T** and **P** with the sequence of registers representing the syntax table and the table of starts of rows, respectively. The L-program is represented in core by the complex *L*.

The complex **N** contains the information about the "quality" of the sentences in the L-program necessary for analysis of the expressions of type (3). This complex is constructed before the operation of the basic algorithm, by a search over the L-program.

The error information accumulated during execution of the algorithm is represented by the complex **R** and is given by the following data:

(1) the number of the sentence in which the error has been detected;
(2) the number of the erroneous word in the sentence; and
(3) the erroneous word and the two words preceding it in the L-program.

The variables **a** and **b** are used to represent the code of the current tested word and the number of the current state, respectively. The variable **d** is used to represent the current number of the tested sentence. As soon as the current tested word in the L-program is $\langle\cdot\rangle$ the complex **R** is printed out and the algorithm stops. The search algorithm for syntactic errors in L-programs has the form

§0	$\bar{\circ}\ a$	Formation of the complex **N**
§1	$\triangle\ a \nrightarrow 15 \Rightarrow \mathbf{a} \oplus \langle\cdot\rangle \circ\!\!\rightarrow 2\mathbf{a} \oplus \langle\S\rangle$ $\mapsto 1 \triangle\ a + 1 \nrightarrow 15 \Rightarrow \mathbf{a} \wedge f_1 \mapsto 1\mathbf{a}$ $\oplus \langle (, !, \downarrow, \uparrow, \triangle, \bar{\triangle}, \circ, \bar{\circ}, \nearrow, \swarrow,$ $', '', ''', '''', \#\rangle \circ\!\!\rightarrow 1a \nrightarrow 15 \wedge 177{:}40$ $\Rightarrow b\mathbf{n}_{я} \vee c_b \Rightarrow \mathbf{n}_{я} \triangle\ a \rightarrow \mathbf{1}$	
§2	$\circ\ \mathbf{b}\ \bar{\circ}\ a \circ \mathbf{d}\ \bar{\circ}\ i \circ \mathbf{a}$	Detection of the next word in the L-program
§3	$\mathbf{a} \oplus \langle\cdot\rangle \circ\!\!\rightarrow 17 \triangle\ \mathbf{d} \triangle\ a \nrightarrow 15 \Rightarrow \mathbf{a}$ $\Rightarrow \mathbf{c} \wedge f_1 \circ\!\!\rightarrow 4{:}100\ я + 77 \Rightarrow \mathbf{c}$	
§4	$a_{16} \Rightarrow a_{14}\ \mathbf{c} \nrightarrow 16 + a_{22} \Rightarrow a_{14}$	Determination and analysis of the current state number
§5	$\circ\ f\ \bar{\circ}\ c$	
§6	$c + 2 \Rightarrow c \nrightarrow 16 + \mathbf{f} \Rightarrow \mathbf{f} - \mathbf{b} \circ\!\!\rightarrow 6$ $\circ\!\!\rightarrow 6\ \bar{\triangle}\ c \nrightarrow 16 \Rightarrow \mathbf{b} \circ \mathbf{f}\,\mathbf{b}$ $\oplus \langle 25, 11, 22\rangle \circ\!\!\rightarrow 14\mathbf{b} \oplus 124$ $\circ\!\!\rightarrow 10\mathbf{b} \oplus 123 \circ\!\!\rightarrow 7\mathbf{b} \oplus \langle 36, 3\rangle$ $\circ\!\!\rightarrow 12\mathbf{b} \oplus \langle 13, 6, 17, 37\rangle \circ\!\!\rightarrow 11\mathbf{b}$ $\mapsto 3 \nrightarrow 13 \rightarrow 5$	
§7	$a + 4 \Rightarrow Q\ \mathbf{d} + 4 \Rightarrow \mathbf{d}\ 10 \Rightarrow \mathbf{b} \rightarrow 3$	Analysis of expressions of type (1), type (2), and type (3)
§10	$\triangle\ \mathbf{d} \triangle\ a \nrightarrow 15 \oplus \langle //\rangle \mapsto 10\ 4 \Rightarrow \mathbf{b} \rightarrow 3$	
§11	$\mathbf{a} \wedge 177{:}40 \Rightarrow b\,\mathbf{n}_{я} \wedge c_b \circ\!\!\rightarrow 3 \nrightarrow 13 \rightarrow 5$	
§12	$\mathbf{a} \wedge 177{:}40 \Rightarrow b\,\mathbf{n}_{я} \wedge c_b \mapsto 3 \nrightarrow 13 \rightarrow 5$	Storage of error information
§13	$\triangle\ ia - 2 \nrightarrow 15 \Lleftarrow 56 \vee \mathbf{f} \Rightarrow \mathbf{f}a - 1$ $\nrightarrow 15 \Lleftarrow 45 \vee \mathbf{f} \Rightarrow \mathbf{f}a \Lleftarrow 34 \vee \mathbf{f} \Rightarrow \mathbf{fd}$ $\Lleftarrow 67 \vee \mathbf{f} \Rightarrow \mathbf{r}_i!$	
§14	$\mathbf{a} \Lleftarrow 51 + 1 \Rightarrow \mathbf{d} \rightarrow 3$	
§15	$:5 \times 11 \Rightarrow \mathbf{m}\ l_{я} \Lleftarrow \mathbf{m} \wedge \mathbf{f}_7!$	Determination of codes in complexes **T** and **P** from a given number
§16	$:6 \times 7 + 1 \Rightarrow \mathbf{m}\mathbf{m}_{я} \Lleftarrow \mathbf{m} \wedge 177!$	
§17	$\mathbf{R}^*$.	Printout of error information

REFERENCES

1. Zakrevskii, A. D., Description of LYaPAS. This volume.
2. Chomsky, N., and Miller, G. H., Finite State Languages, *Information and Control* **1**, 91–112 (1958).

CHECKING OUT L-PROGRAMS ON URAL-1

A. D. Zakrevskii

1. THE PROBLEM OF AUTOMATIC CHECKOUT AND TWO TYPES OF CHECKOUT PROGRAM

Mathematical exploitation of computers shows that the lion's share of the working time, both for the programmer and for the machine, is absorbed by the process of program checkout. Hence arises the need for automatic checkout, which is particularly urgent in the use of automatic programming systems constructed on the basis of languages other than machine language.

It must be kept in mind that the automatic programming system presupposes, as a rule, that the programmer may ignore many characteristics of specific computers; may ignore, for example, the internal language of these machines and certainly may ignore how the program written by him in the input language of the automatic programming system is transformed to the internal language of one or another concrete machine. In this connection it is natural that the programmer wishes to maintain contact with the machine only at the level of the language he directly utilizes. Therefore the program-checkout process must be represented in this language.

Two basic types of checkout program for automatic programming systems can be noted. Programs of the first type are intended to find so-called syntactic errors in the checked out program, i.e., certain illegal combinations of symbols (in the simplest case, adjacent symbols). Programs constructed according to this principle are well known; for example, programs for testing expressions in ALGOL-60.

It is obvious that checkout programs of this type may fail to catch a large number of the errors committed by the programmer. Errors may be present in the algorithm for solution of a problem when there are no syntactic errors in the program, so that the execution of the program does not lead to the desired result. To find this kind of error the participation of the programmer himself is required, and the task of the checkout program is to facilitate this search, listing in clear and convenient form the information needed by the programmer to localize the errors. This work is performed by checkout programs of the second type, which, on the basis of certain brief indications supplied by the programmer, collect the needed information during the execution of the checkout program, first ensuring the most convenient mode of execution for this purpose.

The criterion of effectiveness of a checkout program can be the factor of reduction of the time expended on the average to eliminate all errors from the checked-out programs.

The program considered below, intended for checking out the programs of logical problems on URAL-1, is of the second type. This program, which we shall call OPLU [Russian acronymic for "Checkout Program for Logical Problems"], operates simultaneously with the first-level LYaPAS translator and supplies information on the trajectory of execution of the checked-out program and the sequences of values taken on by the variables, indexes, and complexes of interest to the programmer. This information can be printed and also put out on a signaling panel; the volume of information is kept within reasonable limits by a special control system.

In the development of OPLU the specific characteristics of URAL-1 were taken into account—the small volume of core storage, the low speed, and the nature of the input and output devices. OPLU was written directly in machine language, a knowledge of which on the part of the reader is assumed. The structure of the checkout program has a fairly general character, and it can easily be adapted to other computers.

Experience in using OPLU shows that although the operating speed of URAL-1 is clearly inadequate for the solution of many problems of practical interest, the corresponding programs can be checked out on this machine with complete success.

2. RESTRICTIONS OF "URAL" LYaPAS

Since URAL-1 has in all only 1024 36-bit registers, the following restrictions had to be introduced into first-level L-programs, prepared for checkout on this machine, and divided for coding into pairs of symbols, each of which is located in one incomplete (18-bit) memory register.

(1) The L-program may not contain the operators $\nearrow$, $\swarrow$, $\ddot{\mathbf{X}}$, $\overset{*}{*}$, $\overline{\times}$.
(2) The variable **я** must be considered to be the same as the other vari-

ables (the generation of random values of this variable is not provided), and the index a is the same as the other indexes.

(3) The operator $\Rightarrow$ is not generalized (i.e., expressions of the type $(\mathbf{a}, \mathbf{b}, c) \Rightarrow \mathbf{C}$ are not admitted).

(4) The use of the symbols of the special complexes A and B (the complexes of the starts and cardinalities of the complexes) is not admitted.

(5) Operations are admitted on only eight variable complexes (**A**, **B**, **C**, **D**, **E**, **F**, **G**, and **H**).

(6) The program that prints out the values of complexes must contain indication of the cardinalities of the complexes by means of a constant, variable, or index, whose symbol is placed after the symbol $*$ (for example, $\mathbf{A} * 7$ signifies that the first seven elements of the complex **A** must be listed).

(7) The symbol $\uparrow$ plays an auxiliary role, separating the L-program into parts that are sequentially processed by the translator.

(8) The symbol $\downarrow$ of passage to machine language has a specific meaning: all pairs of symbols following it (up to the first zero) are taken directly to be machine instructions, which are written with the understanding that the positions of the operands in machine memory are taken into account and are used only internally for a given set of transfers, adopting the convention that the composed set of machine instructions is stored beginning from the register with address 0.

(9) Only those L-programs can be directly executed in which the sentence number does not exceed 37 (consequently, the maximum number of sentences in the program is equal to 40—where here and below we use the more convenient octal notation, except for decimal numbers, which are tagged with the subscript $_{10}$). It is possible to check out large L-programs by means of the segmentation of L-programs into several "pieces," and the process of program execution into corresponding stages. The procedure for such a segmentation will be described.

(10) The maximum number of symbols in the L-program in one pass through the translator is equal to 408_{10}. The capability is provided for dividing each piece of L-program into two parts, separated by the symbol $\uparrow$, and translated in two passes.

Otherwise the rules for writing L-programs and their coding remain unchanged.

3. INFORMATION THAT CAN BE EXTRACTED IN THE MAXIMAL CHECKOUT MODE

Several modes of realization of the machine program (MP) assembled by the translator are provided. We shall begin their discussion with the maximal checkout mode, consisting, as do the other modes, of a preparatory stage and a stage of MP execution.

The results of the checkout are printed in the form of tables of 8-position numbers, composed of the digits 0, 1, 2, 3, 4, 5, 6, 7 and the empty symbol (the corresponding positions remain free). Certain of the numbers in the table can be tagged at the left by the symbol "$-$."

In the preparatory stage, consisting of the "adjustment" of the entire checkout system to the prescribed regime, no information is given out. The composition (and simultaneous printing) of tables is carried out during the execution of MP. This is done in the following way.

At each access to the start of some sentence of the L-program the number of this sentence, representing the current element of the trajectory of execution of the L-program, is written at the end of the table, using the two central columns. The trajectory over the L-program can also be observed visually, bringing into the monitor register the contents of register number 4032. In this case only a single lamp will light in the register, namely the one that coincides with the sentence being executed at the given moment (an adjustment must be made for the difference of numbers in the decimal and octal number systems).

In planning checkout the programmer must prescribe the list of variables and indexes of interest to him, in the form of 36-bit codes α_1' and α_2', placing unity in those positions that correspond to the operands in question (in contradistinction to the 18-bit codes occupying the half registers of URAL memory, we shall tag the 36-bit codes located in full registers by primes). The fourth bit from the left in α_1' corresponds to the variable **a**, the fifth to **b**, etc., and analogously for the list of indexes represented by the code α_2'. The volume of information output in checkout is limited in advance by the code β, in which the nine leftmost bits represent a certain number n_1, the nine rightmost, n_2.

When during the execution of MP a value is assigned to any index or variable, this value is printed

(1) if the corresponding operand appears in the list prescribed by the code α_1' or α_2';

(2) if the number of values of this operand already printed does not exceed n_1 (if the given operand is a variable) or n_2 (if it is an index).

The values of the operand are listed in the following form. The 32 binary components of an operand are represented by 11 octal positions, located in two adjacent 8-position lines. The first three positions are identified with the three rightmost positions of a first line; the eight remaining positions form the next line (if all these eight positions have the value 0, the printing of the second line is blocked). The three left positions of the first line indicate the number of the operand whose value is being printed: the variable **a** is coded by the number 0, **b** by 1, **c** by 2, etc.; the index a by 40, b by 41, c by 42, etc. If the last (n_1st or n_2nd) value of an operand is printed, it is tagged by the sign "—" placed in front of the number of the operand.

For example, the value

0000 0000 0000 0000 0000 0000 1010 0010

of the variable **a** is listed by the code

000 242

and the last of the listed values of the index d

1010 0011 1110 0000 0000 0000 0000 0000

is represented by the pair of codes

−043 000
24370000.

The sequence of values of the operands can be observed visually on the signal panel, effecting stop by the address 455; the value of the operand is represented by the arithmetic register, its code by the accumulator.

The output of information in the indicated form, characterized by certain positions remaining blank during printing, is provided by a special subroutine utilizing certain characteristics of printing in the decimal system (which must be taken into account in setting up the printing regime).

In visual observation of the values of the operands it must be recalled that the machine operates with codes shifted by one position to the left, so that in storing 32-bit LYaPAS codes in the 36-bit machine codes, the three left and the extreme right bits are not used.

The execution of MP in this mode is stopped if the planned volume of information is completely printed out (i.e., n_1 values are given out for each of the variables and n_2 values for each of the indexes required by the programmer) or if the L-program has been completely realized.

Aside from this information, at the start and finish of MP execution the complete working file is listed; i.e., the values of all the operands (variables, indexes, and complexes) are listed. The values of the variables and indexes are listed in the above form, the values of the elements of complexes are represented in analogous form, and their number is defined as $(k - 5000)/2$, where k is the number of the complete register occupied by the element. Only the nonzero values of the operands are listed.

In the preparation of the L-program for checkout it may be found expedient to abandon the principle of minimizing the number of variables and indexes used, which is obligatory in the writing of programs. In place of this it is possible to recommend the specialization of each of the operands, so that its meaning does not vary during the entire course of realization of the program. The subsequent minimization in the number of operands in an already checked-out L-program presents no particular difficulties.

By way of example of the method of output of information in the maximal checkout mode, we shall present the following program of the permutation of elements in the complex **A**, whose cardinality is prescribed by the value of the index f and which is located in the registers 5200, 5202, 5204, and 5206. The auxiliary complex **B**, whose elements play the role of counters, is stored in memory beginning with register number 5300.

$$
\begin{array}{ll}
\S 0 & f \Rightarrow e \\
\S 1 & \circ\ \mathbf{b}_e\ \bar{\triangle}\ e \mid\rightarrow 1\ 1 \Rightarrow d \\
\S 2 & \circ\ \mathbf{b}\ \circ\ a \\
\S 3 & \mathbf{b} \times 10\ \mathrm{V}\ \mathbf{a}_a \Rightarrow \mathbf{b}\ \triangle\ a \oplus f \mid\rightarrow 3\mathbf{b} \Rightarrow \mathbf{c} \\
\S 4 & \circ b\ \circ c\ \mathbf{a}_0 \Rightarrow \mathbf{a} \\
\S 5 & \triangle\ b\mathbf{a}_b \Rightarrow \mathbf{a}_c\ \triangle\ c \oplus d \mid\rightarrow 5\mathbf{a} \Rightarrow \mathbf{a}_c\ \triangle\ d \\
 & \triangle\ \mathbf{b}_d \oplus d\ \circ\!\rightarrow 6\ \bar{\triangle}\ d \oplus 1\ \circ\!\rightarrow 2\ \bar{\triangle}\ d \rightarrow 2 \\
\S 6 & \circ\ \mathbf{b}_d d \oplus f \mid\rightarrow 4.
\end{array}
$$

Assuming that the values of the operands **a**, **b**, **c**, a, b, c, d, e are to be printed during execution of the program and that the volume of information printed is prescribed by the values $n_1 = 4$, $n_2 = 3$, we obtain the following results of checking out:

045 004	01	04	04	04	00000003
100 001	043 001	041 000	−041 000	−000 002	040 004
101 002	02	042 000	−042 000	05	041 001
102 003	001 000	000 001	000 002	02	042 001
103 004	040 000	05	05	03	043 001
	03	041 001	06	−002 214	045 004
00	001 001	042 001	04	00000003	100 003
044 004	040 001	043 002	000 001		101 002
01	001 012	−043 001	05	000 002	102 001
044 003	−040 002	02	02	001 214	103 004
01	−001 123	03	03	00000003	142 001
−044 002	002 234	002 134	002 314	002 214	143 001
01	00000001	00000002	00000002		

4. FUNDAMENTAL COMPUTATION MODE

Aside from the maximal checkout mode, a fundamental computation mode and various combinations of these two modes are provided for, set up by the operator by means of keys on the control panel of the machine.

In the fundamental computation mode the machine program and its working field are defined by the maximum possible volume of core from which all the blocks of OPLU except the minimum necessary can be erased. The possibility is provided of binary computation and also (in the realization of large programs, not entirely contained in core) the separation of the computation process into stages, between which information is exchanged by external storage, on magnetic tape. In the fundamental computation mode all blocks of the programming system are called in as needed from magnetic tape, and the parts of the L-program corresponding to the various stages and the initial information representing the conditions of the problem to be solved, from punched tape. In this the correction operations (if needed), translation to machine language, and execution (with binary computation, if provided) alternate in the successive sections of the L-program. After passage of several stages in the computation mode it is possible to pass to checking out of the next piece of L-program in one or another mode, brought in according to the positions of the keys on the control panel.

5. MEMORY ALLOCATION AND REPRESENTATION OF INPUT INFORMATION

During the operation of OPLU and the execution of MP the working store is divided as follows:

The register 0 always contains 0.

Registers 1–33 serve as working registers of OPLU.

Registers 34–77 are also used as working registers in the checkout mode; in the fundamental computation mode they contain the minimally necessary blocks of OPLU (the check-sum block, the block for calling up the entire OPLU, etc.).

The following complete registers are used to store constants: 4100 $-$ c_0, 4102 $-$ c_1, . . ., 4176 $-$ c_{37}, 4200 $-$ d_0, 4202 $-$ d_1, . . ., 4210 $-$ d_4, 4212 $-$ e_0, 4214 $-$ e_1, . . ., 4222 $-$ e_4, 4224 $-$ f_0, 4226 $-$ f_1, . . ., 4246 $-$ f_{11}. Registers 250–277 are used for OPLU auxiliary constants.

Registers 300–1700 are charged in the introduction of OPLU and then, after OPLU has been adjusted to a given switch mode, information associated with the input data appears in registers 746–773, and the complete registers 5000–5176 are freed for the variables and indexes of the L-program, allocated as follows:

5000 $-$ **a**, 5002 $-$ **b**, . . ., 5076 $-$ **я**, 5100 $-$ *a*, . . ., 5176 $-$ **я**.

The next region (registers 1200–3777) is allocated by the programmer to the complexes and the machine program by the setting of the values of the parameters: j_0, the start of MP, $\omega_0, \omega_1, \ldots, \omega_7$, the addresses of the initial elements of the complexes **A**, **B**, $\cdots$, **H**, respectively. At the same time it is necessary to take into account the following: registers 3470–3477 are automatically assigned to the table of starts of sentences (for example, in the execution of MP the register 3742 contains the code $22x$,[1] where x is the address of the register corresponding to the start of sentence number 2); if the operator $\nleftrightarrow$ appears in the L-program, registers 3700–3737 are occupied by the subroutine that corresponds to this operator; further (or, rather, more closely) it is necessary to check MP, whose start must be indicated by the parameter j_0 as the number of an even half-register. If the volume of the L-program does not exceed 630 codes, it can be translated in one pass. In this case it is recommended to put $j_0 \geq 2530$, taking into account the following property of the translator: if $j_0 < 2530$, the translator perceives this as the sign that the L-program is divided into two parts, and after translation of the first part it begins to seek the second.

The region from register 1200 to $j_0 - 1$ is divided among the complexes by the assignment of the parameters $\omega_0, \omega_1, \cdots, \omega_7$ as the addresses of half-registers in which the storage of the corresponding complexes begins (for example, if the complex **C** is located in the full registers 5400, 5402, 5404, ..., the parameter ω_2 is given the value 1400). The programmer must not commit overlap of complexes.

In the fundamental computation mode the region assigned to complexes can be extended through assignment of the registers 300–740, in which are stored OPLU blocks not taking part in the given mode. This additional region can be used only for those complexes whose initial values are not input.

In the checkout of L-programs the LYaPAS-to-URAL-1 machine language translator (TRALU) always takes part, containing the correction block, the checkout program (OPLU) and, in the performance of a multistage process, the block for matching stages (BLOSÉ), which provides for the communication of the required data between stages. All of these programs are stored on magnetic tape in zones 3 and 4 (TRALU), 5 (OPLU), and 1 (BLOSÉ). Further, the possibility of reading TRALU and OPLU from punched tape is provided, from zones with the same numbers (if the magnetic-tape deck is inoperative).

The L-program and its input information are brought in from punched tape, where they are arranged in the following manner.

If the L-program is small (containing not more than 630 codes), it is located in zone 10, containing the sequence

$$\zeta,\ \kappa,\ \vartheta,\ \beta,\ j_0,\ \alpha_1',\ \alpha_2',\ \omega_0,\ \sigma_0,\ \omega_1,\ \sigma_1,\ \ldots,\ \omega_7,\ \sigma_7,\ \xi_1',\ \xi_2',\ \xi_3',\ \ldots,$$

[1] Translator's note: 22 is the URAL-1 unconditional transfer instruction.

where $\zeta = 000040$ if the L-program contains the symbol $\nleftrightarrow$, and $\zeta = 000000$ otherwise; $\kappa = 1000 + 2k$, where k is the total number of values of variables, indexes, and elements of complexes input; $\vartheta = -00\,0014$; the significance of the values β, j_0, α_1', α_2', ω_0, ω_1, $\cdots$, ω_7 have been defined above; and σ_0, σ_1, $\cdots$, σ_7 are the numbers of input elements for the complexes **A**, **B**, . . . , **H**, respectively.

It is necessary that the input elements for each complex have a continuous numbering, beginning with zero, and the input complexes themselves occupy the memory in order, so that a complex "at the right" will be located in registers with higher addresses than those in which a complex "at the left" is stored (we assume that the extreme left complex is **A**, to its right is **B**, etc.). The codes ξ_1', ξ_2', ξ_3', $\cdots$ represent the successive pairs of symbols of the L-program.

Zone 7 must contain the initial information in the form of a sequence ρ_1', ρ_2', ϵ_1^1, ϵ_2', ϵ_3', $\cdots$, where ρ_1', ρ_2' are the lists of variables and indexes whose values are input (these lists are made, as in the utilization of the codes α_1' and α_2'), and ϵ_1', ϵ_2', ϵ_3', $\cdots$ are a continuous array of initial values of operands input before realization of the L-program. The rules for packing these operands into a single array are determined by the values of the parameters ρ_1', ρ_2', ω_0, σ_0, ω_1, σ_1, $\cdots$, ω_7, σ_7, κ, and can be clarified by the following example. If $\rho_1' = 000016\,000000$, $\rho_2' = 000001\,140000$, $\omega_0 = 001200$, $\omega_1 = 001240$, $\sigma_0 = 000004$, $\sigma_1 = 000003$, $\sigma_2 = \sigma_3 = \cdots = \sigma_7 = 0$, $\kappa = 001032$ and the initial array contains the codes ϵ_1', ϵ_2', $\cdots$, ϵ_{15}', this signifies that the following correspondence is established between certain variables, indexes, and elements of complexes, on the one hand, and the initial values of the codes ϵ_1', ϵ_2', . . . , ϵ_{15}' which prescribe them, and which must be supplied by a special block of OPLU called "packaging" in full registers, whose addresses are listed in the right-hand column, on the other hand:

$$
\begin{aligned}
\mathbf{l} &\sim \epsilon_1' \Rightarrow 5026\\
\mathbf{m} &\sim \epsilon_2' \Rightarrow 5030\\
\mathbf{n} &\sim \epsilon_3' \Rightarrow 5032\\
p &\sim \epsilon_4' \Rightarrow 5134\\
s &\sim \epsilon_5' \Rightarrow 5142\\
t &\sim \epsilon_6' \Rightarrow 5144\\
\mathbf{a}_0 &\sim \epsilon_7' \Rightarrow 5200\\
\mathbf{a}_1 &\sim \epsilon_{10}' \Rightarrow 5202\\
\mathbf{a}_2 &\sim \epsilon_{11}' \Rightarrow 5204\\
\mathbf{a}_3 &\sim \epsilon_{12}' \Rightarrow 5206\\
\mathbf{b}_0 &\sim \epsilon_{13}' \Rightarrow 5240\\
\mathbf{b}_1 &\sim \epsilon_{14}' \Rightarrow 5242\\
\mathbf{b}_2 &\sim \epsilon_{15}' \Rightarrow 5244
\end{aligned}
$$

In the preparation of the initial information the 32 positions at the right of the 36-bit URAL codes must be used.

Zone 11 of the punched tape contains information for corrections that are carried out by TRALU when key 2 is set. This information is represented in the form of a sequence of LYaPAS symbols of length not exceeding 216_8, composed according to the rules presented in [1].

Zone 12 contains the second part of the L-program, beginning directly with the symbols ξ_n', ξ_{n+1}', . . . ; zone 13 contains the correction information concerning this part (when corrections are scheduled).

If the L-program is divided into several pieces, the second piece is disposed analogously in the zones 14, 15, 16, 17, the third in the zones 20, 21, 22, 23, etc. For the second piece ϑ will take on the value 00 0020 or −00 0020 (if this piece is the last), for the third piece 00 0024 or −00 0024, etc. The sense of the other parameters listed in zones 14, 20, 24, · · · is conserved and their values are defined in the same way as for zone 10. An exception is formed by the parameters σ_0, σ_1, · · ·, σ_7, which are assigned zero values (in all zones except the tenth). The point is that the initial information for a subsequent stage is obtained by the completion of the preceding one, and the cardinalities of the complexes transmitted between stages are calculated by the program. It is only necessary to provide in the program piece that at the end of the realization of the stage the cardinalities of those complexes **A**, **B**, . . . , **H** which are transmitted to the next stage be represented by the values of the indexes *y*, *z*, ж, л, ф, ш, э, ю, respectively. The block for matching the stages BLOSÉ transmits this information, in particular.

Before the realization of any stage except the first, the initial information for the stage is read, not from the punched tape, but from magnetic tape (from zone 1003), where it is stored by BLOSÉ in packed form.

6. DESCRIPTION OF OPLU

An idea of the character of the various checkout modes can be obtained from the flow chart of OPLU shown in Fig. 1.

OPLU consists of blocks between which the transfer conditions are indicated in the circles: the exits marked by points correspond to the upper position of the key whose number is given in the circle or the value 1 of the binary variable whose symbol is given in the circle. If a certain relation is inscribed in the circle, the marked exit is effected when the relation is satisfied.

An expression of the type 0*a*, 1*b* denotes assignment of the values 0 or 1 to the indicated binary variables. The change of checkout mode can be

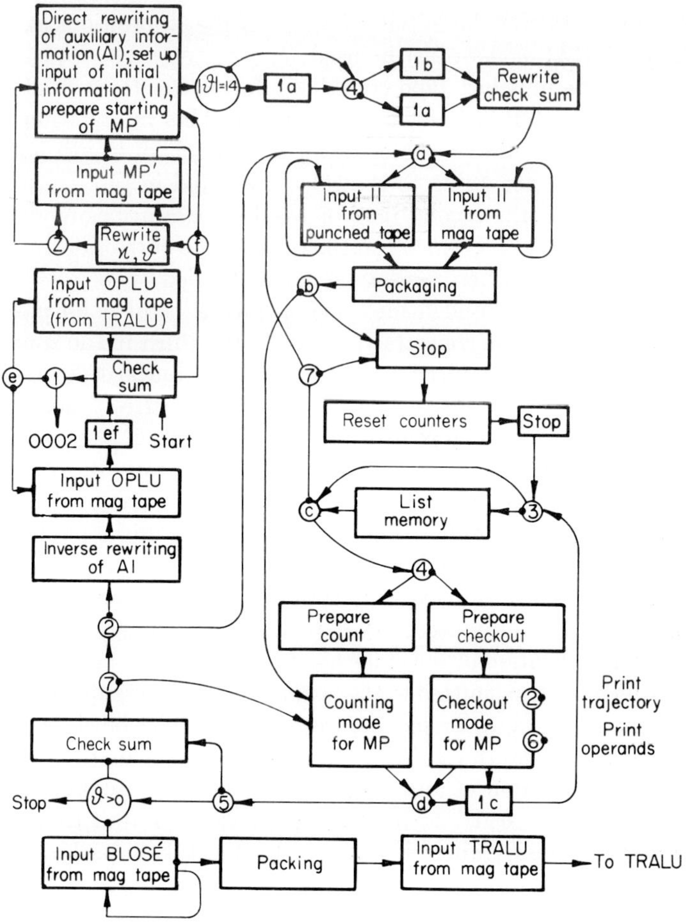

FIG. 1. Flow chart of OPLU. Key: (1) magnetic-tape operation; (2) print trajectory, strict-computation mode; (3) listing of memory; (4) checkout mode; (5) double computation; (6) print values of variables and indexes; (7) internal initialization of memory.

accomplished by changing the positions of the keys when the machine is stopped (block *stop*).

Preparation of the checkout mode begins with the operation of the translator (and therefore key 4 must already be raised). This is done by introducing the pair of codes

00 0000
22 0410

after each $16x$ instruction[2] in the machine program synthesized by the

[2] Translator's note: 16 is the transfer-to-working-storage (magnetic drum) instruction.

translator—these codes assign a new value to a variable or index, where $x \in \{5000, 5002, \cdots, 5176\}$— and by introducing the instruction triplet

16 4022
00 0000
22 0502,

at the start of each sentence of the L-program. Let us recall that the register 4022 is intended for the storage of the implicit variable τ. At the same time, in the preparations for checkout the printout of information by the operator $*$ is blocked—the translator does not synthesize the corresponding instructions.

In preparation of the computation mode the translator does not insert the above pairs and triplets of codes into the machine program, and the listing of data by the operator $*$ is programmed.

The preparation of checkout is continued by the special block included in OPLU. Its action reduces to matching the checked-out program with the block for listing the trajectory (T) and the block for listing the values of the variables and indexes (O), which are not indicated individually in the diagram, and which print out the corresponding information when keys 2 and 6 are raised. The block for checkout preparation locates the codes 22 0410 (instructions for transfer to block O) in the machine program, and places before them the codes $20k$,[3] where k is the address of the register to which the code is to be sent. Thus, when control is transferred to block O during realization of the checked-out program, the accumulator indicates the number k containing sufficient information for operation of the block O and resuming execution of the checked out program. To match the checked-out program with the block T at the start of each sentence, the code 00 0000 is replaced by the code $20n$, where n is the number of the sentence. Directly behind this code the translator places the code 22 0502, representing the instruction for transfer of control to the block T. Return from block T to the checked-out program takes place through the table of starts of sentences stored in memory.

When key 4 is down, in place of preparation of checkout the block for preparation of computation operates, inserting before the codes 22 4010 and 22 0502 the "jumper," the code $22k + 2$, where k is the number of the register in which the code is stored. By this means the block T and O are disconnected from the checked-out program, whose realization is thereby speeded up.

The next block of OPLU in complexity is the "packaging" block, which allocates in memory the initial information for the checked-out program. We recall that this information arrives in working storage in the form of a continuously packed array.

[3] Translator's note: 20 is the transfer-to-accumulator instruction.

The blocks of OPLU described above provide for the realization of the maximal checkout mode, corresponding to the right side of the flow chart of Fig. 1. The realization of the checked-out program can be repeated, as is evident from the flow chart. For this it is required to restore the initial values of the operands and the mechanism for counting the listed values. If the program is self-restoring, key 7 is raised and the repeated input of the initial information from punched tape (block "Input II from punched tape") or from magnetic tape (block "Input II from magnetic tape") and the packaging are blocked, and the machine is stopped directly before the block "Reset counters."

The remaining blocks of OPLU provide for the input of OPLU itself and its preliminary setting to a particular mode of realization of the checked-out program; certain of the blocks operate only in the fundamental computation mode, considered above. When registers 300–740 are used for intermediate information, obtained in the double-computation mode, there are two ways in which the initial data can be reset before computation is repeated. When key 7 is raised the program is self-restoring; in the contrary case the initial information is reentered.

The blocks "Input of II" and "Packaging" are located in registers 566–745; therefore, if these registers are not occupied by intermediate information, it is possible to drop key 2 so that control will be transmitted directly to these blocks. Otherwise key 2 must be raised, which corresponds to the "strict-computation mode," for which initially the entire OPLU is brought in, somewhat adjusted, and only then is control passed to the block "Input II."

After the realization of the current stage of the L-program has been completed, the entire transmitted information is packed into the previously described form and sent to magnetic tape. This is done by the block BLOSÉ, after it has been called in from magnetic tape.

The same block calls in TRALU, transfers the necessary information to it, and starts it.

7. THE OPLU PROGRAM

[Translator's note: In the original this section consists of two sentences of presentation and two pages of URAL-1 machine code—ten columns of about 60 10-place octal numbers. As explained in the translator's preface, this part of the book is omitted.]

REFERENCE

1. Tovshtein, M. Ya., The representation of Input Information to PS-LYaPAS. This volume.

TRANSLATOR FOR URAL-1

N. R. Toropov

The translator described in general form in [1] is submitted to the attention of the reader familiar with the language LYaPAS [2] and basic programming for the machine URAL-1 [3], in the form required by the concrete machine, having limited memory volume (1024 36-bit words) and low-speed performance (100 operations per second).

These machine characteristics have predetermined the basic purpose of the translator-checkout program [4], which is to check out algorithms on test problems that do not require a large amount of machine time.

In the construction of the given translator (TRALU) the program variants of the individual blocks were evaluated primarily for memory volume and secondarily for their execution time.

The process of translator construction can be divided into the following stages:

(1) composition of the translator algorithms (L-programs) in LYaPAS;

(2) minimization of the L-programs with regard to the particularities of the machine; and

(3) translation of the L-program to machine language and minimization of the machine programs.

The utilization of LYaPAS for writing TRALU offers the following advantages over direct machine-language programming.

The entire program is easily grasped. The minimization problem is facilitated, since it is carried out on a compact language, and the translation of the program to machine language is realized with regard to the particularities of the concrete machine. The probability of error in the writing of the program is reduced, since the translation of the compact L-program to machine language is realized by fairly simple, preestablished, and carefully worked-up rules. The checkout process is substantially shortened and facilitated.

In order to offer the largest possible memory volume for the L-program (LP) and machine program (MP) obtained by the translation of the former, the operation of the system TRALU-OPLU is divided into stages that are separated in time, coupled by a minimum of information.

In distribution of machine memory between LP and MP the experience obtained in the exploitation of TRALU for the first variant of LYaPAS has been taken into account.

Descriptions of the individual blocks of TRALU are given in Sections 4–7.

The list of blocks, their abbreviated names, and their location in TRALU are presented in the memory-distribution table (Section 3).

The basic operating principles of TRALU are described in Sections 1 and 2.

The list of working registers and MP constants used in the execution of MP, as well as the basic operands and the text of TRALU, divided into blocks, are given in the appendix. The designations of the blocks are indicated at the heads of the corresponding programs. The sentence numbers corresponding to the L-programs are indicated at the left, and at the right are given the comments, in symbolic form, of the operators and phrases of the L-programs corresponding to the given instructions.

1. BASIC OPERATING PRINCIPLES OF THE TRANSLATOR

The input information for TRALU is an L-program and its accompanying information, in particular the initial address j_0 of the machine program, the initial addresses of the complexes, the tag ζ of the presence in LP of the operators $\nleftrightarrow$ and !, and the number of the zone ϑ in which the LP of the next stage will be found. The output information of TRALU is the machine program.

Every LP is divided into sentences, consisting in turn of phrases. A phrase is a set of LP symbols including an operator and the operands subject to it. Frequently the left operand in a phrase is the implicit variable τ.

All of the LP symbols are encoded by 9-bit binary numbers; i.e., exactly two LP codes are stored in one half-register of URAL-1 memory. Sequentially, code by code, the translator analyzes a phrase and prepares the information needed to synthesize the instructions that will carry out this phrase. Since a strictly defined working register is assigned for each variable, index, and standard constant, the initial addresses of the complexes are given in the information accompanying LP; the address of a natural constant is determined in the process of its formation; the translator directly determines the true addresses of the operands during the analysis of the codes in a phrase. From the code of the operator the corresponding *macroinstruction* is selected, a previously composed program for the execution of one or another operator in machine language with undetermined addresses.

During synthesis the macroinstruction is filled in with addresses obtained by phrase analysis and is added to the machine program already composed. At the same time the number of instructions in MP is kept track of.

Then the next group of codes (phrase) is analyzed, etc., until the symbol of the end of the LP is encountered.

At the start of the synthesis of each sentence the corresponding line of the table starts of sentences (TSS) is filled in with an instruction that unconditionally transfers control to the instruction whose execution begins the execution of the given sentence. The instructions that function as transfer operators in this case are control-transfer instructions in the corresponding row of TSS, which makes possible "single pass" translation.

2. TRANSLATOR OPERATING MODES

In the translation of the LP of a single stage it may be found that LP, and therefore MP, will not fit into the memory allocated to them. In this case it becomes necessary to carry out translation in two steps. The LP and MP corresponding to this stage are divided into two parts (LP′, LP″ and MP′, MP″, respectively).

After translation of LP′ the MP′ obtained is transferred to magnetic tape (MT) and for checking purposes is read into a different memory location, where previously certain translator blocks had been located. Therefore, before translation of LP″ it is necessary to restore these translator blocks, and at the same time the blocks operating before translation and replaced by the machine program. In this connection TRALU is divided into two parts and stored in two zones. Below we shall refer to the first part of the translator program as TP′, the second part as TP″, and the entire translator program as TP.

Let us consider the operation of the translator in the various modes

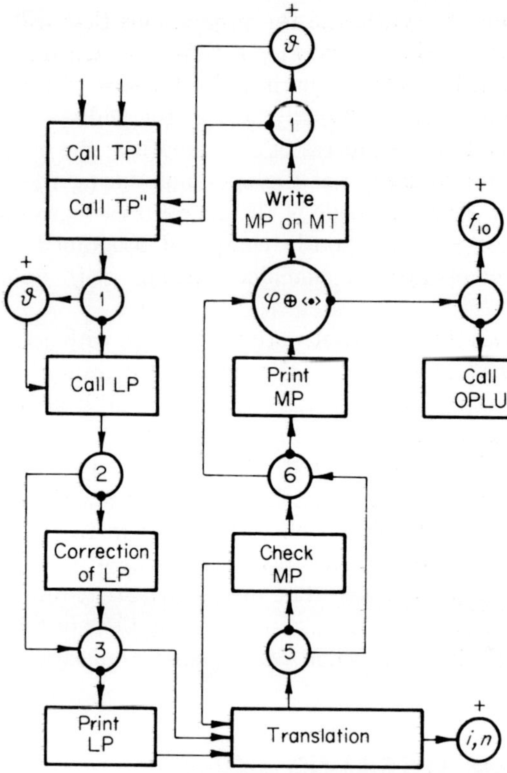

FIG. 1.

according to the consolidated flowchart presented in Fig. 1, where the circles and numbers represent the keys for transfer of control (the output marked by a point denotes transfer with the switch "up," i.e., "set"). The circles with a cross above them denote stop of the machine with transfer to the accumulator of the information designated inside the circle. The circles with logical conditions inscribed denote transfer over the arrow with the point when the condition is fulfilled, and over the arrow without the point otherwise. The check sum block also has two outputs (output "with point" when the check sums agree, and output "without point" otherwise).

Table I lists the operating modes defined by the positions of the control-transfer switches.

The system TRALU–OPLU is "started" by depressing the START button on the control panel. At this moment the reading device of the machine must contain the tape LP (in mode A) or the tape TP (in mode B). Both of these tapes have a zone 2, containing the instructions for input of

TABLE I

Key number	Key on	Key off
1	TP and OPLU read from magnetic tape (mode A)	TP and OPLU read from punched tape (mode B)
2	LP correction	LP correction blocked
3	Print LP	LP not printed
4	Translation for checkout mode	Translation for computing mode
5	Translation and correction repeated	"Ordinary" translation and correction
6	Print MP	MP not printed

TP′ from the corresponding tape. In mode B the machine stops after input of TP for replacement of tape TP by tape LP. Depressing the START button brings in LP. In mode A the machine does not stop. LP is brought in before coincidence of the check sums. If key 2 is "up," LP is corrected; otherwise the correction stage is omitted. Similarly, key 3 controls the LP printing mode. We remark, by the way, that although LP″ is brought in without accompanying information, the latter is printed just the same.

After LP has been printed, if printing has not been blocked, the process of translation proper is carried out. The rectangle marked "Translation" in Fig. 1 is composed of several blocks, which will be discussed below.

If the volume of MP exceeds the admissible volume, the machine stops and sends to accumulator the address i of the last instruction in MP and the address n of the last pair of codes of LP at which translation stopped.

A repeated-translation mode is provided, controlled by key 5. In this mode the check sum of the resultant MP is memorized and the translation process repeated. The second sum is compared with the stored one, and in the case of agreement control is transferred to the MP printing block; otherwise the translation is repeated.

If key 6 is "up," MP is printed. Together with MP″, the table of constants (TC) and the table of starts of sentences (TSS) are printed, as well as the subroutine for execution of the operators $\nleftrightarrow$ and ! if these are found in LP of the given stage. If the stage has terminated (i.e., if LP terminates with the symbol "·"), in mode B the machine stops, while in mode A OPLU is called in.

If the stage has not terminated (i.e., if LP terminates with the symbol "$\uparrow$"), MP′ is transferred to magnetic tape. After check reading from MT, the machine stops in mode B for replacement of tape LP by tape TP. Then, the overrun blocks of TP″ are restored by depressing the START button, and the process is repeated with the next part of LP. In mode A the machine does not stop.

After execution of MP, if this has not been the last stage, TP is brought in by the block for matching stages (BLOSÉ).

3. MEMORY DISTRIBUTION DURING TRANSLATION AND PRELIMINARY REMARKS ON THE DESCRIPTION OF THE BLOCKS OF TRALU

For the purposes of optimal memory utilization, the same locations are used for various purposes in various periods of TRALU operation. For example, the location where the blocks for callup, correction, and printing of LP were found during the preparation for translation is used during translation for the storage of the machine program. The registers occupied by the program for input of TP′ are used as working registers during translation. The check reading of MP′ from magnetic tape is carried out in the location of TP″.

The memory allocation during translation is given for TRALU blocks (abbreviated names are in parentheses).

0002–0010 0024–0051 3145–3156	Call up TP (*calltp*)
0052–0140	Initialize (*init*)
0141–0216	Code recognition (*reco*)
0217–0235	Select next code (*selco*)
0236–0304	Formation of table of constants (*const*)
0305–0421 0426–0437	Analysis and processing of operands (*anop*)
0440–0503	Synthesis of machine instructions (*synth*)
0504–0520	Write machine instructions in memory (*write*)
0521–0546	Prepare MP for output (*promp*)
0547–0567	Block for repeated translation (*repeat*)
0570–0606	Print MP (*primp*)
0607–0657	Write MP on magnetic tape (*tape*)
1133–1141	Call up OPLU (*coplu*)
1540–2310 2475–2527	Block for controlling operator synthesis (*bosc*)
3157–3241	Call up LP (*culp*)
2530–3114	Correction of LP (*corrector*)
3115–3137	Print LP (*prilp*)

Working registers

0002–0033	Used during correction and translation of LP
1142–1146	

Operator macroinstructions

0660–0777	↑, §, △, $\overline{\triangle}$, ⇒, $\dot{\mathbf{X}}$, →, ⇻, ○→, |→, |←, ⊢, ⌉
1116–1132	*
1517–1537	·, :
2311–2474	∇, **I**, ⇇, >, ←, ⊻, +, −, ∧, ∨, ⊕, ×
1000–1115	Table of control transfers (TCT)
1147–1513	LP and accompanying information
3146–3461	LP during correction
3462–3667	INCOR
3242–3270	Program for execution of the operators ⇻ and ! (at the moment of TP″ input)
3700–3737	Working storage for the subroutine for the operators ⇻, !, and the complex of levels **W**
2530–3737	Machine program and table of constants
3740–3777	Table of starts of sentences
1200–···	Location for check reading of MP′ from magnetic tape

For conciseness and clarity, all of TRALU and its individual blocks will be described in LYaPAS. We shall call this an L-description, in contradistinction to an M-description (in machine language, i.e., the text of the program).

For maximum utilization of the machine features, certain trivial changes have been introduced into LYaPAS:

(1) The values of all operands encountered in the L-description of the blocks are represented by 18 bits, one of which is the sign; i.e., the values of the operands can also be negative.

(2) To distinguish the operands of the L-description from LP operands, greek letters are sometimes used.

(3) Various types of constants encountered in the L-description will be enclosed in square brackets; for example, [220000].

(4) Expressions of the type $\varphi \oplus$ <·, !, §, ↓, ⇐>, frequently encountered in L-descriptions, signify the following: If the value of an operand to the left of the symbol ⊕ coincides with the code of one of the operators included between the angular brackets, the implicit variable τ takes on the value 0; otherwise $\tau \neq 0$.

4. PREPARATION FOR TRANSLATION

A. CALL UP TRALU (*calltp*)

The block *calltp* performs the callup of TP′ and TP″ in zones 3 and 4 on magnetic tape (M) in mode A and in punched tape (P) in mode B.

Let us define the conventional designations of certain operators.

The operator $\swarrow$ M α, β, γ causes information to be read from magnetic tape and inscribed in working storage: a volume of β registers are read from zone number α and inscribed in working storage, beginning with register number γ. The modification of this operator for punched tape is the operator $\swarrow$ P α, β, γ.

The operator Σ (α) computes the check sum of the information whose designation is given in the parentheses. The operator α $\overset{+}{\circ}$ stops the machine and sends to the accumulator the information whose designation is given at the left.

calltp

§1 $\mathbf{p}_1 \circ\!\rightarrow 2 \swarrow$ M3, 1120, 0024 $\rightarrow 3$
§2 $\swarrow$ P3, 1120, 0024
§3 $\Sigma\ (\mathrm{TP}') \oplus \Sigma_1 \mid\!\rightarrow 1 \circ x \circ z$
§4 $\mathbf{p}_1 \circ\!\rightarrow 5 \swarrow$ M4, 1556, 1516 $\rightarrow 6$
§5 $\swarrow$ P4, 1556, 1516
§6 $\Sigma\ (\mathrm{TP}'') \oplus \Sigma_2 \mid\!\rightarrow 4$
$\mathbf{p}_1 \mid\!\rightarrow$ *culp* $1\ \vartheta\ \overset{+}{\circ} \rightarrow$ *culp* 1

Here and below $\mathbf{p}_i$ signifies the ith key for transfer of control on the control panel ($i = 1, 2, \ldots, 7$); ($\mathbf{p}_i = 1$ if the ith key is "up"; $\mathbf{p}_i = 0$ if not). Σ_1 and Σ_2 are the known check sums of TP′ and TP″, respectively. The indexes x and z will be discussed below. The value of ϑ is the number of the zone of LP of the next stage.

B. CALL UP LP (*culp*)

The block *culp* performs the preparation for input and the input of LP and INCOR. This block transfers to the required location the complex of cardinalities Ω, prepared by BLOSÉ, at the start of each stage except the first. If LP contains the operators $\nrightarrow$ and !, *culp* transcribes the subroutine P ($\nrightarrow$) for the execution of these operators in the "working" location.

culp

§1 $x \circ\!\!\rightarrow 2\vartheta + 2 \rightarrow 3$	Calculation of the zone numbers of LP and INCOR
§2 ϑ	
§3 $\Rightarrow u + 1 \Rightarrow v$	
§4 $\Sigma \Rightarrow \Sigma_3 \swarrow Pu, 0345 + x, 1147 - x$	Reading of LP
$p_2 \circ\!\!\rightarrow 5 \swarrow Pv, 0216, 3462$	Reading of INCOR
Σ (INCOR)	Check sum of INCOR, LP
§5 $\Sigma (LP) \oplus \Sigma_3 \mid\!\!\rightarrow 4x \mid\!\!\rightarrow 7$	
$\vartheta \oplus \vartheta_0 \circ\!\!\rightarrow 6\Omega' \Rightarrow \Omega$	Transcribe the complex of cardinalities
§6 $\zeta \circ\!\!\rightarrow 7P' (\nrightarrow) \Rightarrow P (\nrightarrow)$	Transcribe subroutine P $(\nrightarrow)$
§7 $\rightarrow$ *corrector* 1	

Σ and Σ_3 are the current and preceding check sums of LP.

The index x takes on the two following values:

$$x = \begin{cases} 0 \text{ during translation of LP'}; \\ -31 \text{ during translation of LP''}. \end{cases}$$

These values are adopted to facilitate the calculation of the initial address for input of LP and the realization of transfers of the type $x \circ\!\!\rightarrow n$ and $x \mid\!\!\rightarrow n$ in machine language. LP′ is input together with the accompanying information, starting in register $1147 - x = 1147$ $(x = 0)$. LP″ is input without accompanying information, starting in register $1147 - x = 1200$ $(x = -31)$.

C. CORRECTION OF LP (*corrector*)

Before correction, LP is transcribed to the working location and its former location is cleaned up. Correction is carried out on the basis of information about the required corrections (INCOR), described in [5]. The LP subject to correction is represented by the complex **A**, INCOR by the complex **C**, and the "corrected" LP by complex **B**. Briefly, the correction principle consists of the following.

(1) The current code in complex **C** is analyzed. If the code is not the code of the symbol $\odot$ or $\otimes$, the next m codes are transcribed to **B**. (The number m is indicated in INCOR before an inserted group of codes.)

(2) If the current code is the code of the symbol ⊗, the position of the left boundary of the transcribed part of LP is sought in LP. The boundary code and all succeeding codes are transcribed into **B** until the sentence number is encountered that has been indicated in INCOR as the right boundary. Then n more codes are transcribed from **A** to **B** (n is indicated by the next code in **C**) and this procedure is repeated until the code of the symbol ⊙ is encountered, signifying the end of operation of *corrector*.

corrector

§1	$\mathbf{p}_2 \circ\!\!\to prilp\ 1\ \mathrm{LP} \Rightarrow (\mathrm{LP})^0$	Rewrite LP in the working field
§2	$\mathbf{O}(1200 \div 1513)\ \bar{\circ}\ b\ \bar{\circ}\ c \to 6$	Clean up a place for LP (as below, the operator for clean-up is denoted by $\mathbf{O}$ $(\alpha \div \beta)$, for the array of registers beginning with α and terminating with β)
§3	$\triangle\ c\mathbf{c}_c \Rightarrow p!$	Extraction of the next code from **C**
§4	$\Rightarrow w \triangle\ bw \Rightarrow \mathbf{b}_b!$	Write corrected codes in LP
§5	$\triangle\ a\mathbf{a}_a \Rightarrow y!$	Extraction of next code in **A**
§6	$\nrightarrow 3 \oplus \langle \odot \rangle \circ\!\!\to 26p \oplus \langle \otimes \rangle$ $\circ\!\!\to 10p + c \Rightarrow q$	Analysis of current code in INCOR
§7	$c \oplus q \circ\!\!\to 6 \nrightarrow 3 \nrightarrow 4 \to 7$	Write codes in **B** inserted from INCOR
§10	$\bar{\circ}\ a \nrightarrow 3 \Rightarrow u$	Drop codes belonging to the operator ⇐ and the M-piece in the search for the left boundary of the part of LP to be transcribed
§11	$\nrightarrow 5 \oplus \langle \Leftarrow \rangle \mapsto 12a + 4 \Rightarrow a \to 11$	
§12	$y \oplus \langle \downarrow \rangle \mapsto 14 \triangle a$	
§13	$\triangle\ a\mathbf{a}_a \to \mathbf{x} \triangle\ a\mathbf{a}_a \vee \mathbf{x} \mapsto 13 \to 11$	
§14	$y \oplus \langle \S \rangle \mapsto 11 \nrightarrow 5 \oplus u \mapsto 11$ $\nrightarrow 3a \Rightarrow g + p - 2 \Rightarrow a \nrightarrow 3 \oplus u$ $\circ\!\!\to 24 \Rightarrow h$	Search for position of left boundary
§15	$\nrightarrow 5h \circ\!\!\to 23y \oplus \langle \Leftarrow \rangle \mapsto 17y$ $\nrightarrow 4\ \bar{\circ}\ z$	Transcribe codes belonging to the operator ⇐

§16	$\triangle\, z \oplus 4 \circ\!\!\rightarrow 15 \nrightarrow 5 \nrightarrow 4 \rightarrow 16$	
§17 §20	$y \oplus \langle \downarrow \rangle \mid\!\!\rightarrow 21y \nrightarrow 4$ $\triangle\, b' \triangle\, a'\mathbf{a}_{a'} \Rightarrow \mathbf{b}_{b'} \circ\!\!\rightarrow 15 \rightarrow 20$	Transcribe the M-piece. The prime on the indexes indicates that transcription is carried out by pairs of codes
§21 §22 §23	$y \oplus \langle \S \rangle$ $\Rightarrow hy \nrightarrow 4 \rightarrow 15$ $y \oplus \boldsymbol{p} \mid\!\!\rightarrow 22a \Rightarrow g - 1 \Rightarrow a$	Transcribe codes from **A** to **B** up to the symbol of the right boundary sentence
§24 §25	$\nrightarrow 3 + g \Rightarrow g$ $a + 1 - g \mid\!\!\rightarrow 6 \nrightarrow 5 \nrightarrow 4w \oplus \langle \downarrow \rangle$ $\mid\!\!\rightarrow 25 \triangle\, a' \triangle\, b' \rightarrow 25$	Transcribe codes from **A** after the symbol of the right boundary sentence (the M-piece begins with a new line)
§26	$\mathbf{p}_5 \circ\!\!\rightarrow \mathit{prilp}\ 1$ $\Sigma\ (\mathrm{LP}) \oplus \Sigma_3 \circ\!\!\rightarrow \mathit{prilp}\ 1$ $\Sigma \Rightarrow \Sigma_3 \rightarrow 2$	Control of the repeated correction of LP

The complexes **A**, **B**, and **C** contain LP and INCOR in "packed" form, where each element of the complex contains two codes.

The presence of the operators $\nrightarrow$ and "!" permits the machine program of the algorithm to be shortened substantially and the algorithm to be simplified; since the operations with the indexes a, b, c, and with the elements of complexes **A**, **B**, and **C** are fairly complicated, it is more convenient to use them as subroutines with the operators $\nrightarrow$ and ! than to repeat them in the program several times.

To number the codes in the elements of the complexes, aside from the indexes a, b, and c, the auxiliary indexes α, β, and σ are introduced, taking on the value 1 for left codes in an element of a complex, and 0 for right.

Let us present typical examples in the given algorithm for manipulating indexes and elements of complexes, in connection with the above.

1. $\triangle\, c\mathbf{c}_c \Rightarrow p \sim 1 \oplus \sigma \Rightarrow \sigma + c \Rightarrow c\mathbf{c}_c \Leftarrow \varphi_\sigma \wedge \mathbf{z}_0 \Rightarrow p$;
2. $\triangle\, bw \Rightarrow \mathbf{b}_b \sim 1 \oplus \beta \Rightarrow \beta + b \Rightarrow b0 - \beta \Rightarrow \gamma w \Leftarrow \varphi_\gamma \wedge \mathbf{z}_\beta \vee \mathbf{b}_b \Rightarrow \mathbf{b}_b$ (the complex **B** is first cleaned up);
3. $\bar{\circ}\ a \sim \circ\ \alpha\ \bar{\circ}\ a$;
4. $a \Rightarrow g - 1 \Rightarrow a \sim 1 \oplus \alpha + a + a \Rightarrow ga - \alpha \Rightarrow a1 \oplus \alpha \Rightarrow \alpha$.

Here $\boldsymbol{\Phi} \equiv \{\varphi_{-1} = 000011,\ \varphi_0 = 000000,\ \varphi_1 = -000011\}$ is the complex of shift constants for extracting LYaPAS codes.

$\mathbf{Z} \equiv \{Z_0 = 000777,\ z_1 = 777000\}$ is the complex of constants for extracting LYAPAS codes.

D. PRINT LP (*prilp*)

The function of the block *prilp* is to print LP′ and LP″ together with the accompanying information. Together with the contents of a register is printed its address, separated by a space.

prilp

§1	$\mathbf{p}_3 \circ\!\!\rightarrow init\ 1[001146] \Rightarrow a$	LP is printed up to the appearance of two empty rows in LP
§2	$\triangle\, a - [001200] \mid\!\!\rightarrow 3a - [001514] \mid\!\!\rightarrow init\ 1$ $k_a \mid\!\!\rightarrow 3a + 1 \Rightarrow bk_b\ \circ\!\!\rightarrow init\ 1$	
§3	$k_a > 21 \Rightarrow \gamma' a < 3 \vee \gamma' * \rightarrow 2$	The complex K, the working complex

The shift of the address by three binary positions is made in order to separate the contents of the register from its address through blocking of the corresponding bar of the printer.

E. INITIALIZATION (*init*)

The purpose of the block *init* is to initialize the counters of constants (k), of instructions (i and j), of LP codes (m and n), and the index c, which will be discussed below. The block also clears space in memory for MP and inserts at the start of MP the instructions for resetting the register ρ of the depth of program realization if the operators ⧻ and ! appear in LP.

In the translation of LP in two steps, the first part of MP is not written in that place in memory to which it is intended. For this two MP instruction counters are introduced. The counter i fixes the address of the memory register in which the current instruction of MP is written, and the counter j fixes the true address of the MP instruction (in single-step translation and in the translation of LP″ the contents of these counters agree).

The number of constants in TC is registered by the counter k, which takes as its initial value $\varsigma + k_0$, where $k_0 = 40$ is the number of rows in TSS; $\varsigma = 0$ if there are no operators $\rightarrow$, !, and $\varsigma = 40$ otherwise—i.e., if P (⧻) is included in TC. The initial value of the counter i is fixed by the value of $\bar{\imath}_0$ in the case of repeated translation.

At the start of the stage, i.e., in translation of LP′ ($x = 0$), a test is made to determine whether LP can be translated in a single step; the test consists of comparing the value of j_0 with the initial address i_0 assigned to the place for MP. If $j_0 < i_0$, the index z takes on a nonzero value; counter j is reset to the value j_0, and the counter i to the value i_0. If $j_0 \geq i_0$, the counters i and j are reset to the value j_0 and LP is translated in a single step. At the start of the stage the entire field reserved for MP and the table of starts of sentences is cleared.

In the translation of LP″ $(x \neq 0)$ a test is made to determine whether MP′ has "grown" to the limit i_0; if not, the counters i and j are restored to the value i_0, otherwise to j. If $x \neq 0$ only part of the MP field is cleared (from i_0 to $i_{max} - k$, where $i_{max} = 3777$). If LP contains the operator $\nrightarrow$, then the instructions for resetting the register ρ governing the depth of program realization are placed at the head of MP′.

init

§1	$x \mid\!\rightarrow 2\zeta + k_0 \Rightarrow k j_0 - i_0 \mid\!\rightarrow 4$	Initialize counters i, j, k,
	$\bar{\circ}\ z j_0 \Rightarrow j i_0 \rightarrow 6$	m, n
§2	$j - i_0 \mid\!\rightarrow 3 i_0 \rightarrow 5$	
§3	$j \rightarrow 6$	
§4	j_0	
§5	$\Rightarrow j$	
§6	$\Rightarrow i \Rightarrow \bar{i}_0 \circ m\ \bar{\circ}\ n \circ c$	
	$x \mid\!\rightarrow 7\mathbf{O}\ (i_{max} - k_0 \div i_{max})$	Clear location of MP
§7	$\mathbf{O}\ (i_0 \div i_{max} - k)\zeta \circ\!\rightarrow$ *reco* $1x \mid\!\rightarrow$ *reco* 1	
	$[200000] \nrightarrow$ *write* 1	Inscription of instructions
	$[163727] \nrightarrow$ *write* $1 \rightarrow$ *reco* 1.	for restoring register ρ

5. THE ANALYSIS OF L-PROGRAMS

A. CHARACTERISTICS OF THE OPERANDS

Before describing the next blocks, let us consider in greater detail the principles of translation adopted in TRALU.

As has been noted, information about the operands and the operator constituting a phrase is prepared in the process of phrase analysis. This work is basically carried out by *reco* and *anop* together with the blocks *selco* and *const.*

Depending on the operator code, one or another subblock of *bosc*, which prepares the information about macroinstructions and transfers control to the block *synth*, operates. The block *synth*, utilizing information about the operands and the macroinstruction, defines its necessary addresses and transfers control to the block *write*, which adds the synthesized instructions to the already composed MP, increases by unity the values of the counters i and j, and also monitors "overflow" of the field reserved for MP.

If we take into consideration that the addresses of transfers are defined outside the block *synth*, all phrases can be considered to be of the same type for it, containing not more than two operands (left and right). Sometimes one or both of the operands may be absent from the phrase. The operands are divided into the three following groups, depending on their type: (a) element of a complex with variable index; (b) element of a complex with constant index; and (c) variable, index, or constant.

The information about the operands is written with the following characteristics: The value $\boldsymbol{\alpha}_r$ represents information about the address of operands of type (b) and (c) or the initial address of a complex for an operand of type (a) ($r = 0$ for the left operand and $r = 1$ for the right). The value $\boldsymbol{\beta}_r$ represents information about the address of an index for an operand of type (a). $\boldsymbol{\gamma}_r \neq 0$ if the rth operand is present in the phrase; otherwise, $\boldsymbol{\gamma}_r = 0$. $\boldsymbol{\delta}_r \neq 0$ if the rth operand is of type (c); otherwise, $\boldsymbol{\delta}_r = 0$. $\boldsymbol{\varepsilon}_r \neq 0$ if the rth operand is of type (a); otherwise, $\boldsymbol{\varepsilon}_r = 0$.

All the characteristics of operands are joined in the complex

$$\mathbf{H} \equiv \{\boldsymbol{\alpha}_0, \boldsymbol{\alpha}_1, \boldsymbol{\beta}_0, \boldsymbol{\beta}_1, \boldsymbol{\gamma}_0, \boldsymbol{\gamma}_1, \boldsymbol{\delta}_0, \boldsymbol{\delta}_1, \boldsymbol{\varepsilon}_0, \boldsymbol{\varepsilon}_1\}.$$

The values of the operands in LP are stored in full memory registers, which makes it possible to avoid checking the parities of the addresses in the execution of operations "12*a*," "13*a*," and "14*a*." [1]

To represent the values of the operands, 32 of the 36 bits of the full register are utilized (3 bits at the left and 1 at the right are not used); i.e., the values of the operands in the machine are doubled with respect to their true value. This makes possible the utilization of the instruction "30*a*" [2] for the elements of complexes (since the values of the indexes are doubled, while the elements of the complexes are located in full registers).

In order that the addresses of operands be obtained from their code, it is necessary to increase the codes of the operands by a factor of two. This is achieved by the block *selco*, which shifts the codes to the left by one position.

[1] Translator's note: URAL codes for fetch, bitwise conjunction, and comparison, respectively.

[2] Translator's note: Instruction modification in URAL-1; the constant a is added to the next instruction before execution.

B. LP CODE RECOGNITION (*reco*)

The purpose of *reco* is the recognition of LP codes supplied by the block *selco*, and the control of the block *anop* in the analysis of phrases and the processing of operands.

reco

§1	$\circ\ \gamma_0\ \circ\ \gamma_1\ \bar{\circ}\ r \nrightarrow selco\ 1$ $\lambda \wedge [001600] \mapsto anop\ 1$	Call up next LP code, fixed by the value of λ, and analyze it
§2	$\lambda > 1 \Rightarrow \varphi$ $\oplus \langle \nabla, \mathrm{I}, \leftarrow\!\mid, \vdash, \urcorner, \uparrow, * \rangle$ $\circ\!\!\rightarrow bosc\ (00)\ 1$	The value of the operator code is fixed by the value of the index φ
	$\varphi \oplus < \cdot, !, \S, \downarrow, \Leftarrow >$ $\circ\!\!\rightarrow const\ 1$	The operator $\Leftarrow$ is taken by the block *reco* as unary (its further analysis is carried out by the block *const*)
	$\varphi \oplus \langle \dot{\mathbf{X}}, \rightarrow, \nrightarrow, \circ\!\!\rightarrow, \mapsto, \rangle \mapsto 3$ $\nrightarrow selco\ 1 \lambda > 1$ $- [000400] + [003740]$ $\Rightarrow \eta$ $\varphi \oplus \langle \rightarrow, \nrightarrow \rangle \circ\!\!\rightarrow bosc\ (00)\ 1$ $\varphi \oplus \langle \circ\!\!\rightarrow, \mapsto \rangle \circ\!\!\rightarrow bosc\ (46)\ 1$	Fixation of a row in TSS in the analysis of transfer operators
	$\gamma_0 \Rightarrow d \mapsto 3 \nrightarrow anop\ 13$	Change of operands by characteristics in the case of phrases of the type $\cdots \Rightarrow \xi_i \dot{\mathbf{X}}\, n\xi_j$ and $\cdots \Leftrightarrow \xi_i \dot{\mathbf{X}}\, n\xi_j$
§3	$\triangle\ r \nrightarrow selco\ 1 \rightarrow anop\ 1$	Call up next operand code

C. SELECTION OF NEXT CODE (*selco*)

The next code in LP is fixed by the doubled value of the index λ.

Two indexes are used to extract the code in LP (the index n to fix the number of the row in LP containing two LP codes, and the index m for numbering the codes in the row).

selco

§1	$1 \oplus m \Rightarrow m$ $+ n \Rightarrow n$	Calculate position of next code
	$\mathbf{b}_n \leftrightarrows \mathbf{m}_m \wedge [001776]$ $\Rightarrow \lambda \circ\!\!\rightarrow 1!$	Extract next code. If it is empty, take following one

Here **B** is a complex each element of which contains two LP codes

$\mathbf{M} \equiv \{\mathbf{m}_0 = 000001, \mathbf{m}_1 = -000010\}$ The complex of shift constants

Let us note that $m = 0$ for the right code of a line in LP, and $m = 1$ for the left code.

D. FORMATION OF THE TABLE OF CONSTANTS (*const*)

The natural constants of LP not appearing after transfer operators and not constituting numbers of sentences, as well as constants in the case of the operator $\Leftarrow$, are inscribed in the table of constants as they appear in the course of analysis. If a constant similar to the current one already exists in TC, the latter is not inscribed in TC. The address of a constant, obtained in *const*, is fixed by the block *anop* by the characteristic α_r. In the case of the operator $\Leftarrow$, *const* marks the presence of a second operand in the phrase by the characteristic γ_1.

const

§1	$\varphi \oplus \langle\Rightarrow\rangle \mid\!\rightarrow bosc\ 1$ $\circ\ h' \circ \gamma_1 \circ a$	Formation of a constant in the case of the operator $\Leftarrow$. The shift constants are obtained from the complex **M**, described above, and the complex $\mathbf{N} \equiv \{\mathbf{n}_0 = 000022, \mathbf{n}_1 = 000000, \mathbf{n}_2 = -000022\}$
§2	$\mathbf{m}_m + \mathbf{n}_a \Rightarrow b$ $\mathbf{b}_n \leftrightarrows b \vee h' \Rightarrow h'$ $\triangle\ a \oplus 4 \circ\!\!\rightarrow 3 \triangle\ n1 \Rightarrow r \rightarrow 2$	
§3	$h' \wedge f_{10}$	
§4	$\Rightarrow h'\ \bar{\circ}\ a$	Search for a constant in TC similar to that formed ($\mathbf{Q} \sim$ TC)
§5	$\triangle\ a \oplus k \circ\!\!\rightarrow 6\mathbf{q}_a \oplus h'$ $\mid\!\rightarrow 5 \rightarrow 7$	
§6	$\triangle\ kh' \Rightarrow \mathbf{q}_a$	Inscription of the formed constant in TC
§7	$a \rightarrow anop\ 11$	Output of the address of the constant

Here and below the prime on an operand signifies that its value is stored in a full register.

E. ANALYSIS AND PROCESSING OF OPERANDS (*anop*)

anop

§1	$\bar{\circ}\ \delta_r\ \bar{\circ}\ \gamma_r\lambda \wedge [001400] \oplus [000400] \mid\!\rightarrow 2$	Test the next operand for membership in the class of variables or indexes
	$\lambda + [000300] \rightarrow 11$	Obtain address of operand when the latter is a variable or index ([000300] is a constant that must be added to the code of the operand to obtain its address)
§2	$\lambda \wedge [001600] \oplus [000200] \mid\!\rightarrow 7 \circ \delta_r$	If the operand is an element of a complex, it is tagged by the characteristic $\boldsymbol{\delta}_r$
	$\lambda \wedge [001700] \oplus [000300] \mid\!\rightarrow 3 b_\lambda \rightarrow 4$	If the complex is special, its start (b_λ) is fixed in $\boldsymbol{\alpha}_r$
§3	$b_{[\lambda+\omega]}$	Obtain the initial address of the complex and fix it in $\boldsymbol{\alpha}_r$
§4	$\Rightarrow \boldsymbol{\alpha}_r$	
	$\not\mapsto$ *selco* $1\lambda \oplus \langle * \rangle \mid\!\rightarrow 5$ $\lambda > 1 \Rightarrow \varphi 1 \Rightarrow r \not\mapsto$ *selco* $1 \rightarrow 1$	Process phrases of type $\mathbf{A} * \xi_j$
§5	$\lambda \wedge [001400] \oplus [000400] \circ\!\rightarrow 6$ $\lambda - [000400] + \boldsymbol{\alpha}_r \rightarrow 11$	Formation of the characteristic $\boldsymbol{\alpha}_r$ for an operand of type (b) (see text, Section 5A)
§6	$\lambda + [000300] \Rightarrow \boldsymbol{\beta}_r\ \bar{\circ}\ \boldsymbol{\varepsilon}_r \rightarrow 12$	Formation of the characteristic $\boldsymbol{\beta}_r$
§7	$\lambda \wedge [001400] \oplus [001000] \mid\!\rightarrow 10$ $\lambda - [000400] > 22 \rightarrow$ *const* 4	If the operand is a natural constant, the latter is formed and put out in block *const*
§10	$\lambda - [001300]$	Obtain the address of a standard constant
§11	$\Rightarrow \boldsymbol{\alpha}_r \circ \boldsymbol{\varepsilon}_r$	Formation of characteristics $\boldsymbol{\alpha}_r$ and $\boldsymbol{\varepsilon}_r$ in the case of operand types (b) and (c)

§12 $r \mid\!\rightarrow$ *bosc* (00) 1 $\not\mapsto$ *selco* 1 $\rightarrow$ *reco* 2	If all the operands have been processed, control is transferred to *bosc*; otherwise, the next code is taken
§13 $\circ\ a \circ b$ §14 $\triangle\ b\mathbf{h}_a \Rightarrow e\mathbf{h}_b \Rightarrow \mathbf{h}_a$ $e \Rightarrow \mathbf{h}_b \triangle\ a \oplus 6 \mid\!\rightarrow 14!$	Exchange of characteristics among operands

6. THE SYNTHESIS OF MACHINE INSTRUCTIONS

A. MACROINSTRUCTIONS

The synthesis of an operator (by which we mean "the synthesis of machine instructions realizing an operator") is carried out by means of the macroinstruction of this operator, composed so as to minimize the number of undetermined addresses.

In the synthesis of the operators !, $\downarrow$, $\Leftarrow$, $\circ$, $\bar{\circ}$, $\Leftrightarrow$, >, $\uparrow$, which do not have particular macroinstructions, the macroinstructions of other operators are used entirely or in part, sometimes with slight modification.

The location of the macroinstruction in the text of TRALU is indicated in the memory-distribution table (Section 3).

In the analysis of binary operators a common structure can be seen for them. There exists a group of instructions relating to the left operand (we shall call it a group of type I) and a group of instructions for the second operand (group of type II). Each of these groups may consist of subgroups. The first subgroup (I_a or II_a) consists of the instructions

$$\begin{array}{ll} 30 & 0001 \\ \theta & 4000, \end{array}$$

where in place of the symbol θ there stands the code of some machine operation. The final form of these instructions is completed in the process of synthesis by means of the operand characteristics (the addresses are indicated in the characteristics without the completeness tag). For example, if $\boldsymbol{\alpha}_r = 1000$ and $\boldsymbol{\beta}_r = 1100$, the instructions of the first subgroup after completion take the form

$$\begin{array}{ll} 30 & 1101 \\ \theta & 5000; \end{array}$$

i.e., the odd halves of the full registers are used for the indexes of elements of complexes.

The second subgroup (I_b or II_b) consists of instructions not requiring completion in the block *synth* (e.g., the macroinstruction of the operator ":").

The basic function of the subblocks of *bosc* is the preparation of the following data:

s—the initial address of the macroinstruction;
s_0—the final address of group I;
s_1—the final address of group II;
γ—the number of the register from which control is transferred from the block *synth* (in the block *synth*, transfer on index γ is used, which signifies transfer to the sentence whose number is indicated by the value of γ).

Let us consider, for example, the data for the macroinstruction of the operator $\leftarrow$:

$$s = 2372, \quad s_0 = 2374, \quad s_1 = 2412, \quad \gamma = [\mathit{reco}\ 1].$$

Let us dwell a little on the particularities of the macroinstructions of certain operators. The macroinstructions of the operators $\Leftarrow$, $\triangle$, $\bar{\triangle}$, $\circ$, $\bar{\circ}$, $\Rightarrow$, $\Leftrightarrow$, $\dot{\mathbf{X}}$, contain instructions for the checkout mode, which are omitted in the composition of MP if the operand changing its value is of type (a) or (b) or if key 4 is "up."

In the macroinstructions of unary operators one of the groups I or II is omitted.

The synthesis of the operators $\circ\!\!\rightarrow$ and $|\!\!\rightarrow$ is realized differently, depending on the preceding operator. For this the list of macroinstructions, aside from the proper macroinstructions for the operators $\circ\!\!\rightarrow$ and $|\!\!\rightarrow$, contains their common macroinstruction with the symbolic designation "$\pm \circ\!\!\rightarrow$, $|\!\!\rightarrow$," by means of which these operators are synthesized if preceded by $+$ or $-$.

The operator $*$ also has two macroinstructions: one for the case where the values of individual operands are listed, the other for the case of "list complex," denoted by the symbol $\bar{*}$.

The macroinstructions for the operators $\dot{\mathbf{X}}$, $\rightarrow$, $\nleftrightarrow$, $\circ\!\!\rightarrow$, $|\!\!\rightarrow$, $*$, ∇, and $\mathbf{I}$ contain control-transfer instructions whose addresses are completed by the value of the index η and the counters i and j.

The details of the macroinstructions will be explained in the descriptions of the corresponding subblocks of *bosc*.

B. BLOCK FOR OPERATOR SYNTHESIS CONTROL (*bosc*)

The block *bosc* is divided into subblocks according to operator, which prepare the information about the macroinstructions for the block *synth* and control the synthesis of the corresponding operators.

A number is assigned to each subblock, numerically equal to the operator code. For example, the subblock that controls the synthesis of the operator $\Leftrightarrow$ has the number 15. The sentence numbers in the L-description of the subblocks have two components: the first indicates the number of the block, the second, the number of the sentence in its L-description. For example, the number of sentence 5 of subblock 15 of block *bosc* is written as follows: *bosc* 15.5. Below, for conciseness, the designation *bosc* will be omitted and the first component enclosed in parentheses; e.g., in our example (15)5. In order to organize exit from *synth* the index γ is assigned the value of the address of the start of the corresponding sentence. For example, $(01)4 \Rightarrow \gamma$ signifies that control will be transferred from *synth* to sentence 4 of subblock (01).

To select the corresponding subblock by the operator code it is convenient to introduce transfer by variable, which is particularly simple to accomplish in URAL-1, which has an operation for modification of instruction (30a). The choice of subblock is performed by means of the table of control transfers (TCT). Each row, associated with one of the operators, contains the unconditional transfer instruction to the subblock for control of the synthesis of this operator.

The choice of row in TCT by operator code is realized by subblock (00) of *bosc*. This subblock initializes the value of the index c:

$$c = \begin{cases} 0 & \text{at the start of synthesis of all operators except } \circ\!\rightarrow \text{ and } |\!\rightarrow; \\ 1 & \text{during the synthesis of the operator } +; \\ 2 & \text{during the synthesis of the operator } -. \end{cases}$$

The table TCT begins in register 1000, where the rows corresponding to operator codes not encountered in LP are used to store additional constants. The use of TCT permits the block *reco* to be simplified, since in this case the operator codes are not recognized in isolation.

(00)

§1 $\circ c \rightarrow \mathbf{r}_\varphi$, where the complex $\mathbf{R}$ is TCT

The subblocks of *bosc* will be given the designation of the *operator* whose synthesis they control. (In the descriptions of certain subblocks that do not require comment, only their L-programs will be given).

Operator §

The operators §, ↓, ↑ will be assigned to the class of unary operators with left operand, since, aside from their basic functions, they must conserve and transmit the value of the implicit variable τ.

(01)

§1	$(01)4 \Rightarrow \gamma \mathbf{t}_1 \Rightarrow s\mathbf{p}_4 \circ\!\!\rightarrow 2\ 4 \rightarrow 3$	Preparation for synthesis. If key 4 is on at the start of the sentence in MP, three checkout-mode instructions are inserted; otherwise, they are omitted
§2	1	
§3	$+\, s \rightarrow (05)3$	
§4	$\not\mapsto$ *selco* $1\lambda > 1 - [000400] + [003740] \Rightarrow a$ $\mathbf{p}_4 \circ\!\!\rightarrow 5j - 3 \rightarrow 6$	Formation of a row of TCT
§5	j	
§6	$+\, [220000] \Rightarrow k_a \rightarrow$ *reco* 1	

Here and below the complex **T** is the complex of "starts" of macroinstructions. For example, $t_{14} = 0700$ is the start of the macroinstruction of operator $\Rightarrow$.

Operator ↓

(05)

§1	$(05)4$	Selection of macroinstruction of operator $\wedge$, since the given operator does not have its proper macroinstruction
§2	$\Rightarrow \gamma \mathbf{t}_{42} \Rightarrow s + 1$	
§3	$\Rightarrow \mathbf{s}_0 \rightarrow (54)4$	
§4	$j \Rightarrow f$	Memorization of the start of the M-piece (group of machine language instructions inserted in MP from LP)
§5	$\triangle\, n\mathbf{b}_n \circ\!\!\rightarrow$ *reco* $1 \circ a$	If the current instruction of the M-piece is not a transfer instruction, it is inscribed in MP without modification; otherwise, its address is added to the initial address of the M-piece
§6	$\mathbf{b}_n \wedge [370000] \oplus \mathbf{u}_a \circ\!\!\rightarrow 7$ $\triangle\, a \oplus 3 \mid\!\rightarrow 6$ $\mathbf{b}_n \rightarrow 10$	
§7	$\mathbf{b}_n + f$	
§10	$\not\mapsto$ *write* $1 \rightarrow 5$	The complex $\mathbf{U} \equiv \{210000, 220000, 240000\}$

Operator "·"

(06)

§1 *repeat* $1 \Rightarrow \gamma\mathbf{t}_6 - 2$ $\rightarrow (46)4$ — Synthesis of control-transfer instructions in OPLU for listing of a "final memory state"

Since the operator "·" does not have a left operand and is synthesized by the overall block *synth*, the initial address of the macroinstruction will be given as two less than the true one. This will also be the practice below.

Operators ⇐, ⇒

This subblock controls the synthesis of the operator ⇐ (§ 1), operator ⇒ (§ 2), and other operators that assign values to operands (§ 4).

(07)

§1 $\bar{\circ}\ \boldsymbol{\gamma}_0 \not\leftrightarrow$ *anop* 13 — Exchange of characteristics of operands in the case of operator ⇐

§2 *reco* 1 — Preparation for synthesis of operators ⇐, ⇒

§3 $\Rightarrow \gamma\mathbf{t}_{14} \Rightarrow s + 1$

§4 $\Rightarrow \mathbf{s}_0 + 2 \Rightarrow \mathbf{s}_1$
$\delta_1 \circ\!\rightarrow$ *synth* 1
$\mathbf{p}_4 \circ\!\rightarrow$ *synth* 1
$\mathbf{s}_1 \rightarrow (54)4$
— Sentence 4 controls the translation modes (checkout, computation), depending on the position of key 4, during the synthesis of operators ⇐, ⇒, ○, $\bar{\circ}$, △, $\bar{\triangle}$, $\dot{\mathbf{X}}$

Operator △

(10)

§1 $\mathbf{t}_{10} + 3 \Rightarrow \mathbf{s}_1\mathbf{t}_{10} - 2$
§2 $\Rightarrow s(07)2$
§3 $\Rightarrow \gamma s + 1$
§4 $\Rightarrow \mathbf{s}_0 \rightarrow$ *synth* 1

Operator $\bar{\triangle}$

(11)

§1 $\mathbf{t}_{11} + 4 \Rightarrow \mathbf{s}_1$
$\mathbf{t}_{11} - 2 \rightarrow (10)2$

Operators $\circ$, $\bar{\circ}$

(12)

§1	(07)2	Preparation for return from block *write* after inscription in MP of first instruction realizing operator $\circ$ or $\bar{\circ}$
§2	$\Rightarrow \varepsilon \mathbf{d}_\varphi \rightarrow$ *write* 1	

The groups of machine instructions realizing these operators differ only in the first instruction ("200000" for operator $\circ$ and "024244" for operator $\bar{\circ}$), which is taken according to the operator code from the table of auxiliary constants **D**, distributed among the rows of TCT; the remaining instructions are synthesized as for the operator $\Rightarrow$.

The last instructions of operator $\Leftrightarrow$ are synthesized in the same way, differing from the above in a single instruction (024022), which is also selected in table **D** by means of subblock (12), but in this case the exit from the block *write* must be to (07)1 and not (07)2, as in the case of the operators $\circ$ and $\bar{\circ}$.

Therefore, exit from the block *write* is organized by means of an index ϵ.

Operator $\Leftrightarrow$

The program realizing this operator is written thus:

$$\xi_i \Leftrightarrow \xi_j \sim \xi_i \Rightarrow \tau\xi_j \Rightarrow \xi_i\tau \Rightarrow \xi_j.$$

(15)

§1	(15) 2 $\Rightarrow \gamma \mathbf{t}_{60} \rightarrow$ (44) 2	Synthesis of instructions realizing the expression $\xi_i \Rightarrow \tau\xi_j$
§2	$\nleftrightarrow$ *anop* 13 $\circ\ \gamma_0$ (15) 3 $\rightarrow$ (07) 3	Synthesis of instructions realizing the phrase $\Rightarrow \xi_i$, with account to the checkout mode. For this the operands exchange characteristics, and the macroinstruction of the operator $\Rightarrow$ is used without the instructions of group I ($\gamma_0 = 0$)
§3	$\circ\ \gamma_1$ (07) 1 $\rightarrow$ (12) 2	Synthesis of instructions realizing the phrase $\tau \Rightarrow \xi_j$, for which the operands again exchange characteristics

This procedure for the synthesis of the operator $\Leftrightarrow$ has permitted us to avoid the creation of a special macroinstruction for it.

Operator $\dot{\mathbf{X}}$

$$(\xi_i \dot{\mathbf{X}}\, n\xi_j \sim \xi_i \circ\!\!\rightarrow n \vdash \Rightarrow \xi_j c_{[\xi_j]} \oplus \xi_i \Rightarrow \xi_i).$$

The left operand of this operator can be the right operand of the preceding operator if the latter is of the class that changes values of operands ($\Rightarrow$, $\Leftrightarrow$).

The block *reco* detects the criterion of this situation by the value of the index d ($d \neq 0$ if the left operand is "intrinsic," $d = 0$ otherwise).

(42)

§1 $d \mid\!\!\rightarrow 2 \Rightarrow \gamma_0$ §2 (42) $3 \Rightarrow \gamma \mathbf{t}_{42} \Rightarrow s$ $\mathbf{t}_{42} + 6 \rightarrow (07)\ 4$	If $d = 0$, instructions are synthesized which realize the phrase $\circ\!\!\rightarrow n \vdash \Rightarrow \xi_j$; otherwise, $\xi_i \circ\!\!\rightarrow n \vdash \Rightarrow \xi_j$. (The instruction "300001" is in the macroinstruction in front of the instruction for assignment of value to the index ξ_j, in order to "fit" the macroinstruction to standard form.)
§3 (42) $7 \Rightarrow \gamma \mathbf{p}_4 \circ\!\!\rightarrow 4$ $-\ 6 \rightarrow 5$ §4 $-\ 4$ §5 $+\ i \Rightarrow a$ §6 $\circ\ \gamma_0 \eta + k_a \Rightarrow k_a$	Completion of the transfer-instruction address (the address of the row in TCT, prepared in *reco*, is added to the instruction written in the register with address $[i - 6]$ in the checkout mode or the instruction in register $[i - 4]$ in the computation mode)
$\nRightarrow$ *anop* 13 $\mathbf{t}_{42} + 12 \Rightarrow s$ $\mathbf{t}_{42} + 15 \rightarrow (07)\ 4$	Synthesis of instructions realizing the phrase $c_{[\xi_j]} \oplus \xi_i \Rightarrow \xi_j$, for which the operands exchange characteristics
§7 $[02\ 4000] + \alpha_0$ $\nRightarrow$ *write* $1 \rightarrow$ *reco* 1	Synthesis of instructions for storing the value of τ in the accumulator (the value of the implicit variable τ always remains in the accumulator at the end of the realization of operators)

Operator $\rightarrow$

§1 (44) $3 \Rightarrow \gamma \mathbf{t}_{44}$ §2 $\Rightarrow s + 2 \rightarrow (05)\ 3$	Synthesis of instructions transmitting the value of τ and not the incomplete instruction "220000"
§3 η §4 $+\ k_{[i-1]} \Rightarrow k_{[i-1]}$ $\rightarrow$ *reco* 1	Completion of the unconditional-transfer instruction

Operators ⇻, !

To reduce the volume of MP the operators ⇻ and "!" are executed by means of a subroutine P (⇻), whose algorithm is presented below. In MP, instructions are synthesized for the execution of these operators which transmit to the subroutine P (⇻) the "return" address ν and the value of τ. The subroutine has two entries (§ 1 for the operator ⇻ and § 2 for the operator "!").

The depth of realization of the program registering the value of the index ρ must not exceed 10; otherwise, the machine stops with the transfer of the value of ρ to the accumulator.

To exercise control transfer by the operator ! to the corresponding "depth," the complex of levels $\mathbf{W} = \{\mathbf{w}_0, \mathbf{w}_1, \ldots, \mathbf{w}_7\}$ is introduced, each of whose elements contains the address of transfer to the corresponding level. If $\rho = 0$ (the program is realized at the "zero depth"), the operator ! is omitted.

P (⇻)

§1	$\Rightarrow \nu + 1 \Rightarrow \mathbf{w}_\rho$	Formation of complex **W**
	$\triangle\ \rho - 10\ \circ\!\!\rightarrow 3$ $\rho\ \overset{+}{\circ}$	Change of value of program-execution-depth register
§2	$\Rightarrow \nu\rho\ \circ\!\!\rightarrow 3$ $\bar{\triangle}\ \rho\tau \rightarrow \mathbf{w}_\rho$	Execution of transfer by operator "!"; if $\rho \neq 0$, change of value of ρ
§3	$\tau \rightarrow \nu$	Return to execution of MP with conservation of the value of τ

(45)

§1	$(45)\ 2 \Rightarrow \gamma\mathbf{t}_{45} \Rightarrow s$ $\mathbf{t}_{45} + 5 \rightarrow (05)\ 3$	Synthesis of instructions that fix τ and transmit the return address to the subroutine P (⇻)
§2 §3	$j - 1$ $+\ k_{[i-3]} \Rightarrow k_{[i-3]}$ $\varphi \oplus \langle ! \rangle \mid\!\!\rightarrow (44)\ 3$	Definition of the address of the transfer instruction and the instruction transmitting the return address in the case of the operator ⇻
§4	$\circ\ k_{[i-1]}\ \bar{\triangle}\ i\ \bar{\triangle}\ j$	Erasure of last instruction in MP, not required in the case of the operator "!"
	$\varphi \oplus \langle ! \rangle \mid\!\!\rightarrow reco\ 1$ $12 \rightarrow (44)\ 4$	Definition of the address in the instruction transferring control to the subroutine P (⇻)

Operators $\circ\!\rightarrow$, $|\!\rightarrow$

Two cases are distinguished in the synthesis of the operators $\circ\!\rightarrow$ and $|\!\rightarrow$:

(a) the preceding operator was $+$ or $-$ $(c \neq 0)$
(b) any other operator $(c = 0)$

In case (b) instructions are synthesized for logical addition with zero for the operator $\circ\!\rightarrow$, or addition modulo 2 with zero for the operator $|\!\rightarrow$. These operations conserve the value of τ.

In case (a) the last instruction of MP (extraction of the 32-bit result) is erased and instructions are added that correspond to the expression

(1)	$\cdots \Rightarrow \alpha \wedge \mathbf{f}_{10} \Rightarrow \tau \oplus \alpha \mid\!\rightarrow 1\tau \rightarrow 2$	for the combinations of operators
§1	$\tau \rightarrow n$	$\cdots + \cdots \circ\!\rightarrow n, \ \cdots - \cdots |\!\rightarrow n,$
§2	$\cdots$	
(2)	$\cdots \Rightarrow \alpha \wedge f_{10} \Rightarrow \tau \oplus \alpha \mid\!\rightarrow n\tau \cdots$	for the combinations $\cdots - \cdots \circ\!\rightarrow n, \ \cdots + \cdots |\!\rightarrow n.$

The subblock (46), using the macroinstruction for the operators $\circ\!\rightarrow$, $|\!\rightarrow$ and the special macroinstruction "$\pm \circ\!\rightarrow, |\!\rightarrow$," depending on the cases listed above, controls the synthesis of the operators $\circ\!\rightarrow$, $|\!\rightarrow$.

(46)

§1	$(44)\ 3 \Rightarrow \gamma c \circ\!\rightarrow 2$ $(46)\ 5 \Rightarrow \gamma\, \overline{\triangle}\, i\, \overline{\triangle}\, j$ $\mathbf{t}'_{47} - 2 \Rightarrow s$ $\mathbf{t}'_{47} + 10 \rightarrow (05)\ 3$	Formation of data on the macroinstruction "$\pm \circ\!\rightarrow, |\!\rightarrow$" in case (a) and erasure of last instruction in MP
§2	$\varphi \oplus \langle \circ\!\rightarrow \rangle \mid\!\rightarrow 3$ $\mathbf{t}_{46} \rightarrow 4$	Synthesis of the operators $\circ\!\rightarrow$ and $|\!\rightarrow$ in case (b). [The definition of the address in the transfer instruction is realized by subblock (44).]
§3	$\mathbf{t}_{47}$	
§4	$\Rightarrow s + 3 \rightarrow (05)\ 3$	
§5	$j - 2 + k_{[i-5]} \Rightarrow k_{[i-5]}$ $\varphi \oplus \langle \circ\!\rightarrow \rangle \mid\!\rightarrow 7$ $c \oplus 1 \mid\!\rightarrow 10$	Completion of the control-transfer instructions for the combination operators $\cdots + \cdots \circ\!\rightarrow n, \ \cdots - \cdots |\!\rightarrow n$
§6	$\circ\ cj \rightarrow (45)\ 3$	
§7	$c \oplus 1 \mid\!\rightarrow 6$	
§10	$\circ\ c\eta + k_{[i-3]}$ $\Rightarrow k_{[i-3]}$ $\rightarrow (45)\ 4$	Completion of the control-transfer instructions for the combinations $\cdots - \cdots \circ\!\rightarrow n,$ $\cdots + \cdots |\!\rightarrow n$ and removal of the last instruction in MP, which is no longer necessary

Operator ∇

The phrase $\xi_i \nabla$ is performed by means of the program

$\xi_i \Rightarrow \alpha\ \bar{\circ}\ f \circ \beta$

§1 $\triangle f \oplus 40 \circ\!\rightarrow 2c_f \wedge \alpha \circ\!\rightarrow 1 \triangle \beta \rightarrow 1$
§2 $\beta\ldots$

(50)

§1 $10 \Rightarrow a\mathbf{t}_{50} \Rightarrow s$ — Synthesis of instructions that execute the expressions given above
$\mathbf{t}_{50} + 15$
§2 $\Rightarrow \mathbf{s}_0(50)\ 3 \rightarrow (54)\ 3$

§3 $i - a \Rightarrow bj - 2$ — Completion of addresses in transfer instructions
$+ k_b \Rightarrow k_b$
§4 $j - a + k_{[i-2]} \Rightarrow k_{[i-2]}$
$\rightarrow$ *reco* 1

Operator I

The subblock (51) controls the synthesis of the operation ξ_i I, performed by the program

$\xi_i \Rightarrow \alpha\ \bar{\circ}\ f40 \Rightarrow g \circ \beta$

§1 $\bar{\triangle} g \triangle f \oplus 40 \circ\!\rightarrow 2\alpha \wedge c_f \circ\!\rightarrow 1$
$c_g \vee \beta \Rightarrow \beta \rightarrow 1$
§2 $\beta\ldots$

(51)

§1 $12 \Rightarrow a\mathbf{t}_{51} \Rightarrow s$ — Synthesis of instructions corresponding to the operator and completion of the addresses of the internal transfers by means of subblock (50), transferring to it the corresponding value of the index a
$\mathbf{t}_{51} + 17 \rightarrow (50)\ 2$

Subblocks (52), (53), (54), (55) control the synthesis of the operators $|\!\leftarrow$, $\vdash$, $\urcorner$, $\uparrow$ and do not require particular explanation, nor do the programs that realize them. It need only be noted that the operator $\uparrow$ in this translator carries out a "separation" function between LP′ and LP″, and transmission of the value of τ.

Operator $|\!\leftarrow$

(52)

§1 $\mathbf{t}_{52}$
§2 $\Rightarrow s + 4 \rightarrow (54)\ 2$

Operator $\vdash$

(53)

§1 $\mathbf{t}_{53} \rightarrow (52)\ 2$

Operator $\urcorner$

(54)

§1 $\mathbf{t}_{54} \Rightarrow s + 2$	This fragmenting of the sentence is required because the given program is common to many subblocks
§2 $\Rightarrow \mathbf{s}_0$ *reco* 1	
§3 $\Rightarrow \gamma \mathbf{s}_0$	
§4 $+ 2$	
§5 $\Rightarrow \mathbf{s}_1 \rightarrow$ *synth* 1	

Operator $\uparrow$

(55)

§1 *promp* $2 \rightarrow (05)\ 2$

Operator $*$

The operator $*$ has two modifications and, correspondingly, two macro-instructions: (a) to list the values of "individual" operands and (b) to list complexes. In view of the absence of the complex of cardinalities in the simplified version of LYaPAS for TRALU, when the values of complexes are listed their cardinalities are indicated at the right of the operator by an index, variable, or constant.

In the translation of LP in the checkout mode the given operator is completely left out in case (b); in case (a) it retains one of its functions—the transmission of the value of τ.

The values of the operands are listed in the same significant positions as in input.

(56)

§1 $\gamma_1 \mid\!\rightarrow 5\mathbf{p}_4 \circ \rightarrow 4$	Control of synthesis of operator $*$ in the listing of "individual" operands
§2 $\mathbf{t}_{72}$	
§3 $\Rightarrow s + 1 \rightarrow (54)\ 2$	
§4 $\mathbf{t}_{56} \rightarrow (52)\ 1$	
§5 $\mathbf{p}_4 \mid\!\rightarrow$ *reco* 1 $\nleftrightarrow$ *anop* $13\alpha_0 \Rightarrow \beta_1\ \bar{\circ}\ \epsilon_1\ 0 - \boldsymbol{\alpha}_1 \Rightarrow \boldsymbol{\alpha}_1$	In checkout mode the operator $*$ is omitted. Exchange of operands by their characteristics; correction of operands, and addition of missing ones, in order to put them in "standard" form and use block *synth*

$6 \Rightarrow a(50)\ 4 \Rightarrow \gamma \mathbf{t}_{57} - 1$
$\Rightarrow s\mathbf{t}_{57} + 7$
§6 $\Rightarrow \mathbf{s}_1$
§7 $s + 2 \rightarrow (10)\ 4$

Preparation for synthesis of the operator * in the case of listing of complexes and completion of the address of internal transfer by means of subblock (50)

Operator $\leftrightarrows$

(60)

§1 $\mathbf{t}_{60} \Rightarrow s\mathbf{t}_{60} + 11 \rightarrow (63)\ 2$

Operator $<$

(61)

§1 $(61)\ 2 \rightarrow (62)\ 2$
§2 $[070134] \Rightarrow k_{[i-4]} \rightarrow reco_1$

In the synthesis of the given operator the macroinstruction of the operator $>$ is used, with a certain correction

Operator $>$

§1 *reco* 1

§2 $\Rightarrow \gamma \mathbf{t}_{62} \Rightarrow s\mathbf{t}_{62} + 1 \rightarrow (56)\ 6$

Operator $\leftarrow$

(63)

§1 $\mathbf{t}_{63} \Rightarrow s\mathbf{t}_{63} + 20$
§2 $\Rightarrow \mathbf{s}_1\ reco\ 1 \Rightarrow \gamma$
$\rightarrow (56)\ 7$

Control of the synthesis of instructions realizing the operator $\leftarrow$ $(\xi_j \leftarrow \xi_j \sim \xi_i < \xi_j \Rightarrow \beta 40 - \xi_j \Rightarrow \delta \xi_i > \delta \vee \beta \wedge f_{10})$

Operator $\veebar$

Subblock (64) controls synthesis of the operation $\xi_i \veebar \xi_j$, performed by the program

$$\xi_i \Rightarrow \alpha \xi_j \Rightarrow \gamma c_{37} < \gamma \Rightarrow \delta\alpha < \delta \wedge d_\gamma \Rightarrow \beta\alpha > \delta \wedge e_\gamma \vee \beta$$

(64)

§1 $\mathbf{t}_{64} \Rightarrow s\mathbf{t}_{64} + 25 \rightarrow (63)\ 2$

Operator :

Division $\xi_i : \xi_j$ is executed by means of a program carrying out a sequence of subtractions of the divisor ξ_j from the dividend ξ_i.

$\xi_i \Rightarrow \tau \circ я$ §1 $\tau - \xi_j \circ\!\!\rightarrow 2 \Rightarrow \tau \triangle я \rightarrow 1$ §2 $\tau \ldots$	The quotient is "accumulated" in the index я, and the remainder remains in the accumulator

(65)

§1 (65) $2 \Rightarrow \gamma \mathbf{t}_{65} \Rightarrow s$ $\mathbf{t}_{65} + 4 \Rightarrow \mathbf{s}_0 \mathbf{t}_{65} + 17 \rightarrow (54)\ 5$	Synthesis of instructions that carry out the division program
§2 $j + k_{[i-6]} \Rightarrow k_{[i-6]}\, j - 5 + k_{[i-10]}$ $\Rightarrow k_{[i-10]} \boldsymbol{\delta}_1 \circ\!\!\rightarrow 3j - 14 \rightarrow (44)\ 4$	Completion of addresses in the internal transfers
§3 $j - 13 \rightarrow (44)\ 4$	

Operator ×

Because of the absence of a multiplication operation without roundoff in URAL-1, it would be necessary to use a rather cumbersome algorithm to execute the operator × as it is described in [2]; this is undesirable.

Most often in practice the need arises to multiply natural numbers. Therefore for memory economy we limit the range of the multiplied numbers to the segment $[0\text{–}2^{16}]$, which permits an uncomplicated algorithm to be applied for multiplication, the sense of which is clear from the macroinstruction.

(66)

§1 $\mathbf{t}_{66} \Rightarrow s + 2 \rightarrow (54)\ 2$

Operator +

(70)

§1 $1 \Rightarrow c\mathbf{t}_{70}$
§2 $\Rightarrow s + 4$
§3 $\Rightarrow \mathbf{s}_1\ rcco\ 1 \rightarrow (10)\ 3$

Operator −

(71)

§1 $2 \Rightarrow c\mathbf{t}_{71} \Rightarrow s\mathbf{t}_{71} + 5 \rightarrow (70)\ 3$

Control of the synthesis of the operator ∧ is exercised by subblock (56) (§ 2 and 3)

Operator $\vee$

§1 $t_{73} \rightarrow (56)\ 3$

Operator $\oplus$

§1 $t_{74} \rightarrow (56)\ 3$

C. SYNTHESIS OF MP INSTRUCTIONS (*synth*)

The block *synth*, utilizing data about the macroinstruction and the operand characteristics, constructs the MP instructions, leaving the addresses in transfer instructions undetermined. The final form of the transfer instructions is formulated by the corresponding subblocks of *bosc*.

synth

§1	$\bar{o}\ a$	If the synthesis is terminated of instructions of group II, exit is realized from *synth*
§2	$\triangle\ a \oplus 2 \circ\!\!\rightarrow \gamma$	
	$\gamma_a \mid\!\!\rightarrow 3\ \triangle\ s \rightarrow 5$	If an operand is absent from the phrase, the corresponding subgroup of instructions I_a or II_a is omitted
§3	$\varepsilon_a \circ\!\!\rightarrow 4k_s + \beta_a \nleftrightarrow$ *write* 1	Synthesis of instructions 30*a*, when the operand is an element of a complex
§4	$\triangle\ sk_s + \alpha_a \nleftrightarrow$ *write* 1	Synthesis of instructions completed by means of the characteristic α_a.
§5	$\triangle\ ss_a - s \circ\!\!\rightarrow 2k_s \nleftrightarrow$ *write* $1 \rightarrow 5$	Transcription from the macroinstruction of instructions for the subgroups I_b and II_b

D. INSCRIPTION OF MACHINE INSTRUCTIONS IN WORKING STORAGE (*write*)

The functions of writing instructions in MP, counting their number, and monitoring for overflow of the memory field reserved for MP are concentrated in a single block whose program is presented below.

write

§1	$\Rightarrow k_i$	Write current instruction
§2	$\triangle j \triangle i$ $- i_{\max} \mid\!\!\rightarrow 3!$	Change values of counters and check for overflow of MP field
§3	(i, n) ♁	Stop machine on overflow and transfer the values of counters i and n to the accumulator

7. BLOCKS THAT COMPLETE THE OPERATION OF TRALU

A. PREPARATION FOR OUTPUT OF MP (*promp*)

The block *promp* prepares information on the volume of MP subject to listing or transfer to magnetic tape (MP′), "combines" parts of MP, and checks the parity of final address in transfer to magnetic tape (in tape instructions only full addresses may appear).

The block *promp* has two entries: § 1 after synthesis of the operator "•" and § 2 after synthesis of the operator ↑ (i.e., after translation of LP′).

promp

§1	$i_{\max} \Rightarrow \beta \rightarrow 5$	Preparation for listing of MP after "•". If after synthesis of the operator ↑ it is found that $j < i_0$, then the instruction [22 2530] is written at the end of MP, since in this case MP″ is written starting from register $i_0 = 2530$, and $i = j$ in the composition of MP″
§2	$j - i_0 \mid\!\!\rightarrow 3[22\ 2530]$ ⇻ *write* 1	
§3	$i \wedge [00\ 0001] \circ\!\!\rightarrow 4$ ⇻ *write* 2	If the address of the last instruction in MP′ is even, then zero is appended to the end of MP to make it terminate with a full register, since the addresses in magnetic-tape-manipulation instructions in the given computer must be full
§4	$j \Rightarrow \beta i$	Fixation of the last values of counters i and j and the volume of MP for listing and transfer to tape
§5	$\Rightarrow \alpha - \bar{i}_0 \Rightarrow \mu$ $\rightarrow$ *primp* 1	

B. BLOCK FOR REPEATED TRANSLATION (*repeat*)

Repetition of the translation of LP begins with § 1 of the block *init* in translation of LP′ and from § 5 of the same block (i.e., without initialization of the counter of constants and without clearing the memory zone occupied by the constants, subroutine, and TSS) in the translation of LP″.

Repeated translation is blocked by key 5.

repeat

§1	$\mathbf{p}_5 \circ\!\!\rightarrow$ *primp* 1	Block repeated translation by key 5
	$\Sigma\ (\mathrm{MP}) \oplus \Sigma_4 \circ\!\!\rightarrow$ *primp* 1 Σ $\Rightarrow \Sigma_4\, x \circ\!\!\rightarrow$ *init* 1 $\bar{\imath}_0 \rightarrow$ *init* 5	If the check sums do not agree, the sum obtained is stored and translation is repeated

C. PRINT MP (*primp*)

From the information prepared by *promp, primp* prints the instructions of MP together with their addresses. To reduce the volume of the material listed, zeros are suppressed. Together with the instructions of MP″ are printed the table of constants, subroutine P ($\nleftrightarrow$) if it is used, and the table of starts of sentences.

primp

§1	$\mathbf{p}_6 \circ\!\!\rightarrow$ *tape* 1 $\bar{\imath}_0 \Rightarrow a\beta \;-\; \mu \Rightarrow b$	Block printing of MP by key 6
§2	$k_a \circ\!\!\rightarrow 3 > 21 \Rightarrow \gamma' b < 3 \vee \gamma' *$	Print out MP instructions together with their addresses and zero suppression
§3	$\triangle\, b \triangle\, a \oplus \alpha \mid\!\!\rightarrow 2 \rightarrow$ *tape* 1	

D. BLOCK FOR STORAGE OF MP ON MAGNETIC TAPE (*tape*)

In the translation of LP in two steps, MP′ is transcribed to magnetic tape (MT) in order to free space for MP″. After composition of the latter, MP′ is read back from MT and written in memory in front of MP″, beginning from register j_0.

The transcription of MP′ to MT is carried out with check reading into another location. If the check sums do not agree, the transfer is repeated.

Information about MP′ written on tape (its cardinality μ_0 and the check sum Σ') are sent to OPLU. The signal of presence of MP′ on tape is the value $z \neq 0$.

tape

§1	$\varphi \oplus \langle \uparrow \rangle \mid\!\rightarrow$ *coplu* 1 Σ (MP′) $\Rightarrow \Sigma'$	MP″ is not written on tape. Storage of check sum of MP′
	$[004000] - 2 + i \Rightarrow a$ $[004000] - 2 + \bar{\jmath}_0 + \mu \Rightarrow b$	Preparation of addresses of the tape-manipulation instructions
§2	$\nearrow \bar{\imath}_0, \mu,$ M 1005	Inscription of MP′ (registers $\bar{\imath}_0$ to $i - 2$) in zone 1005 of MT
	$\swarrow$ M1005, $\mu, \bar{\jmath}_0$	Reading of MP′ from MT into registers $\bar{\jmath}_0$ to $\bar{\jmath}_0 + \mu - 2$ (here $\bar{\jmath}_0 = 1200$)
	$\Sigma\ (\mathrm{MP}')' \oplus \Sigma' \mid\!\rightarrow 2$	Check inscription on MT
	$\mu \Rightarrow \mu_0[-000031] \Rightarrow x$ $\mathbf{p}_1 \mid\!\rightarrow$ *calltp* 4 (ϑ, ζ) ♁ $\rightarrow$ *calltp* 4	Preparation for translation of the next part of LP. (In mode B the machine stops for change of tape in the machine reading device, since for translation of LP″ it is necessary to dispose TP″ in memory, while the tape of LP is in the reading device.)

E. CALL OPLU (*coplu*)

The block *coplu* terminates the work of TRALU in the current stage. TRALU will be restored in memory in succeeding stages of translation by BLOSÉ.

The task of the block *coplu* is to call in OPLU in mode *A* or to stop the machine in mode *B*.

coplu

§1	$\mathbf{p}_1 \mid\!\rightarrow 2f_{10}$ ♁	In mode *B* the machine stops for change of tape in the reading device before input of OPLU
§2	$\swarrow$ M5, 1035, 0034 $\rightarrow$ OPLU	

8. CONCLUSION

TRALU is an example of a translator from LYaPAS for small digital computers with small memory volume, having words of not less than 32 bits.

The translator was written in LYaPAS with maximal utilization of the particularities of the concrete machine. Since LYaPAS is not intended for these purposes, it was necessary to "adapt" it for the description of concrete machine programs by the introduction of certain changes. Other procedures known at the present time for writing such large programs, copiously supplied with transfers, are unsuitable.

APPENDIX

BASIC OPERANDS USED IN L-DESCRIPTIONS OF BLOCKS

Address of register	Symbol	Basic function
0002	α_0	Characteristics of operands
0003	α_1	
0004	β_0	
0005	β_1	
0006	γ_0	
0007	γ_1	
0010	δ_0	
0011	δ_1	
0012	ε_0	
0013	ε_1	
0014	s	Initial address of the macroinstruction in *bosc* and current address in *synth*
0015	$\mathbf{s}_0$	Final address of group I in macroinstruction (*bosc–synth*)
0016	$\mathbf{s}_1$	Final address of group II in macroinstruction (*bosc–synth*)
0017	r	Operand counter in phrase analysis
0020	m	Counter modulo 2 of LP codes (*selco*)
0021	δ	Index of exit from block *selco*
0022	h'	Working register
0023		
0024	Σ_4	Check sum of MP (*repeat*)
0025		
0026	φ	Operator code (*reco, anop, bosc*)
0027	k	Constant counter in TC

APPENDIX—*Continued*

Address of register	Symbol	Basic function
0030	λ	Next code in LP (*selco, reco, anop, const, bosc*)
0031	γ	Index of exit from block *synth*
0032	i	Counter for MP instructions stored in memory
0033	n	LP line counter (*selco, const, bosc*)
1025	c	Criterion of type of operators in synthesis
1027	x	Counter of parts of LP in a single stage
1041	d	Criterion of a phrase of type $\Rightarrow \dot{\xi}_i \dot{\mathbf{X}} n \dot{\xi}_j$
1043	η	Row number in TSS in analysis and synthesis of transfer operators
1047	$i_{\max}$	Maximal address of the memory field reserved for MP
1067	ϵ	Index of exit from block *write*
1101	ξ	Index of exit from block *anop*
1103	$\bar{\imath}_0$	Address of register in which storage of MP begins
1105	j	Counter of true addresses of MP instructions
1142	μ_0	Cardinality of MP′, transcribed on MT′
1143	z	Tag indicating that translation of LP is carried out in two steps
1144 1145	Σ'	Check sum of MP′ written on MT′
1146	ϑ	Number of zone of LP in next stage
1147	ζ	Tag indicating presence of the operators $\nleftrightarrow$, ! in LP
1153	j_0	True initial address of MP
3727	ρ	MP depth-of-realization register
3730 . . . 3737	**W**	Complex of MP depths of realization

(Constants, used in the L-description, are mainly situated between the rows of table TCT.)

REFERENCES

1. Zakrevskii, A. D., Description of LYaPAS. This volume.
2. Smol'nikov, N. Ya., "Basic Programming for URAL Digital Computer" [in Russian]. Sovetskoe radio, Moscow, 1961.
3. Tovshtein, M. Ya., Translator for High-Speed Computers. This volume.
4. Zakrevskii, A. D., Checking out L-Programs on URAL-1. This volume.
5. Tovshtein, M. Ya., The Representation of Input Information to PS-LYaPAS. This volume.

PART II: APPLICATIONS

Section A: ABSTRACT PROBLEMS IN SYNTHESIS THEORY

PROGRAMMING BOOLEAN COMPUTATIONS

A. D. Zakrevskii

The present chapter is written for the purpose of expanding certain concepts and introducing the corresponding symbols, utilized in the subsequent chapters in the discussion of various problems of a logical character.

The constructions are based on the elementary theory of finite sets [1–4] or, more exactly, Russell's elementary theory of types, whose basic ideas are presented, for example, in [5].

The basic material for the present article is taken from earlier publications by the same author [6, 7].

1. THE HIERARCHY OF BOOLEAN SPACES

Let us consider a certain finite set $X \equiv \{x_0, x_1, \ldots, x_{n-1}\}$, calling it the space of first rank. By $\tilde{X}$ we shall denote the set of all 2^n of its subsets and call it the space of second rank over X. By $\tilde{\tilde{X}}$ we denote the set of all 2^{2^n} subsets of $\tilde{X}$ and call it the space of third rank over X, etc. Passing to a more convenient symbolism, we shall denote by $X^{(k)}$ the space of kth rank over X, which is also called the Boolean space over $X^{(k-1)}$. For example, $X = X^{(1)}$, $\tilde{X} = X^{(2)}$, $\tilde{\tilde{X}} = X^{(3)}$.

We shall bring into consideration set-theoretical variables, representing them by the symbols α, β, and γ, which can be accompanied by superscripts and subscripts (in the representation of concrete sets we shall also operate with letters of the latin alphabet). By the superscript we shall indicate, if useful, one important characteristic of the variable, called its rank. We shall assume that the values of the variables of the kth rank are subsets of the space of kth rank, and only such subsets: $\alpha^k, \beta^k, \gamma^k \subseteq X^{(k)}$. We shall also adopt an alternative convention for the representation of the rank of a variable (or constant set): $\rho(\alpha)$ denotes the rank of the variable α. From these definitions it follows that if $\alpha \in X^{(k)}$, then $\rho(\alpha) = k - 1$. Conversely, if $\rho(\alpha) = k$, then α is an element of the space $X^{(k+1)}$; for example, $\alpha^0 \in X$, $\alpha^1 \in \tilde{X}$, $\alpha^2 \in \tilde{\tilde{X}}$. Subscripts will be used to distinguish variables.

The cardinality of set α, defined as the number of its elements, will be denoted by $\sigma(\alpha)$. If $\sigma(\alpha) = 0$, the set α is called empty, and is denoted by the symbol $\varnothing$: $\sigma(\alpha) = 0 \leftrightarrow \alpha = \varnothing$. To avoid possible confusion, we remark that the empty set $\varnothing$ can be a "legitimate" element of any other set; for example, if it is known that the unique element of a set α is the empty set, it must be considered that $\sigma(\alpha) = 1$. At the same time, the empty set is subject to special rules; it can be assigned arbitrary rank, and therefore may be manipulated on the same level as sets of arbitrary ranks.

The point is that such relations among sets as equality ($=$), inclusion ($\subseteq$), strict inclusion ($\subset$), and exclusion ($\not\subset$) are defined only for sets of identical rank, constituting elements of the same space. A similar remark holds with respect to the application of the set-theoretical operations of union ($\cup$), intersection ($\cap$), and difference ($\setminus$) of sets; for example, the expression $\alpha^1 \cup \alpha^2$ has no meaning, while the expression $\{\alpha^1\} \cup \alpha^2$ is perfectly meaningful, since $\{\alpha^1\}$ denotes the set of second rank whose unique element is α^1. Expressions of the type $\alpha^k = \varnothing$, $\alpha^k \cup \varnothing$, etc., are always admissible, i.e., have a defined sense. We shall adopt the convention that $\{\alpha_i\}$ is a set containing the single element α_i, the expression $\{\alpha_i / i \in I\}$ is the set of all elements α_i for which $i \in I$.

The set $\bar{\alpha}^k$ is the complement of the set α^k in the space $X^{(k)}: \bar{\alpha}^k = X^{(k)} \setminus \alpha^k$. Let us also introduce the operator $+$, the summation of sets, defined in the following way: $\alpha + \beta = (\alpha \cup \beta) \setminus (\alpha \cap \beta)$. From the definition it follows that $\alpha + \beta = \bar{\alpha} \cap \beta \cup \alpha \cap \bar{\beta}$. It is assumed that the operator $\cap$ operates before the operator $\cup$.

Let us adopt a convention on the order of action of the operators. By convention, if two adjacent unary operators are present in the formula, the one at the right acts before the one at the left: if ψ_i and ψ_2 are unary operators, then $\psi_1 \psi_2 \alpha = \psi_1(\psi_2 \alpha)$. We divide binary operators and relation symbols into four sets:

$$\{\cap\}, \quad \{\cup, \setminus, +\}, \quad \{\in, \subset, \subseteq, \notin, \not\subset, \not\subseteq, =, \neq\}, \quad \{\rightarrow, \leftrightarrow\},$$

considering that the order of action of the operators occurring in the different sets coincides with the order of the sets. If this order is violated or if we wish to make explicit the order of action of the operators in a single set, we shall introduce additional parentheses. Thus, the expression

$$\alpha \cap \beta \cup \bar{\alpha} \cap \bar{\beta} \in A \rightarrow \alpha + \beta \in B$$

is equivalent to the expression

$$(((\alpha \cap \beta) \cup \cup(\bar{\alpha} \cap \bar{\beta})) \in A) \rightarrow ((\alpha + \beta) \in B).$$

A useful element of the model being constructed is the existence quantifier $\exists$; by $(\exists \alpha)[\varphi(\alpha)]$ we agree to denote the proposition that there exists a certain α that satisfies the condition $\varphi(\alpha)$ [4].

For example, let us define by means of the existence quantifier the sense of the symbol $\in^2$: $\alpha \in^2 \beta \leftrightarrow (\exists \gamma)[\alpha \in \gamma \in \beta]$, i.e., $\alpha \in^2 \beta$ if and only if there exists a γ such that $\alpha \in \gamma$ and $\gamma \in \beta$. By way of generalization we obtain the expression

$$\alpha \in {}^k\beta \leftrightarrow (\exists \gamma_1, \gamma_2, \ldots, \gamma_{k-1})[\alpha \in \gamma_1 \in \gamma_2 \in \cdots \in \gamma_{k-1} \in \beta],$$

read in the following way: $\alpha \in {}^k\beta$ if and only if there exist $\gamma_1, \gamma_2, \cdots, \gamma_{k-1}$ such that $\alpha \in \gamma_1, \gamma_1 \in \gamma_2, \ldots, \gamma_{k-1} \in \beta$. We obtain a similar generalization for the relation $\notin$ ($\alpha \notin \beta$ signifies that α is not an element of the set β):

$$\alpha \notin {}^k\beta \leftrightarrow (\overline{\exists \gamma_1, \gamma_2, \ldots, \gamma_{k-1}})\,[\alpha \in \gamma_1 \in \gamma_2 \in \cdots \in \gamma_{k-1} \in \beta],$$

where $(\overline{\exists\ \gamma_1, \gamma_2, \ldots, \gamma_{k-1}})$ signifies that "there do not exist such $\gamma_1, \gamma_2, \ldots,$ γ_{k-1} that ..."

As for the universal quantifier [4], which in our case must necessarily be bounded, it is possible to avoid it, since an expression "for all α, where $\alpha \in \beta, \ldots$" has the following equivalent: "$\alpha \in \beta \rightarrow \cdots$"

Let us introduce an abbreviated notation for the operation of union of elements of a set

$$\alpha \equiv \{\alpha_0, \alpha_1, \ldots, \alpha_{m-1}\} : \bigcup [\alpha] \equiv \alpha_0 \bigcup \alpha_1 \bigcup \cdots \bigcup \alpha_{m-1}.$$

Analogously,

$$\bigcap [\alpha] \equiv \alpha_0 \bigcap \alpha_1 \bigcap \cdots \bigcap \alpha_{m-1}$$

and

$$+ [\alpha] \equiv \alpha_0 + \alpha_1 + \cdots + \alpha_{m-1}.$$

Obviously,

$$\rho(\bigcup [\alpha]) = \rho(\bigcap [\alpha]) = \rho(+ [\alpha]) = \rho(\alpha) - 1.$$

By way of generalization of these operations we shall assume that if

$$\alpha \equiv \{\alpha_0, \alpha_1, \ldots, \alpha_{m-1}\},$$

then

$$\begin{aligned} \bigcup^{(2)}[\alpha] &\equiv \{\bigcup[\alpha_0], \bigcup[\alpha_1], \ldots, \bigcup[\alpha_{m-1}]\}, \\ \bigcap^{(2)}[\alpha] &\equiv \{\bigcap[\alpha_0], \bigcap[\alpha_1], \ldots, \bigcap[\alpha_{m-1}]\}, \\ +^{(2)}[\alpha] &\equiv \{+[\alpha_0], +[\alpha_1], \ldots, +[\alpha_{m-1}]\}, \end{aligned}$$

while

$$\bigcup^2[\alpha] \equiv \bigcup[\bigcup[\alpha]], \quad \bigcap^2[\alpha] \equiv \bigcap[\bigcap[\alpha]], \quad +^2[\alpha] \equiv +[+[\alpha]].$$

The meaning of the symbols $\bigcup^k$, $\bigcup^{(k)}$, $\bigcap^k$, $\bigcap^{(k)}$, $+^k$, $+^{(k)}$ will be defined by induction.

For example, let $\alpha = \{\alpha_1, \alpha_2\}$, where $\alpha_1 = \{\beta_1, \beta_3\}$ and $\alpha_2 = \{\beta_2, \beta_3\}$, while $\beta_1 = \{b\}$, $\beta_2 = \{a, d\}$, $\beta_3 = \{b, c\}$. Then

$$\begin{aligned} &\bigcup[\alpha] = \{\beta_1, \beta_2, \beta_3\}, \quad \bigcap[\alpha] = \{\beta_3\}, \quad +[\alpha] = \{\beta_1, \beta_2\}, \\ &\bigcup^2[\alpha] = \{a, b, c, d\}, \quad \bigcap^2[\alpha] = \beta_3, \\ &\bigcup^{(2)}[\alpha] = \{\{b, c,\} \{a, b, c, d\}\}, \quad \bigcap^{(2)}[\alpha] = \{\{b\}, \varnothing\}, \end{aligned}$$

and so on.

The relations among the ranks of the sets under consideration for arbitrary k must satisfy the following conditions:

(a) $\rho(\alpha) \geq k + 1$,
(b) $\rho(\bigcup^k[\alpha] = \rho(\bigcap^k[\alpha]) = \rho(+^k[\alpha]) = \rho(\alpha) - k$, and
(c) $\rho(\bigcup^{(k)}[\alpha]) = \rho(\bigcap^{(k)}[\alpha]) = \rho(+^{(k)}[\alpha]) = \rho(\alpha) - 1$.

We also generalize the operation of complementing of sets $\alpha \equiv \{\alpha_0, \alpha_1, \ldots, \alpha_{m-1}\}$, putting $\neg\,\alpha \equiv \bar{\alpha}$:

$$\begin{aligned} \neg\,[\alpha] &\equiv \{\neg\,\alpha_0, \neg\,\alpha_1, \ldots, \neg\,\alpha_{m-1}\}, \\ \neg^{(2)}[\alpha] &\equiv \{\neg\,[\alpha_0], \neg\,[\alpha_1], \cdots, \neg\,[\alpha_{m-1}]\} \end{aligned}$$

and so on. Obviously, the expression $\neg^{(k)}[\alpha]$ has meaning only if $\rho(\alpha) \geq k + 1$. Note that $\neg^2[\alpha] = [\alpha]$.

Let us introduce the operators *inf* and *sup*, obtaining the lower and upper bounds of a set α (the operations have meaning for $\rho(\alpha) \geq 2$):

$$\begin{aligned} \beta \in \mathit{inf}\,\alpha &\leftrightarrow (\beta \in \alpha) \wedge (\overline{\exists\gamma})[\gamma \in \alpha) \wedge (\gamma \subset \beta)], \\ \beta \in \mathit{sup}\,\alpha &\leftrightarrow (\beta \in \alpha) \wedge (\overline{\exists\gamma})[(\gamma \in \alpha) \wedge (\gamma \supset \beta)]. \end{aligned}$$

In connection with the use in these expressions of the operators of the propositional calculus, note that the order of their execution is defined by the inclusion of the operator of conjunction $\wedge$ in the same set as the operator $\cap$, and the operator of disjunction $\vee$ and the operator of addition modulo 2 $\oplus$ in the same set as the operators $\cup$ and $+$.

Let us introduce an operation of multiplication of sets, which will be very essential for further constructions, consisting of obtaining a set called the product of the initial sets. In the present chapter we shall, as a rule, consider only sets with distinct elements, and furthermore, for the meantime, we shall neglect ordering of elements in sets, assuming, for example, $\{a, b\} \equiv \{b, a\}$.

Therefore this operation, which we shall denote by the symbol *cart*, differs from Cartesian multiplication of sets and has commutative properties: $\alpha \; cart \; \beta = \beta \; cart \; \alpha$.

For the case $\alpha \cap \beta = \varnothing$ the product $\alpha \; cart \; \beta$ of the sets α and β is defined as the set of all sets containing exactly one element from each of the factor sets. For example,

$$\{a, b\} \; cart \; \{c, d\} = \{\{a, c\}, \{a, d\}, \{b, c\}, \{b, d\}\}.$$

The case $\alpha \cap \rho \neq \varnothing$ is somewhat more complicated. The multiplication operation corresponding to this case is defined as follows: associating the set indexes with the elements of the factor sets, we first consider them as distinct elements and obtain the product of sets corresponding to the case $\alpha \cap \beta = \varnothing$; then we eliminate the indexes and "reduce like terms," retaining in each set only distinct elements. For example, if $\alpha = \{a, b, c\}$ and $\beta = \{a, b, d\}$, then

$$\alpha \; cart \; \beta = \{\{a\}, \{b\}, \{a, b\}, \{a, c\}, \{a, d\}, \{b, c\}, \{b, d\}, \{c, d\}\}.$$

From this example it is evident that in the case $\alpha \cap \beta \neq \varnothing$ the sets that are elements of the product of the sets α and β may contain more than one element of each of the factor sets. Thus, in the example considered above, the set $\{a, b\}$ contains two elements of the factor $\{a, b, c\}$ and two elements of the factor $\{a, b, d\}$.

Let us put $Cart[\alpha] \equiv \alpha_0 \; cart \; \alpha_1 \; cart \cdots cart \; \alpha_{m-1}$ if $\alpha \equiv \{\alpha_0, \alpha_1, \ldots, \alpha_{m-1}\}$ [obviously, the given operation has meaning for $\rho(\alpha) \geq 2$]. For example, if

$$\alpha = \{\{a\}, \{b\}, \{a, b\}, \{c\}, \{c, d\}\},$$

then

$$Cart[\alpha] = \{\{a, b, c\}, \{a, b, c, d\}\}.$$

We formulate a pair of theorems whose simple proofs we omit.

Theorem 1.

$$\bigcup [Cart\, [\alpha]] = \bigcup [\alpha].$$

Theorem 2.

$$\alpha \cup \beta = X^{(k)} \rightarrow \bigcup^{(2)} [\tilde{\alpha}\ \mathrm{cart}\ \tilde{\beta}] = X^{(k+1)}.$$

Let us introduce the operator $\min_\varphi$, the φ-minimization of a set, defined by:

$$\alpha \in \min_\varphi \beta \leftrightarrow (\alpha \in \beta) \wedge (\overline{\exists \gamma}) [(\gamma \in \beta) \wedge (\varphi(\gamma) < \varphi(\alpha))],$$

where $\varphi(\alpha)$ is some numeric function, defined on the set β. Analogously we introduce the operator $\max_\varphi$, φ-maximization of a set:

$$\alpha \in \max_\varphi \beta \leftrightarrow (\alpha \in \beta) \wedge (\overline{\exists \gamma}) [(\gamma \in \beta) \wedge (\varphi(\gamma) > \varphi(\alpha))].$$

For example, if

$$\alpha = \{\{a, b, c\}, \{a, d\}, \{a, e, k\}, \{c, d\}, \{b, e, m, t\}\},$$

then

$$\min_\sigma \alpha = \{\{a, d\}, \{c, d\}\}, \quad \text{and} \quad \max_\sigma \alpha = \{\{b, e, m, t\}\}.$$

2. OBJECTS IN BOOLEAN SPACE AND CERTAIN FORMS OF THEIR REPRESENTATION

The set $X^{(k+1)}$, containing, in particular, the empty element, is called a Boolean space over $X^{(k)}$. The consideration of the properties of various Boolean spaces reduces in an obvious way to the investigation of the Boolean space $M \equiv \tilde{X}$.

Practical forms of representation of objects in Boolean space require a definite ordering of the elements of the space. Since positional binary coding of numbers is widely used in modern computers, it is convenient to adopt as the number of the element σ_i of a Boolean space $\tilde{X}$, where $X \equiv \{x_0, x_1, \cdots, x_{n-1}\}$, the number

$$i = \sum_{j=0}^{n-1} \tau_j \cdot 2^j$$

where $\tau_j \in \{0, 1\}$ and $\tau_j = 1 \leftrightarrow x_j \in \sigma_i$.

On this method of numbering is based, in particular, the representation of the subsets M_i in M, given by the row matrix

$$|| \{M_i\} \ni M ||.$$

For example, if $X = \{x_0, x_1, x_2\}$ and $M_i = \{\emptyset, \{x_0, x_1\}, \{x_1, x_2\}\}$, then

$$|| \{M_i\} \ni M || = [10010010]$$

(it must be taken into account that the components are numbered from the left).

We shall say that the subset $M_i \subseteq M$ represents the Boolean function $f_i(x_0, x_1, \cdots, x_{n-1})$, taking on the value 1 on elements of M_i and the value 0 on elements of $\bar{M}_i$. For convenience we shall adopt the convention that the symbols M_i and $M(f_i)$ have the same meaning.

The nonempty subset $I \subseteq M$, satisfying the condition

$$\gamma \in I \leftrightarrow \alpha \subseteq \gamma \subseteq \beta,$$

where $\alpha, \beta, \gamma \in M$ will be called an interval of the Boolean space, and α and β are its lower and upper elements, denoted by $I^{(\alpha)}$ and $I^{(\beta)}$, respectively. The interval I represents a Boolean function that can be expressed algebraically by a single elementary conjunction, in which the number of letters is equal to the rank of the interval $\rho(I)$. The relation

$$\rho(I) = n - \sigma(\beta \backslash \alpha),$$

where σ is the symbol of the cardinality of the set, is obvious.

We define the reference sets V and W of intervals of the first rank, assuming that the intervals forming the set V correspond to single-letter conjunctions with the inversion symbol, and the intervals in W are single-letter conjunctions without the inversion symbol. In other words,

$$I \in V \leftrightarrow I^{(\alpha)} = \emptyset, \quad I^{(\beta)} = X \backslash (x_j\};$$
$$I \in W \leftrightarrow I^{(\alpha)} = \{x_j\}, \quad I^{(\beta)} = X,$$

where $j = 0, 1, \cdots, n-1$ and $\emptyset$ is the symbol of the empty set.

Since any Boolean function f_i may be expressed as a disjunction of elementary conjunctions, it can be represented by the corresponding set of intervals $Int_i \in \{I_i, I_i^2, \cdots, I_i^k\}$, where k is the number of conjunctions in the given disjunction, while

$$M_i = \cup [Int_i] \equiv I_i^1 \cup I_i^2 \cup \cdots \cup I_i^k.$$

Boolean functions can have diverse forms of representation. We shall limit our consideration to three fundamental forms, calling them the vector, elementary, and interval forms (by convention).

(1) In the vector form the Boolean function f_i is represented by a vector $\| \{M_i\} \ni M \|$ of dimensionality $\sigma(M) = 2^n$.

(2) In the elementary form the Boolean function f_i is represented by the matrix $\| M_i \ni X \|$ of dimensionality $\sigma(M_i)$ on $n = \sigma(X)$ or by the matrix $\| \bar{M}_i \ni X \|$, with corresponding sign.

(3) In the interval form the function f_i is represented by two matrices $\| Int_i \subseteq V \|$ and $\| \mathrm{Int}_i \subseteq W \|$, each of dimensionality $k = \sigma(Int_i)$ on $n = \sigma(X)$.

Expressed in bits, the memory capacity N required to represent a Boolean function is 2^n for the vector form, $\sigma(M_i) \cdot n$ (or $\sigma(\bar{M}_i) \cdot n$) for the elementary form, and $2n \cdot \sigma(Int_i)$ for the interval form. Starting from these relations and considering the particularities of the function represented, the form of representation in each concrete case is to be chosen, bearing in mind that the working memory of a modern computer contains 10^4–10^6 bits.

For example, the vectors form of representation of Boolean functions, which is particularly convenient for machine computations, is practically inapplicable for $n > 15$. The elementary form may be used up to n of the order of several hundred, if the cardinality of the set M_i is severely limited (for example, if $\sigma(M_i) \leq 100$).

It is frequently expedient to consider spaces of a more restricted character. For example, suppose that in the execution of the algorithm for solving a certain problem it is possible to analyze only a subset of some set P which, in turn, is only a subset of a Boolean space M. In this case it is not useful to code the subset of P as a subset of M, since this coding may be too complicated. Instead, it is convenient to renumber in sequence the elements of the set P and represent the set by the vector $\Psi(P)$, in which $\sigma(P)$ of the left components have the value 1, and of the others 0, expressing any subset P_i in P by the vector $\| \{P_i\} \ni P \|$. This method of coding is fairly economical.

REFERENCES

1. Aleksandrov, P. S., "Introduction to the General Theory of Sets and Functions" [in Russian]. Moscow-Leningrad, 1948.
2. Kurosh, A. G., "Lectures on General Algebra" [in Russian]. Moscow, 1962.
3. Loev, M., "Theory of Probability" [in Russian]. Moscow, 1962.
4. Uspenskii, V. A., "Lectures on Computable Functions" [in Russian]. Moscow, 1960.
5. Van Hao and MacNaughton, R., "Axiomatic Systems of Set Theory" Moscow, International Literature [Translated from English], 1963.
6. Zakrevskii, A. D., Elements of a Metalanguage for First Level LYaPAS [in Russian], *Tr. SFTI* **48** (1966).
7. Zakrevskii, A. D., Operations in Boolean Space, *Tr. SFTI* **48** (1966).

OPTIMAL COVERAGE OF SETS

A. D. Zakrevskii

1. COVERAGES OF SETS

One of the very general concepts used both in finite automata theory and beyond it is the concept of coverage of a set or, more precisely, coverage of a set by elements of another. We shall define this concept, using the language described in [1].

We shall call α-coverage of a set β every subset γ of a set α for which $\bigcup [\gamma] \supseteq \beta$. The set of all α-coverages of the set β is denoted by $cov_\alpha \beta$.

From the definition it follows that

$$\gamma_1 \subset \gamma_2 \rightarrow \bigcup [\gamma_1] \subseteq \bigcup [\gamma_2];$$
$$(\gamma_1 \subset \gamma_2 \subseteq \alpha) \wedge (\gamma_1 \in cov_\alpha \beta) \rightarrow \gamma_2 \in cov_\alpha \beta;$$
$$\bigcup [\alpha] \not\supseteq \beta \leftrightarrow cov_\alpha \beta = \varnothing;$$
$$\rho(\alpha) = \rho(\beta) + 1.$$

There exists a convenient matrix interpretation of the concept of coverage. Indeed, this concept is based on the relation $\ni$ between elements of the sets α and β which can be represented by the Boolean matrix $\| \alpha \ni \beta \|$. To each α-coverage of a set β there will correspond a certain subset of rows of a matrix having together ones in all the columns of this matrix.

Let, for example,

$$\| \alpha \ni \beta \| = \begin{bmatrix} 0 & 1 & 0 & 1 & 0 \\ 1 & 0 & 1 & 0 & 0 \\ 1 & 0 & 1 & 1 & 0 \\ 0 & 1 & 0 & 0 & 1 \end{bmatrix},$$

i.e., $\alpha = \{\alpha_1, \alpha_2, \alpha_3, \alpha_4,\}$; $\beta = \{\beta_1, \beta_2, \beta_3, \beta_4, \beta_5\}$; $\alpha_1 \ni \beta_2$; $\alpha_1 \ni \beta_4$, $\alpha_2 \ni \beta_1$ etc. The set $cov_\alpha \beta$ (ordered in some arbitrary manner) is expressed by the matrix

$$|| cov_\alpha \beta \ni \alpha || = \begin{bmatrix} 1 & 1 & 1 & 1 \\ 0 & 1 & 1 & 1 \\ 1 & 0 & 1 & 1 \\ 1 & 1 & 0 & 1 \\ 0 & 0 & 1 & 1 \end{bmatrix}.$$

Theorem 1.

$$cov_\alpha \beta = Cart[\alpha^*],$$

where $|| \alpha^* \ni \alpha || = T || \alpha \ni \beta ||$ and T is the matrix transpose operator.

Since the set $cov_\alpha \beta$ is convex [2] with respect to the relation $\subset$ (i.e., $(\gamma_1 \subset \gamma_2 \subseteq \alpha) \wedge (\gamma_1 \in cov_\alpha \beta) \rightarrow \gamma_2 \in cov_\alpha \beta)$, the lower bound of this set $inf\ cov_\alpha \beta$, which we shall call the set of irredundant coverages, is of interest. In the example considered above, this set is expressed by the matrix

$$|| inf\ cov_\alpha \beta \ni \alpha || = \begin{bmatrix} 1 & 1 & 0 & 1 \\ 0 & 0 & 1 & 1 \end{bmatrix}.$$

On the set $\alpha \equiv \{\alpha_1, \alpha_2, \ldots, \alpha_n\}$ a certain positive function e can be defined, which can be considered as a vector $\mathbf{e} = [e_1 e_2 \ldots e_n]$, where $e_1 = e(\alpha_1)$, $e_2 = e(\alpha_2), \ldots, e_n = e(\alpha_n)$. We use the term cost of coverage γ for the value of the scalar product $\varphi(\gamma) = \mathbf{e} \cdot || \{\gamma\} \ni \alpha ||$. The problem of finding a coverage of minimal cost or of optimizing the coverage (the optimality criterion in this case is the minimum cost) arises in may practical applications. The best-known interpretations of this problem are the minimization of Boolean functions in the class of disjunctive normal forms [3, 4, 5], finding the externally stable set of nodes of a graph [6], and expansion of parentheses in the conjunctive normal form of a Boolean function [7].

Since the cost function $\varphi(\gamma)$ is linear in γ, the problem of optimization of a coverage can be solved by means of linear programming [8, 9], although this involves certain difficulties arising in the search for integer solutions—the values of the vector $|| \{\gamma\} \ni \alpha ||$.

On the other hand, there is a basis for hoping that the consideration of the specifics of the problem will permit more efficient algorithms to be constructed for its solution, whose execution will require smaller outlays of time. In this chapter we present some results of investigation along these lines.

2. SOME WAYS TO FIND OPTIMAL COVERAGES

Let us consider the set $\min_\varphi cov_\alpha \beta$ of the φ-minimal α-coverages of the set β, i.e., coverages that correspond to minimal values of the function φ.

Theorem 2.

$$cov_\alpha \beta \supseteq inf\ cov_\alpha \beta \supseteq \min_\varphi cov_\alpha \beta.$$

The theorem follows from the positiveness of the function e.

Particular attention is merited by the cases $e \equiv 1$, or $\mathbf{e} = [11\ldots1]$. In these cases $\varphi(\gamma) = \sigma(\gamma)$ and the φ-minimal coverage will have the minimal number of elements, so that these coverages can be called shortest, and the set formed by them will be denoted by $\min_\sigma cov_\alpha \beta$. The search for shortest coverages was followed, in particular [8, 9].

It is obvious that in the search for optimal coverages a certain sifting of the subsets in α must be realized. This sifting can be shortened on the basis of a preliminary division of the set α into equivalence classes with respect to the given optimality criterion. If the criterion of optimality of the coverage is the minimum cost, then two elements α_i and α_j of the set α are considered to belong to the same equivalence class if the substitution in any coverage of the element α_i by α_j (or vice versa) does not increase the cost or violate the coverage. More exactly, $e(\alpha_i) = e(\alpha_j)$ and $\alpha_i \cap \beta = \alpha_j \cap \beta$. The set $\alpha_i \cap \beta$ is called the projection of the set α_i on the set β, or the β-projection of the set α_i (it can, of course, also be called the α_i-projection of the set β, but we wish particularly to distinguish the set β, projecting onto it various sets $\alpha_i, \alpha_j, \ldots$).

Theorem 3.

$$(\alpha_i, \alpha_j \in \alpha) \wedge (\alpha_i \cap \beta = \alpha_j \cap \beta) \rightarrow (\alpha_i \in^2 \min_\sigma cov_\alpha \beta \leftrightarrow \alpha_j \in^2 \min_\sigma cov_\alpha \beta).$$

Theorem 4.

$$(\alpha_i, \alpha_j \in \alpha) \wedge (\alpha_i \cap \beta = \alpha_j \cap \beta) \wedge (\varphi(\alpha_i) = \varphi(\alpha_j))$$
$$\rightarrow (\alpha_i \in^2 \min_\varphi cov_\alpha \beta \leftrightarrow \alpha_j \in^2 \min_\varphi cov_\alpha \beta).$$

The proof of these theorems is based on the already observed possibility of using the interchangeability of the elements α_i and α_j to satisfy the formulated conditions; such a substitution effectuates passage between equivalent coverages. We note that by virtue of Theorem 2 the symbols $\min_\sigma cov_\alpha$ and $\min_\varphi cov_\alpha$ in the formulations of Theorems 3 and 4 can be replaced by either the symbol $inf\ cov_\alpha$ or cov_α.

Taking σ-minimality as the criterion of optimality of coverage, we separate the set α into equivalence classes with respect to the relation $\alpha_i \cap \beta = \alpha_j \cap \beta$; i.e., we shall include in one class the elements α_i and α_j if and only if their β-projections are equal. We denote by $\alpha_{(\gamma)}$ the class for which

$$\alpha_i \in \alpha_{(\gamma)} \leftrightarrow \alpha_i \cap \beta = \gamma.$$

We denote by α^* the set of distinct β-projections of the elements of the set α. Obviously,

$$\alpha^* = \cap^{(2)} [\alpha \; Cart \; \{\beta\}].$$

Denoting by A_i the set $\{\alpha_{(\gamma_j)})/\gamma_j \in \gamma_i\}$, we formulate the following theorems from the above considerations.

Theorem 5.

$$\gamma_i \in cov_{\alpha^*} \beta \leftrightarrow Cart [A_i] \subseteq cov_\alpha \beta.$$

Theorem 6.

$$\gamma_i \in inf \; cov_{\alpha^*} \beta \leftrightarrow Cart [A_i] \subseteq inf \; cov_\alpha \beta.$$

Theorem 7.

$$\gamma_i \in \min_\sigma cov_{\alpha^*} \beta \leftrightarrow Cart [A_i] \subseteq \min_\sigma cov_\alpha \beta.$$

Theorem 8.

$$cov_\alpha \beta = \cup [\{Cart [A_i]/\gamma_i \in cov_{\alpha^*} \beta\}].$$

Theorem 9.

$$inf \; cov_\alpha \beta = \cup [\{Cart [A_i]/\gamma_i \in inf \; cov_{\alpha^*} \beta\}].$$

Theorem 10.

$$\min_\sigma cov_\alpha \beta = \cup [\{Cart [A_i]/\gamma_i \in \min_\sigma cov_\alpha \beta\}].$$

These theorems permit the problem of finding the α-coverage of a set β to be reduced to the problem of finding the α^*-coverages of a set β, which is of practical interest if $\sigma(\alpha^*) \ll \sigma(\alpha)$. Indeed, according to the theorems, to find all α-coverages of the set β it is sufficient to find the α^*-coverages of the set β, to obtain for each of these coverages the product of the corresponding elements of the coverage of the classes in α, and to unite these products.

These theorems can be used, for example, to construct algorithms for the minimization of weakly defined Boolean functions [10]. To find the set of φ-minimal coverages of the set β it is useful to reduce the set $\alpha_{(\gamma_j)}$ to $\min_e \alpha_{(\gamma_j)}$ and then assume that the function e takes on the same value on the element γ_j of the set α^* as on the elements of the set $\min_e \alpha_{(\gamma_j)}$.

Theorem 11.

$$\gamma_i \in \min_\varphi cov_{\alpha^*} \beta \leftrightarrow Cart\,[\{\min_\varphi \alpha_{(\gamma_j)}/\gamma_j \in \gamma_i\}] \subseteq \min_\varphi cov_\alpha \beta.$$

Theorem 12.

$$\min_\varphi cov_\alpha \beta = \cup\,[\{Cart\,[\{\min_\varphi \alpha_{(\gamma_j)}/\gamma_j \in \gamma_i\}]/\gamma_i \in \min_\varphi cov_{\alpha^*} \beta\}].$$

3. FINDING IRREDUNDANT COVERAGES

Passing to the consideration of concrete coverage-optimization algorithms, we introduce the operator R of passage from the matrix $\|\alpha \ni \beta\|$ to the matrix $\|inf\ cov_\alpha \beta \ni \alpha\|$:

$$\|inf\ cov_\alpha \beta \ni \alpha\| = R\,\|\alpha \ni \beta\|.$$

As recalled above, the set $cov_\alpha \beta$ is convex with respect to $\subset$. Therefore it is completely defined by its lower bound $inf\ cov_\alpha \beta$, to find which is of course equivalent to finding the entire set $cov_\alpha \beta$.

The utilization of the property of convexity of the set $cov_\alpha \beta$ permits the volume of calculation in the execution of the operator R [11] to be reduced. The basic idea of this reduction is to traverse the boundary of the convex set, attemping to analyze the smallest possible number of elements of the space $\tilde{\alpha}$ (for membership in the set $cov_\alpha \beta$ or $inf\ cov_\alpha \beta$). It is obvious that the volume of calculation carried out in this will depend primarily on the dimensions of the boundary.

We denote by $k_{\max}^n$ the maximal value of the cardinality $\sigma(inf\ \alpha)$ of the lower bound of the convex set α, given in the n-dimensional Boolean space M. This value corresponds to the convex set α, defined in the following way: $\gamma \in \alpha \leftrightarrow \sigma(\gamma) \geq n/2$, i.e., is equal to the number of combinations of $n/2$ out of n (for even n):

$$k_{\max}^n = C_n^{n/2} = \frac{n!}{(n/2)!(n/2)!}.$$

Approximating this value for large n by Stirling's formula, we obtain

$$k_{\max}^n \cong (2/\pi n)^{1/2} \cdot 2^n.$$

As we see, the reduction is not particularly great. For example, operating with convex sets in a space $\tilde{X}$, where $\sigma(X) = 60$, it must be recalled that since $(2/60\pi)^{1/2} > 10^{-1}$, there exist convex sets α in $\tilde{X}$ whose boundaries will contain one tenth of all $2^{60} \approx 10^{18}$ elements of the space $\tilde{X}$, i.e.,

$$\sigma(inf\ \alpha) \approx 10^{17}.$$

It is possible to hope, however, that in the solution of many practical problems such that the set of solutions of each of them is a convex set, the boundary of this set will be found to be substantially smaller and the method of boundary traversal developed here will be substantially more efficient—all the more since the traversal of the boundary can be organized with minimal connectivity of the algorithm [11]; there is no need to store the part of the set $inf\ cov_\alpha\ \beta$ already obtained, and the result can be listed element by element.

A substantial acceleration of the traversal of the boundary can be obtained by a preliminary ordering of the elements of the set α (the rows of the matrix subject to the operator R), which leads, in particular, to the extraction of the kernel of the solution—the intersection of all coverages [12, 13]. Ordering can also be carried out during the execution of the algorithm, in which case it is carried out with respect to an uncovered part of the set [14].

Below we describe the subroutine which realizes the operator R, effecting passage from the matrix $\| A \ni B \|$ to the matrix $\| inf\ cov_A\ B \ni A \|$. The operator is given the code *fisirco*, since it finds the set of irredundant coverages.

The reduction of search volume obtained by the algorithm is expressed in the sequence of values of the variable set A^*.

The proposed algorithm for executing the operator R is based on reduced search (the idea of which is given in [11]) and the analysis of the subsets A_i of the set A for membership in the sets $cov_A\ B$ and $inf\ cov_A\ B$. In the process of direct incrementation of the set A_i the cardinality of the set progressively increases through inclusion of those elements of the set A, taken in order of increasing number, that are not absorbed by the set P—the union of the elements $(P \subseteq B)$ already included in A_i. Together with the current value of the set P is also determined the current value of the set Q which contains those elements of B that belong to not less than two elements of A_i.

Theorem 13.

$$(\exists \alpha)\ [\alpha \in A_i \in cov_A\ B) \wedge (\alpha \subseteq Q)] \leftrightarrow A_i \in inf\ cov_A\ B.$$

Some reduction of execution time for the algorithm is achieved by the utilization of the auxiliary set $S \equiv \{S_1, S_2, \cdots, S_n\}$, where the element $S_j \equiv a_{j+1} \cup a_{j+2} \cup \cdots \cup a_n$ represents the possibility of a remainder $A_j^* \equiv \{a_{j+1}, a_{j+2}, \cdots, a_n\}$; if the regime of direct incrementation is connected with the inclusion in A_i of elements taken from A_j^* and $P \cup S_j \neq B$, then passage is effected to a special regime of incrementation of the set A_i, presented in sentence 4 of the subroutine, after which the values of the sets P and Q are determined anew.

The program is (2, 2)-terminal. The basic terminals correspond to the start and finish of sifting of all the irredundant coverages. In finding the current irredundant coverage, represented by the current value of the set A_i, exit from the subroutine is through the auxiliary exit terminal α, and to obtain the next coverage it is necessary to return to the subroutine through the auxiliary entry terminal, sentence 4.

$$\begin{aligned} &\beta :: \| A \ni B \|, \\ &\gamma :: \psi (B), \\ &\delta :: \| \{A_i\} \ni A \|, \\ &\epsilon :: \| S \ni B \|. \end{aligned}$$

The current values of the sets P and Q are the variables **b** and **c**:

$$\mathbf{b} :: \| \{P\} \ni B \|, \mathbf{c} :: \| \{Q\} \ni B \|.$$

fisirco αч, βк +, γп +, δп, ϵк/**d**, $b/(\beta\gamma\epsilon)$
5 247 10 (β**abc**) (δ**d**)

	$b_\beta - 1 \Rightarrow a \ \bar{\circ} \ b \circ \mathbf{a} \circ \mathbf{b} \circ \mathbf{c} \circ \delta$	Initialization
§1	$\beta_a \vee \mathbf{a} \Rightarrow \mathbf{a} \ \bar{\triangle} \ a\mathbf{a} \Rightarrow \epsilon_a a \mid\!\rightarrow 1$	Obtain set S
§2	$\triangle \ b\mathbf{b} \ \urcorner \ \wedge \ \beta_b \circ\!\rightarrow 2\mathbf{b} \wedge \beta_b \vee \mathbf{c} \Rightarrow \mathbf{c}$ $c_b \vee \delta \Rightarrow \delta\beta_b \vee \mathbf{b} \Rightarrow \mathbf{b} \oplus \gamma \mid\!\rightarrow 2\delta \Rightarrow \mathbf{d}$	Direct incrementation of the sets A_i, P, Q; find $A_i \in cov_A \, B$
§3	$\mathbf{d} \ \ddot{\mathbf{X}} \ \alpha a\mathbf{c} \ \urcorner \ \wedge \ \beta_a \mid\!\rightarrow 3$	Test $A_i \in inf \ cov_A \, B$
§4	$\delta \ \urcorner + 1 \wedge \delta \vdash \Rightarrow bc_b \oplus \delta \Rightarrow \delta \Rightarrow \mathbf{d} \mid\!\rightarrow 5$ $\epsilon_b \oplus \gamma \mid\!\rightarrow 10$	Remove the last element from A_i
§5	$\circ \ \mathbf{b} \circ \mathbf{c}$	
§6	$\mathbf{d} \ \ddot{\mathbf{X}} \ 7a\beta_a \wedge \mathbf{b} \vee \mathbf{c} \Rightarrow \mathbf{c}\beta_a \vee \mathbf{b} \Rightarrow \mathbf{b} \rightarrow 6$	Reevaluate sets P and Q
§7	$\mathbf{b} \vee \epsilon_b \oplus \gamma \circ\!\rightarrow 2 \rightarrow 4$	Analysis of the possibilities of remainder in A
§10	.	

To illustrate the algorithm we present three matrices: the initial matrix $\| A \ni B \|$, and the matrix $\| S \ni B \|$ corresponding to it and obtained during execution of sentence 1, and $\| D \ni A \|$ where D is the set of those subsets A_i of A that are considered for the execution of sentence 2 and are represented by the value of the variable δ.

The rows of the matrix $\| D \ni A \|$ marked by an asterisk represent the elements of the required set $inf\ cov_A\ B$.

$$\| A \ni B \| = \begin{bmatrix} 0 & 0 & 1 & 1 & 0 & 1 & 1 \\ 1 & 1 & 1 & 0 & 0 & 0 & 0 \\ 0 & 1 & 0 & 1 & 1 & 0 & 0 \\ 0 & 1 & 0 & 0 & 0 & 1 & 0 \\ 1 & 0 & 0 & 0 & 0 & 0 & 1 \\ 0 & 0 & 1 & 0 & 1 & 0 & 0 \end{bmatrix}$$

$$\| S \ni B \| = \begin{bmatrix} 1 & 1 & 1 & 1 & 1 & 1 & 1 \\ 1 & 1 & 1 & 1 & 1 & 1 & 1 \\ 1 & 1 & 1 & 0 & 1 & 1 & 1 \\ 1 & 0 & 1 & 0 & 1 & 0 & 1 \\ 0 & 0 & 1 & 0 & 1 & 0 & 0 \\ 0 & 0 & 0 & 0 & 0 & 0 & 0 \end{bmatrix}$$

$$\| D \ni A \| = \left[\begin{array}{cccccc} 1 & 0 & 0 & 0 & 0 & 0 \\ 1 & 1 & 0 & 0 & 0 & 0 \\ 1 & 1 & 1 & 0 & 0 & 0 \\ 1 & 1 & 0 & 0 & 0 & 1 \\ 1 & 0 & 1 & 0 & 0 & 0 \\ 1 & 0 & 1 & 0 & 1 & 0 \\ 1 & 0 & 0 & 1 & 0 & 0 \\ 1 & 0 & 0 & 1 & 1 & 0 \\ 1 & 0 & 0 & 1 & 1 & 1 \\ 0 & 1 & 0 & 0 & 0 & 0 \\ 0 & 1 & 1 & 0 & 0 & 0 \\ 0 & 1 & 1 & 1 & 0 & 0 \\ 0 & 1 & 1 & 1 & 1 & 0 \\ 0 & 0 & 1 & 0 & 0 & 0 \\ 0 & 0 & 1 & 1 & 0 & 0 \\ 0 & 0 & 1 & 1 & 1 & 0 \\ 0 & 0 & 1 & 1 & 1 & 1 \end{array}\right] \begin{array}{c} \\ \\ * \\ * \\ \\ * \\ \\ \\ * \\ \\ \\ \\ * \\ \\ \\ \\ * \end{array}$$

The initial information for this example has been taken from [15].

4. FINDING THE SHORTEST COVERAGE

Here we present an algorithm for obtaining the set $\min_\sigma cov_A B$ of the shortest A-coverage of the set B, proposed by Novoselov [16].

The algorithm utilizes the method described for reducing the sifting of the sets A_i in A with the following additions, permitting the speed of sifting to be increased.

(1) In the course of execution of the algorithm all coverages that have a minimal (among the coverages considered) cardinality are memorized. These coverages form a certain variable set U. In the search for a new coverage only those subsets A_i in A are subject to analysis whose cardinalities do not exceed the cardinality of the coverages of U.

(2) The coverages found are not tested for redundancy, since the method of sifting the subsets in A guarantees that all shortest coverages are directly obtained.

In this connection the algorithm operates with the sets A, B, A_i, P, S, whose significance is defined in the description of the operator *fisirco*, and with the cited set U. The set $\min_\sigma cov_A B$ is obtained by the operation of *fishco* $\alpha\beta\gamma//$; the initial data are given by the complex α:

$$\alpha::\| A \ni B \|.$$

The complex β is used to store the intermediate information:

$$\beta::\| S \ni B \|.$$

The complex γ represents the set U during execution of the algorithm:

$$\gamma::\| U \not\ni A \|;$$

upon termination of the computation the complex γ is the result of the operation

$$\gamma::\| \min_\sigma cov_A B \ni A \|.$$

The current value of the sets A_i and P are represented by the variables **c** and **f**:

$$\mathbf{c}::\| \{A_i\} \ni A \|, \quad \mathbf{f}::\| \{P\} \ni B \|.$$

fishco αк $+$, βк??, γк?/f, $c/(\alpha\beta)$
11 260 7 (α**df**) (γ**ac**)

	$\circ\ \mathbf{c}b_a \Rightarrow a \Rightarrow b_\beta??\ \circ\ \mathbf{d}\ \circ\ \mathbf{f}$	Initialization
§1	$\bar{\triangle}\ a\alpha_a \vee \mathbf{d} \Rightarrow \mathbf{d} \Rightarrow \beta_a a \mid\!\rightarrow \mathbf{1}\ \bar{\circ}\ \mathbf{e}$ $\Rightarrow a\ \circ\ \mathbf{a}\ \circ\ \mathbf{b}$	Obtain set S
§2	$\mathbf{a}\ \dot{\mathbf{X}}\ 3b\alpha_b \vee \mathbf{f} \Rightarrow \mathbf{f} \rightarrow 2$	Obtain set P
§3	$\triangle\ a \oplus b_\alpha \circ\!\rightarrow 6\mathbf{b} - \mathbf{e} \mid\!\rightarrow 6\beta_a \vee \mathbf{f} \oplus \mathbf{d} \mid\!\rightarrow 6$ $\mathbf{f}\ \urcorner\ \wedge\ \alpha_a \circ\!\rightarrow 3\alpha_a \vee \mathbf{f} \oplus \mathbf{d} \mid\!\rightarrow 5$ $1 + \mathbf{b} \oplus \mathbf{e} \circ\!\rightarrow 4\ \circ\ c1 + \mathbf{b} \Rightarrow \mathbf{e}$	Analysis of current element a_j in A for utility of its inclusion in A_i
§4	$c_a \vee \mathbf{c} \Rightarrow \gamma_c?\ \triangle\ c \Rightarrow b_\gamma \rightarrow 3$	Inclusion of A_i in U
§5	$c_a \vee \mathbf{c} \Rightarrow \mathbf{c}\ \triangle\ \mathbf{b}\alpha_a \vee \mathbf{f} \Rightarrow \mathbf{f} \rightarrow 3$	Inclusion of a_j in A_i
§6	$\mathbf{c} \circ\!\rightarrow 7\ \urcorner + 1 \wedge \mathbf{c} \vdash \Rightarrow a\ \bar{\triangle}\ \mathbf{b}\ \circ\ \mathbf{f}\ c_a$ $\oplus\ \mathbf{c} \Rightarrow \mathbf{c} \Rightarrow \mathbf{a} \rightarrow 2$	Exclusion of an element from A_i
§7		

In the search for the set of shortest coverages it is possible to utilize also the method of testing all subsets A_i of cardinality k, obtained by sifting the C_n^k combinations of k out of n, where n is the cardinality of the set A [17]. The initial value of k is defined as the minimum possible [18]. The coverages found as a result of the inspection of all C_n^k combinations form the set $\min_\sigma cov_A\, B$; if none of the analyzed subsets of A_i is a coverage, the parameter k is increased by unity and the combinations corresponding to its new value are sifted.

This idea is embodied in the operator *fishort*, performed by the following program where, as before,

$$\alpha::\ \|A \ni B\|,\ \beta::\ \|S \ni B\|,\ \gamma::\ \|\min_\sigma cov_A\, B \ni A\|.$$

fishort αк $+$, βк??, γк?, δп $+$ /**e**, $c/(\alpha\beta)$
14 300 10 (γ**dac**) (α**be**)

	$\mathbf{b}_\alpha \Rightarrow a \Rightarrow b_\beta??\ c_a \Rightarrow \mathbf{d}\ \circ\ \mathbf{b}\ \circ\ b$	
§1	$\bar{\triangle}\ a\alpha_a \vee \mathbf{b} \Rightarrow \mathbf{b} \Rightarrow \beta_a a \mid\!\rightarrow 1\ \circ\ \mathbf{a}\delta$ $\Rightarrow c\ \bar{\circ}\ a \rightarrow 4$	Construct set S
§2	$\circ\ c\mathbf{a} \wedge \mathbf{d} \circ\!\rightarrow 3\mathbf{a} + \mathbf{d} \wedge \mathbf{a} \Rightarrow \mathbf{c} \oplus \mathbf{a}\ \nabla$ $\Rightarrow c\mathbf{c} \Rightarrow \mathbf{a}$	Sift combinations

§3 $\mathbf{a} - \mathbf{d} \wedge \mathbf{a} \Rightarrow \mathbf{ca} - \mathbf{d} \oplus \mathbf{c} \vdash \Rightarrow a\mathbf{c} \vee c_a$
$\Rightarrow \mathbf{ac} \circ\!\!\rightarrow 6$

§4 $\triangle\, a \oplus b_a \circ\!\!\rightarrow 10c_a \vee \mathbf{a} \Rightarrow \mathbf{a}\, \bar{\triangle}\, c \mid\!\!\rightarrow 4\mathbf{a}$
$\Rightarrow \mathbf{c} \circ \mathbf{e}$

§5 $\mathbf{c} \vdash \Rightarrow ac_a \oplus \mathbf{c} \Rightarrow \mathbf{c} \circ\!\!\rightarrow 6\alpha_a \vee \mathbf{e}$
$\Rightarrow \mathbf{e} \rightarrow 5$

§6 $\mathbf{e} \vee \beta_a \oplus \mathbf{b} \circ\!\!\rightarrow 7 \triangle\, c\mathbf{a} - 1 \wedge \mathbf{a} \Rightarrow \mathbf{a} \rightarrow 3$

Test current A_i and include new elements in $\min_\sigma cov_A B$

§7 $\mathbf{e} \vee \alpha_a \oplus \mathbf{b} \mid\!\!\rightarrow 2\mathbf{a} \Rightarrow \gamma_b?\ \triangle b \rightarrow 2$

§10 $b \Rightarrow b_\gamma.$

5. OBTAINING φ-MINIMAL COVERAGES

The operator *fiminco* described below for finding φ-minimal A-coverages of the set B is also due to Novoselov [19]. It executes an operator similar to *fishco*, having only two distinguishing features.

(1) The criterion of the quality of the coverage of A_i is not the cardinality of the coverage but the value of the function φ:

$$\varphi = \sum_{a_j \in A_i} e(a_j),$$

where e is some positive function, given on the set A. Of the coverages considered, only those for which φ takes on its minimal value are included in the set U.

(2) The algorithm provides for finding both a set $\min_\sigma cov_A B$ and a sequence of irredundant coverages A_i' (which offers substantial economy in the volume of certain programs).

The operands in the operation *fiminco* $\alpha\beta\gamma\delta\epsilon\zeta\eta//$ are given the following sense:

α is an auxiliary exit terminal, corresponding to the finding of all sets $\min_\sigma cov_A B$ (the basic output terminal is reached upon obtaining the current irredundant coverages; to obtain the next, it is necessary to return to an auxiliary entry terminal, § 3).

The initial information is given by the complexes β and δ:

$$\beta :: \| A \ni B \|,$$

$[\delta_i] = e(a_i)$ (δ is the complex of weights of the elements in A).

The index η defines the operating mode of the subroutine: for $\eta = 1$ the mode of obtaining the entire set $\min_\varphi cov_A B$; for $\eta = 0$ the mode of obtaining the current irredundant coverage.

The result of operation of the subroutine is given by the values of the variable ζ, constituting the current coverage A_i',

$$\zeta::|| \{A_i'\} \ni A ||$$

and the complex ϵ:

$$\epsilon::|| \min_\varphi cov_A B \ni A ||.$$

Intermediate information is represented by the same complex ϵ:

$$\epsilon::|| U \ni A ||$$

and the complex γ:

$$\gamma::|| S \ni B ||.$$

fiminco αч, βк +, γк??, δк +, ϵк?, ζп,
ηи + / **g**, *b* / $(\beta\gamma)$ $(\zeta\eta)$
12 315 7 (β**df**) (ϵ**ac**)

	$\circ$ $\mathbf{c}b_\beta \Rightarrow a \Rightarrow b_\gamma$?? $\circ$ **b** $\circ$ **d** $\circ$ **f**	Initialization
§1	$\bar{\triangle}\ a\beta_a \vee \mathbf{d} \Rightarrow \mathbf{d} \Rightarrow \gamma_a a \mid\!\rightarrow 1\ \bar{\circ}\ \mathbf{g} \Rightarrow a \circ \mathbf{a}$	Obtain set S
§2	$\mathbf{a}\ \dot{\mathbf{X}}\ 3b\beta_b \vee \mathbf{f} \Rightarrow \mathbf{f}\eta \circ\!\rightarrow 2\delta_b + \mathbf{b} \Rightarrow \mathbf{b} \rightarrow 2$	Obtain set P
§3	$\triangle\ a \oplus b_\beta \circ\!\rightarrow 6\gamma_a \vee \mathbf{f} \oplus \mathbf{d} \mid\!\rightarrow 6f\ \urcorner$ $\wedge\ \beta_a \circ\!\rightarrow 3$ $\eta \circ\!\rightarrow 7\delta_a + \mathbf{b} \Rightarrow \mathbf{e} - 1 - \mathbf{g} \mid\!\rightarrow 3\beta_a \vee \mathbf{f}$ $\oplus\ \mathbf{d} \mid\!\rightarrow 5\mathbf{e} \oplus \mathbf{g} \circ\!\rightarrow 4\mathbf{e} \Rightarrow \mathbf{g} \circ c$	Analysis of the current element $a_j \in A$ for expediency of its inclusion in A_i
§4	$c_a \vee \mathbf{c} \Rightarrow \epsilon_c?\ \triangle\ c \Rightarrow b_\epsilon \rightarrow 3$	Inclusion of A_i in U
§5	$c_a \vee \mathbf{c} \Rightarrow \mathbf{ce} \Rightarrow \mathbf{b}\beta_a \vee \mathbf{f} \Rightarrow \mathbf{f} \rightarrow 3$	Inclusion of a_j in A_i
§6	$\mathbf{c} \circ\!\rightarrow \alpha\ \urcorner + 1 \wedge \mathbf{c} \vdash \Rightarrow a \circ \mathbf{f} \circ \mathbf{b}c_a$ $\oplus\ \mathbf{c} \Rightarrow \mathbf{c} \Rightarrow \mathbf{a} \rightarrow 2$	Exclusion of an element from A_i
§7	$\mathbf{a} \vee \mathbf{f} \oplus \mathbf{d} \mid\!\rightarrow 5c_a \vee \mathbf{c} \Rightarrow \zeta.$	Output of the current coverage

6. OBTAINING A SINGLE SHORTEST COVERAGE

In the solution of certain practical problems it is sufficient to find one of the shortest coverages (it is not important which one). In such cases it is possible to use the algorithm described below, in which the enumeration of the variants considered is strongly reduced. This algorithm is based on the following two theorems, whose proofs are simple.

Theorem 14.

$$(\exists a_i, a_j \in A)[a_i \subset a_j] \rightarrow \min_\sigma cov_{A \setminus \{a_i\}} B \subseteq \min_\sigma cov_A B.$$

Theorem 15.

$$b_i \in B \rightarrow (\exists a_j)[b_i \in a_j \in^2 \min_\sigma cov_A B.$$

From Theorem 14 it follows that in the search for only a single shortest A-coverage of the set B it is expedient to reduce the set A to its upper bound. Theorem 15 represents the expansion of the problem of finding the shortest A-coverages of the set B into a collection of analogous problems for sets of smaller cardinalities. This expansion is preferably carried out with respect to that element $b_i \in B$ to which corresponds the column of the matrix $\| A \in B \|$ containing the minimal number of ones; in this case the sifting of variants is, as a rule, reduced.

This expansion leads to a repeated replacement of A and B by certain $A_i \subset A$ and $B_j \subset B$, for which $\min_\sigma cov_{A_i} B_j$ is sought. In this the operation of reducing the set A_i alternates with the operation of finding the minimal column in the matrix $\| A_i \in B_j \|$.

The sifting of coverage variants, which can be represented by a search tree, is also reduced by stopping the examination of paths of length t or more, where t is the cardinality of the shortest of the coverages examined.

The operator of finding one shortest coverage is expressed, in particular, by the following three L-operators.

We define the operation *redminor* $\alpha\beta\gamma//$ or reduction of the minor of a Boolean matrix, represented by the complex α. The minor is defined by the values of the variables β and γ:

$$\beta{::}\| \{P'\} \in P \|,$$
$$\gamma{::}\| \{Q'\} \in Q \|,$$

where P' is a distinguished subset of rows in the set P of rows of the matrix, and Q' is a distinguished subset of columns of the set Q of columns of the matrix. As a result of the given operation the set of rows P' must be reduced to a certain $P'' \subseteq P$ through removal from P' of the rows absorbed by other rows of the minor, and cancelling of like rows. The result is represented by the variable β:

$$\beta\!::\| \{P''\} \ni P \|.$$

redminor αк $+$, βп, γп $+$ / **b**, b / $(\alpha\gamma)$
1 62 3 $(\beta$**ab**$)$

$\beta \Rightarrow$ **b**
§1 **b** $\dot{\mathbf{X}}$ $3bc_b \oplus \beta \Rightarrow$ **a**
§2 **a** $\dot{\mathbf{X}}$ $1a\alpha_a \urcorner \wedge \gamma \wedge \alpha_b \mid\!\rightarrow 2c_b \oplus \beta \Rightarrow \beta \rightarrow 1$
§3 .

In the execution of the operator *mincol* $\alpha\beta\gamma\delta//$ a minimal column of the minor of a Boolean matrix is found, given by the complex α and the variables β and γ, exactly as in the case of the operation *redminor* $\alpha\beta\gamma//$. The result of the operation, the determination of a minimal column, is represented by the variable δ:

$$\delta\!::\| \{A^*\} \ni A \|,$$

where A^* is the subset of rows of the minor having ones in a distinguished column. If there are several minimal columns, the rightmost is taken.

mincol αк $+$, βп $+$, γп $+$, δп/**c**, d/$(\alpha\gamma)$ $(\beta\delta)$
2 107 4$(\beta$**ac**$)$ $(\gamma$**b**$)$

$\gamma \Rightarrow$ **b** $\bar{\circ}$ d
§1 **b** $\dot{\mathbf{X}}$ $4b\beta \Rightarrow$ **a** $\circ$ c $\circ$ **c**
§2 **a** $\dot{\mathbf{X}}$ $3a\alpha_a \wedge c_b \circ\!\!\rightarrow 2 \triangle cc_a \vee$ **c** $\Rightarrow$ **c** $\rightarrow 2$
§3 $c - d \mid\!\rightarrow 1$ $c \Rightarrow d$**c** $\Rightarrow \delta \rightarrow 1$
§4 .

The examination of different variants of coverage in this method is carried out by means of the operator *tree*, intended for application in the following situation.

Let the process of solution of a certain problem have a multistep character, and be representable by a tree whose root corresponds to the start of solution and each of whose nodes q_i corresponds to a branching of the process, represented by a subset T_i of a certain set T. In the realization of the current step of the solution process one element in T_i is selected and passage is effected to the next node q_j of the tree, located deeper by unity (i.e., farther from the root). Then T_j is determined, where the method of determination can differ in different problems. The depth of transversal of the tree increases until one of the required solutions is obtained or the current set T_k becomes empty.

The operator *tree* provides a complete traversal of the tree under these conditions, which may be reduced by auxiliary considerations. For this the subroutine has an additional entry terminal, sentence 2, upon entry to which the inspection of any given variant in depth is blocked.

To accelerate the process of solution the information used in the solution corresponding to the nodes q_i, located along the path joining the current node to the root of the tree, is memorized. This information is given by the values of the operand

$$\delta::\| \; T_i \ni T \; \|,$$

where T_i corresponds to the set of variants not yet examined by the continuation of the calculation, and by the operand ζ, whose meaning is made precise in the programming of a concrete problem, and which can represent, in particular, the list of certain variables enclosed in parentheses. The memorization of this information is carried out on the complex ϵ. The number of the element of T selected in the current step is the value of the index γ, and the depth of the step, the value of the index β.

The program has two exit terminals. The basic terminal is attained after termination of the planned enumeration of variants, the auxiliary (α) after the current, but not last traversal.

tree αч, βи, γи, δп, ϵк, ζв/—, —
3 41 2

$\triangle\,\beta$
§1 $\delta\,\dot{\mathbf{X}}\,2\gamma(\delta\zeta) \Rightarrow \epsilon \rightarrow \alpha$
§2 $\bar{\triangle}\,\beta\epsilon \Rightarrow (\delta\zeta)\,\beta \mid\!\rightarrow 1.$

We pass, finally, to the description of the operation *oshco* $\alpha\beta\gamma\delta\epsilon//$ of obtaining one shortest A_i-coverage of the set B_j, where A and B are certain sets, $\alpha::\| \; A \ni B \; \|$, $\beta::\| \; \{A_i\} \ni A \; \|$, $\gamma::\| \; \{B_j\} \ni B \; \|$, the variable ϵ is its result after completion of the operation ($\epsilon::\| \; \{A_k\} \ni A \; \|$, where A_k is the found shortest coverage), and the complex δ is intended for storage of the intermediate information in traversal of the tree of variant enumeration.

$$\mathit{oshco}\ (1, 2, 3)\ \alpha\text{к}\ +,\ \beta\text{п}\ +,\ \gamma\text{п},\ \delta\text{к},\ \epsilon\text{п}/\mathbf{e},\ e/(\alpha\gamma\delta)\ (\beta\epsilon)$$
$$\mathbf{4\ 125\ 2}\ (\beta\mathbf{cde})$$

$$\bar{\circ}\ e\ \circ\ c\beta \Rightarrow \mathbf{c}\ \circ\ \mathbf{e} \rightarrow 2$$
$$\S 1\quad c_d \vee \mathbf{e} \Rightarrow \mathbf{e}\alpha_d\ \urcorner\ \wedge\ \gamma \Rightarrow \gamma\ |\!\rightarrow 2\mathbf{e} \Rightarrow \epsilon c - 1 \Rightarrow e \rightarrow \mathit{tree}\ 2$$
$$\S 2\quad c \oplus e\ \circ\!\rightarrow \mathit{tree}\ 2\ \mathit{redminor}\ \alpha\mathbf{c}\gamma//\mathit{mincol}\ \alpha\mathbf{c}\gamma\mathbf{d}//\mathit{tree}\ 1\ cd\mathbf{d}\delta(\mathbf{c}\gamma\mathbf{e})//.$$

Figure 1 illustrates the method: a tree traversed in execution of the above subroutine seeking the shortest A-coverage of the set B if

$$\| A \ni B \| = \begin{bmatrix} 0 & 0 & 1 & 1 & 0 & 1 & 1 \\ 1 & 1 & 1 & 0 & 0 & 0 & 0 \\ 0 & 1 & 0 & 1 & 1 & 0 & 0 \\ 0 & 1 & 0 & 0 & 0 & 1 & 0 \\ 1 & 0 & 0 & 0 & 0 & 0 & 1 \\ 0 & 0 & 1 & 0 & 1 & 0 & 0 \end{bmatrix},$$

where $a_1 = a$; $a_2 = b$; $a_3 = c$; $a_4 = d$; $a_5 = e$; $a_6 = f$.

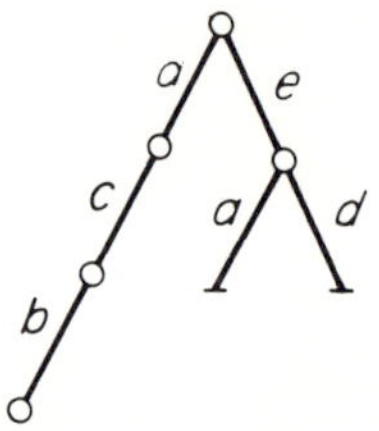

FIG. 1.

The tree has only three branches. The leftmost represents the first of the coverages found $\{a, c, b\}$, the next two, $\{e, a\}$ and $\{e, d\}$, are examined only to depth 2 (since the already found coverage has three elements) and are dropped, since they are not coverages.

Comparison of this operator with the operator *fisirco* which was illustrated by the same example, shows the substantial reduction of solution time obtained with the operator *oshco*.

7. AN APPROXIMATION METHOD FOR FINDING A SHORTEST COVERAGE

The approximation method in question was described by way of example in the description of LYaPAS [20]. Therefore we shall limit ourselves here to brief presentations of the corresponding subroutines.

The operation *approshco* $\alpha\beta\gamma\delta//$ consists of finding one of the A_i-coverages of the set B_j where, as in the definition of the operator *oshco*,

$$\alpha::\| A \ni B \|, \beta::\| \{A_i\} \ni A \|, \gamma::\| \{B_j\} \ni B \|.$$

The coverage A_k found is considered as an approximation to the shortest, and is represented in the result by the value of the variable δ:

$$\delta::\| \{A_k\} \ni A \|.$$

The algorithm is very rapid, but the degree of approximation may be rather low.

approshco (7, 6)/αк +, βп, γп +, δп/**c**, d/($\alpha\gamma$) ($\beta\delta$)
10 56 1 (β**c**)

$\circ\ \delta$
§1 *mincol* $\alpha\beta\gamma$**c**//*maxrow* α**c**γc//
$c_c \vee \delta \Rightarrow \delta \urcorner \wedge \beta \Rightarrow \beta\alpha_c \urcorner \wedge \gamma \Rightarrow \gamma \mid\rightarrow 1.$

The operator *mincol* has been presented earlier. The operation *maxrow* $\alpha\beta\gamma\delta$//, described in [20], consists of finding the maximal row of a minor prescribed, as above, by the representative matrix of the complex α and the variables β and γ, defining the rows and columns of the matrix, respectively.

maxrow αк +, βп, γп +, δи/−, b/($\alpha\gamma$)
6 42 2

$\circ\ \delta\ \circ\ b$
§1 $\beta\ \dot{\mathbf{X}}\ 2a\alpha_a \wedge \gamma \nabla - b \circ\rightarrow 1 + b \Rightarrow ba \Rightarrow \delta \rightarrow 1$
§2 .

In the solution of certain problem—for example, in the synthesis of functionally stable circuits—the need appears for finding h-fold coverages. By definition, an h-fold coverage of the set B is a subset A_i of A such that the following condition is satisfied: Each element of B enters into not less than h elements of A_i.

An algorithm for finding h-fold coverages was considered in [21].

REFERENCES

1. Zakrevskii, A. D., Programming Boolean Computations. This volume.
2. Kurosh, A. G., "Lectures on General Algebra" [in Russian]. Fizmatgiz, Moscow, 1962.

3. Quine, W. V., The Problem of Simplifying of Truth Functions, *Amer. Math. Monthly*, **59**, No. 8, 521–531 (1952).
4. Gavrilov, M. A., Minimization of Boolean Functions Characterizing Switching Networks, *Avtomat. i Telemek.* No. 9, 1217–1238 (1959).
5. Waligorsky, S., Calculation of the Quine's Table for Truth Functions, *Prace ZAM PAN* **2**, A 15 (1961).
6. Berge, C., "Theory of Graphs and Its Applications" [Russian translation from French]. IL, Moscow, 1962.
7. Nelson, R. L., Simplest Normal Truth Functions, *J. Symbolic Logic* **20**, No. 2, 105–108 (1955).
8. Pyne, J. B., McCluskey, E. J., Jr., An Essay on Prime Implicant Tables Fundamental Product Table of Combinations or Truthtable, *SIAM J. Appl. Math.* **9**, No. 4 (1961).
9. Maistrova, T. L., Linear Programming and the Problem of Minimization of Normal Forms of Boolean Functions, *in* "Problems of Information Transmission," No. 12, pp. 5–15 [in Russian]. Izd. Akad. Nauk SSSR, 1963.
10. Zakrevskii, A. D., Algorithms for the Minimization of Weakly-Defined Boolean Functions, *Kibernetika* (*Kiev*) No. 2, 53–60 (1965).
11. Zakrevskii, A. D., On Reduced Sifting in the Solution of Certain Problems in the Synthesis of Discrete Automata, *Izv. Vuzov, Rsdiofiz.* **8**, No. 1, 166–174 (1964).
12. Quine, W. V., On Cores and Prime Implicants of Truth Functions. *Amer. Math. Monthly* **66**, No. 9, 755–760 (1959).
13. Zhuravlev, Yu. I., Set-Theoretical Methods in Boolean Algebra, *in* "Problems of Cybernetics," No. 8, pp. 5–44 [in Russian]. Izd. Akad. Nauk SSSR, 1962.
14. Choudhury, A. K., and Basu, M. S., A Mechanized Chart for Simplification of Switching Functions, *IRE Trans., Electron. Computers* **11**, 376–418 (1962).
15. Pyne, J. B., and McCluskey, E. J., Jr., The Reduction of Redundancy in Solving Prime Implicant Tables, *IRE Trans., Electron. Computers* **11**, 473–482 (1962).
16. Novoselov, V. G., Finding Shortest Coverages, *Tr. SFTI, Tomsk* **48** (1966).
17. Roth, J., Algebraic Topological Methods in Synthesis, *Proc. Intern. Symp. Theory Switching*, Pt. I. Harvard Univ. Press, 1959.
18. Kazakov, V. D., Finding Minimal Normal Forms of a Logical Function by a Method of Bounded Search, *in* "Structural Theory of Switching Devices," pp. 145–147 [in Russian]. Izd. Akad. Nauk SSSR, Moscow, 1963.
19. Novoselov, V. G., Finding φ-Minimal Coverages, *Tr. SFTI, Tomsk* **48** (1966).
20. Zakrevskii, A. D., Description of LYaPAS. This volume.
21. Agibalov, G. P., Finding Optimal Multiple Coverages of Sets, *Tr. SFTI, Tomsk* **48** (1966).

THE SOLUTION OF SYSTEMS OF LOGICAL EQUATIONS

A. D. Zakrevskii and A. Yu. Kalmykova

1. STATEMENT OF THE PROBLEM

We shall consider that to solve a system of logical equations means to obtain all sets of values of the logical variables satisfying the entire system as a whole, i.e., each of the equations in the system. Sometimes the conditions of the problem admit of only several such sets, called roots of the system, in particular, a single one.

An arbitrary logical equation is easily brought to the form $F_i = 1$, where F_i is some Boolean function (for example, $a = b \sim ab \vee \bar{a}\bar{b} = 1$), and a system of m logical equations can be brought to the form

$$F_i = 1, \quad i = 1, 2, \ldots, m,$$

which we shall use in this chapter.

It is obvious that the above system is equivalent to a single equation

$$F_1 \cdot F_2 \cdot \cdots \cdot F_m = 1,$$

where the Boolean function $F \equiv F_1 \cdot F_2 \cdot \cdots \cdot F_m$ represents all the solutions of the system; i.e., the set of roots of the equation $F = 1$ coincides with the set of roots of the initial system. Therefore it is possible to formulate the problem of solving a system of logical equations as the problem of finding the function F, i.e., as the problem of carrying out the logical multiplication of the functions $F_1, F_2, \ldots, F_m$. The essential factors determining the degree of complexity of the resolution of the given problem are the form of representation of the Boolean functions $F_1, F_2, \ldots, F_m$ and the form in which the result is to be expressed.

In the examination of systems with a small number n of variables (for example, for $n \leq 10$) the problem is solved fairly simply, but as n increases it becomes far from trivial, requiring enormous expenditures of time to solve (even using computers, which it is impossible to avoid in the general case).

At the same time, in any practical statements of problems of a logical character the initial volume of information, as a rule, is not large. This means, for example, that systems of Boolean equations whose solutions are of practical interest are prescribed by Boolean functions of limited class, and it is possible to develop efficient algorithms for solution of these systems by taking into account their particularities.

In the present chapter we present the results of some studies in this direction, expressed by algorithms in LYaPAS. Reference [1] is devoted to the exploitation of the particularities of search for only one solution of a system.

2. THE FORMS USED FOR REPRESENTING THE INFORMATION

We shall use the forms of representation of Boolean functions introduced in [2]: the vector, elementary, and interval forms.

By analogy with the forms of representation of the ordinary Boolean functions it is also possible to define the form of representation of a system of Boolean functions $F_1, F_2, \ldots, F_m$: the vector form $\| M^* \ni M \|$, where $M^* \equiv \{M_1, M_2, \ldots, M_m\}$, the elementary form consisting of a set of matrices $\| M_i \ni X \|$ for all $M_i \in M^*$, and the interval form, consisting of a set of pairs of matrices $\| Int_i \subseteq V \|$ and $\| Int_i \subseteq W \|$ for all $i = 1, 2, \ldots, m$.

Aside from these, we bring into consideration a limited vector form of representation of a system of Boolean functions, whose application is useful in those situations in which the total number n of system variables can be fairly large, but the number of essential variables of each of the functions in the system is bounded by a certain small number l (for example, $n = 50$ and $l = 5$). Completing the number of variables in each of the functions to l (by the introduction of certain fictitious arguments, if needed), we represent the entire system by a pair of matrices A and B: $B :: \| X^* \ni X \|$, where $X^* \equiv \{X_1, X_2, \ldots, X_m\}$ and X_i is the set of arguments of the function F_i,

$$A_i \equiv \| \{M_i'\} \ni \tilde{X}_i \|,$$

where $i = 1, 2, \ldots, m$; A_i is the ith row of the matrix A; $\tilde{X}_i$ is the set of all subsets in X_i, forming a Boolean space over X_i; and M_i' is a subset of $\tilde{X}_i$ representing the function F_i.

Of course, aside from the above, it is possible to apply other forms of representation of Boolean functions (for example, superpositions, giving Boolean functions in the form of the superposition of certain reference functions). In certain situations it is expedient to transform the Boolean functions, i.e., to pass from one form to another, to increase the efficiency of information processing.

For example, in the interval form of representation of a system of Boolean functions and for $n < 15$ it is useful to pass to the vector form, which is executed by the operator *passinvec*. The interval form of a system of Boolean functions is given by complexes α, β and a complex γ, representing the required division of the complexes α and β by the enumeration of the numbers of those elements of the complexes α and β that correspond to the last elements of the subcomplexes, in the form of individual matrices

$$\| Int_i \subseteq V \| \quad \text{and} \quad \| Int_i \subseteq W \|,$$

for each $i = 1, 2, \ldots, m$.

The vector form obtained in the result is represented by the complex

$$\delta :: \| M^* \ni M \|$$

passinvec αк, βк, γк +, δк / **b**, c / (α, β)
24 115 3 (δ**a**b)

$\bar{\circ}$ **a** $\circ$ a $\circ$ **b** $\circ$ b
§1 α_a $\dot{\mathbf{X}}$ 2c**a** $\wedge$ $d_c \Rightarrow$ **a** $\rightarrow 1$
§2 β_a $\dot{\mathbf{X}}$ 3c**a** $\wedge$ $e_c \Rightarrow$ **a** $\rightarrow 2$
§3 **a** $\vee$ **b** $\Rightarrow$ **b** $\bar{\circ}$ **a** $\triangle$ $a\gamma_b - a \mid\!\rightarrow 1$
b $\Rightarrow \delta_b \circ$ **b** $\triangle$ $b \oplus b_\gamma \mid\!\rightarrow 1$.

3. SIMPLE ALGORITHMS

One of the simplest algorithms for solving system of logical equations consists of enumerating all elements in the Boolean space M, substituting them successively in the system equations, and finding among them those that satisfy each equation. As soon as some one of the equations is not satisfied, the tested element of the Boolean space is eliminated from further consideration.

This algorithm is carried out by the operator *sisle* for solving a system of logical equations by the method of sifting the elements of the Boolean space, which is performed by the following program:

$$sisle\ (*)\ /\ \alpha\text{ч},\ \beta\text{к}\ +,\ \gamma\text{к}\ +,\ \delta\text{к},\ \epsilon\text{п}\ /\ -,\ b\ /\ (\beta\delta)$$
62 64 2

$$\bar{\triangle}\ \epsilon \Rightarrow b \circ \epsilon \circ a$$
§1 $\gamma_a \Rightarrow a_\delta \triangle a \oplus b_\gamma \circ\!\rightarrow \alpha\gamma_a - a_\delta \Rightarrow b_\delta$
$calvaboof\ 1\epsilon\delta\ //$
§2 $\circ\ a\epsilon + c_b \Rightarrow \epsilon \mid\!\rightarrow 1.$

All the information about the system of logical equations is represented by the complex β, which is divided by the elements of the complex γ into parts, each of which represents one of the functions F_i. The sifting of the elements σ_i of the Boolean space is reflected by the sequence of values of the variable ϵ: $\epsilon{::}\|\ \{\sigma_i\} \ni X\ \|$, for each of which the operator *calvaboof* defines the values of the functions F_i, represented in sequence by the complex δ (depending on the form of representation of the Boolean function, one or another operator of the type *calvaboof* can be taken). The operator *calvaboof* has one additional exit terminal, corresponding to the unit value of the calculated function and returning to the program in § 1. The operator *sisle* also has an additional exit terminal α, attained when a root of the system is found, which will be represented at that moment by the value of the variable ϵ [at the start of execution of the program $[\epsilon] = n \equiv \sigma(X)$]. To continue the search for roots it is necessary to return to the auxiliary entry terminal of the subroutine *sisle*, § 2.

In utilizing this subroutine it is first necessary to match its internal operands (a, b) with the corresponding set of the operator *calvaboof*.

The solution of a system of logical equations given in vector form is carried out exceedingly simply. For example, if $\mathbf{C}{::}\|\ M^* \ni M\ \|$, after execution of the program

$$\circ\ a\ \bar{\circ}\ \mathbf{a}$$
§1 $\mathbf{c}_a \wedge \mathbf{a} \Rightarrow \mathbf{a} \triangle a \oplus b_2 \mid\!\rightarrow 1.$,

the set of all roots of the system will be represented by the function F, given by means of the variable $a{::}\|\ M(F) \ni M\ \|$.

It is not difficult to find a root of the system if it is represented in elementary form. Below we give a subroutine *solsysgri*, corresponding to a method proposed by Grigor'yan [3].

The initial information is given by a complex α, whose initial portion represents a sequence of matrices $\|\ M_i \ni X\ \|$ for $i = 1, 2, \ldots, k$ and last part the sequence of matrices $\| \bar{M}_j \ni X \|$ for $j = k + 1, k + 2, \ldots, m$. The parameter k is given by the value of the index γ: $[\gamma] = k$, and the division of the complex α into subcomplexes is determined by the complex β, constituting the ordered list of their cardinalities. The result of the operation of the subroutine is represented by the complex α, $\alpha :: \| M' \ni X \|$, where M' is the set of roots of the system.

The idea of the algorithm is to search for elements common to the first k sets of M_i and absent from the remaining sets $\bar{M}_j$.

$$solsysgri\ (26, 46)\ /\ \alpha\text{к}, \beta\text{к}, \gamma\text{и}\ /\ \mathbf{b}, i\ /$$
32 206 4 $(\alpha\mathbf{a})$

$$\circ\ \mathbf{b}\ \circ\ a\ \circ\ b\ \circ\ d\ \circ\ f\beta_a \Rightarrow c \rightarrow 2$$

§1 $f \mid\!\rightarrow 4b_\beta \Rightarrow \gamma\ \bar{\triangle}\ ad - \beta_a \Rightarrow dc - 1 \Rightarrow b\mathbf{b} - c_b \Rightarrow \mathbf{b}1 \Rightarrow f \nleftrightarrow 2c_h\ \urcorner \wedge \mathbf{b}$
$\Rightarrow \mathbf{b} \rightarrow sequel\ 2$

§2 $\circ\ e\beta_a + d \Rightarrow d \triangle a \oplus \gamma\ \circ\!\rightarrow 1\beta_a \Rightarrow i\ sequel\ 3\alpha\alpha icgh\mathbf{a}de \,\|\, f \mid\!\rightarrow 2b \Rightarrow c$
$\circ\ b \rightarrow 2$

§3 $a \Rightarrow \alpha_b \triangle b \rightarrow sequel\ 2$

§4 $compress\ \alpha\mathbf{b}\alpha\ //.$

This subroutine uses the operator *sequel*, which performs the following function. In the complexes β and γ certain subcomplexes are defined by the indexes κ and λ, giving the positions of their initial elements, and by the indexes δ and ϵ, giving their cardinalities, respectively.

The operator *sequel* finds in these subcomplexes equal elements and supplies information in the following form: the variable ϑ represents the values of the elements found, the indexes ζ and η their number in the complexes β and γ. Exit from *sequel* is realized by the auxiliary terminal α, and to obtain the next pair of equal elements it is necessary to return to the subroutine by the auxiliary entry § 2.

$$sequel\ \alpha\text{ч}, \beta\text{к}\,+, \gamma\text{к}\,+, \delta\text{и}, \epsilon\text{и}, \zeta\text{и}, \eta\text{и}, \vartheta\text{и}, \kappa\text{и}, \lambda\text{и}/-, -/(\beta\gamma\vartheta)$$
26 62 2

$$\kappa \Rightarrow \zeta + \delta \Rightarrow \delta\lambda \Rightarrow \eta + \epsilon \Rightarrow \epsilon$$

§1 $\beta_\zeta \oplus \gamma_\eta \mid\!\rightarrow 2\gamma_\eta \Rightarrow \vartheta \rightarrow \alpha$

§2 $\triangle\ \zeta \oplus \delta \mid\!\rightarrow 1\kappa \Rightarrow \zeta \triangle \eta \oplus \epsilon \mid\!\rightarrow 1.$

The case where the system is prescribed in interval form is more complicated, corresponding, as noted above, to the representation of the system functions in the form of the disjunction of elementary conjunctions. We offer an algorithm that models the abbreviated process of direct multiplication of these disjunctions and is carried out by the subroutine *roots cdc* (finding the roots of a Boolean function given in the form of the conjunction of disjunctions of elementary conjunctions). The multiplication process is carried out by the rules of traversal of a logical tree with storage of intermediate results obtained in operations along the path connecting the current node of the tree with the root. The initial information is represented by the complexes δ and ϵ, containing the complete list of all conjunctions of the system and the complex ζ, expressing the division of the complexes δ and ϵ into subcomplexes, representing the equations of the system individually; the value of the element ζ_c is the number of those elements in the complexes δ and ϵ that are last in the subcomplexes representing the [c]th equations of the system in interval form.

The resultant function F, representing all the roots of the system, is also given out in interval form, as a sequence of intervals, represented by the pair of variables β and γ at the moments of exit by the auxiliary terminal α (to obtain the next interval it is necessary to return to § 2 of the subroutine). The values of these variables β and γ express the process of sifting over the intervals obtained in the intersections of the elements belonging to various sets Int_i. The complex η is used to store the intermediate information obtained in the traversal of the tree.

roots cdc αч, βп, γп, δк +, ϵк +, ζк +, ηк/**b**, $c/(\beta\gamma\delta\epsilon)$
17 136 3 (β**ab**)

$\bar{\circ}\ a \circ b \circ \mathbf{a} \circ \mathbf{b} \circ b_\eta$
§1 $\triangle\ a\delta_a \vee \mathbf{a} \Rightarrow \beta\epsilon_a \vee \mathbf{b} \Rightarrow \gamma \wedge \beta \mid\!\rightarrow \mathbf{3}\zeta_b \Rightarrow c \triangle b \oplus b_\zeta \circ\!\rightarrow \alpha(a\mathbf{ab}) \Rightarrow \eta$
§2 $\bar{\triangle}\ b\beta \Rightarrow \mathbf{a}\gamma \Rightarrow \mathbf{b}c \Rightarrow a \rightarrow 1$
§3 $a \oplus \zeta_b \mid\!\rightarrow 1\eta \Rightarrow (a\mathbf{ab})\ \bar{\triangle}\ b \oplus f_{10} \mid\!\rightarrow 3.$

Before the application of the operator *roots cdc* it is useful to order the equations of the system in the order of increasing cardinalities of the sets Int_i. Ordering is carried out by the operator *ord cdc*, for which the initial information is given by the complexes α, β, and γ, whose meanings coincide with the meanings of the complexes δ, ϵ, and ζ, respectively, of the subroutine *roots cdc*. The result supplied by the operator, representing the ordered system of equations, is fixed by the values of the complexes δ, ϵ, and ζ (α is transformed to δ, β to ϵ, and γ to ζ). The complex η is auxiliary.

ord cdc (22)/αк +, βк +, γк +, δк, ϵк, ζк, ηк/**a**, $f/(\alpha\beta\delta\epsilon)(\gamma\zeta\eta)$
36 136 4

$\circ\ d \circ b \circ g \circ f$
§1 $\gamma_b + 1 - g \Rightarrow \eta_b + g \Rightarrow g \triangle b \oplus b_\gamma \mid\!\rightarrow 1$
§2 $\circ$ *e fiminel* $3\eta c\mathbf{a}$ //
§3 $\gamma_c - \eta_c + 1 \Rightarrow b$
§4 $\alpha_b \Rightarrow \delta_d\beta_b \Rightarrow \epsilon_d \triangle d \triangle b \triangle e \oplus \mathbf{a} \mid\!\rightarrow 4\ \bar{\circ}\ \eta_c d - 1 \Rightarrow \zeta_f \triangle f \oplus b_\gamma \mid\!\rightarrow 2.$

The operator *fiminel* finds the minimal element in the complex β, which is interpreted in the given case as the code of a natural number; its value is represented by the variable δ, its number by the index γ. On finding the minimal element, exit is realized over terminal α; on a new entry to the subroutine *fiminel* (to § 2) a different element is found (if there are no more, exit is realized over the basic terminal).

fiminel αч, βк +, γи, δп/ −, $a/(\beta\delta)$
22 51 2

$\circ\ a \Rightarrow \gamma$
§1 $\beta_a - \beta_\gamma\ \circ \!\rightarrow 2\ \triangle\ a \oplus b_\beta \mid\!\rightarrow 1 \beta_\gamma \Rightarrow \delta \rightarrow \alpha$
§2 $\circ\ a\ \triangle\ \gamma \oplus b_\beta \mid\!\rightarrow 1.$

4. ABBREVIATED SIFTING

Cherry and Vaswani [4] have proposed a method for reducing the sifting over elements of a Boolean space in the search for the roots of a system of equations in which each one depends on a small number of variables. Its difference from the full sifting method described (the operator *sisle*) consists of the following.

In the analysis of the current element σ_i of the Boolean space the first unsatisfied equation is found, and the number j of the highest-order argument of this equation is determined. Then the code representing the element σ_i is incremented by transfer to the mode of arithmetic addition of unity in the jth position from the left, and the assignment of zero to all positions to the right.

For example, if σ_i = 00101011 and it is found that this value does not satisfy the equation whose arguments are x_2, x_3, and x_5, the following value of σ_i is taken to be 00110000. As is evident from the example, the abbreviation is achieved through the elimination from consideration of certain elements in the Boolean space (in the present case, the elements 00101100, 00101101, 00101110, 0010111).

The effectiveness of the method is increased by a preliminary ordering of the system and the renumbering of the variables, as a result of which the equations are ordered by increasing index j of the highest-order argument.

The operator *solsysche* for the solution of a system of logical equations, given in limited vector form by the complexes $\beta::A$ and $\gamma::B$, is based on this idea. The operator *solsysche* is realized by the following subroutine, giving the current root of the system on exit over the auxiliary terminal α

in the form of the value of the variable δ, and writing the number of variables of the system of equations in ϵ.

solsysche $(30)/\alpha\text{ч}, \beta\text{к}+, \gamma\text{к}+, \delta\text{п}, \epsilon\text{и}/\mathbf{c}, f/(\gamma\delta)$
31 120 3 $(\beta\mathbf{b})(\gamma\mathbf{ac})$

$\circ\ \mathbf{c}\epsilon - 1 \Rightarrow e$
§1 $\bar{\circ}\ d\mathbf{c} + \mathbf{c}_e \Rightarrow \mathbf{c}\ \circ\!\rightarrow 3 \Rightarrow \delta$
§2 $\triangle\ d \oplus b_\beta\ \circ\!\rightarrow \alpha\gamma_d \Rightarrow \mathbf{b}$ *packva* $\mathbf{cb}c//\beta_d \wedge c_c \mid\!\rightarrow 2\mathbf{b}\ \urcorner + 1 \vee \mathbf{b}\ \urcorner\ \vee\ \mathbf{c}$
$+ 1 \Rightarrow \mathbf{c}\ \circ\!\rightarrow 3\ \bar{\circ}\ d\mathbf{c} \Rightarrow \delta \rightarrow 2$
§3 .

The operator *packva* appearing in the subroutine "packs" the components of the variable α, separating the unit components of the variable β, preserving their order, and locating them at the extreme right positions of the index γ. For example, if $\alpha = 00101011$ and $\beta = 01101000$, then γ takes on the value 00000011.

packva $\alpha\text{п}+, \beta\text{п}+, \gamma\text{и}/\mathbf{a}, b/(\alpha\beta)$
30 65 2 $(\alpha\mathbf{a})$

$\circ\ \gamma\ 40 \Rightarrow b\beta \Rightarrow \mathbf{a}$
§1 $\bar{\triangle}\ b\mathbf{a}\ \circ\!\rightarrow 2\ \urcorner + 1 \wedge \mathbf{a} \vdash \Rightarrow ac_a \oplus \mathbf{a} \Rightarrow \mathbf{a}\alpha \wedge c_a\ \circ\!\rightarrow 1c_b \vee \gamma$
$\Rightarrow \gamma \rightarrow 1$
§2 .

The inverse operation is performed by the operator *unpack*, which distributes the right-hand components of the index γ over the positions of the variable α marked by unit components of the variable β (the values of the remaining components of the variable α in this do not change). For example, if $\alpha = 10110100$, $\beta = 01100101$, and $\gamma = 00001011$, the result of the operator *unpack* is that the variable α takes the value 11010101.

unpack $\alpha\text{п}, \beta\text{п}+, \gamma\text{и}+/\ \mathbf{a}, b\ /\ (\alpha\beta)$
27 70 2 $(\alpha\mathbf{a})$

$\beta \Rightarrow \mathbf{a}\ \urcorner \wedge \alpha \Rightarrow \alpha 40 \Rightarrow b$
§1 $\bar{\triangle}\ b\mathbf{a}\ \circ\!\rightarrow 2\ \urcorner + 1 \wedge \mathbf{a} \vdash \Rightarrow ac_a \oplus \mathbf{a} \Rightarrow \mathbf{a}c_b \wedge \gamma\ \circ\!\rightarrow 1c_a \vee \alpha$
$\Rightarrow \alpha \rightarrow 1$
§2 .

5. SYSTEMS OF LOGICAL EQUATIONS IN LIMITED VECTOR FORM

In [5] it was pointed out that in a number of cases a system of logical equations can be subjected to a preliminary simplification, leaving unchanged the set of roots of the system but substantially reducing the subsequent sifting. Simplification is carried out in pairwise comparison of the system equations by the removal of all those roots of one equation that are not roots of the second. The simplification process terminates by obtaining some unsimplifiable (in the framework of the method) form.

Let us consider here an algorithm for simplification of a system in limited vector form, and below, for a system given in interval form.

The simplification of the limited vector form of a system of logical equations, consisting of the removal of components of unity from the matrix, is carried out by the operator *simpliseq*, for which the initial information is given by the complexes $\gamma::A$ and $\delta::B$, and the index β, whose value is the number of variables on which each one of the equations of the system depends.

The application of the operator results in the complex $\gamma::A'$, where A' is the matrix obtained from A in which certain 1's are replaced by zeros, and the variables ζ and ϵ, containing information that can substantially reduce further search for the solution: $\zeta::\|\{P\} \ni X\|$ and $\epsilon::\|\{Q\} \ni X\|$, where P is a subset of X whose elements are known not to enter into any of the solutions of the system, and the elements of Q, on the contrary, must appear in a root of the system if any exist. In particular, the overlap of sets P and Q is a criterion of incompatibility of the system.

A detailed description of the algorithm of the subroutine *simpliseq* is given in [5]. We remark only that in its execution all the equations of the system are examined, and intervals given on $\tilde{X}_i$ are sought that do not contain roots of the function F_i (we therefore call them empty intervals). The process begins with the search for intervals of lowest rank. Then the empty intervals are transformed to the space $\tilde{X}_j$ in which are given the functions F_j of the other equations, the roots of the functions F_j covered by them are removed, and the corresponding elements of the matrix A are given the value zero. This simplification process has a chain character, since the removal of a single unity from the matrix A increases the probability of removing others. The effectiveness of the process increases with increase of the number m of system equations. If the system is incompatible, exit from *simpliseq* is realized over terminal α.

simpliseq (7, 27, 30, 23)/αч, βи, γк, δк, ϵп, ζп / **g**, *n* / ($\delta\epsilon\zeta$)
37 436 21 (δ**edfg**)

	$\circ\ \xi\ \circ\ \epsilon$	
§1	$\cup\ m\ 40 - \beta \Rightarrow d \Rightarrow fc_d \Rightarrow e \Rightarrow g \rightarrow 4$	
§2	$g \urcorner\ h\vee + 1 \wedge g \Rightarrow h < d \Rightarrow jh \oplus g$ $\Rightarrow k < d \Rightarrow i \not\nrightarrow 12 \circ c\mathbf{c} \Rightarrow \mathbf{h} \rightarrow 5$	Search for empty intervals
§3	$k \mapsto 2$ *gencombva* $4g\ //\ g \nabla \oplus 1 \mapsto 13m$ $\circ\!\rightarrow 14 \rightarrow 1$	
§4	$g \Rightarrow h \rightarrow 2$	
§5	$\bar{\circ}\ l\gamma_c \wedge \mathbf{h} \mapsto 7k \Rightarrow n$	
§6	$\delta_c \Rightarrow \mathbf{e}$ *unpack* $\mathbf{ee}n\ //!\ h \Rightarrow n\mathbf{e} \Rightarrow \mathbf{d} \not\nrightarrow 6\mathbf{d}$ $\wedge \zeta \mapsto 7\mathbf{e} \wedge \epsilon \mapsto 7\mathbf{d} \vee \mathbf{e} \Rightarrow \mathbf{f} \nabla \oplus 1$ $\mapsto 10\mathbf{d} \vee \zeta \Rightarrow \zeta\mathbf{e} \vee \epsilon \Rightarrow \epsilon \wedge \zeta \circ\!\rightarrow 10$ $\rightarrow \alpha$	Transformation of empty intervals to other spaces
§7	$\triangle\ c \oplus b_\gamma \mapsto 5 \rightarrow 3$	
§10	$\triangle\ l \oplus b_\delta \circ\!\rightarrow 7l \oplus c \circ\!\rightarrow 10\delta_l \wedge \mathbf{f} \oplus \mathbf{f}$ $\mapsto 10\mathbf{e} \Rightarrow \mathbf{g}$	Removal of roots covered by empty intervals
§11	*packva* $\mathbf{g}\delta_l c\ //!\ \mathbf{d} \Rightarrow \mathbf{g}i \Rightarrow j < d \Rightarrow j \not\nrightarrow 11i$ $< d \Rightarrow i \nrightarrow 12\gamma_l \wedge \mathbf{c} \circ\!\rightarrow 10\mathbf{c} \urcorner \wedge \gamma_l$ $\Rightarrow \gamma_l 1 \Rightarrow m \rightarrow 10$	
§12	*vectin* $ij\mathbf{c}//!$	
§13	$m \mapsto 1$	Organization of the order of sifting the equations and their analysis
§14	$\triangle f \oplus 40 \circ\!\rightarrow 15e \vee c_f \Rightarrow e \Rightarrow g \circ m \rightarrow 4$	
§15	.	

This subroutine uses the operator *gencombva*, an operator for sifting the values of the variable β with fixed number of 1's. Exit is realized over the terminal α on obtaining the current value, and over the basic terminal after all combinations have been generated.

gencombva αч, βп / **b**, *a* /
7 66 1 (β**ab**)

$\beta + 1 \wedge \beta \Rightarrow \mathbf{a} \circ\!\rightarrow 1\beta + 1 \oplus \beta < 1 + 1 \Rightarrow \mathbf{ba} - 1 \oplus \mathbf{a} \vdash \Rightarrow a\mathbf{b}$
$\vdash - a \Rightarrow a\mathbf{b} < a \oplus \mathbf{a} \Rightarrow \beta \rightarrow \alpha$

§1 .

The operator *vectin* is also used; it finds the vector form of an interval, given in an interval form by the variables α and β, and represents the result by the variable γ.

vectin αп +, βп +, γп / **b**, *a* / (α, β)
23 52 3 (α**ab**)

$\bar{o}$ $\gamma\alpha \Rightarrow$ a$\beta \Rightarrow$ b
§1 a $\dot{\mathbf{X}}$ $2ad_a \wedge \gamma \Rightarrow \gamma \rightarrow 1$
§2 b $\dot{\mathbf{X}}$ $3ae_a \wedge \gamma \Rightarrow \gamma \rightarrow 2$
§3

Directly after the operator *simpliseq* for simplifying the system, it is possible to apply the operator *solsysche* for solution of the system, giving the pair of matrices A' and B for its initial information.

It is also possible to take another path, preparing the information in interval form suitable for use by the operator *roots cdc*. The corresponding transformation consists of finding the shortest disjunctive forms of the functions F_i of the simplified system.

In this minimization those unit elements of the matrix A to which correspond zero elements of the matrix A' are considered to be elements whose values can be defined arbitrarily.

6. SIMPLIFICATION OF SYSTEMS OF LOGICAL EQUATIONS IN INTERVAL FORM

The above general idea for the preliminary simplification of the system of logical equations also can be carried out in the case of the interval form of representation of the system.

We submit an algorithm corresponding to this case, based on the following elementary transformations of the system (we denote by I_s^i and I_t^j the intervals that are elements of the sets Int_i and Int_j, representing the ith and jth equations of the system, respectively):

(a) if $I_s^i \cap I_t^j = \varnothing$ for all $I_t^j \in Int_j$, the interval I_s^i is removed from the set Int_i;

(b) if $I_s^i \cap I_t^j = \varnothing$ for all $I_t^j \in Int_j$ except for one I_r^j, then the interval I_s^i is replaced by the interval $I_s^i \cap I_r^j$, which increases, as a rule, the rank of the interval and increases the probability of further reduction;

(c) if there exists an interval I satisfying the condition

$$J_s^i \supset J \supseteq \bigcup [\{I_s^i \cap I_t^j / I_t^j \in Int_j\}],$$

then the interval I_s^i may be replaced by the interval I (this transformation includes, in particular, the two preceding ones).

The algorithm is performed by the subroutine *sinseq* for the simplification of the interval form of a system of logical equations, carrying out simplification in the following order. For a given pair of equations i and j the set $\{I_s^i \cap I_t^j / I_t^i \in Int_j\}$ is found for one of the values of s, after which a minimal covering interval I is found for this set and the condition $I_s^i \supset I$ is verified. When the condition is satisfied, I_s^i is replaced by I (in particular, instead of I the empty set may sometimes be taken), and when the possibility exists, the new interval is absorbed by some other interval of Int_i. If absorption cannot be effected, the possibility of joining the new interval to other intervals of Int_i is examined and, if this is possible, a junction is made and the new interval obtained is again analyzed for the possibility of joining it to others.

This process of analysis is carried out in three nested cycles; in the inner cycle the various intervals of Int_i are sifted, in the next the value of the parameter j is changed, i.e., the second of the equations in the pair under investigation is changed, and in the external cycle the first of the equations is changed by change of the value of i.

The process of sifting the pairs of equations is stopped only if all the possibilities of simplification by the given method have been exhausted.

The initial information for the subroutine *sinseq* is given in interval form by the complexes β and γ and by the complex ζ, separating the preceding complexes into subcomplexes by listing their cardinalities. The complexes ϵ and δ are auxiliary, in which the sets $\{I_s^i \cap I_t^j / I_t^j \ni Int_j\}$ obtained during execution of the algorithm are represented.

The result, the simplified system of equations, is represented by the same complexes β, γ, and ξ, in the same form.

sinseq (21, 20, 25, 33, 35)/αч, βк, γк, δк, ϵк, ζк/$\mathbf{d}$, l/($\beta\gamma\delta\epsilon$)
24 442 14 (α**dc**)

§1 $\circ\ e\ \bar{\circ}\ h\ \circ\ n\zeta_e \Rightarrow f \Rightarrow l1 \Rightarrow d \rightarrow 3$
§2 $1 \Rightarrow n\ \circ\ j\ \circ\ k$ *setin* $(\beta_h)(\gamma_h)\beta\gamma\delta\epsilon f(\zeta_d)//b_\gamma\ \circ\!\rightarrow 11$
mincovin $\delta\epsilon\mathbf{cd} // \mathbf{c} \vee \mathbf{d} \Rightarrow i\beta_h \vee \gamma_h \oplus i\ \circ\!\rightarrow 3$
$1 \Rightarrow m\mathbf{c} \Rightarrow \beta_h\mathbf{d} \Rightarrow \gamma_h \rightarrow 6$
§3 $\triangle\ h \oplus l\ \circ\!\rightarrow 4\beta_h \vee \gamma_h\ \circ\!\rightarrow 3 \rightarrow 2$
§4 $n\ \circ\!\rightarrow \alpha\ \circ\ nl - \zeta_e - 1 \Rightarrow h$
§5 $\zeta_d + f \Rightarrow f\ \triangle\ d \oplus e\ \circ\!\rightarrow 5d \oplus b_\zeta \mid\!\rightarrow 3\ \circ\ d\ \circ\ fl - 1 \Rightarrow h\ \triangle\ el + \zeta_e$
$\Rightarrow le \oplus b_\zeta \mid\!\rightarrow 3m \mid\!\rightarrow 1 \rightarrow 13$

§6 $l - \zeta_e - 1 \Rightarrow g$
§7 $\triangle\, g \oplus h \circ\!\!\rightarrow 7\beta_g \vee \gamma_g \circ\!\!\rightarrow 7g \oplus l \circ\!\!\rightarrow 10j \mid\!\!\rightarrow 12$
absin 11**cd**$(\beta_g)(\gamma_g)$ // $\rightarrow 7$
§10 $1 \Rightarrow jk \circ\!\!\rightarrow 6 \rightarrow 3$
§11 $1 \Rightarrow m \circ \beta_h \circ \gamma_h j \circ\!\!\rightarrow 3g \Rightarrow h \rightarrow 6$
§12 *joinin* 11**cd**$(\beta_g)(\gamma_g)(\beta_g)(\gamma_g)$ // $1 \Rightarrow k \rightarrow 7$
§13 *condinf* $\beta\gamma\zeta$ //.

If in the operation of *sinseq* it is found that the system of equations is incompatible, exit is carried out over terminal α.

The operator *setin* finds the set $\{I_s^i \cap I_t^j / I_t^j \in Int_j\}$ for the interval I_s^i, given by the variables α and β, of the set Int_j, represented by subcomplexes of the complexes γ and δ, separated by indexes η and ϑ, giving the starts of the distinguished subsets and their cardinalities, respectively. The result is represented by the complexes ϵ and ζ.

setin αп +, βп +, γк +, δк +, ϵк, ζк, ηи, ϑи/**b**, *a*/
21 72 2 (α**ab**)

$\eta + \vartheta \Rightarrow \vartheta \circ a$
§1 $\alpha \vee \gamma_\eta \Rightarrow \mathbf{a}\beta \vee \delta_\eta \Rightarrow \mathbf{b} \triangle\, \eta \oplus \vartheta \circ\!\!\rightarrow 2\mathbf{b} \wedge \mathbf{a} \mid\!\!\rightarrow 1\mathbf{a} \Rightarrow \epsilon_a\mathbf{b}$
$\Rightarrow \zeta_a \triangle\, a \Rightarrow b_\epsilon \rightarrow 1$
§2 .

The operator *mincovin* finds the minimal covering interval for the set of intervals represented by the complexes α and β, representing the result by the variables γ and δ.

mincovin αк +, βк +, γп, δп / −, *c* / ($\alpha\beta\gamma\delta$)
20 61 3

$\circ\; a1 \Rightarrow b1 \Rightarrow c$
§1 $\alpha_a \wedge \alpha_b \Rightarrow \gamma \circ\!\!\rightarrow 2 \triangle\, b \oplus b_\alpha \mid\!\!\rightarrow 1$
§2 $\beta_a \wedge \beta_c \Rightarrow \delta \circ\!\!\rightarrow 3 \triangle\, c \oplus b_\beta \mid\!\!\rightarrow 2$
§3 .

The operator *absin* has two exit terminals, and realizes exit through the auxiliary terminal α if the interval given by the intervals β and γ is absorbed by an interval given by the variables δ and ϵ. Otherwise exit is through the basic terminal.

absin αч, βч +, γп +, δп +, ϵп + / −, − / ($\beta\gamma\delta\epsilon$)
25 26 1

$\delta \vee \beta \oplus \delta \mid\!\!\rightarrow 1\epsilon \vee \gamma \oplus \epsilon \mid\!\!\rightarrow 1 \rightarrow \alpha$
§1 .

The operator *joinin* joins, if possible, the interval given by the variables β and γ with the interval given by the variables δ and ϵ and represents the result of joining by the variables ζ and η, realizing exit through terminal α. If joining is not possible, the basic exit terminal is used.

joinin αч, βп +, γп +, δп +, ϵп +, ζп, ηп/**b**, $-/(\beta\gamma\delta\epsilon\zeta\eta)$
33 57 1 (β**ab**)

$\beta \oplus \delta \Rightarrow \mathbf{a} \circ\!\!\rightarrow 1\epsilon \oplus \gamma \Rightarrow \mathbf{b} \circ\!\!\rightarrow 1\mathbf{a} \vee \mathbf{b} \bigtriangledown \oplus 1 \mid\!\!\rightarrow 1\mathbf{b} \urcorner \wedge \beta$
$\Rightarrow \zeta\mathbf{b} \urcorner \wedge \gamma \Rightarrow \eta \rightarrow \alpha$
§1 .

In the course of execution of the subroutine *sinseq* some of the Int_i are reduced through elimination of individual intervals in simplification, and the elements of the complexes α and β representing them take on zero values. When the simplification process terminates, the complexes α and β are condensed through the elimination of the elements with zero values, and the values of the elements of the complex γ are altered correspondingly. This operation is carried out by the subroutine *condinf*.

condinf αк, βк, γк $\mid -, d \mid (\alpha\beta)$
35 75 2

$\circ\ c \circ a \circ b\gamma_b \Rightarrow d$
§1 $\alpha_c \vee \beta_c \circ\!\!\rightarrow 2\alpha_c \Rightarrow \alpha_a\beta_c \Rightarrow \beta_a \bigtriangleup a$
§2 $\bigtriangleup c \oplus d \mid\!\!\rightarrow 1a \Rightarrow \gamma_b \bigtriangleup b\gamma_b + d \Rightarrow db \oplus b_\gamma \mid\!\!\rightarrow 1.$

REFERENCES

1. Zakrevskii, A. D., Testing for Identities in Boolean Algebra. This volume.
2. Zakrevskii, A. D., Programming Boolean Computations. This volume.
3. Grigor'yan, Yu. G., Algorithms for the Solution of a System of Logical Equations. *ZhVM i MF* **2**, No. 1, 186–189 (1962).
4. Cherry, C., and Vaswani, P. K., A New Type of Computer in Propositional Logic, with Greatly Reduced Scanning Procedures, *Information and Control* **4**, 155–168 (1964).
5. Zakrevskii, A. D., On the Solution of Systems of Logical Equations, *in* "Principles of Network and Control System Design," pp. 48–55 [in Russian]. Izd. Nauka, Moscow, 1964.

TESTING FOR IDENTITIES IN BOOLEAN ALGEBRA

A. D. Zakrevskii

1. STATEMENT OF THE PROBLEM AND VARIOUS INTERPRETATIONS

Let us formulate the problem of testing the relation

$$A_1 \vee A_2 \vee \cdots \vee A_m = 1$$

for identity, where A_i are certain elementary conjunctions, containing the symbols of binary variables (with or without inversion) in the set $X \equiv \{x_1, x_2, \ldots, x_n\}$.

This problem, which becomes far from trivial as n and m increase, is of particular interest already because certain other problems, of broad significance, reduce to it. For example, the identity of a given relation can be considered as the necessary and sufficient condition for absorption $(A' \to N)$ of an elementary conjunction A' by a disjunction

$$N \equiv A_1' \vee A_2' \vee \cdots \vee A_m'$$

of elementary conjunctions not orthogonal to the conjunction A' [1, 2]. For this it is sufficient to utilize the theorem formulated in [2]

$$(A' \to N) \leftrightarrow (\bigvee_{i=1}^{m} (A_i' : A') = 1),$$

where $A_i' : A'$ is a conjunction obtained from A_i' by the elimination of the symbols entering into A' and the replacement of $A_i' : A'$ by A_i.

We take a second example from the field of machine search for a logical derivation [3, 4]. In this field the following problem is typical: for a given system of propositions $B_1, B_2, \ldots, B_k, C$ it is required to establish whether the truth of C follows from the truth of $B_1, B_2, \ldots, B_k$. An algorithm that answers this question may be called "oracle" [4]. In situations presenting practical interest the number of essential variables in each of the propositions considered is, as a rule, small, but the total number of variables can be fairly great, in which connection computational difficulties arise. Considering $B_1, B_2, \ldots, B_k, C$ as Boolean functions taking on the value 1 for truth of the corresponding proposition, it is possible to formulate the problem of "Oracle" as the problem of testing the identity of the relation

$$\bar{B}_1 \vee \bar{B}_2 \vee \cdots \vee \bar{B}_k \vee C = 1.$$

Since each of the functions $B_1, B_2, \cdots, B_k$ and C depends on a small number of variables, it is not difficult to obtain the disjunctive normal forms of the functions $\bar{B}_1, \bar{B}_2, \ldots, \bar{B}_k, C$ and, uniting them, pass to a formula of the type

$$A_1 \vee A_2 \vee \cdots \vee A_m = 1.$$

A third example can be the problem of finding one of the solutions of a system of logical equations. In effect, in [5] it was shown that the general solution of equations characterizing the Boolean functions $f_1, f_2, \ldots, f_k$ (each of which takes on the value 1 if and only if the corresponding equation is satisfied) can be reduced to the inversion of the Boolean function $F \equiv \bar{f}_1 \vee \bar{f}_2 \vee \cdots \vee \bar{f}_k$ in the class of disjunctive normal forms. In seeking only one (arbitrary) solution the problem is substantially simplified, and becomes the problem of testing the identity of the relation

$$\bar{f}_1 \vee \bar{f}_2 \vee \cdots \vee \bar{f}_k = 1:$$

if this relation is satisfied identically, the system of equations has no solution; if for a certain combination of values of the arguments the relation is not satisfied, then the given combination is one of the solutions of the initial system. In the satisfaction of conditions similar to those considered in the preceding example, the disjunctive form of the function F is fairly easy to obtain. Thus, we again come to the problem of testing the identity of the relation

$$A_1 \vee A_2 \vee \cdots \vee A_m = 1,$$

where A_i are certain elementary conjunctions, and this problem is solved much faster than the problem of finding all solutions of the system of logical equations.

2. BASIC RELATIONS

We denote by K the set of all elementary conjunctions and by $a \perp A$ the relation of orthogonality between a ($a \in K$) and all conjunctions entering into the set A ($A \equiv \{A_1, A_2, \ldots, A_m\} \subset K$):

$$a \perp A \leftrightarrow (A_i \in A \rightarrow A_i \wedge a = 0).$$

It is obvious that the problem of testing the identity of the relation $A_1 \vee A_2 \vee \cdots \vee A_m = 1$ reduces to finding a conjunction a with the indicated property.

Let us consider the set A/b, where $b \in K$ and $A \subset K$, obtained from A by removal of all conjunctions orthogonal to the conjunction b and removal from the remaining conjunctions of the symbols occurring in b. For example,

$$\{abc, ad, ac\bar{d}, bde\}/ad = \{bc, 1, be\}.$$

By definition, it follows that

$$a \perp A \leftrightarrow (A/a = \varnothing),$$

where $\varnothing$ is the symbol of the empty set.

The following theorem is easily proved.

Theorem 1.

$$(A/c = B) \wedge (B \perp a) \rightarrow ac \perp A.$$

Assuming that $a, b, c \in K$, $A \subset K$, and $\gamma \in \{x_1, \bar{x}_1, x_2, \bar{x}_2, \ldots, x_n, \bar{x}_n\}$, we formulate a corollary of Theorem 1.

Corollary. If there exists a symbol γ that does not enter into any of the conjunctions of the set A, and if there exists a conjunction a orthogonal to all conjunctions of set A, there also exists a certain conjunction b containing the symbol γ and also orthogonal to all conjunctions of the set A.

In effect, it is sufficient to put $b = a\gamma$, since

$$(A/\gamma = A) \wedge (A \perp a) \rightarrow a\gamma \perp A.$$

We note that from the definition of orthogonality there follows the proposition

$$(\gamma \in A) \wedge (A \perp a) \rightarrow \bar{\gamma}a = a.$$

In other words, if in A there exists a certain single-literal conjunction γ and if there exists a conjunction a orthogonal to all conjunctions in A, then a contains the literal $\bar{\gamma}$.

The relations formulated permit the symbols entering into one of the conjunctions, which can serve as a solution, to be found successively. Only when the conditions represented in the left sides of the above formulas are not satisfied does the necessity arise to sift variants, which can be organized, for example, by the values of the remaining free variables in their order of occurrence.

3. SOLUTION ALGORITHM

Let us formulate the algorithm for solution of the above problem on the basis of the relations obtained.

The set A is analyzed for the possibility of applying the above relations; i.e., conjunctions are sought that contain only one literal, and literals of the type x_i or $\bar{x}_i$ that are not contained in a single conjunction of the set A are sought. The corresponding conjunction a is constructed, all conjunctions orthogonal to a are removed from A, and the remainder is again analyzed. This process is continued until further reduction of A becomes impossible. If the set A has become empty, the current value of the conjunction a is a conjunction orthogonal to all conjunctions of the set A, and the problem is considered to be solved. Otherwise some sifting of variants is carried out to finish the construction of the conjunction a, represented by a logical tree, whose nodes (which are not terminal leaves) are branch points with two exits, one of which corresponds to the symbol x_i, the other to $\bar{x}_i$, where x_i is one of the binary variables not figuring in the current value of the conjunction a. The basic idea of the algorithm is that the application of the described rules permits the dimensions of the tree to be substantially reduced and thereby its traversal to be accelerated. If in the course of calculation 1 appears in the current remainder of A, a backtracking movement is carried out over the current branch of the tree (since there does not exist a conjunction orthogonal to 1); if this has not occurred and at the same time the remainder of the set A is no longer subject to reduction by means of the formulated relations, a movement upward along the tree is effected. Information is memorized about the values of the constructed conjunction a corresponding to those nodes of the tree to which it is possible to return, i.e., those nodes lying on the path joining the current node to the root of the tree.

4. EXAMPLE

We shall illustrate the application of the given algorithm by the example of testing the identity

$$a\bar{f} \vee ace \vee a\bar{b}ef \vee a\bar{e}f \vee ab\bar{c}f \vee \bar{a}bf \vee \bar{a}\bar{e}f \vee \bar{a}\bar{b}e \vee \bar{a}\bar{d}e\bar{f} \vee \bar{a}bde\bar{f} \vee \\ \vee \bar{a}b\bar{e}\bar{f} \vee \bar{a}c\bar{e}\bar{f} \vee \bar{a}\bar{b}\bar{c}\bar{e}\bar{f} = 1,$$

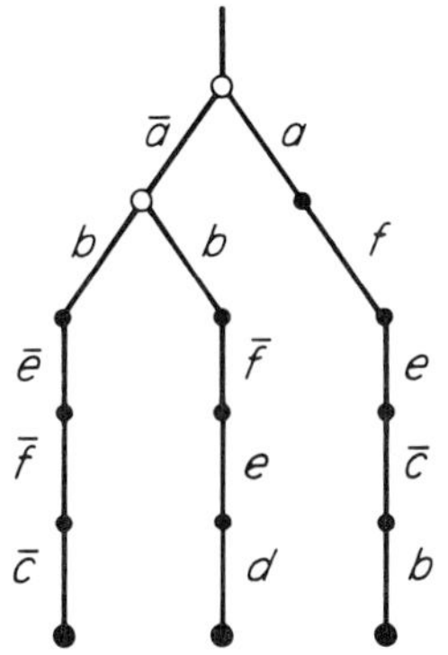

FIG. 1.

in which the tree of variants considered (shown in Fig. 1) is traversed, and it is established that the given disjunction is identically equal to 1. The initial set A of conjunctions is subjected during this to the following sequence of reductions:

$$\begin{aligned}
A_1 &= A/\bar{a} = \{bf,\ \bar{e}f,\ \bar{b}e,\ \bar{d}e\bar{f},\ bde\bar{f},\ b\bar{e}\bar{f},\ c\bar{e}\bar{f},\ \bar{b}\bar{c}\bar{e}\bar{f}\};\\
A_2 &= A_1/\bar{b} = \{\bar{e}f,\ e,\ \bar{d}e\bar{f},\ c\bar{e}\bar{f},\ \bar{c}\bar{e}\bar{f}\};\\
A_3 &= A_2/\bar{e} = \{f,\ c\bar{f},\ \bar{c}\bar{f}\};\\
A_4 &= A_3/\bar{f} = \{c,\ \bar{c}\};\\
A_5 &= A_4/\bar{c} = \{1\};\\
A_6 &= A_1/b = \{f,\ \bar{e}f,\ \bar{d}e\bar{f},\ de\bar{f},\ \bar{e}\bar{f},\ c\bar{e}\bar{f}\};\\
A_7 &= A_6/\bar{f} = \{\bar{d}e,\ de,\ \bar{e},\ c\bar{e}\};\\
A_8 &= A_7/e = \{\bar{d},\ d\};\\
A_9 &= A_8/\bar{d} = \{1\};\\
A_{10} &= A/a = \{\bar{f},\ ce,\ \bar{b}ef,\ \bar{e}f,\ b\bar{c}f\};\\
A_{11} &= A_{10}/f = \{ce,\ \bar{b}e,\ \bar{e},\ b\bar{c}\};\\
A_{12} &= A_{11}/e = \{c,\ \bar{b},\ b\bar{c}\};\\
A_{13} &= A_{12}/\bar{c} = \{\bar{b},\ b\};\\
A_{14} &= A_{13}/\bar{b} = \{1\}.
\end{aligned}$$

5. L-PROGRAM

The algorithm described is carried out by the L-program *testidendisc* (test of the identity to unity of a certain disjunction of conjunctions). The program corresponding to it will now be described.

A. THE EXTERNAL OPERANDS

α is an auxiliary exit terminal of the program attainable when a certain conjunction a is being found, orthogonal to all conjunctions of A (the basic exit terminal is reached when the truth of the tested identity is found).

$$\beta::\|\, A \ni X \,\|, \text{ where } X \equiv \{x_1, x_2, \ldots, x_n\}.$$
$$\gamma::\|\, A \ni X' \,\|, \text{ where } X' \equiv \{\bar{x}_1, \bar{x}_2, \ldots, \bar{x}_n\}.$$
$$\delta::\not\psi(A).$$
$$\epsilon::\not\psi(X).$$

ζ is a complex for storing intermediate information in traversal of the tree; its cardinality is $3k$, where k is the maximum height of the tree.

$$\eta::\|\, \{a\} \ni X \,\|,$$
$$\vartheta::\|\, \{a\} \ni X' \,\|,$$

where a is the constructed conjunction, which tends to become orthogonal to all conjunctions of A (the current subset A^* of those conjunctions in A that remain nonorthogonal to a is given by the value of the variable **f**, where $\mathbf{f}::\|\, \{A^*\} \ni A \,\|$).

Motion away from the root of the tree of sifting variants in the construction of the conjunction a is taken as motion upward, that toward the root as motion downward.

testidendisc $\alpha\text{ч}, \beta\text{к} +, \gamma\text{к} +, \delta\text{п} +, \epsilon\text{п} +, \zeta\text{к}, \eta\text{п}, \vartheta\text{п}/\mathbf{h}, a/(\beta\gamma\epsilon\zeta\eta\vartheta)$
13 351 11 (δ**bf**) (β**acdegh**)

$\circ\ \eta \circ \vartheta\ \delta \Rightarrow \mathbf{f} \circ b_\zeta$

§1 $\mathbf{f} \Rightarrow \mathbf{b} \circ\!\rightarrow \alpha \circ \mathbf{c} \circ \mathbf{d} \circ \mathbf{g} \circ \mathbf{h}\epsilon \oplus \eta$
$\oplus\ \vartheta \Rightarrow e$ — Find in A a single-literal conjunction

§2 $\mathbf{b}\ \ddot{\mathbf{X}}\ 3a\beta_a \vee \mathbf{c} \Rightarrow \mathbf{c}\gamma_a \vee \mathbf{d} \Rightarrow \mathbf{d}\beta_a \vee \gamma_a \wedge$
$e \Rightarrow \mathbf{a} \circ\!\rightarrow 11\ \nabla - 2 \,|\!\rightarrow 2\beta_a \vee \mathbf{g} \Rightarrow \mathbf{g}$
$\gamma_a \vee \mathbf{h} \Rightarrow \mathbf{h} \rightarrow 2$

§3	$\mathbf{c} \oplus \mathbf{d} \Rightarrow \mathbf{a} \wedge \mathbf{c} \vee \mathbf{g} \wedge \mathbf{e} \Rightarrow \mathbf{ga} \wedge \mathbf{d} \vee$ $\mathbf{h} \wedge \mathbf{e} \Rightarrow \mathbf{h} \vee \mathbf{g} \circ\!\!\rightarrow 7\mathbf{g} \wedge \mathbf{h} \mid\!\!\rightarrow 11\mathbf{h} \vee$ $\eta \Rightarrow \eta\mathbf{g} \vee \vartheta \Rightarrow \vartheta$	Find literals not entering into conjunctions in A
§4	$\mathbf{f} \Rightarrow \mathbf{b}$	Remove from A conjunctions orthogonal to a
§5	$\mathbf{b}\ \ddot{\mathbf{X}}\ 1a\beta_a \wedge \vartheta \mid\!\!\rightarrow 6\gamma_a \wedge \eta \circ\!\!\rightarrow 5$	
§6	$c_a \oplus \mathbf{f} \Rightarrow \mathbf{f} \rightarrow 5$	Upward along the tree
§7	$\epsilon \oplus \eta \oplus \vartheta \Rightarrow \mathbf{a}(\eta\vartheta\mathbf{f}) \Rightarrow \zeta\mathbf{a} \vdash \Rightarrow ac_a \vee$ $\vartheta \Rightarrow \vartheta \rightarrow 4$	
§10	$\zeta \Rightarrow (\eta\vartheta\mathbf{f})\epsilon \oplus \eta \oplus \vartheta \vdash \Rightarrow ac_a \vee$ $\eta \Rightarrow \eta \rightarrow 4$	Downward along the tree
§11	$b_\zeta \mid\!\!\rightarrow 10.$	

REFERENCES

1. Zhuravlev, Yu. I., Set-Theoretical Methods in Boolean Algebra, *in* "Problems of Cybernetics," No. 8, pp. 5–44. [in Russian]. Fizmatgiz, Moscow, 1962.
2. Zakrevskii, A. D., The Problem of Simplification of the Shortened dnf of a Boolean Function, *Comm. Siberian Conf. Math. Mech., 3rd,* pp. 266–267 [in Russian]. Izd. Tomskogo Univ., 1964.
3. Van Hao. On the Road to Mechanical Mathematics [Russian translation], *in* "Cybernetic Collection," No. 5, pp. 114–165. IL, Moscow, 1962.
4. Shanin, N. A., General Description of the Problem of Machine Search for Logical Derivation, *Comm. All-Union Symposium on the Problem of Machine Search for Logical Derivation, 1st* [in Russian]. Trakai, July 1–7, 1964.
5. Zakrevskii, A. D., and Kalmykova, A. Yu., The Solution of Systems of Logical Equations. This volume.

PART II: APPLICATIONS

Section B: STRUCTURAL SYNTHESIS

DETERMINATION OF THE CONNECTIVITY OF A GRAPH

A. D. Zakrevskii

Basically utilizing the terminology of Berge [1], we shall call a symmetrical graph k-connected if it remains connected on removal of any $k - 1$ branches. The solution to the problem of determining the connectivity number k of a graph involves great computational difficulties. In the general case it is possible to propose the sifting of the subsets of branches of the graph, testing whether they are cutsets, i.e., whether their removal from the graph induces loss of connectivity. It is obvious that the set of all cutsets of a graph is a convex set with respect to inclusion, and therefore the search can be reduced as was done in the realization of the operator *fisirco*. On the other hand, since the connectivity number of a graph is defined by the minimal (in cardinality) section, the search can be shortened still more, excluding from consideration all subsets of cardinality not less than p, where p is the minimal cardinality of the cutsets already found.

The maximum reduction of the computational process can be achieved for the solution of particular cases—analysis of the graph for 1-connectivity (or simple connectivity) or 2-connectivity.

The analysis of a graph for 1-connectivity can be carried out on the basis of the algorithms presented in [2] or [3].

The program presented below in LYaPAS is distinguished by high speed of machine execution; α is an auxiliary exit terminal of the subroutine corresponding to the detection of nonconnectivity of the graph; $\beta::\|F \diamond F\|$, $\gamma::\psi(F)$, F is the set of nodes of the analyzed graph; $\diamond$ is the relation of adjacency between nodes.

ansigrac αч, βк, γп/**b**, $b/(\beta\gamma)$
15 61 3 (β**ab**)

$\bar{\text{o}}\ b$
§1 $\triangle b \oplus b_\beta \circ\!\!\rightarrow 3\beta_b \Rightarrow \mathbf{a} \Rightarrow \mathbf{b}$
§2 $\mathbf{a}\ \ddot{\mathbf{X}}\ 1a\mathbf{b} \vee \beta_a \Rightarrow \beta_a \rightarrow 2$
§3 $\beta_0 \oplus \gamma \mid\!\!\rightarrow \alpha.$

The analysis of the graph for 2-connectivity is somewhat more complex. We define a symmetrical graph G by the set of nodes F and the set of branches U: $G = (F, U)$. We say that the graph (F, U) is 2-connected on $C \subseteq F$ if its subgraph $(C, U_{(c)})$, corresponding to the set C, is 2-connected.

We submit the following algorithm, whose idea consists of the successive extension of the set C on which 2-connectivity of the graph (F, U) is established through the inclusion in C of nodes located in elementary circuits of the partial graph $(F, U \backslash U_{(c)})$ closing on C, i.e., circuits whose boundary nodes belong to C. We shall assume that $\sigma(F) \geq 2$.

The expression of the algorithm contains six sentences, corresponding to various stages of the algorithm. In stage 1 a branch identical to the set C is found. In stage 2 motion is simulated over a certain elementary circuit, whose start is the branch found in stage 1, and which passes outside C. In this motion the matrix β, representing the tested graph, is changed so as to exclude the possibility of backward motion over the branches already traversed, and the nodes traversed are included in a set B, the set of nodes of the graph (F, U) subjected to analysis, beginning with the analysis of the zero node (stage 0). As soon as the continuation of the elementary circuit outside C becomes impossible, execution of stages 3 and 4 begins, in which it is clarified whether there exists in the graph (F, U), in which backward motion over the already traversed branches is forbidden, a path connecting the final node of the circuit with the set C. In the presence of such a circuit stage 5 is carried out; all nodes belonging to B are included in C. If C coincides with F, the 2-connectedness of the graph is taken to be established and the basic exit terminal of the subroutine is attained; otherwise, stage 1, etc., is again executed. The absence of 2-connectivity is found if in the execution of stage 1 no branch is found identical to the set C or if in the execution of the last stages the absence of the sought path is established. The auxiliary terminal α is then attained.

As in the earlier subroutine for the operator *ansigrac*, $\beta :: \| F \diamond F \|$, where $\diamond$ is the relation of adjacency between nodes of the graph and

$\gamma::\psi(F)$. The variable sets B and C represent the current values of the internal variables of the subroutines **b** and **c**:

$$\mathbf{b}::\|\ \{B\} \ni F\ \|, \quad \mathbf{c}::\|\ \{C\} \ni F\ \|.$$

ansigrac two αч, βк, γп/**e**, $b/(\beta\gamma)$
16 166 5 (β**abcde**)

$c_0 \Rightarrow \mathbf{b} \to 5$

§1	$\mathbf{a}\ \dot{\mathbf{X}}\ \alpha a\mathbf{c}\ \urcorner \wedge \beta_a \circ\!\!\to 1 \Rightarrow \mathbf{a}$	Search for exit from C
§2	$a \Rightarrow b\mathbf{a} \vdash \Rightarrow ac_a \vee \mathbf{b} \Rightarrow \mathbf{b}c_b \oplus \beta_a \Rightarrow$ $\beta_a\mathbf{b}\ \urcorner \wedge \beta_a \Rightarrow \mathbf{a} \mid\!\!\to 2c_a \Rightarrow \mathbf{e} \Rightarrow \mathbf{a} \Rightarrow \mathbf{d}$	Motion over the elementary network outside C (with inhibition of return)
§3	$\mathbf{a}\ \dot{\mathbf{X}}\ 4a\beta_a \vee \mathbf{d} \Rightarrow \mathbf{d} \to 3$	Attempt at closure of the elementary circuit obtained on C
§4	$\mathbf{d} \wedge \mathbf{c} \mid\!\!\to 5e\ \urcorner \wedge \mathbf{d} \Rightarrow \mathbf{a} \circ\!\!\to \alpha \vee e \Rightarrow e \to 3$	
§5	$\mathbf{b} \Rightarrow \mathbf{c} \Rightarrow \mathbf{a} \oplus \gamma \mid\!\!\to 1.$	Inclusion of B in C and test for equality of $C = F$

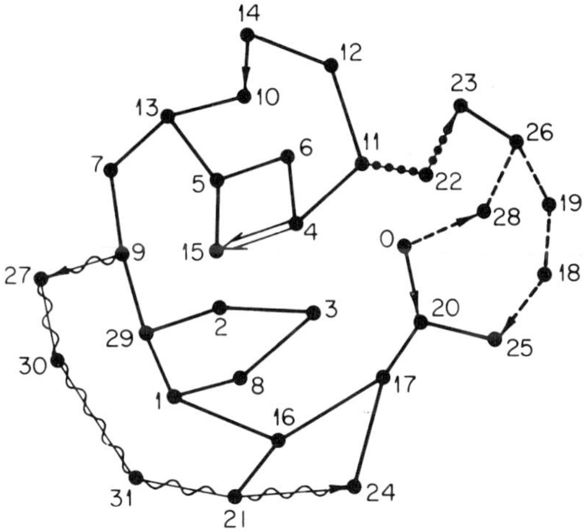

FIG. 1.

The functioning of the algorithm is illustrated by the example of analysis of the graph (Fig. 1), where 0–31 are the nodes. It is expressed in the successive inclusion in C of those nodes that enter into elementary circuits, obtained in the execution of sentence 2, indicated in the figure by the heavy, broken, double, wavy, and dotted lines.

REFERENCES

1. Berge, C., "Theory of Graphs and Its Applications" [Russian translation]. IL, Moscow, 1962.
2. Warshall, S., A Theorem on Boolean Matrices, *J. Assoc. Comput. Mach.* **9**, 11–12 (1962).
3. Martynyuk, V. V., Economical Construction of a Transitive Closure of a Binary Relation, *ZhVM i MF* **2**, No. 4, 723–725 (1962).

ALGORITHMS FOR THE MINIMIZATION OF BOOLEAN FUNCTIONS

V. G. Novoselov

The theory of the minimization of Boolean functions in the class of disjunctive normal forms is one of the most highly developed divisions of the general theory of minimization of Boolean functions. However, in engineering practice its application encounters the following difficulties:

(1) The weak knowledge on the part of engineering and technical workers of the mathematical apparatus and language in which the minimization algorithms are presented.
(2) The large expenditures of time for the execution of computations connected with the solution of minimization problems.

One of the methods for avoiding these difficulties is the following. Each minimization method or even system of methods is represented in the form of a computer program, and each program is characterized by the form of presentation of the input information. Many of the methods described in the literature are theoretically applicable to the minimization of any Boolean function, but in practice it is found that there does not exist a single method that would be universal in this sense. In other words, each method is applicable only for the minimization of a certain class of Boolean functions that is characteristic of it. The limits of the class are established, starting from the requirement of limited volume of the information presented at any one time and limited time for solution of the problem. This means that a system of programs is necessary to carry out various methods of minimization.

Thus, in the representation of minimization methods in the form of computer programs, all the calculations can be carried out by the machine, and it is only required that the engineer have knowledge of the form of representation of the input information and the possibilities of each program, expressed in a finite number of characteristics.

1. SOME DEFINITIONS

An *interval* I of a Boolean space is a subset such that it is completely defined by its *minimal* α and *maximal* β elements, namely,

$$\gamma \in I \leftrightarrow \alpha \subseteq \gamma \subseteq \beta.$$

The prescription of the interval partitions the set X of binary variables into classes of *internal* and *external* variables for the given interval. The external variables of the set X are those whose values coincide for all elements of the interval. The remaining variables are internal. The set of internal and external variables is easily defined by the minimal and maximal elements of the interval. The *rank* of the interval is equal to the number of its external variables. An interval *covers* itself and any of its subsets.

We introduce in a Boolean space the concept of symmetry. Two intervals are *symmetrical* if they have the same external variables and if the values of identical external variables differ for one and only one variable. We say that the intervals are symmetrical with respect to this variable. Two intervals I_1 and I_2, symmetrical with respect to x_i, are *joined* by this variable, forming the new interval $I_3 = I_1 \cup I_2$.

Let $X' \subseteq X$ be the set of external variables of the interval I. An interval I is *extended* by the variable $x_i \in X'$ if the extended interval I' is the result of the joining of the interval I and the interval symmetrical to it in x_i. The interval I is *extended by the set of variables* $X'' \subset X'$ if it is extended by the variable $x_i \in X''$, then the result is extended by the variable $x_j \in X''$ $(i \neq j)$, etc., for all the variables of the set X''.

If on the set of elements formed by the union of two intervals I_1 and I_2 there exists overlapping, but not equal sets of external variables, there exists a maximal interval I_3, distinct from the initial intervals, and the intervals I_1 and I_2 admit a *generalized joining*, whose result is the interval I_3. It is obvious that the values of the common external variables of the intervals I_1 and I_2 must differ only for one variable.

An interval formed by elements of some subset M' in the Boolean space M is called an *interval on the set* M'. The interval I is a *maximal interval* of the set $M' \subseteq M$ if on M' there exists no interval I' that completely covers the interval I and does not coincide with it.

2. ALGORITHM FOR THE MINIMIZATION OF BOOLEAN FUNCTIONS IN THE INTERVAL REPRESENTATION

One possible initial representation of Boolean functions $f(X)$ is the disjunctive normal form (dnf) which, in interval representation, is given by the set of intervals $Int = \{I_1, I_2, \ldots, I_m\}$.

For the given form of representation of a Boolean function there exists the well-known method of construction of the minimal form developed by Blake. Blake's method consists of the following two parts:

(1) Construction of the set of maximal intervals of a Boolean function by applying to the initial set of intervals the operation of generalized joining and removal of the completely covered intervals.

(2) Construction of the set of irredundant sets of maximal intervals by applying an operation inverse to generalized joining, and selecting in it the minimal form.

A machine algorithm based on Blake's method is described in [1]. Using the results of this article, however, we introduce changes in certain phases of the algorithm that permit the volume of search for the construction of the set of irredundant sets of maximal intervals to be reduced.

On the basis of the operators of the algorithm for construction of the minimal form is constructed an approximate algorithm, based on the sequential extraction of kernels from the residues; specifically, the kernel on the set Int_{max} of maximal intervals of the Boolean function is found. The intervals of the kernel and the antikernel are removed, as well as the intervals covered by the kernel intervals and one of the remaining intervals, of rank not greater than the rank of the interval removed. On the remaining set of intervals Int^* a kernel of this set is again removed, etc. If on the set Int^* there is no part belonging to a kernel, one of the intervals of minimal rank is distinguished, which is introduced into the resultant form of the Boolean function and removed from the set Int^*; the algorithm for extraction of a kernel is again applied to the new set Int^* of intervals.

A. ALGORITHM FOR THE CONSTRUCTION OF ONE MINIMAL dnf OF A BOOLEAN FUNCTION

(1) All possible generalized joinings among the elements of the set *Int* are carried out; the results distinct from the elements of the set *Int* are included in it. This process is repeated over the new set *Int* as long as the results of the generalized joining are different from the elements of the set *Int*.

(2) From the set *Int* obtained, all intervals are removed that are completely covered by other elements of this set. A new set Int_{max} of maximal intervals of the Boolean function is constructed.

(3) In stage (1), together with the construction of the set *Int* there are associated with each element of this set values of the variables $A, B \in \{0, 1\}$, where

$$A = \begin{cases} 1, & \text{if the interval can be generally joined with other elements of the set } Int; \\ 0, & \text{otherwise;} \end{cases}$$

$$B = \begin{cases} 1, & \text{if the interval is the result of a generalized joining of elements of } Int; \\ 0, & \text{otherwise.} \end{cases}$$

The set Int_{max} of maximal intervals is separated according to this information: $\text{sep}^{(4)}\ Int_{max} = \{Int_1, Int_2, Int_3, Int_4\}$. In the class Int_1 (or Int_2, Int_3, Int_4) are included the intervals of the set Int_{max} for which $A = 0$, $B = 0$ (or $A = 1, B = 0; A = 0; B = 1, A = 1; B = 1$). Intervals of the classes Int_1 and Int_2 are included in the kernel of the set Int_{max} and, as a result, enter into each minimal form of the Boolean function. Intervals of the class Int_3 are included in the antikernel of the set Int_{max} and do not appear in any of the minimal forms. Further investigation concerning the membership of the intervals of the fourth class Int_4 in minimal forms is necessary, and is carried out in the next stages of the algorithm.

(4) Up to this point our algorithm coincides completely with the algorithm given in [1]. In the next stages all irredundant subsets are found by sifting various subsets of the elements of the class Int_4, and from them one minimal subset is chosen. In many cases it is possible that the set Int_4 can be substantially reduced before sifting is carried out.

Intervals that are completely covered by the union of kernel intervals are removed. These intervals belong to the antikernel. Intervals (also in the antikernel) each of which is completely covered by the union of the intervals of the kernel and one of the elements of the set Int_4 of lower rank are removed. Since we are looking for one minimal form, it is also possible to remove intervals covered by the union of kernel intervals and one of the elements of Int_4 whose rank is equal to the rank of the interval removed.

It is clear that the removal of these intervals from Int_4 can lead to a substantial reduction of the subsequent search. We denote the set of intervals thereby formed by $Int_4{}^*$. The reduction of Int_4 to $Int_4{}^*$ is accomplished in the following way.

(a) The part of the interval $I_i \in Int_4$ that is not covered by kernel intervals is found. It is represented by the set of intervals Int_i'.

(b) A search is carried out for an interval $I_j \in Int_4$ $(i \neq j)$, covering all intervals of the set Int_i' and with rank not exceeding the rank of the interval I_i.

(c) If such an interval $I_j \in Int_4$ has been found, the interval I_i is removed from the set Int_4. If all the intervals $I_i \in Int_4$ have been examined, the process terminates; otherwise, step (a) is repeated for the next interval.

(5) In this stage the algorithm branches.

(a) If it is necessary to find a minimal form, pass to stage (6).

(b) If it is sufficient to find a certain (on the average good) approximation to the mdnf, second- and higher-order kernels are found, and specifically, the removal of certain intervals from the fourth class changes the conditions of mutual generalized joining among the intervals of this class, as a result of which certain intervals may be permitted to enter into the kernel and antikernel of the set $Int_4{}^*$. For this purpose stages (1) and (3) of the algorithm are carried out, but now on the set $Int_4{}^*$. Sets $Int_1^{(*)}$, $Int_2^{(*)}$, $Int_3^{(*)}$, and $Int_4^{(*)}$ are found. The elements of the sets $Int_1^{(*)}$, $Int_2^{(*)}$, and $Int_3^{(*)}$ are included in the corresponding sets Int_1, Int_2, and Int_3, while the set $Int_4^{(*)}$ is renamed Int_4. The operations of stage (4) of the algorithm are applied to the set Int_4 as long as the result is not the identity of the sets Int_4 and $Int_4{}^*$, or that $Int_4{}^*$ is empty. In the former case passage to stage (10) takes place; in the latter the process of construction of the resultant form has terminated.

(6) Operations similar to those described in [1] are performed on the set of intervals $Int_4{}^*$, except that in stage (8) a more general criterion of termination of the sifting of irredundant subsets of the set $Int_4{}^*$ is introduced before all the irredundant subsets have been taken. A set X' of literals (variables and their inversions) external to the intervals of the set $Int_4{}^*$ is found, as well as the lower bound k of the number of intervals in $Int_4{}^*$ whose union covers all the intervals of the sets $Int_4{}^*$ and which are called in the ensemble a *coverage*.

(7) The next set (of cardinality k) of intervals in Int_4* is found, which contains all the literals in X', and is tested if it is a coverage. For this stages (1) and (2) are executed over the set Int_2 and the current set of intervals in Int_4*.

Then stage (8) of the algorithm is performed.

(8) The set of intervals of the fourth class, formed by the application of the operations of stages (1) and (2) to the set Int_2 and the given set of intervals from Int_4*, is compared with the set Int_4*. If the union of intervals of the new fourth class contains all the intervals of the set Int_4*, one of the irredundant sets of maximal intervals of the Boolean function has been obtained. On subsequent entries into the stage (7) all other irredundant sets are obtained. Only one of the sets with minimal sum of ranks of the intervals entering into it is retained. The current irredundant set is either eliminated, if the sum of ranks of its intervals is not smaller than the sum of ranks of the intervals of the set in memory, or (in the contrary case) stored as the minimal. The process of obtaining a minimal form terminates if the sum of ranks of intervals of the set in memory is not greater than the sum of ranks k of the intervals in the set Int_4*; otherwise, the next stage is carried out.

(9) Stage (7) of the algorithm is executed, and if all coverages of cardinality k have been obtained, k is increased by unity.

(10) In the set Int_4* an interval of minimal rank is found, which is introduced into the resultant form of the Boolean function. The interval found is removed from the set Int_4* and stage (4) is applied to the new set Int_4*.

B. LYaPAS REPRESENTATION OF THE ALGORITHM

We first present a number of subroutines of the general L-program corresponding to the algorithm. The L-program is also represented in the form of a subroutine, and executes the operator *mifboof*. Let us introduce the general notation. By definition, each set Int of intervals is equivalent to the assignment of two sets Y and Z, where Y contains the minimal elements and Z the maximal elements of the corresponding elements of the set Int. According to this the set of intervals is represented in successive subroutines by the complexes $\alpha::\| Y \ni X \|$ and $\beta::\| Z \ni X \|$, where X is the set of variables on which the Boolean function depends. To represent the elements of the intervals the zero component of the elements of complexes α and β is not used, and therefore the set X is defined in the following way: $X = \{x_1, x_2, \ldots, x_n\}$.

genadj. Perform generalized adjunction (joining) of intervals of a set.

The set of intervals *Int* of the Boolean function is represented by the complexes α and β. Each interval $I_i(\alpha_i, \beta_i)$ is tested for the possibility of generalized joining with all the elements of the set *Int*. If the result of the generalized adjunction differs from the elements of the set *Int*, they are added to the set *Int*. The values of the variables A and B introduced in stage (3) of the algorithm are represented in the zero positions of the elements of complexes α and β, respectively. The result of the subroutine is represented by the changed complexes α and β.

$genadj\ \alpha\text{к}?,\ \beta\text{к}?/\mathbf{b},\ d/(\alpha\beta)$
170 233 4 $(\alpha\mathbf{ab})$

$\circ\ a \Rightarrow bb_\alpha \Rightarrow dc_0\ \urcorner \Rightarrow \mathbf{b}$

§1 $\triangle\ a \oplus d\ \circ\!\!\rightarrow 4\alpha_a \oplus \alpha_b \wedge \mathbf{b} \Rightarrow \mathbf{a}\beta_a$
$\oplus\ \beta_b \wedge \mathbf{a} \Rightarrow \mathbf{a}\ |\!\!\leftarrow\ \oplus\ c_0\ |\!\!\rightarrow 1\alpha_a$
$\vee\ \alpha_b \oplus \mathbf{a} \Rightarrow \alpha_d?\beta_a \wedge \beta_b \vee \mathbf{a}$
$\vee\ c_0 \Rightarrow \beta_d?c_0 \vee \alpha_a \Rightarrow \alpha_a c_0 \vee \alpha_b$
$\Rightarrow \alpha_b\ \circ\ c\ \circ\ \mathbf{a}$

Test for generalized adjunction of intervals I_a and I_b, formation of the interval I_d—the result of the generalized adjunction—and setting of tags

§2 $\alpha_c \oplus \alpha_d \wedge \mathbf{b}\ |\!\!\rightarrow 3\beta_c \oplus \beta_d \wedge \mathbf{b}$
$|\!\!\rightarrow 3c_0 \vee \beta_c \Rightarrow \beta_c\ \bar{\circ}\ \mathbf{a}$

Test for presence of interval I_d in *Int*

§3 $\triangle\ c \oplus d\ |\!\!\rightarrow 2\mathbf{a}\ |\!\!\rightarrow 1\ \triangle\ d \rightarrow 1$
§4 $\bar{\circ}\ a\ \triangle\ b \oplus d\ |\!\!\rightarrow 1d \Rightarrow b_\alpha \Rightarrow b_\beta.$

redcoin. Reduce the set of intervals by removal of completely covered intervals.

The lower bound of the set of intervals represented by the complexes α and β is found. The unit components of the variable γ represent the essential variables of the Boolean function.

The result of the subroutine is represented by the same complexes α and β.

$redcoin\ \alpha\text{к},\ \beta\text{к},\ \gamma\text{п} + /\mathbf{a},\ c/(\alpha\beta\gamma)$
171 133 3 $(\alpha\mathbf{a})$

$\circ\ a \Rightarrow bb_\alpha \Rightarrow c$

§1 $\triangle\ a \oplus c\ \circ\!\!\rightarrow 2\beta_a \oplus \alpha_a\ \urcorner \wedge \gamma \Rightarrow \mathbf{a}\alpha_b$
$\oplus\ \beta_b \wedge \mathbf{a}\ |\!\!\rightarrow 1\alpha_b \oplus \alpha_a \wedge \mathbf{a}\ |\!\!\rightarrow 1a$
$\oplus\ b\ \circ\!\!\rightarrow 1\ \bar{\triangle}\ c\alpha_c \Rightarrow \alpha_b\beta_c \Rightarrow \beta_b \rightarrow 3$

Test for coverage of interval I_b by I_a and removal of the covered interval

§2 $\triangle\ b$
§3 $\bar{\circ}\ ab \oplus c\ |\!\!\rightarrow 1c \Rightarrow b_\alpha \Rightarrow b_\beta.$

sepic. Separate the set of intervals by classes.

The operator *sepic* separates the set of intervals $Int_{\max}$ by classes, according to the description of stage (3) of the algorithm. The intervals of classes Int_1 and Int_2 are represented by the complexes β and γ, the intervals of class Int_4 by complexes δ and ϵ. If it is found that the set of intervals of the fourth class is empty, passage over the auxiliary exit terminal to sentence α of the external program is executed.

$$sepic - \alpha\text{ч},\ \beta\text{к},\ \gamma\text{к},\ \delta\text{к}?,\ \epsilon\text{к}?/\mathbf{a},\ c/(\beta\gamma\delta\epsilon)$$

172 161 4 $(\beta\mathbf{a})$

$$\circ\ a\ \circ\ b\ \circ\ cc_0 \urcorner \Rightarrow \mathbf{a}$$

§1 $\beta_a \wedge c_0 \mid\!\rightarrow 2\gamma_a \wedge c_0 \mid\!\rightarrow 4\beta_a \Rightarrow \beta_b\gamma_a \Rightarrow \gamma_b \triangle b \rightarrow 4$

§2 $\gamma_a \wedge c_0 \mid\!\rightarrow 3\beta_a \Rightarrow \beta_b\gamma_a \Rightarrow \gamma_b \triangle b \rightarrow 4$

§3 $\beta_a \wedge \mathbf{a} \Rightarrow \delta_c?\gamma_a \wedge \mathbf{a} \Rightarrow \epsilon_c? \triangle c$

§4 $\triangle a \oplus b_\beta \mid\!\rightarrow 1b \Rightarrow b_\beta \Rightarrow b_\gamma c \circ\!\rightarrow \alpha \Rightarrow b_\delta \Rightarrow b_\epsilon.$

consinew. Construct from an initial set of intervals a new set not overlapping a certain interval I.

The initial set of intervals is represented by the complexes β and γ, and the interval I by its minimal element δ and maximal element ϵ. Each interval I_i of the initial set that overlaps the interval I is decomposed into a set Int_i' of intervals, each of which does not overlap I and whose union, together with the elements of the interval I, covers the interval I_i. The interval I_i is removed from the initial set of intervals and the new set Int_i' is added to the initial set. If the interval I is completely covered by the initial set of intervals, then control is transferred to sentence α of the external program. The new set of intervals is represented by the complexes β and γ.

$$consinew - \alpha\text{ч},\ \beta\text{к}?,\ \gamma\text{к}?,\ \delta\text{п} +,\ \epsilon\text{п} + /\mathbf{b},\ c/(\beta\gamma\delta\epsilon)$$

200 213 5 $(\gamma\mathbf{ab})$

$$\bar{\circ}\ bb_\beta \Rightarrow c\delta \oplus \epsilon \urcorner \Rightarrow \mathbf{a}$$

§1 $\triangle b \oplus c \circ\!\rightarrow 5\beta_b \oplus \delta \Rightarrow \mathbf{b}\gamma_b \oplus \epsilon \wedge \mathbf{b} \mid\!\rightarrow 1\beta_b \oplus \gamma_b \wedge \mathbf{a} \Rightarrow \mathbf{b}$

§2 $\mathbf{b}\,\dot{\mathbf{X}}\,4ac_a \wedge \delta \circ\!\rightarrow 3c_a \urcorner \wedge \gamma_b \Rightarrow \gamma_c?c_a \urcorner \wedge \beta_b \Rightarrow \beta_c? \triangle c \rightarrow 2$

§3 $c_a \vee \beta_b \Rightarrow \beta_c?c_a \vee \gamma_b \Rightarrow \gamma_c? \triangle c \rightarrow 2$

§4 $\bar{\triangle} c \circ\!\rightarrow \alpha\beta_c \Rightarrow \beta_b\gamma_c \Rightarrow \gamma_b \rightarrow 1$

§5 $c \Rightarrow b_\beta \Rightarrow b_\gamma.$

depinc. Detect a part of an interval not covered by a set of intervals.

An interval I_i is represented in the zero elements of the complexes δ and ϵ [in the initial assignment $\sigma(\delta) = \sigma(\epsilon) = 1$]. The set $Int_1 \cup Int_2$ is represented in the complexes β and γ. The subroutine successively determines the parts of the interval I_i not covered by all intervals of the set $Int_1 \cup Int_2$. The result of the subroutine, the set Int_i', is represented by the complexes δ and ϵ. When the interval I_i is completely covered by the union of intervals of the set $Int_1 \cup Int_2$, control is transferred to sentence α of the external program.

depinc (200)/αч, βк +, γк +, δк?, ϵк?/**b**, d/($\beta\gamma\delta\epsilon$)
201 35 1

$\circ\ d$
§1 *consinew* $\alpha\delta\epsilon(\beta_d)(\gamma_d)//\ \triangle\ d \oplus b_\beta \mid\!\rightarrow 1.$

tecosin. Test the coverage of a set of intervals.

The set $Int_4{}^*$ is represented by the complexes δ and ϵ. The set Int_i' of intervals, represented by the complexes β and γ, is tested for coverage by the elements of the set $Int_4{}^*$. The interval $I_\zeta \in Int_4{}^*$, covering the set Int_i', is tagged by the index ζ. When none of the elements of the set $Int_4{}^*$ covers the set Int_i' control is transferred over the auxiliary exit terminal to sentence α of the external program.

tecosin (174)/αч, βк +, γк +, δк +, ϵк +, ζи/**b**, a/($\beta\gamma\delta\epsilon$)
173 51 1 (β**ab**)

inmir $\beta\gamma$**ab**// $\bar{\circ}\ \zeta$
§1 $\triangle\ \zeta \oplus b_\delta \circ\!\rightarrow \alpha$**a** $\urcorner \wedge \delta_\zeta \mid\!\rightarrow 1\epsilon_\zeta \urcorner \wedge$ **b** $\mid\!\rightarrow 1.$

inmir. Find an interval of minimal rank that completely covers a set of intervals.

Minimal and maximal elements γ and δ of an interval of set Int_i', represented by the complexes α and β, are found.

inmir αк +, βк +, γп, δп/—, a/($\alpha\beta\gamma\delta$)
174 37 1

$\circ\ a\ \bar{\circ}\ \gamma \circ \delta$
§1 $\alpha_a \wedge \gamma \Rightarrow \gamma\beta_a \vee \delta \Rightarrow \delta$
$\triangle\ a \oplus b_\alpha \mid\!\rightarrow 1.$

reman. Remove the antikernel from a set of intervals.

The intervals of the set $Int_4{}^*$, belonging to the antikernel, represented by the complexes β and γ are removed. The complexes δ and ϵ are working complexes (their cardinalities are defined by the operator *depinc*). The set of intervals of the kernel is represented by the complexes ζ and η. If the set of intervals of the antikernel is empty, control is transferred over the auxiliary exit terminal to sentence α of the external program.

reman (172, 173)/αч, βк, γк, δк?, ϵк?, ζк +, ηк + /**c**, e/($\beta\gamma\delta\epsilon\zeta\eta$)
175 152 3

○ e ○ **c**
§1 $\beta_e \Rightarrow \delta_0\gamma_e \Rightarrow \epsilon_0 1 \Rightarrow b_\delta$ *depinc* $2\zeta\eta\delta\epsilon$//*tecosin* $3\delta\epsilon\beta\gamma a$//$a \oplus e$ ○→ *tecosin* $1\beta_a$
$\oplus\ \gamma_a$ ⌉ ∇ ⇒ $d\beta_e \oplus \gamma_e$ ⌉ ∇ − d ○→ *tecosin* 1
§2 $\bar{○}$ **c** △ $b_\beta \Rightarrow b_\gamma \Rightarrow a\beta_a \Rightarrow \beta_e\gamma_a \Rightarrow \gamma_o$ $\overline{△}$ e
§3 △ $e - b_\beta$ ○→ 1**c** ○→ α.

finexli. Find a set of literals external to the elements of a set of intervals.

A set of literals (variables and their inversions) is found, external to the elements of the set Int_4*, represented by the complexes α and β. The set of external variables encountered without inversion is represented by the variable γ, and the set of external variables encountered with inversion, by the variable δ. The number of variables on which the Boolean function depends is represented by the index ϵ.

finexli αк +, βк +, γп, δп, ϵи + /**b**, a/($\alpha\beta\gamma\delta$)
176 70 1 (α**ab**)

○ a ○ γ ○ $\delta c_0 - c_\epsilon \Rightarrow$ **b**
§1 $\alpha_a \oplus \beta_a$ ⌉ ∧ **b** ⇒ **a** ∧ α_a ∨ $\gamma \Rightarrow \gamma\alpha_a$ ⌉ ∧ **a** ∨ $\delta \Rightarrow \delta$ △ $a \oplus b_\alpha$ |→ 1.

extrank. Find an interval of extremal rank.

The set of intervals is represented by the complexes γ and δ. The extremal interval found is characterized by an index ϵ and rank ζ. The operator *extrank* finds the interval of minimal rank (α = $\bar{○}$, β = |→) or maximal rank ($\alpha = 0$, β = ○→).

The number of variables on which the Boolean function depends is represented by the index η.

extrank α ○, β ○, γк +, δк +, ϵи, ζп, ηи +/**a**, a/($\gamma\delta$)
177 71 2 (γ**a**)
○ $a\alpha\zeta c_0 - c_\eta \Rightarrow$ **a**
§1 $\gamma_a \oplus \delta_a$ ⌉ ∧ **a** ∇ − $\zeta\beta 2 + \zeta \Rightarrow \zeta a \Rightarrow \epsilon$
§2 △ $a \oplus b_\gamma$ |→ 1.

copoco. Construct possible coverages.

The operator *copoco* finds on a set $Int_4{}^*$, represented by the complexes β and γ, one of the sets (of cardinality δ) of intervals that contain in their ensemble all external literals of the elements of the set $Int_4{}^*$ (the variables are represented by the unit components of the variable ϵ and their inversions by the unit components of the variable ζ). The set is represented by the variable η, so that $\eta^i = 1$ if the interval $I_i \in Int_4{}^*$, and I_i occurs in the set; otherwise, $\eta^i = 0$ (η^i is the ith component of the variable η). The current (except the first) sets are obtained by entry to § 1 of the subroutine. After all coverages of cardinality δ have been obtained, control is transferred to sentence α of the external L-program.

$copoco\ \alpha\text{Ч},\ \beta\text{К}+,\ \gamma\text{К}+,\ \delta\text{П},\ \epsilon\text{П},\ \zeta\text{П},\ \eta\text{П},\ \vartheta\text{П},\ \kappa\text{П}\ /\ \mathbf{f},\ b\ /\ (\beta\gamma\epsilon\zeta\vartheta\kappa)$
202 267 3 $(\beta\mathbf{abd})$ $(\eta\mathbf{ecf})$

$\delta - 1 \Rightarrow ac_a \urcorner + 1 \Rightarrow \eta \Rightarrow \mathbf{c}b_\beta - 1 \Rightarrow ac_a \Rightarrow \mathbf{e} \circ \mathbf{f} \circ \vartheta \circ \kappa \rightarrow 2$

§1 $\eta + \mathbf{e} \wedge \eta \Rightarrow \mathbf{c} \circ\!\rightarrow \alpha\mathbf{c} - \mathbf{e} \oplus \mathbf{c} \vdash \Rightarrow a\eta$
$+ \mathbf{e} \oplus \eta < 1 + \mathbf{e} \Rightarrow \mathbf{c} \vdash - a \Rightarrow a\mathbf{c}$
$< a \oplus \eta \Rightarrow \eta\mathbf{f} \Rightarrow \mathbf{c}\eta \wedge \mathbf{e} \Rightarrow \mathbf{fc} \circ\!\rightarrow 3\eta$
$- \mathbf{e} \wedge \eta \Rightarrow \mathbf{c} \circ \vartheta \circ \kappa$

Construction of the next set

§2 $\mathbf{c}\,\dot{\mathbf{X}}\,3a\beta_a \oplus \gamma_a \urcorner \Rightarrow \mathbf{d} \wedge \beta_a \vee \vartheta \Rightarrow \vartheta\beta_a \urcorner$
$\wedge \zeta \wedge \mathbf{d} \vee \kappa \Rightarrow \kappa \rightarrow 2$

§3 $\eta - \mathbf{e} \oplus \eta \vdash \Rightarrow a\beta_a \oplus \gamma_a \urcorner \Rightarrow \mathbf{d} \wedge \beta_a$
$\vee \vartheta \oplus \epsilon \mid\!\rightarrow 1\beta_a \urcorner \wedge \zeta \wedge \mathbf{d} \vee \kappa \oplus \zeta$
$\mid\!\rightarrow 1.$

Test that in their ensemble the intervals of the set contain all external literals of the set $Int_4{}^*$

mifboof. Minimize the interval form of a Boolean function.

The operator *mifboof* performs the algorithm for the construction of one mdnf, described above. The Boolean function is represented by the complexes α and β. The complexes γ and δ represent the intervals of the fourth class; ϵ and η are intermediate complexes. The complexes ζ and ϑ represent the intervals of the fourth class that belong to the minimal form. The number of variables on which the Boolean function depends is represented by the index κ.

mifboof (170, 171, 172, 175, 176, 177, 202) $\alpha\text{К}??,\ \beta\text{К}??,\ \gamma\text{К}?,\ \delta\text{К}?,\ \epsilon\text{К}?,$
$\zeta\text{К}?,\ \eta\text{К},\ \vartheta\text{К},\ \kappa\text{И}+,\ \lambda\text{И}/1,\ k/(\alpha\beta\gamma\delta\epsilon\zeta\eta\vartheta)$
203 1071 22 $(\alpha\mathbf{abghikl})$

$\circ\ k \circ e\ \bar{\circ}\ j \circ b_\zeta c_0 - c_\kappa \Rightarrow \mathbf{i}.$

§1 $genadj\ \alpha\beta//redcoin\ \alpha\beta\mathbf{i}// \circ bk \mid\!\rightarrow 15e \circ\!\rightarrow 3b_\alpha \Rightarrow ab_\alpha + b_\epsilon \Rightarrow b_\alpha??$
$\Rightarrow b_\beta??$

§2 $\epsilon_b \Rightarrow \alpha_a\eta_b \Rightarrow \beta_a \triangle b \triangle a \oplus b_\alpha \mid\!\rightarrow 2$

§3 $\textit{sepic}\ 22\alpha\beta\gamma\delta//$

§4 $b_\alpha \circ\!\rightarrow 6\ \textit{reman}\ 6\gamma\delta\epsilon\eta\alpha\beta//\lambda \circ\!\rightarrow 6a_\alpha \Leftrightarrow a_\epsilon b_\alpha \Rightarrow b_\epsilon a_\gamma \Leftrightarrow a_\alpha b_\gamma \Rightarrow b_\alpha a_\beta$
$\Leftrightarrow a_\eta b_\beta \Rightarrow b_\eta a_\delta \Leftrightarrow a_\beta b_\delta \Rightarrow b_\beta \circ b_\gamma \Rightarrow b_\delta\ \bar{\circ}\ e \rightarrow 1$

§5 $b_\alpha \Rightarrow a\gamma_c \Rightarrow \alpha_a?\delta_c \Rightarrow \beta_a \triangle b_\alpha \Rightarrow b_\beta b_\gamma \Rightarrow a\gamma_a \Rightarrow \gamma_c\delta_a \Rightarrow \delta_c\ \bar{\triangle}\ b_\gamma \Rightarrow b_\delta \rightarrow 4$

§6 $\lambda \mid\!\rightarrow 13\ \textit{finexli}\ \gamma\delta\mathbf{hg}\kappa//\mathbf{h} \nabla \Rightarrow i\mathbf{g} \nabla + i \Rightarrow ib_\gamma \Rightarrow g \circ hb_\alpha \Rightarrow f$

§7 $\textit{extrank} \circ \circ\!\rightarrow \gamma\delta c\mathbf{c}\kappa//\bar{\triangle}\ b_\gamma \Rightarrow a\gamma_c \Leftrightarrow \gamma_a\delta_c \Leftrightarrow \delta_a\mathbf{c} + h \Rightarrow h - i$
$\circ\!\rightarrow 7g \Rightarrow b_\gamma - a \Rightarrow g\ \bar{\circ}\ k \circ h$

§10 $\textit{copoco}\ 12\gamma\delta g\mathbf{hgjkl}//f \Rightarrow b \circ c\mathbf{j} \Rightarrow \mathbf{a}$

§11 $\mathbf{a}\ \dot{\mathbf{X}}\ 1a\gamma_a \Rightarrow \alpha_b? \Rightarrow \epsilon_c?\delta_a \Rightarrow \beta_b? \Rightarrow \eta_c? \triangle c \Rightarrow b_\epsilon \Rightarrow b_\eta \triangle b \Rightarrow b_\alpha \Rightarrow b_\beta$
$\rightarrow 11$

§12 $\triangle g \Rightarrow i \circ hb_\gamma \Rightarrow b$

§13 $\textit{extrank}\ \bar{\circ} \mid\!\rightarrow \gamma\delta c\mathbf{c}\kappa//\lambda \mid\!\rightarrow 5\ \bar{\triangle}\ b_\gamma \Rightarrow a\gamma_a \Leftrightarrow \gamma_c\delta_c \Leftrightarrow \delta_a\mathbf{c} + h \Rightarrow h\ \bar{\triangle}\ i$
$\mid\!\rightarrow 13b \Rightarrow b_\gamma \rightarrow 10$

§14 $h - j \mid\!\rightarrow 21 \rightarrow \textit{copoco}\ 1$

§15 $\gamma_b \Rightarrow \mathbf{a}\delta_b \Rightarrow \mathbf{b}f \Rightarrow a$

§16 $\alpha_a \oplus \mathbf{a} \mid\!\rightarrow 17\beta_a \oplus \mathbf{b} \mid\!\rightarrow 17 \triangle b \oplus b_\gamma \mid\!\rightarrow 15 \circ a \circ i \rightarrow 20$

§17 $\triangle a \oplus b_\alpha \mid\!\rightarrow 16 \rightarrow \textit{copoco}\ 1$

§20 $\epsilon_a \oplus \eta_a \urcorner \wedge \mathbf{i} \nabla + i \Rightarrow i \triangle a \oplus b_\epsilon \mid\!\rightarrow 20i - j \mid\!\rightarrow \textit{copoco}\ 1i$
$\Rightarrow ja_\epsilon \Leftrightarrow a_\zeta b_\epsilon \Rightarrow b_\zeta a_\eta \Leftrightarrow a_\vartheta b_\eta \Rightarrow b_\vartheta \circ b_\epsilon \Rightarrow b_\eta$

§21 $\zeta * \vartheta * f \Rightarrow b_\alpha \Rightarrow b_\beta$

§22 $\alpha * \beta *.$

C. EXAMPLE

The Boolean function $f(x_1, x_2, \ldots, x_6)$ is prescribed by the following set of intervals: $\{\bar{x}_1\bar{x}_2\bar{x}_3\bar{x}_4\bar{x}_5,\ \bar{x}_1\bar{x}_2\bar{x}_3x_5x_6,\ x_1\bar{x}_2\bar{x}_3\bar{x}_5\bar{x}_6,\ x_1x_2\bar{x}_3x_4\bar{x}_6,\ \bar{x}_1x_2\bar{x}_3x_4x_5,\ \bar{x}_1\bar{x}_2\bar{x}_4\bar{x}_5\bar{x}_6,\ x_1x_2x_4\bar{x}_5\bar{x}_6,\ \bar{x}_1x_2x_4x_5\bar{x}_6,\ \bar{x}_1\bar{x}_3\bar{x}_4\bar{x}_5x_6\}$.

After operation of the operators *genadj* and *redcoin* a set $Int_{\max}$ of maximal intervals is obtained:

		A	B	
$\vee$	$\bar{x}_2\ \bar{x}_3\ \bar{x}_4\ \bar{x}_5\ \bar{x}_6$	1	1	
	$x_1\ \bar{x}_2\ \bar{x}_3\ \bar{x}_5\ \bar{x}_6$	1	1	
$\vee$	$x_1\ \bar{x}_3\ x_4\ \bar{x}_5\ \bar{x}_6$	1	1	
	$x_2\ \bar{x}_3\ x_4\ x_5\ \bar{x}_6$	1	1	
$\vee$	$x_1\ x_2\ \bar{x}_3\ x_4\ \bar{x}_6$	1	1	Int_4
$\vee$	$\bar{x}_1\ x_2\ \bar{x}_3\ x_4\ x_5$	1	1	
	$\bar{x}_1\ \bar{x}_3\ x_4\ x_5\ x_6$	1	1	
	$\bar{x}_1\ \bar{x}_2\ \bar{x}_3\ x_4\ x_6$	1	1	
$\vee$	$\bar{x}_1\ \bar{x}_2\ \bar{x}_3\ \bar{x}_5\ x_6$	1	1	
*	$\bar{x}_1\ \bar{x}_2\ \bar{x}_3\ \bar{x}_4\ \bar{x}_5$	1	1	

$$\left.\begin{array}{lllllcc} \bar{x}_1 & \bar{x}_2 & \bar{x}_4 & \bar{x}_5 & \bar{x}_6 & 1 & 0 \\ x_1 & x_2 & x_4 & \bar{x}_5 & \bar{x}_6 & 1 & 0 \\ \bar{x}_1 & x_2 & x_4 & x_5 & \bar{x}_6 & 1 & 0 \\ \bar{x}_1 & \bar{x}_3 & \bar{x}_4 & \bar{x}_5 & x_6 & 1 & 0 \end{array}\right\} Int_2$$

The operator *sepic* extracts the sets Int_4 and Int_2. The application of the operator *reman* permits the interval $\bar{x}_1\bar{x}_2\bar{x}_3\bar{x}_4\bar{x}_5$, belonging to the antikernel, and marked by the $*$, to be determined. Since only one minimal form is found, it is possible to remove from Int_4 the intervals marked by the symbol $\vee$. The remaining unmarked four intervals in $Int_4{}^*$, after processing by *genadj* and *redcoin*, enter into the sets $Int_2^{(*)}$ and $Int_1^{(*)}$. Since the set $Int_4^{(*)}$ is empty, one of the minimal forms of the Boolean function is written in the following form:

$$\begin{aligned} f_{\mathrm{mdf}} = {} & \bar{x}_1\bar{x}_2\bar{x}_4\bar{x}_5\bar{x}_6 \vee x_1x_2x_4\bar{x}_5\bar{x}_6 \vee \bar{x}_1x_2x_4x_5\bar{x}_6 \vee \\ & \vee \bar{x}_1\bar{x}_3\bar{x}_4\bar{x}_5x_6 \vee x_1\bar{x}_2\bar{x}_3\bar{x}_5\bar{x}_6 \vee x_2\bar{x}_3x_4x_5\bar{x}_6 \vee \\ & \vee \bar{x}_1\bar{x}_3x_4x_5x_6 \vee \bar{x}_1\bar{x}_2\bar{x}_3x_4x_6. \end{aligned}$$

3. ALGORITHM FOR SIMPLIFICATION OF THE INTERVAL FORM OF AN INCOMPLETELY DEFINED BOOLEAN FUNCTION

An incompletely defined Boolean function $f(X)$ is prescribed by the sets Int_1 and Int_0 of intervals on which the function takes on the unit and zero values, respectively.

There does not exist in the literature an exact method for the minimization of Boolean functions defined in this way. The proposed algorithm for finding a form of the Boolean function approaching the shortest is based on the coverage of the set Int_1 by the smallest possible number of maximal intervals.

A. DESCRIPTION OF THE ALGORITHM

(1) All possible joinings of the elements of the set Int_1 are carried out. The intervals that are completely covered by other elements of the set Int_1 are removed from the set Int_1.

(2) Finding an interval I_i^*. For the interval $I_i \in Int_1$ a set of intervals $Int_i' \subseteq Int_1$ exists, the union of each of which with the interval I_i is included in an interval not overlapping with the elements of the set Int_0. An interval I_i^* is found (if it exists), not overlapping with the elements of the set Int_0 and covering the interval I_i and all elements of the set Int_i'. The interval I_i^* is the one sought, since it can be extended to a maximal interval entering into at least one of the shortest forms of the Boolean function. If no interval I_i^* exists for the given Boolean function, the cardinality of the set Int_i' and the index i of the interval I_i are stored as the characteristics of the interval I_i.

This procedure is repeated for the next interval $I_i \in Int_1$, and either there is found for it the corresponding interval I_i^*, which is then given as the sought, or again the cardinality $\sigma(Int_i')$ of the corresponding set Int_i' is calculated. The characteristics of the new interval are stored if $\sigma(Int_i')$ for the new interval is less than the corresponding cardinality of the previously stored one; otherwise the next interval $I_i \in Int_1$ is examined.

If all intervals have been examined in this way, then that interval whose characteristics have been stored, i.e., the interval I_j with minimal value of $\sigma(Int_j')$, is taken as the one sought. The criterion of choice of the interval with minimal cardinality of the corresponding set Int_j' is intuitive, and analogous to the criterion of choice of elements in the matrix method [2].

(3) A set X' is found of those external variables of the interval I_i^* for each of which the interval I_i^* can be extended without overlapping the elements of the set Int_0.

(4) The interval I_i^* is extended over all sets of variables, defined by combinations of m out of $\sigma(X')$ elements of the set X' [in the initial state $m = \sigma(X')$]. If all intervals obtained after such extension overlap the elements of the set Int_0, then m is decreased by unity; otherwise, of the intervals not overlapping the elements of the set Int_0 is taken an interval I_i' which absorbs the maximal number of elements of the set Int_1. If there are several such intervals, then the interval I_i'' is chosen with maximal overlapping with the elements of the set Int_1.

(5) The interval I_i'' is included in the resultant form of the Boolean function. All elements completely covered by I_i'' are removed from the set Int_1. If the remaining set of intervals Int_1 is empty, the process of constructing the resultant form has terminated; otherwise the set Int_1 is altered in the following manner. In each interval $I_j \in Int_1$ is found a subset, represented by a set Int_j' of intervals, not covered by the interval I_i''. The interval I_j is removed from the set Int_1, and the elements of the set Int_j' are included in Int_1. Stage (1) of the algorithm is realized.

B. REPRESENTATION OF THE ALGORITHM IN LYaPAS

We first present a group of subroutines used in the general L-program, which program is also executed in the form of a subroutine and realizes the operator *mincodefin*. We shall assume that in the subroutines the sets of intervals are represented by two complexes, where the first will represent the minimal elements, the other the maximal elements of the intervals of the corresponding sets.

adjin. Perform possible adjunctions in the set of intervals.

All possible adjunctions of intervals in the set Int_1, represented by complexes α and β, are carried out. The adjoined intervals are removed from the set Int_1, and the results of adjunction are included in Int_1.

$adjin - \alpha\text{к}, \beta\text{к}/\mathbf{a}, b/(\alpha\beta)$
205 132 3 $(\alpha\mathbf{a})$

§1 $\circ\ a \Rightarrow b$
§2 $\triangle\ a \oplus b_\alpha\ \circ\!\!\rightarrow 3\alpha_a \oplus \alpha_b \Rightarrow \mathbf{a}\ |\!\!\leftarrow \oplus\ c_0\ |\!\!\rightarrow 2\beta_a \oplus \beta_b \oplus \mathbf{a}\ |\!\!\rightarrow 2\mathbf{a}\ \urcorner \wedge \alpha_a$
$\Rightarrow \alpha_a$
$\mathbf{a} \vee \beta_a \Rightarrow \beta_a\ \overline{\triangle}\ b_\alpha \Rightarrow b_\beta \Rightarrow a\beta_a \Rightarrow \beta_b\alpha_a \Rightarrow \alpha_b \rightarrow 1$
§3 $\triangle\ b \Rightarrow a \oplus b_\alpha\ |\!\!\rightarrow 2.$

finin. Find the next interval.

In the subroutine (according to stage (2) of the algorithm) the interval I_i^* is found. The set Int_1 is represented by the complexes α and β. The complexes γ and ζ represent the minimal and maximal elements of the intervals of the set Int_0. The variable η represents the essential variables of the Boolean function. The sought interval I_i^* is represented by its minimal and maximal elements δ and ϵ.

$finin\ \alpha\text{к}+, \beta\text{к}+, \gamma\text{к}+, \delta\text{п}, \epsilon\text{п}, \eta\text{п}+, \zeta\text{к}+/\mathbf{e}, f/(\alpha\beta\gamma\delta\epsilon\eta)$
206 246 11 $(\alpha\mathbf{abcde})$

$\bar{\circ}\ c\ \bar{\circ}\ e$
§1 $\triangle\ c \oplus b_\alpha\ \circ\!\!\rightarrow 10\alpha_c \Rightarrow \mathbf{a} \oplus \beta_c\ \urcorner \wedge \eta \Rightarrow \mathbf{b}\ \bar{\circ}\ \mathbf{c}\ \bar{\circ}\ b\ \circ\ g\ \circ\ d$
§2 $d\ |\!\!\rightarrow 6\ \triangle\ b \oplus b_\alpha\ \circ\!\!\rightarrow 5\alpha_b \oplus \beta_b\ \urcorner \wedge \mathbf{b} \Rightarrow \mathbf{d}\alpha_b \oplus \mathbf{a}\ \urcorner \wedge \mathbf{d} \Rightarrow \mathbf{d}\ \circ\!\!\rightarrow 2$
§3 $\circ\ a$
§4 $\gamma_\alpha \oplus \zeta_\alpha\ \urcorner \Rightarrow \mathbf{e}\gamma_\alpha \oplus \mathbf{a} \wedge \mathbf{d} \wedge \mathbf{e}\ \circ\!\!\rightarrow 2\ \triangle\ a \oplus b_\gamma\ |\!\!\rightarrow 4d\ |\!\!\rightarrow 7\mathbf{d} \wedge \mathbf{c}$
$\Rightarrow \mathbf{c}\ \circ\!\!\rightarrow 1\ \triangle\ g \rightarrow 2$
§5 $\bar{\circ}\ d\mathbf{c} \Rightarrow \mathbf{d} \rightarrow 3$
§6 $g - e\ |\!\!\rightarrow 1g \Rightarrow ec \Rightarrow f \rightarrow 1$
§7 $\mathbf{a} \wedge \mathbf{c} \Rightarrow \delta\mathbf{c}\ \urcorner \wedge \mathbf{b} \vee \beta_c \Rightarrow \epsilon \rightarrow 11$
§10 $\alpha_f \Rightarrow \delta\beta_f \Rightarrow \epsilon$
§11 .

sevinex. Find the set of variables in which an interval can be extended.

For the interval I_i^*, represented by its minimal and maximal elements, γ and δ, the possibility of extending it without overlapping with the elements of the set Int_0, represented by the complexes α and β, is tested. Extension is tested for each external variable of the interval I_i^*. The set of variables ϵ is formed for each of which the interval I_i^* can be extended. The set of essential variables of the Boolean function is represented by η.

sevinex $\alpha\text{к}+,\ \beta\text{к}+,\ \gamma\text{п}+,\ \delta\text{п}+,\ \epsilon\text{п},\ \eta\text{п}+/\mathbf{c},\ \mathbf{c}/(\alpha\beta\gamma\delta\epsilon\eta)$
207 110 4 $(\gamma\mathbf{abc})$

$\circ\ \epsilon\gamma \oplus \delta\ \urcorner \wedge \eta \Rightarrow \mathbf{a}$
§1 $\mathbf{a}\,\dot{\mathbf{X}}\,4bc_b\ \urcorner \Rightarrow \mathbf{c}\ \bar{\circ}\ a$
§2 $\triangle\ a \oplus b_\alpha\ \circ\!\rightarrow 3\alpha_a \oplus \gamma \Rightarrow \mathbf{b}\beta_a \oplus \delta \wedge \mathbf{c} \wedge \mathbf{b}\ \circ\!\rightarrow 1 \rightarrow 2$
§3 $c_b \vee \epsilon \Rightarrow \epsilon \rightarrow 1$
§4 .

extin. Extension of an interval.

According to the stage (4) of the algorithm the interval I_i^* is extended by the set of variables of the set X', for each of which it can be extended without overlapping the elements of the set Int_0. The subroutine finds all such sets that have maximal cardinality. The set Int_0 is represented by the complexes β and γ, the interval I_i^* by the variables δ and ϵ. The set X' is prescribed by the unit components of the variable ζ. The current extension of the interval is represented by the variables η and ϑ. After all sets of maximal cardinality have been obtained, control is transferred to sentence α of the external program. To obtain the next interval, § 3 of the subroutine is entered.

extin $(212)/\alpha\text{ч},\ \beta\text{к}+,\ \gamma\text{к}+,\ \delta\text{п}+,\ \epsilon\text{п}+,\ \zeta\text{п}+,\ \eta\text{п},\ \vartheta\text{п},\ \kappa\text{и}/\mathbf{e},\ c/$
$(\beta\gamma\delta\epsilon\zeta\eta\vartheta)$
210 120 4 $(\zeta\mathbf{abe})$

$77 \Rightarrow \kappa\ \circ\ \mathbf{e}$
§1 $\mathbf{e} \oplus \zeta\ \urcorner \Rightarrow \mathbf{a}\ \bar{\circ}\ a$
§2 $\triangle\ a \oplus b_\beta\ \circ\!\rightarrow 4\beta_a \oplus \delta \Rightarrow \mathbf{b}\gamma_a \oplus \epsilon \wedge \mathbf{b} \wedge \mathbf{a}\ \circ\!\rightarrow 3 \rightarrow 2$
§3 *genco* $\alpha e\zeta\kappa//\rightarrow 1$
§4 $\mathbf{e}\ \nabla \Rightarrow \kappa\mathbf{e}\ \urcorner \wedge \delta \Rightarrow \eta\mathbf{e} \vee \epsilon \Rightarrow \vartheta.$

abovin. Calculate the cardinalities of subsets of elements that are absorbed by or overlap a certain interval.

The number ϵ of elements is counted that appear in the set Int_1, represented by the complexes α and β, and are completely covered by the interval I_i', represented by its minimal and maximal elements δ and ϵ. The number of elements of Int_1 overlapping the interval I_i' is given by the index ζ.

$\mathit{abovin}\ \alpha\text{к}\ +,\ \beta\text{к}\ +,\ \gamma\text{п}\ +,\ \delta\text{п}\ +,\ \epsilon\text{и},\ \zeta\text{и}/\mathbf{a},\ a/(\alpha\beta\gamma\delta)$
$\mathbf{211\ 77\ 3}\ (\alpha\mathbf{a})$

$\bar{\circ}\ a\ \circ\ \epsilon\ \circ\ \zeta$
§1 $\triangle\ a \oplus b_\alpha\ \circ\!\rightarrow 3\alpha_a\ \urcorner\ \wedge\ \gamma\ |\!\rightarrow 2\delta\ \urcorner\ \wedge\ \beta_a\ |\!\rightarrow 2\ \triangle\ \epsilon \rightarrow 1$
§2 $\gamma \oplus \alpha_a \Rightarrow \mathbf{a}\delta \oplus \beta_a \wedge \mathbf{a}\ |\!\rightarrow 1\ \triangle\ \zeta \rightarrow 1$
§3 .

genco. Generate codes of combinations.

The subroutine operates in two modes. If $\delta = k$, then $(\sigma(\gamma), k)$ combinations of β unit components of the variable γ are found. If $\delta = 77$, then all $2^{\sigma(\gamma)}$ combinations of β are found. After all the required combinations have been found, control is transferred to sentence α of the external program.

$\mathit{genco}\ \alpha\text{ч},\ \beta\text{п},\ \gamma\text{п}\ +,\ \delta\text{и}\ +/\mathbf{c},\ b/(\beta\gamma)$
$\mathbf{212\ 177\ 5}\ (\gamma\mathbf{bac})$

$\gamma \Rightarrow \mathbf{b} - 1\ \urcorner\ \wedge\ \gamma \Rightarrow \mathbf{c}\delta \oplus 77\ \circ\!\rightarrow 1\delta \Rightarrow b\beta\ \circ\!\rightarrow 3$
§1 $\circ\ b\beta \wedge \mathbf{c}\ \circ\!\rightarrow 2\gamma\ \urcorner\ \vee\ \beta + \mathbf{c} \wedge \beta \Rightarrow \mathbf{a} \oplus \beta\ \nabla \Rightarrow b\mathbf{a} \Rightarrow \beta$
§2 $\beta - 1 \wedge \beta \Rightarrow \mathbf{a}\beta - 1 \oplus \mathbf{a} \wedge \gamma \Rightarrow \mathbf{b} \vdash \Rightarrow a\mathbf{a} \vee c_a \Rightarrow \beta b$
$\circ\!\rightarrow 4c_a \oplus \mathbf{b} \Rightarrow \mathbf{b}$
§3 $\mathbf{b}\ \dot{\mathbf{X}}\ \alpha a c_a \vee \beta \Rightarrow \beta\ \bar{\triangle}\ b\ |\!\rightarrow 3$
§4 $\delta \oplus 77\ \circ\!\rightarrow 5\beta\ \nabla \oplus \delta\ \circ\!\rightarrow 5 \rightarrow \alpha$
§5 .

mincodefin. Minimize incompletely defined Boolean functions, given in interval form.

The subroutine performs the algorithm described in 3.A. The set Int_1 is represented by the complexes α and β, the set Int_0 by the complexes γ and δ, and the set of intervals of the resultant form, by the complexes ϵ and ζ. The variable η represents the essential variables of the Boolean function.

$\mathit{mincodefin}$ (205, 171, 206, 207, 210, 211, 200)
$\alpha\text{к}??,\ \beta\text{к}??,\ \gamma\text{к}\ +,\ \delta\text{к}\ +,\ \epsilon\text{к}?,\ \zeta\text{к}?\ \eta\text{п}\ +/\mathbf{j},\ h/(\alpha\beta\gamma\delta\epsilon\zeta\eta)$
$\mathbf{213\ 215\ 4}\ (\alpha\mathbf{afghij})\ \circ\ h$

§1 $\mathit{adjin}\ \alpha\beta//\mathit{redcoin}\ \alpha\beta\eta//\mathit{finin}\ \alpha\beta\gamma\mathbf{fg}\eta\delta//\mathit{sevinex}\ \gamma\delta\mathbf{fgh}\eta//\ \circ\ f\ \circ\ g$
§2 $\mathit{extin}\ 3\gamma\delta\mathbf{fghij}c//\mathit{abovin}\ \alpha\beta\mathbf{ij}bd//f - b\ |\!\rightarrow \mathit{extin}\ 3b \Rightarrow fd - g$
$\circ\!\rightarrow \mathit{extin}\ 3d \Rightarrow g\mathbf{i} \Rightarrow \epsilon_h? \Rightarrow \zeta_h?\mathbf{j} \rightarrow \mathit{extin}\ 3$
§3 $\epsilon_h \Rightarrow \mathbf{d}\zeta_h \Rightarrow \mathbf{e}\ \triangle\ h\ \mathit{consinew}\ 4\alpha\beta\mathbf{de}//\!\rightarrow 1$
§4 $h \Rightarrow b_\epsilon \Rightarrow d_\zeta\epsilon * \zeta *.$

C. EXAMPLE

Let us consider the simplification of the Boolean function given by the following sets of intervals:

$$Int_1 = \{\bar{x}_3\bar{x}_4\bar{x}_5\bar{x}_6,\ x_2\bar{x}_4x_5\bar{x}_6,\ \bar{x}_2\bar{x}_3x_4\bar{x}_5x_6,\ \bar{x}_1\bar{x}_2x_3\bar{x}_5x_6,\ \bar{x}_1x_2x_3x_6\};$$
$$Int_0 = \{\bar{x}_3x_4x_5\bar{x}_6,\ x_1\bar{x}_2x_3\bar{x}_6,\ x_1x_2\bar{x}_3x_6,\ \bar{x}_3\bar{x}_4x_5x_6,\ \bar{x}_2x_3\bar{x}_4x_5x_6\}.$$

The operator *adjin* does not alter the initial sets. The interval $I_0 \in Int_1$ is considered. For this interval the set Int_0' is empty, and the operator *finin* supplies it as the sought set. The extension of the interval I_0^* is possible for each variable in the set $X' = \{x_4, x_5\}$. Of two possible maximal intervals $\bar{x}_3\bar{x}_4\bar{x}_6$ and $\bar{x}_3\bar{x}_5\bar{x}_6$ the former is taken, as it overlaps the interval $x_2\bar{x}_4x_5\bar{x}_6$. The intervals I_0 and I_1 are removed from the set Int_1, and in their place is introduced the interval $x_2x_3\bar{x}_4x_5\bar{x}_6$, representing the subset of elements of the interval I_1 not covered by the interval I_0. The process is repeated for the following interval. The resultant form of the Boolean function is

$$f(X) = \bar{x}_3\bar{x}_4\bar{x}_6 \vee x_2x_3 \vee \bar{x}_2\bar{x}_5x_6.$$

It is one of the shortest forms of the initial Boolean function.

4. ALGORITHM FOR THE MINIMIZATION OF COMPLETELY DEFINED BOOLEAN FUNCTIONS

Here we consider the prescription of a Boolean function by a list of elements of a set M_1, which is equivalent to the perfect disjunctive normal form. The classical method for the minimization of Boolean functions given in the perfect dnf is the method of Quine [3], as perfected by McCluskey [4]. This method can be practically applied for the minimization of Boolean functions of a large number of variables, but with limited cardinality of the set M_1.

The Quine–McCluskey method consists of two parts:

(1) Construction of the set of maximal intervals of the Boolean function by the realization of all possible adjunctions of the elements of the set M_1 and removal of the completely covered elements of this set. The algorithm that performs this part of the method is fairly fast, but for its computer execution it requires a large memory volume.

(2) Construction of tables of coverage by maximal intervals of the elements of the set M_1 and finding from this table the resulting minimized forms of the Boolean function. This part of the method is connected with a large enumeration, to reduce which, depending on the required result, various algorithms have been developed for the construction of coverages. These algorithms are fairly completely described in [5], and therefore here we shall consider only the first part of the Quine–McCluskey method.

5. PROGRAM FOR CONSTRUCTING A SET OF MAXIMAL INTERVALS OF A BOOLEAN FUNCTION

oswep. Order a set by weights of elements and partition it into classes of equal weights.

For each element of the complex α, that represents the initial set, its weight (the number of unit components of the element) is calculated. The set of elements of the complex α, ordered by increasing weight, is the complex β. A complex γ of cardinality δ is constructed, where δ is the maximal weight of the elements of complex α.

The ith element of the complex γ is the number of elements of the complex α whose weights do not exceed $i - 1$. Then it is found that the complex γ defines a partitioning into classes (by equal weights) of the elements of the complex α: the element γ_i is the minimal value, less one, of the index of the elements of the complex β belonging to the class of weight i, and γ_{i+1} is the maximal value of the index of the elements of this class. In the construction of the complex β the above interpretation of the value of the elements of the complex γ is used.

oswep (61)/αк, βк??, γк??, δи +/**a**, b/($\alpha\beta$)
215 132 3

$\delta + 2 \Rightarrow b_\gamma ?? b_\alpha \Rightarrow b_\beta ??$ *cleanup* $\gamma // \circ \ a$
§1 $\alpha_a \nabla \Rightarrow b \triangle \gamma_b \triangle a \oplus b_\alpha \mid\!\rightarrow 1 \ \bar{\circ} \ \mathbf{a} \circ a$
§2 $\mathbf{a} + \gamma_a \Rightarrow \gamma_a \Rightarrow \mathbf{a} \triangle a \oplus b_\gamma \mid\!\rightarrow 2 \circ a$
§3 $\alpha_a \nabla \Rightarrow b\gamma_b \Rightarrow c\alpha_a \Rightarrow \beta_c \bar{\triangle} \gamma_b \triangle a \oplus b_\beta \mid\!\rightarrow 3.$

desec. Delimit a set of elements of a complex.

Those elements of the complex α whose bitwise conjunction with the variable β is equal to 0 are delimited. The delimited elements form a complex γ.

desec αк +, βп +, γк?, δв/−, $b/\gamma\alpha(\alpha\beta\gamma)$
216 50 2

$\circ\ ab_\gamma \Rightarrow b$
§1 $\alpha_a \wedge \beta \mid\!\rightarrow 2\alpha_a \Rightarrow \gamma_b\delta \triangle b$
§2 $\triangle\ a \oplus b_\alpha \mid\!\rightarrow 1b \Rightarrow b_\gamma.$

cosmif. Construct the set of maximal intervals of a Boolean function.

The initial information for the L-program is the set M_1, which is represented by the complex α, and namely, $\alpha{:}{:}\|\ M_1 \ni X\ \|$; the zero component of the elements of the complex α in the representation of the set M_1 is not used, so the set X is given in the following form:

$$X = \{x_1, x_2, \ldots, x_n\}.$$

In the operation of the program the complex α is ordered and partitioned into classes of equal weights of the elements. The ordered complex is represented by the complex β, and the boundaries of the corresponding classes of the partition are represented by the complex γ.

The possible adjunctions of the elements of adjacent classes of the complex β, i.e., those classes whose weights of elements differ by unity, are carried out. The intervals obtained as a result of adjunction in the set M_1 are represented by the two complexes ζ and ϵ. The elements of the complex ζ are minimal elements of the intervals and the elements of the complex ϵ represent sets of their internal variables. Elements of the complex β are considered as minimal elements of the intervals of nth rank, and are associated with elements of the complex α, representing the internal variables of the intervals (with all elements of the complex α equal to 0). We formulate the conditions for adjunction of intervals with respect to the given representation: (1) the sets of their internal variables must coincide, and (2) the minimal elements of the intervals must be adjacent.

Adjunctions of intervals are tagged by the unit value in the zero components of the corresponding elements of the complex β. After all possible adjunctions of intervals represented by the complexes α and β have been realized, the untagged elements of β are found, and the corresponding intervals are written into the complexes η and ν, representing their maximal and minimal elements. If the complexes ζ and ϵ are empty, the process terminates; otherwise, the complexes ζ and ϵ are renamed complexes β and α, respectively, and the adjunction process is renewed. The value of the index κ represents the number of essential variables of the Boolean function.

cosmif (215, 61, 61, 216)/αк, βк??, γк??, δк??, ϵк?, ζк?, ηк?, ϑк?, кИ $+/\mathbf{c}, f/(\alpha\beta\epsilon\zeta\eta\vartheta), (\gamma\delta)$

214 340 6 (α**ab**)

oswep $\alpha\beta\gamma\kappa//c_0 - c_\kappa \Rightarrow \mathbf{b}$ *cleanup* $\alpha//b_\gamma \Rightarrow b_\delta?? \circ b_\vartheta \Rightarrow b_\eta$

§1 $\circ\ e\ \bar{\circ}\ \mathbf{c} \circ f$ *cleanup* $\delta// \circ a$

§2 $\gamma_a \Rightarrow b\mathbf{c} \Rightarrow \delta_a \triangle a \oplus b_\gamma \circ\!\rightarrow 6\mathbf{c} \Rightarrow \delta_a\gamma_a \Rightarrow c \oplus b \circ\!\rightarrow 2 \triangle a \oplus b_\gamma$ $\circ\!\rightarrow 6\gamma_a \oplus c \circ\!\rightarrow 2$

§3 $\triangle bc \Rightarrow d$

§4 $\triangle d\alpha_b \oplus \alpha_d \mid\!\rightarrow 5\beta_b \oplus \beta_d \wedge \mathbf{b} \Rightarrow \mathbf{a} \mid\!\leftarrow \oplus c_0 \mid\!\rightarrow 5\alpha_b \vee \mathbf{a} \Rightarrow \epsilon_f?\mathbf{a} \urcorner \wedge \beta_b$ $\wedge \mathbf{b} \Rightarrow \zeta_f? \triangle f \triangle \mathbf{c}c_0 \vee \beta_b \Rightarrow \beta_b c_0 \vee \beta_d \Rightarrow \beta_d$

§5 $d \oplus \gamma_a \mid\!\rightarrow 4b \oplus c \mid\!\rightarrow 3\ \bar{\triangle}\ a \rightarrow 2$

§6 $f \Rightarrow b_\epsilon \Rightarrow b_\zeta$ *desec* $\beta c_0\vartheta(\alpha_a \vee \beta_a \Rightarrow \eta_b?)//a_\epsilon \Leftrightarrow a_\alpha a_\zeta \Leftrightarrow a_\beta b_\epsilon \Leftrightarrow b_\alpha b_\zeta \Leftrightarrow b_\beta b_\epsilon$ $\mid\!\rightarrow 1b_\vartheta \Rightarrow b_\eta$.

6. ABOUT THE METAPROGRAM

In the theoretical consideration of the problems of minimization, all the algorithms have been classified basically according to the mathematical apparatus applied to the solution (algebraic, graphical, matricial, etc.). From the viewpoint of their computer execution it is necessary to classify the algorithms according to the circle of problems solved by means of these algorithms. We propose the following criteria of classification:

(1) The maximal dimensionality of the Boolean space, expressed in terms of the number of binary variables n on which the Boolean function depends.

(2) The maximal cardinalities of the sets representing the initial prescription of the Boolean function.

(3) The precision of the solution obtained.

(4) The form in which the initial information is presented.

On the basis of these criteria we have considered and classified certain minimization methods and written the corresponding programs. A part of them is represented in this chapter. The system thereby composed contains programs that perform the following algorithms.

(1) Algorithms for the minimization of Boolean functions represented in matricial form. The corresponding program [2] guarantees the construction of an irredundant form fairly close to minimal. The program handles problems in the minimization of Boolean functions with not more than 11 variables.

In the other programs of the system the number of variables can be fairly great (up to 32 and more), but other constraints are imposed.

(2) In Section 4 of the present chapter a program is considered that performs the Quine–McCluskey algorithm, whose essential constraint is the cardinality of the Boolean space in which the function takes on the unit value.

(3) In [6 and 7] are considered algorithms and the corresponding programs for the minimization of weakly-defined Boolean functions, i.e., incompletely defined functions with constrained volume of the Boolean spaces on which the function takes on unit and zero values.

(4) Finally, in Sections 2 and 3 of the present chapter and in [8] are considered algorithms and programs for minimization of Boolean functions, prescribed in interval form, with constraints on the cardinality of the initial set of intervals.

In the description of each program the form in which the initial information is supplied and the precision of the solution are discussed.

As is evident from the characteristics of the above programs, each of them is suitable for the solution of only a limited circle of problems, but at the same time the system as a whole permits a fairly broad range of problems to be solved, to a given degree of precision. The composition of such a system of subroutines is the first approximation of the results obtained to their practical utilization. The next stage is the construction, on the basis of the system of programs, of a metaprogram for the minimization of Boolean functions that uses certain criteria to associate with each problem the sequences of programs for its solution, taken in the existing system. The basic problem here reduces to the choice of a suitable system of algorithms and the development of the corresponding criteria of choice.

The present programs can serve as the basis for such a system. Furthermore, it is possible to develop new programs, using various combinations of operators of the L-programs described. For example, by completing the operator for construction of the reduced dnf described in Section 5 by various operators for construction of a coverage, it is possible to obtain various approximate algorithms. Or, in the operator *mincodefin*, by replacing the operator for finding the next interval *finin* by any other operator with the same functional content (for example, an operator for random selection of the next interval), it is possible to obtain another program with a different degree of approximation and speed. It is obviously desirable to complete the system with various statistical algorithms.

To construct an algorithm for associating definite minimization programs with concrete problems, it is possible to use as the first approximation the criteria proposed above for the classification of the algorithms, completing them with the characteristics of the programs. In particular, work is necessary on the estimation of speed and the possibilities of the programs.

In this sense the representation of the algorithms in the form of programs can give an objective criterion for comparing the effectiveness of the algorithms and for determining the class of problems for which each algorithm can be applied effectively. The literature basically presents *bounds* on the effectiveness of algorithms. For practical use it is more necessary to have statistical, *average* estimates, which can be obtained partly with the utilization of the theoretical results [9] obtained, partly by accumulation of statistical results during the application of the algorithms to the solution of concrete problems.

REFERENCES

1. Stognii, A. A., On the Minimization of Boolean Functions on a Computer, *in* "Problems of Computer Engineering," pp. 65–70 [in Russian]. Kiev, 1961.
2. Novoselov, V. G., Approximate Method for the Minimization of Boolean Functions Given in Matrix Form, *Tr. SFTI* **48** (1965).
3. Quine, W. V., The Problem of Simplifying of Truth Functions, *Amer. Math. Monthly* **59**, No. 8, 521–531 (1952).
4. McCluskey, E. J., Minimization of Boolean Functions, *B. S. T. J.* **35**, No. 6, 1417–1444 (1956).
5. Zakrevskii, A. D., Optimization of Coverage of Sets, *Tr. SFTI* **48** (1965).
6. Novoselov, V. G., Some Algorithms for the Minimization of Weakly Defined Boolean Functions, *Tr. SFTI* **48** (1965).
7. Zakrevskii, A. D., Algorithm for the Approximate Minimization of Boolean Functions, *Tr. SFTI* **48** (1965).
8. Toropov, N. R., Algorithm for the Simplification of the dnf of a Boolean Function, *Tr. SFTI* **48** (1965).
9. Mileto, F., and Putzolu, G., Average Values of Quantities Appearing in Boolean Functions Minimization, *IEEE Trans., Electron. Computers* **13**, 87–92 (1964).

THE DECOMPOSITION PROBLEM FOR BOOLEAN FUNCTIONS

I. L. Fadeev

The branch of switching-circuit theory concerned with the synthesis problem is of major practical interest. In recent years interest in synthesis methods based on the property of decomposability of Boolean functions has grown. There is a basis for supposing that the most general methods of synthesis can be constructed on the basis of the decomposability of Boolean functions.

The present chapter is connected with the decomposition of completely defined Boolean functions and includes the construction of several decomposition algorithms, as well as investigations connected with their construction.

1. ALGORITHM FOR DECOMPOSITION OF A BOOLEAN FUNCTION ACCORDING TO A PRESCRIBED PARTITION

A.

We denote by $f(X)$ a completely defined Boolean function, where $X = \{x_0, x_1, \ldots, x_{n-1}\}$ and by A/B the partition of the set X into three nonoverlapping subsets A, B, C. We shall be concerned only with those partitions for which

$$C = X \backslash (A \cup B), \qquad \sigma(A) > 0, \quad \sigma(B) > 1.$$

The function $f(X)$ is called decomposable by the partition A/B if and only if there exists a pair of Boolean functions $\varphi(A, C, \psi)$ and $\psi(C, B)$ such that

$$f(X) = \varphi(A, C, \psi(C, B)).$$

Whether $f(X)$ is decomposable by the given partition A/B must be determined.

The problem has been stated in this form in [1], and an algorithm for its solution presented. Here the algorithm described in [1] is given by the LYaPAS operator *decodeboom*. In the operator *decodeboom* the operator *dissym* is used and is of independent interest.

B.

The operator *dissym* for the disjunctive symmetrization of a Boolean function with respect to Y, where $Y \subseteq X$, maps the initial function $\eta(X)$ onto a function $\psi(X)$ by the following rules:

$$\eta_i(Y) \neq 0 \leftrightarrow \psi_i(Y) = 1,$$
$$\eta_i(Y) = 0 \leftrightarrow \psi_i(Y) = 0.$$

Here $\eta_i(Y)$ and $\psi_i(Y)$ are the coefficients of expansion with respect to $X \backslash Y$ of the functions $\eta(X)$ and $\psi(Y)$, respectively.

The operator *dissym* in LYaPAS corresponds to the operator $S_Y(\vee)$ described in [1]. It has the form:

dissym αп $+$, βп $+$, γп/**a**, $a/(\beta\gamma)$
76 33 2

$\beta \Rightarrow \gamma\alpha \Rightarrow$ **a**
§1 **a** $\dot{\mathbf{X}}$ $2a\gamma \veebar a \vee \gamma \Rightarrow \gamma \rightarrow 1$
§2 .

$$\alpha::\|\ \{Y\} \ni X\ \|,$$
$$\beta::\|\ \{\tilde{X}_1{}^0\} \ni \tilde{X}\ \|,$$
$$\gamma::\|\ \{\tilde{X}_1{}^1\} \ni \tilde{X}\ \|.$$

Here $\tilde{X}_1{}^0$ and $\tilde{X}_1{}^1$ are the sets of unit vertices of the n-dimensional cube, defining the initial and resultant functions, respectively.

C.

The operator *decodeboom* tests the decomposability of a completely defined Boolean function in matrix form according to a prescribed partition. The LYaPAS program has the following form:

decodeboom (76)/αч, βп +, γп +, δп + /**d**, *a*/(βγ)
77 111 2 (δ**bcd**)

$\gamma \Rightarrow \mathbf{a}\ \bar{\circ}\ \mathbf{b}$
§1 $\mathbf{a}\ \dot{\mathbf{X}}\ 2a\mathbf{b} \wedge e_a \Rightarrow \mathbf{b} \rightarrow 1$
§2 $\mathbf{b} \wedge \delta \Rightarrow \mathbf{b}$ *dissym* $\gamma\mathbf{bc}//\mathbf{c} \oplus \delta \Rightarrow \mathbf{b} \Rightarrow \mathbf{c}\gamma \Rightarrow \mathbf{a} \rightarrow$ *dissym* $1\mathbf{c} \Rightarrow \mathbf{db} \Rightarrow \mathbf{c}\beta$
$\Rightarrow \mathbf{a} \rightarrow$ *dissym* $1\mathbf{c} \oplus \mathbf{b} \wedge \mathbf{d} \circ \rightarrow \alpha.$

Thc described operator has one auxiliary exit α. This exit is realized if and only if the tested function is decomposable in A/B.

The possibility of priming the variables permits functions to be investigated for which $5 \leq \sigma(X) \leq 12$. If $\sigma(X) < 5$ it is possible to add dummy variables to make the total number of variables equal to 5.

The same dimensional considerations apply to the operator *dissym*.

We refer the reader to [1] for fuller detail.

2. STRUCTURAL RELATIONS IN THE SET OF SOLUTIONS OF THE DECOMPOSITION PROBLEM

A.

We introduce the following concepts:

M is the set of all partitions satisfying the definition of partition given in the last paragraph;

M^n is the set of all such partitions in M for which $f(X)$ is not decomposable.

M^d is the set of all partitions in M for which $f(X)$ is decomposable.

It is obvious that $M = M^n \cup M^d$ and therefore, disposing of M^n, it is fairly easy to establish for any $A/B \in M$ whether $f(X)$ is decomposable with respect to A/B or not. We shall call M^n the set of solutions. A trivial enumeration of all partitions in M with test of $f(X)$ for decomposability for each of these partitions offers the possibility of obtaining M^n, but the enumeration of partitions is so extensive that even for functions of a relatively small number of variables it becomes practically impossible.

An important structural relation in M^n was demonstrated in [1].

Theorem 1.

$$A/B \in M^n \rightarrow A'/B' \in M^n \qquad \text{if} \quad A' \supseteq A,\ B' \supseteq B.$$

From Theorem 1 it follows that M^n is convex with respect to the relation given among its elements by the auxiliary condition of the theorem. In this case there exists a lower bound Γ of the convex set M^n that uniquely defines M^n. The relation $\Gamma \subseteq M^n$ holds, and the following relations, connected with the convexity of M^n, also hold.

Let $A/B \in \Gamma$ and A'/B', $A''/B'' \in M \backslash \{A/B\}$, where $A' \subseteq A \subseteq A''$, $B' \subseteq B \subseteq B''$. Then $A'/B' \in M^d$, $A''/B'' \in M^n$.

We shall consider the problem of finding Γ.

The next section is devoted to an algorithm for finding Γ, and here we establish certain structural relations in M^n that aid in reducing the enumeration of partitions in M in the search for Γ.

B.

We introduce the basic concepts. For this we use the set-theoretical model of representation of a Boolean space and related concepts [2]. In this model:

X is a reference set;

$\tilde{X}$ is a Boolean space over X or an n-dimensional Boolean cube, where $n = \sigma(X)$;

$\alpha \in \tilde{X}$ is a vertex of the cube $\tilde{X}$;

$(\alpha_{\min}, \alpha_{\max})$ is the representation of an interval of the cube X by its minimal and maximal elements, i.e., those elements for which $\alpha_{\min} \subseteq \alpha \subseteq \alpha_{\max}$ holds, where α is an arbitrary vertex of the interval concerned.

We call the interval set S_a of the interval $a = (\alpha_{\min}, \alpha_{\max})$ a subset of X which is defined by the expression

$$S_a = \alpha_{\max} \backslash \alpha_{\min}.$$

By V_Y we denote a set of intervals having the internal set $X \backslash Y$, where $Y \subseteq X$.

We define the distance R_{ab} between intervals $a = (\alpha_{\min}, \alpha_{\max})$ and $b = (\beta_{\min}, \beta_{\max})$, where $a, b \in V_Y$, by means of the expression $R_{ab} = \alpha_{\min} + \beta_{\min}$. The operation $+$ between two sets c and d has the sense

$$c + d = (c \cup d) \backslash (c \cap d).$$

The set V_Y may be considered to be a Boolean space over Y. In this case we shall consider any interval in V_Y as an element of a new space $\tilde{Y}$ and thereby we abstract from the internal structure of the intervals in V_Y. In this we shall concentrate our attention on various relations among the elements of V_Y; for example, the relations describable by means of the concept of distance between intervals in V_Y. The following lemmas contain propositions about this kind of relation among the intervals.

Lemma 1.

$$a, b \in V_Y \rightarrow R_{ab} \subseteq Y.$$

Lemma 2. If

$$a, b \in V_{X \setminus U} \quad \text{and} \quad a, b \subseteq p \in V_{X \setminus Y},$$

then

$$R_{ab} \subseteq Y \setminus U.$$

Lemma 3. If $a \in V_Y$ and $U \subseteq Y$, there exists in V_Y a single interval b for which $R_{ab} = U$.

Lemma 4. Let $a, b, c \in V_Y$ and let p be an interval of the cube that absorbs a and b. Then the following relation holds:

$$R_{ab} \supseteq R_{bc} \rightarrow c \subseteq p.$$

We introduce the concept of R-absorption of a pair of intervals by a second pair of intervals.

Let $a, b \in V_Y$ and $c, d \in V_U$, where $U \subseteq Y$, and the relations $a \subseteq c$, $b \subseteq d$ also hold. Then the pair c, d R-absorbs the pair a, b if and only if $R_{ab} = R_{cd}$.

Lemma 5. The necessary and sufficient condition for R-absorption of the pair, a, b in V_Y by the pair c, d in V_U, where $U \subseteq Y$, is the satisfaction of the conditions $a \subseteq c$, $R_{ab} = R_{cd}$.

Lemma 6. Let $a, b, c \subseteq V_Y$ and $a', b', c' \subseteq V_U$, where $U \subseteq Y$. If a', b' R-absorbs the pair a, b and b', c' R-absorbs the pair b, c, then a', c' R-absorbs the pair a, c.

C.

In Section 1.B we described the representation of $f(X)$ by a partition of the set of vertices of the $\sigma(X)$-dimensional cube into classes $\tilde{X}_1$ and $\tilde{X}_0$. Further we shall consider $f(X)$ to be given on the cube by means of a coloring of its vertices by the colors 0 and 1. A vertex of the cube will have color 1 if and only if the function takes on the value 1 at this vertex.

We call an interval with colored vertices a K-interval, where this coloring is carried out by the prescription of $f(X)$ on the vertices of the n-dimensional cube. The K-interval may be represented as the combination of the concepts of the expansion coefficient of the function $f(X)$ and the interval on the cube in which this coefficient is given. If we are concerned with a concrete coefficient α and the interval in which it is defined, we shall speak of a K_α-interval.

Since the difference between the concepts of interval and K-interval reduces simply to the coloring of the vertices of the latter, all concepts connected with an individual interval and concepts characterizing the relations among several intervals hold also for K-intervals.

We denote by K_Y decomposition of the function $f(X)$, where $Y \subseteq X$, the set Z_Y of all K-intervals on the n-dimensional cube having the internal set $X \backslash Y$.

We also denote by Z_Y the set of all K_α-intervals on the n-dimensional cube having the internal set $X \backslash Y$. The difference between the K-interval and the K_α-interval corresponding to it is of a purely formal order: in the case of the K_α-interval there is a direct indication concerning the coefficient that is represented by it; in the case of the K-interval this indication is absent. We shall therefore identify the K_α-interval and the corresponding K-interval. When necessary we shall introduce an index indicating the coefficient or, on the contrary, omit it if this does not cause ambiguity. For example, we shall denote the distance between the K_α-intervals a_α and b_β, where $a_\alpha, b_\beta \in Z_Y$, by R_{ab}.

If $Y = X \backslash \{x_i\}$, where $x_i \in X$, we shall denote Z_Y by Z_i.

We shall call a K-interval with empty internal set an element.

Lemma 7. Let $a_\alpha, b_\beta \in Z_Y$. The necessary and sufficient condition for inequality of α and β is the existence of K-intervals c_γ and d_δ which are R-absorbed by the pair c, d and for which $\gamma \neq \delta$.

D.

Let us investigate the form of the function $f(X)$ for certain assumptions about its decomposability. For proof we require the following two criteria.

Criterion 1. The function $f(X)$ is decomposable with respect to the partition A/B if and only if the following holds for all i, j:

$$f_{ij}(B) \in \{0, 1, \kappa_j(B), \bar{\kappa}_j(B)\}.$$

Here $f_j(A, B)$ is the expansion coefficient of the function $f(X)$ with respect to C, $f_{ij}(B)$ is the expansion coefficient of $f_j(A, B)$ with respect to A, $\kappa_j(B)$ is an arbitrary function and the limits of variation of i and j are defined by the conditions $0 \leq i \leq 2^{\sigma(A)}$, $0 \leq j < 2^{\sigma(C)}$.

Criterion 2. The function $f(X)$ is decomposable with respect to the partition A/B if and only if the following holds for all i, j:

$$f_{ij}(A) \in \{\varphi_i(A), \psi_i(A)\},$$

where $f_i(A, B)$ is the expansion coefficient of $f(X)$ with respect to C, $f_{ij}(A)$ is the expansion coefficient of $f_i(A, B)$ with respect to B, $\varphi_j(A)$ and $\psi_j(A)$ are arbitrary Boolean functions of A, and the values of i and j are defined by the conditions $0 \leq i < 2^{\sigma(C)}$, $0 \leq j < 2^{\sigma(B)}$.

Criterion 1 was formulated in [2], Criterion 2 in [1]. In terms of our new concepts, these two criteria appear as follows.

Criterion 1a. $f(X)$ is decomposable with respect to A/B if and only if there does not exist a pair $a_\alpha, b_\beta \in Z_{X \setminus B}$ such that $R_{ab} \subseteq A$, $\alpha \neq$ const, $\beta \neq$ const, $\alpha \neq \beta$, $\alpha \neq \bar{\beta}$.

Criterion 2a. $f(X)$ is decomposable with respect to A/B if and only if there does not exist in $Z_{B \cup C}$ a triplet $a_\alpha, b_\beta, c_\gamma$ such that $\alpha \neq \beta$, $\alpha \neq \gamma$, $\beta \neq \gamma$, $R_{ab} \cup R_{ac} \subseteq B$.

For the proof of Criteria 1a and 2a we utilize Lemmas 1, 2, and 4.

Lemma 8. Let $A \cup B = X$. If $A/B \in M^n$ and for every partition A'/B in M with respect to which $A' \subset A$, the relation $A'/B \in M^d$ holds, then there exist $a_\alpha, b_\beta \in Z_A$ for which $R_{ab} = A$, $\alpha \neq$ const, $\beta \neq$ const, $\alpha \neq \beta$, $\alpha \neq \bar{\beta}$. For every $c_\gamma \in Z_A \setminus \{a, b\}$ the relation $\gamma =$ const holds.

Proof: On the basis of Criterion 1a and the conditions $A/B \in M^n$ it follows that in Z_A there exists a pair a_α, b_β for which $R_{ab} \subseteq A$, $\alpha \neq$ const, $\beta \neq$ const, $\alpha \neq \beta$, $\alpha \neq \bar{\beta}$. On the basis of this criterion and the condition $A'/B \in M^d$ it follows that in $Z_{A'}$ there exists no pair c_γ, d_δ for which

$$\gamma \neq \text{const}, \ \delta \neq \text{const}, \ \gamma \neq \delta, \ \gamma \neq \bar{\delta}, \ R_{cd} \subseteq A' \subset A,$$

thus $R_{ab} = A$.

We shall assume that in $Z_A \backslash \{a, b\}$ there exists a c_γ for which $\gamma \neq$ const. It is easily seen that at least one of the following conditions holds:

$$1.\ \gamma \neq \alpha, \gamma \neq \bar{\alpha};$$
$$2.\ \gamma \neq \beta, \gamma \neq \bar{\beta}.$$

Let, for example, Criterion 1 be satisfied. Then in Z_A there exists a pair a_α, c_γ for which $\alpha \neq$ constant, $\gamma \neq$ constant, $\gamma \neq \alpha$, $\gamma \neq \bar{\alpha}$ and for which $R_{ac} \neq A$ on the basis of Lemma 3, $R_{ac} \subseteq A$ according to Lemma 1 which implies $R_{ac} \subset A$. According to Criterion 1*a*, $f(X)$ is not decomposable with respect to the partition R_{ac}/B, which contradicts the hypothesis of the lemma to be proved. Consequently, there does not exist a $c_\gamma \in Z_A \backslash \{a, b\}$ for which $\gamma \neq$ constant. This completes the proof of our lemma.

Lemma 9. Let $A \cup B = X$, $\sigma(B) > 2$. If $A/B \in M^n$ and for every partition A/B' in M with respect to which $B' \subset B$ the relation $A/B' \in M^d$ holds, then there exist a_α, b_β in Z_B for which $\alpha \neq \beta$ and $R_{ab} = B$. For any c_γ, $d_\delta \in Z_B \backslash \{a, b\}$ the relations $\gamma = \delta$, $\gamma \neq \alpha$, and $\gamma \neq \beta$ hold.

PROOF: We shall consider Z_B to be a Boolean space over B, and the relations among the coefficients given on the K-intervals in Z_B of interest will be represented on the cube B by means of a special type of coloring: two vertices of the cube will have the same color if and only if the K-intervals in Z_B corresponding to them have identical coefficients. According to Criterion 2*a* and the conditions of the lemma to be proved, the coloring of the cube $\tilde{B}$ should satisfy the requirements that the total number of colors used for coloring the entire cube is greater than two, but no proper subcube contains vertices with more than two colors.

We denote by $\lambda_{(\alpha)}$ the color of the vertex α, where α is an arbitrary vertex of the cube $\tilde{B}$. In the cube $\tilde{B}$ whose coloring satisfies the above conditions there exists at least one triplet of vertices a, b, c for which $\lambda(a) \neq \lambda(b)$, $\lambda(a) \neq \lambda(c)$, $\lambda(b) \neq \lambda(c)$.

Let us consider the vertex $d = (a \cap b) \cup (a \cap c) \cup (b \cap c)$. The minimal interval L_{ab} covering the vertices a, b has the internal set $S_{ab} = R_{ab} = a + b$. It is easily shown that $a + d \subseteq a + b$, which on the basis of Lemma 4 implies $d \in L_{ab}$. Similarly we prove that $d \in L_{bc}$, $d \in L_{ac}$, where L_{bc}, L_{ac} are minimal intervals of the cube $\tilde{B}$ covering b, c and a, c, respectively. We assume that $S_{ab}, S_{ac}, S_{bc} \subset B$, i.e., each of L_{ab}, L_{ac}, L_{bc} is a proper subcube of $\tilde{B}$. In this case there does not exist a color for d that would satisfy the conditions for coloring the cube. This requires that the assumption be abandoned and that one of the distances R_{ab}, R_{ac}, R_{bc} be put equal to B. To be concrete, let $R_{ab} = B$. Then on the basis of Lemma 3 it follows that $R_{ac}, R_{bc} \subset B$.

Let us consider a certain vertex $e \in \tilde{B}\backslash\{a, b, c\}$. We assume that $\lambda(e) \neq \lambda(c)$ and $\lambda(e) \neq \lambda(a)$. Then for the triplet of vertices a, c, e the relations $\lambda(a) \neq \lambda(c)$, $\lambda(c) \neq \lambda(e)$, $\lambda(a) \neq \lambda(e)$ hold. As has been established in the preceding part of the proof, for such a triplet of vertices the distance between one of the pairs of vertices must be equal to B. This condition can be satisfied if and only if $R_{ce} = B$.

Let us take a certain vertex $f \in \tilde{B}\backslash\{a, b, c, e\}$. If $\lambda(f) \in \{\lambda(c), \lambda(a)\}$, then for the triplet of vertices e, f, b the conditions $\lambda(e) \neq \lambda(f)$, $\lambda(e) \neq \lambda(b)$, $\lambda(f) \neq \lambda(b)$, and R_{ef}, R_{eb}, $R_{fb} \subset B$ hold; but such a situation, as we have seen, leads to a contradiction. If $\lambda(f) \in \{\lambda(c), \lambda(a)\}$, then an analogous situation can arise for the triplet a, c, f. The impossibility of coloring f requires us to abandon the assumption that $\lambda(e) \neq \lambda(a)$ and $\lambda(e) \neq \lambda(c)$ and to assume that $\lambda(e) \neq \lambda(c)$ and $\lambda(e) \neq \lambda(b)$. But this assumption cannot give rise to the existence of new possibilities for coloring f, because of the symmetry of the coloring conditions with respect to the colors $\lambda(a)$, $\lambda(b)$ of vertex e. The assumption that $\lambda(e) \neq \lambda(c)$ is abandoned. The coloring of all vertices in $\tilde{B}\backslash\{a, b, c\}$ by the color $\lambda(c)$ does not contradict the coloring conditions. The interpretation of the results of coloring $\tilde{B}$ for the initial model Z_B comprises the proposition stated by the lemma.

Let us introduce the following concept.

We give the name "U-type function with respect to Y" to a function $f(X)$, for which Z_i contains a single pair of K-intervals a_α and b_β for which α and β are not constants and $R_{ab} = X\backslash\{x_i\}$, where $x_i \in Y \subset X$, $\sigma(X) > 3$. Among the remaining K-intervals in Z_i there may exist only one K-interval c_γ such that $\gamma \neq$ constant.

Lemma 10. If $A/B \in \Gamma$, $A \cup B = X$, $\sigma(B) > 2$, then $f(X)$ is a U type function with respect to B.

PROOF: It is easily seen that the hypothesis of Lemma 10 absorbs the hypotheses of Lemmas 8 and 9, and therefore $f(X)$ belongs to the types described in the propositions of Lemmas 8 and 9.

We denote by a_α and b_β that unique pair in Z_A for which α and β are not constants, and by c_γ and d_δ the unique pair in Z_B for which $\gamma \neq \delta$ and $R_{cd} = B$. We find all those K-intervals in Z_i that have nonconstant coefficients. Here we have in mind $x_i \in B$. Each element of Z_i is absorbed by a certain element of Z_A. It is obvious that each $e_\epsilon \in Z_i$ for which $\epsilon \neq$ const can be absorbed by only such $f_\zeta \in Z_A$ for which $\zeta \neq$ const. Thus it is necessary to investigate only the K-intervals in Z_i absorbed by the K-intervals a and b.

On the basis of Lemmas 9 and 7, $\epsilon =$ constant for $e_\epsilon \in Z_i$ if e is a K-interval that has an empty intersection with c or d.

Thus nonconstant coefficients may be possessed only by $p, q, r, s \in Z_i$ such that $p, q \subset a; r, s \subset b; p \cap c \neq \emptyset; q \cap d \neq \emptyset; r \cap c \neq \emptyset; s \cap d \neq \emptyset$.

The number of elements in the set $K = \{p, q, r, s\}$ having nonconstant coefficients must be greater than one, since otherwise there arises a contradiction with Lemma 9. The case in which all elements in K have nonconstant coefficients is also impossible, since in this case the coefficients of the K-intervals a, b are equal or the inverses of each other, which contradicts Lemma 8. Of the remaining variants we must exclude the case in which exactly two elements of K, whose distance is A or $B \backslash \{x_i\}$, have nonconstant coefficients, since in these cases there arise contradictions with Lemma 8 or 9.

The other cases do not contradict Lemmas 8 or 9. They are presented in the formulation of the present lemma.

E.

On the basis of this analysis of the form of the function we shall interest ourselves in the structural relations in the set of solutions, but we shall first formulate the following lemma.

Lemma 11. Let $A \cup B = X$, $x_i \in B$. If there exist $a_\alpha, b_\beta, c_\gamma \in Z_i$, such that $a, b \subset p \in Z_A$, $R_{ac} \subseteq A$, $\alpha =$ constant, $\beta \neq$ constant, $\gamma \neq$ constant, then $A/B \in M^n$.

PROOF: The union of K-intervals m and n in Z_B which absorbs a also absorbs c, as is easily shown with the use of Lemma 4.

Let t and s be two other K-intervals in Z_B whose union absorbs b. Then m, n, t, s in their ensemble absorb a, b, c. Let us consider the set T of elements entering into the union of a and b, and the set $L = \{m, n, t, s\}$. These sets can be put into a one-to-one relation with each other by means of the relation of absorption of elements of T by elements of L. Any pair of K-intervals in L R-absorbs the pair of elements corresponding to it in T. From $\alpha =$ constant and $\beta \neq$ constant it follows on the basis of Lemma 7 that the coefficient t of the K-interval is not equal to any of the coefficients m, n, s of the K-intervals or the coefficient s of the K-interval is not equal to any of the coefficients m, n, t of the K-intervals.

The elements entering into c are R-absorbed by the pair of K-intervals m, n. From the condition $\gamma \neq$ constant and Lemma 7, it follows that the coefficients of the K-intervals m and n are not equal to each other. But this means that the number of different coefficients among the coefficients m, n, s or m, n, t of the K-intervals is not less than 2. Consequently, the number of distinct coefficients m, n, t, s of K-intervals is greater than 2, which implies nondecomposability of the function with respect to the partition A/B on the basis of Criterion 2a.

Theorem 2. If $A/B \in \Gamma$, then $A'/B' \in M^n$, where

$$A \cup B = A' \cup B' = X, \sigma(B) > 2.$$

PROOF: (1) Let us consider partitions A'/B' for which $B' \cap B \neq \emptyset$. Let $x_i \in B' \cap B$. On the basis of Lemma 10 the function $f(X)$ is a U-type function with respect to B; i.e., in Z_i there exists a pair of K-intervals a_α and b_β for which $R_{ab} = X\backslash\{x_i\}$, $\alpha \neq$ constant, $\beta \neq$ constant, and among the other elements in Z_i there may be only a single K-interval having nonconstant coefficient. Let us consider $c, d \in Z_i$ such that $R_{bc} = R_{ad} = A'$. On the basis of Lemma 5, b and c are R-absorbed by the K-intervals $a', b' \in Z_{A'}$ for which $a' \supset a$, $b' \supset b$, and $R_{a'b'} = A'$. From this it follows that $c \subset a'$. Analogous considerations for d show that $d \subset b'$. From the fact that $f(X)$ is a U-type function with respect to B it follows that the coefficient of at least one of $\{c, d\}$ is equal to a constant. Then at least one of the triplets $\{a, b, c\}$, $\{a, b, d\}$ satisfies the conditions of Lemma 11, which implies $A'/B' \in M^n$.

(2) The case $A'/B' = B/A$. Here it is necessary that $\sigma(A) > 1$. On the basis of Lemma 8 and Criterion 2a we conclude that $A'/B' \in M^n$.

(3) The case A'/B' when $B' \subset A$. According to Lemma 8, Z_A contains a pair a_α, b_β for which $R_{ab} = A$, $\alpha \neq \gamma$, $\beta \neq \gamma$, where $c_\gamma \in Z_A\backslash\{a, b\}$.

Let us consider the set P, defined by the expression

$$c \in P \leftrightarrow (c \in Z_A) \wedge (R_{ca} \subseteq B').$$

We have $a \in P$, $b \notin P$, which follows from the definition of the set P, the condition $B' \subset A$, and Lemma 4. Consequently, for a_α and any $c_\gamma \in P\backslash\{a\}$ we have $\gamma \neq \alpha$.

We denote by a' the K-interval in $Z_{B'}$ for which $a' \supset a$. Let

$$c'_{\gamma'} \in Z_{B'}\backslash\{a'_{\alpha'}\}.$$

In P there exists a K-interval c such that $R_{ac} = R_{a'c'}$. On the basis of Lemma 6, the pair a', c' R-absorbs the pair a, c. Since $a' \neq c'$, $R_{a'c'} = R_{ac} \neq \emptyset$ and, consequently, $c \in P\backslash\{a_\alpha\}$. On the basis of Lemma 7 and $\alpha \neq \gamma$ it follows that $\alpha' \neq \gamma'$.

Thus in $Z_{B'}\backslash\{a'_{\alpha'}\}$ there does not exist a single K-interval $c'_{\gamma'}$ for which $\alpha' = \gamma'$. Similarly $\beta' \neq \delta'$, where $\beta'_{\beta'} \in Z_{B'}$, $b' \supset b$, and d' is any K-interval in $Z_{B'}\backslash\{b'\}$. But from the two last propositions it follows that there exists in $Z_{B'}$ a triplet of K-intervals whose coefficients are pairwise unequal, and this, on the basis of Criterion 2a, implies the nondecomposability of $f(X)$ with respect to A'/B'.

Theorem 3. If $A/B \in \Gamma$, then $A'/B' \in M^n$, where

$$A' \cup B' = A \cup B, \sigma(B) > 2.$$

PROOF: On the basis of the definition of decomposability of $f(X)$ with respect to A/B, any K-interval in Z_C, where $C = X\backslash(A \cup B)$ is a certain function of $A \cup B$ that either is decomposable with respect to A/B or has the conditions of Theorem 2 fulfilled for it. There exists in Z_C at least one K-interval representing a function of $A \cup B$ for which the conditions of Theorem 2 are satisfied. But then the truth of the present theorem follows by the definition of the decomposability of $f(X)$ with respect to some partition and on the basis of Theorem 2.

Theorem 4. If $A/B \in M^n$ and for any partition A/B for which $A' \subset A$ we have $A'/B \in M^d$, then every partition A''/B'' in M for which $B'' \supset B$ or $A'' \supseteq B$, and $A'' \cup B'' = A \cup B$ is a partition belonging to M^n.

Theorem 5. If $A/B \in M^n$ and for any partition A/B' for which $B' \subset B$ the relation $A/B' \in M^d$ holds, then every partition A''/B'' in M for which $A'' \cup B'' = A \cup B$ and $A'' \supset A$ belongs to M^n.

The proof of Theorems 4 and 5 follows almost word for word the proofs of Theorems 2 (point 3) and 3.

We now formulate a lemma whose proposition follows from Criterion 1a of the decomposability of $f(X)$ according to the partition A/B when $A \cup B = X$.

Lemma 12. A necessary condition for decomposability of $f(X)$ with respect to A/B, when $A \cup B = X$, is the following. If there exists an $a_\alpha \in Z_i$ for which $\alpha \neq$ const, then from the existence of $b_\beta \in Z_i$ for which $\beta =$ const, $R_{ab} \subseteq A$, there follows $\gamma =$ const, where $c_\gamma \in Z_A$ and $c \supset b$. Here $x_i \in B$.

Theorem 6. If $A/B \in \Gamma$, then $A'/B' \in M^n$, where

$$A \cup B = A' \cup B' = X, \sigma(X) > 3, \sigma(B) = 2, \sigma(A') > 1.$$

PROOF: On the basis of Theorem 4 the function $f(X)$ is nondecomposable with respect to every A'/B' in M for which $B' \supset B$ or $B \subseteq A'$.

Let us consider such A'/B' in M for which $A' \cap B \neq \varnothing$ and $B' \cap \bar{B} \neq \varnothing$. Let $B' \ni x_i \in B$ and $A' \ni x_j \in B$.

Assume that $A'/B' \in M^d$. According to Lemma 8, there exists in Z_A a single pair a_α, b_β for which $\alpha \neq$ constant, $\beta \neq$ constant, $\alpha \neq \beta$, $\alpha \neq \bar{\beta}$, $R_{ab} = A$. Each of a and b absorbs a pair of K-intervals in Z_i, i.e., $a = c \cup d$, $b = e \cup f$, where c_γ, d_δ, e_ϵ, $f_\zeta \in Z_i$. The distance between c and d is $R_{cd} = \{x_j\}$ and between e and f, $R_{ef} = \{x_j\}$. We take such c', d', e', f' in $Z_{A'}$ for which $c' \supset c$, $d' \supset d$, $e' \supset e$, $f' \supset f$. It is easily shown that among c', d', e', f', no two are identical. Hence $R_{c'd'} \neq \varnothing$ and $R_{e'f'} \neq \varnothing$. On the other hand, since $c' \supset c$, $d' \supset d$, it must hold that $R_{c'd'} \subseteq R_{cd}$. For the same reason, $R_{e'f'} \subseteq R_{ef}$. But $R_{cd} = R_{ef} = \{x_j\}$, so $R_{e'f'} = R_{c'd'} = \{x_j\}$. Among c, d, e, f there exists at least one K-interval whose coefficient is not constant, since otherwise the coefficients α and β cannot enter into the above relation. Let, for example, $\gamma \neq$ constant. We choose g, $h \in Z_i$ such that the pair c, g is R-absorbed by the pair c', e' and the pair c, h is R-absorbed by the pair c', f'. By Lemma 6, c', f' are R-absorbed by the pair g, h and, consequently, $R_{gh} = R_{e'f'} = \{x_j\}$. This signifies that $g \cup h \in Z_A$, and

$$g \cup h \in Z_A \backslash \{a, b\}.$$

On the basis of Lemma 7 the coefficients of the K-intervals g and h are equal and constant. The pairs of K-intervals c, g and c, h are such pairs as were defined in the formulation of Lemma 12. The application of this lemma is possible since by hypothesis $A'/B' \in M^d$. According to the lemma the coefficients of e', f' must be constant. Aside from this they must also be equal, since the coefficients g and h of the K-intervals are equal. But this implies $\epsilon =$ const, $\zeta =$ const, and $\epsilon = \zeta$, whence $\beta =$ constant, which contradicts the hypothesis.

This contradiction proves the incompatibility of the conditions

$$A'/B' \in M^d \qquad \text{and} \qquad A/B \in \Gamma,$$

which proves the theorem.

We shall generalize the theorem to the case of A/B for which $A \cup B \neq X$.

Theorem 7. If $A/B \in \Gamma$, then $A'/B' \in M^n$, where

$$\sigma(X) > 3, \sigma(B) = 2, \sigma(A') > 1.$$

The proof of the theorem repeats word for word the proof of Theorem 3, but in place of Theorem 2, Theorem 6 is utilized.

3. COMPLEXITY OF THE ALGORITHM FOR FINDING THE LOWER BOUND OF A SET OF SOLUTIONS

The complexity of the algorithm for the solution of the problem put at the start of the preceding paragraph can be estimated by means of the total number of partitions A/B for which decomposability of $f(X)$ is established directly by the use of the operator *decodeboom*. The algorithm for solution of the decomposition problem will consist of an operator for selection of the next partition and the operator *decodeboom* for testing $f(X)$ according to the given partition. The selection of the next partition to be tested must be performed taking into account the results of the previous tests and our knowledge of the structural relations in the set of solutions. This permits the total number of tests to be reduced. If we use the information about the structural relations in the set of solutions that is contained in Theorems 1, 3, and 7, an algorithm can be constructed whose relative complexity is defined by the following proposition.

Theorem 8. There exists an algorithm for the solution of the problem of decomposition of $f(X)$, constructed on the basis of Theorems 1, 3, and 7 such that the number of entries to the operator *decodeboom* for the purpose of the direct testing of the decomposability of $f(X)$ with respect to the partition $A/B \in M$, for which $\sigma(A \cup B) > 3$ does not exceed

$$2\,(2^n - \sum_{i=0}^{3} C_n{}^i),$$

where $n = \sigma(X)$.

PROOF: Let A/B be such a partition that $A \cup B = E \subseteq X$, where $\sigma(E) > 3$. We shall assume that with respect to every partition A'/B' for which $A' \subseteq A$, $B' \subseteq B$, $\sigma(A' \cup B') < \sigma(E)$, it is known to which of the sets M^n, M^d it belongs. Then on the basis of Theorem 1 for certain $A/B \in M_E$, where M_E is the set of partitions A/B for which $A \cup B = E$, it can be shown that they belong to M^n. We denote the set of all $A/B \in M_E$ for which this can be done by M_E^{n*}

(1) If $M_E^{n*} = M_E$ it is then obvious that it is not necessary to enter *decodeboom*.

(2) Case (1) does not hold. If $M_E \backslash M_E^{n*}$ contains a partition A/B for which $\sigma(B) > 2$, $\sigma(A) > 1$, the test of decomposability of $f(X)$ with respect to A/B answers the question of decomposability for any partition in $M_E \backslash M_E^{n*}$. In fact, if $A/B \in M^n$, then on the basis of Theorem 3 there follows the nondecomposability of $f(X)$ for any partition in M_E. If

$A/B \in M^d$, then no partition in $M_E \backslash M_E^{n*}$ can belong to the set M^n, since in this case we enter into contradiction with either Theorem 3 or Theorem 7.

(3) Cases (1) and (2) do not hold. Let $M_E \backslash M_E^{n*}$ contain a partition A/B for which $\sigma(B) > 2$, $\sigma(A) = 1$. If the entry to *decodeboom* establishes that $A/B \in M^n$, then it is not required to test other partitions in $M_E \backslash M_E^{n*}$. For the opposite outcome it is possible that it will be necessary to test such a partition $A'/B' \in M_E{}^* \backslash M_E^{n*}$, for which $\sigma(B) = 2$. These two entries to *decodeboom* are sufficient to distribute all partitions in M_E over M^d and M^n, as follows from Theorems 3 and 7.

(4) Cases (1), (2), and (3) do not hold. Let $M_E \backslash M_E^{n*}$ not contain a single partition A/B for which $\sigma(B) > 2$. Then the test of any partition in $M_E \backslash M_E^{n*}$ decides the question of decomposability of all partitions in $M_E \backslash M_E^{n*}$. Theorem 7 is used for this.

It is thus possible to make not more than two entries to *decodeboom* in order to decide the question of the decomposability of any partition in M_E. If the algorithm is organized in such manner that *decodeboom* is first entered to test A/B for which $\sigma(A \cup B) = 3$, then, applying one of the rules (1)–(4) for each M_E, to investigate all M_E where $\sigma(E) = 4$, then to pass to M_E for which $\sigma(E) = 5, 6, 7, \ldots$, then in order to investigate any M_E for $\sigma(E) > 3$, it is not necessary to execute more than two entries to *decodeboom*. The algorithm is terminated when either all M_E for which $\sigma(E) = \sigma(X) = n$ have been investigated or when in the investigation of each M_E, where $\sigma(E)$ is some fixed number less than n, a case is found analyzed in point (1), i.e., $M_E^{n*} = M_E$. The demonstration that the number of all possible M_E for $3 < \sigma(E) \leq n$ is equal to $2^n - \sum_{i=0}^{3} C_n{}^i$ terminates the proof of the theorem.

There is reason to believe that this estimate of the complexity of the decomposition algorithm can be improved.

REFERENCES

1. Zakrevskii, A. D., An Algorithm for Decomposition of a Boolean Function, *Tr. SFTI* **44** (1964).
2. Povarov, G. N., On the Functional Decomposability of Boolean Functions, *DAN SSSR* **94**, No. 5 (1954).
3. Zakrevskii, A. D., Operations in Boolean Space, *Tr. SFTI* **48** (1965).

THE CONSTRUCTION OF PARTICULAR MINIMAL NORMAL FORMS OF BOOLEAN FUNCTIONS

V. P. Didenko

The difficulty and volume of the computational process to obtain minimal normal representations of Boolean functions increase sharply as the number of variables increases for the overwhelming majority of Boolean functions. In this connection it is interesting to develop methods for the direct construction (without the construction of tables of coverage and finding all minimal terms) of particular minimal disjunctive (conjunctive) normal forms (mdnf).[1] This makes possible a substantial memory economy and reduction of the number of computer operations for the computation of a particular mdnf. The development of such methods permits a complete solution of the problem of obtaining a minimal normal representation of Boolean functions although not for all cases. The reduction of the sifting operations substantially extends the limits of application of the methods both with respect to the number of variables on which the function may depend and the number of vectors $\tilde{\alpha}$ and $\tilde{\beta}$ on which the function takes on the values $F_i(\tilde{\alpha}) = 1$ and $F_0(\tilde{\beta}) = 0$, respectively.

Below we describe one of the methods of solving of this problem. The method contains the following basic stages.

(1) Find the minimal terms of the kernel (these are terms of every particular mdnf representing the function F_1) from the vectors $\tilde{\alpha} \in M_{F_1}$ and $\tilde{\beta} \in M_{F_0}$, where M_{F_1} and M_{F_0} are the sets of vectors of the values of the variables on which $F_1(\tilde{\alpha}) = 1$ and $F_0(\tilde{\beta}) = 0$, respectively.

(2) Remove from M_{F_1} the vectors covered by the kernel[2] (the minimal

[1] The terminology is adopted from [1].

[2] The vector $\tilde{\alpha}$ is said to be covered by a minimal term if all the values of the literals encountered simultaneously in $\tilde{\alpha}$ and in the minimal term are identical.

terms). If the set M'_{F_1} obtained after removal from M_{F_1} of the covered vectors is empty, the process of constructing the mdnf has terminated. Otherwise pass to stage (3).

(3) Find from the vectors in the set M'_{F_1} and in M_{F_0} the sets M_3, $\bar{M}_3$ of the nonobligatory variables[3] and their negations (the variable x_s is nonobligatory if its values on all vectors $\tilde{\alpha} \in M'_{F_1}$ are nonobligatory).

In the presence of nonobligatory variables the rule of point (4) is applied, in their absence the rule of point (5) or (6).

(4) Cross off in the vectors $\tilde{\alpha} \in M'_{F_1}$ and $\tilde{\beta} \in M_{F_0}$ the values of one of the nonobligatory variables[4] (their order of selection is arbitrary, although the variant of mdnf obtained at the end depends on this order) encountered in M_3 or $\bar{M}_3$. In comparison, or sum modulo 2, the crossed-off values of the variable x_s in the vectors $\tilde{\alpha}$ and $\tilde{\beta}$ are reduced to a form where $\alpha_s \equiv \beta_s$ with subsequent restoration of the previous values of α_s and β_s in the vectors $\bar{\alpha}$ and $\bar{\beta}$, respectively.

We denote the vectors of the set M'_{F_1} with the crossed-off value of x_s by $M_{F_1,\tilde{x}_s}$ and of M_{F_0} by $M_{F_0,\tilde{x}_s}$ (the subscript $\tilde{x}_s$ indicates that in the vectors $\tilde{\beta} \in M_{F_0}$ both occurrences of x_s and $\bar{x}_s$ are crossed off), and we apply to the vectors of the sets $M_{F_1,\tilde{x}_s}$ and $M_{F_0,\tilde{x}_s}$ the rules of points (1)–(4) until either in point (2) the process of formation of the particular mdnf terminates or in point (3) passage to point (5) occurs.

(5) Form the subset $(M'_{F_1})_{\tilde{x}_m}$, composed only of those vectors $\tilde{\alpha} \in M'_{F_1}$ on which the values of the variable $\tilde{x}_m$ are obligatory. Further, apply the rules of points (1)–(4) to the vectors of the sets $(M'_{F_1})_{\tilde{x}_m}$ and $(M_{F_0})_{\tilde{x}_m}$.

If in point (1) no minimal terms are found, passage is effected to point (5), where the subsets $(M'_{F_1})_{\tilde{x}_m,\tilde{x}_{m_1}}$ and $(M_{F_0})_{\tilde{x}_m,\tilde{x}_{m_1}}$ are found.

With repeated passage to point (5) the subsets $(M'_{F_1})_{\tilde{x}_m,\tilde{x}_{m_1},\ldots,\tilde{x}_{m_k}}$ and $(M_{F_0})_{\tilde{x}_m,\tilde{x}_{m_1},\ldots,\tilde{x}_{m_k}}$, are found, where k is smaller than the number of variables n in F, with subsequent passage to point (1) and the replacement of M_{F_1}, M_{F_0} by $(M'_{F_1})_{\tilde{x}_m,\ldots,\tilde{x}_{m_k}}$ and $(M_{F_0})_{\tilde{x}_m,\ldots,\tilde{x}_{m_k}}$. Otherwise in point (3) passage is effected to point (6).

(6) If in passage to point (6) the vector of obligatory variables $\{\tilde{x}_m, \tilde{x}_{m_1}, \ldots, \tilde{x}_{m_k}\}$ is found to be empty, the rule of point (3) with subsequent passage to point (5) is carried out. If in passage to point (6) the

[3] The value α_s of the variable x_s ($s = 1, 2, \cdots, n$) on the vector $\tilde{\alpha} \in M_{F_1}$ is called obligatory if and only if in M_{F_0} is found a vector $\tilde{\beta} \in M_{F_0}$ such that the vectors $\tilde{\alpha}$ and $\tilde{\beta}$ are adjacent, i.e., differ only in the value of a single variable x_s. Otherwise the value of α_s in the vector $\tilde{\alpha}$ is nonobligatory.

[4] By crossing off of the values of a variable x_s in vectors $\tilde{\gamma}$ and $\tilde{\beta}$ we shall mean the temporary reduction (until termination of the computation of one particular mdnf) of these vectors to a form in which for any $\tilde{\gamma} \in M_{F_1}$ and all $\tilde{\beta} \in M_{F_0}$ the identity $\gamma_s \equiv \beta_s$ ($1 \leq s \leq n$) holds after the values of the variable have been crossed off.

vector $\{\tilde{x}_m, \tilde{x}_{m_1}, \ldots, \tilde{x}_{m_k}\}$ is not empty, then it is used to form the subsets $(M'_{F_1})_{\tilde{x}_{i_1},\ldots,\tilde{x}_{i_t}}$ and $(M_{F_0})_{\tilde{x}_{i_1},\ldots,\tilde{x}_{i_t}}$ of the vectors $\tilde{\alpha}$ and $\tilde{\beta}$, where

$$\{\tilde{x}_{i_1}, \ldots, \tilde{x}_{i_t}\} \subseteq \{\tilde{x}_m, \tilde{x}_{m1}, \ldots, \tilde{x}_{m_k}\}.$$

If $\{\tilde{x}_{i_1}, \ldots \ldots, \tilde{x}_{i_t}\} = \{\tilde{x}_m, \tilde{x}_{m1}, \ldots, \tilde{x}_{m_k}\}$, passage is effected to point (1), with replacement of M_{F_1} and M_{F_0} by the sets

$$(M'_{F_i})_{\tilde{x}_{i_1},\ldots,\tilde{x}_{i_t}} \qquad \text{and} \qquad (M_{F_0})_{\tilde{x}_{i_1},\ldots,\tilde{x}_{i_t}}.$$

If $\{\tilde{x}_{i_1}, \ldots, \tilde{x}_{i_t}\} \subset \{\tilde{x}_m, \ldots, \tilde{x}_{m_k}\}$, the vector $\{\tilde{x}_m, \tilde{x}_{m1}, \ldots, \tilde{x}_{m_k}\}$ is corrected to the vector $\{\tilde{x}_{i_1}, \cdots, \tilde{x}_{i_t}\}$ at the expense of those variables that were found to be nonobligatory in the vectors of a subset of the type $M''_{F_{1,\tilde{x}_s}}$, where $M''_{F_{1,\tilde{x}_s}}$ is the subset of vectors obtained from $M'_{F_{1,\tilde{x}_s}}$ by removal of the covered vectors.

After this, passage is effected to point (1) with the above-mentioned substitution of M_{F_1} and M_{F_0}.

The basic component parts of this method are the rules of points (1), (3), and (5), since they determine the possibility of forming the particular mdnf. Therefore, before passing to a description of the working program of the method in LYaPAS [2], we present the basic theoretical premises to show that in using this method the minimal terms entering into some particular mdnf will be found, representing the Boolean function prescribed by the vectors $\tilde{\alpha}$ on which $F_1(\tilde{\alpha}) = 1$.

Let us consider the method for determining the kernel [3, 4].

Let there be given a Boolean function on the vectors $\tilde{\alpha} \in M_{F_1}$ and $\tilde{\beta} \in M_{F_0}$, taking on the value $F_1(\tilde{\alpha}) = 1$ and $F_0(\tilde{\beta}) = 0$, where

$$M_{F_1} \cap M_{F_0} \equiv 0.$$

Furthermore, let the values of the n variables in the vectors $\tilde{\alpha}$ and $\tilde{\beta}$ be strictly ordered; i.e., the values of α_1, β_1 correspond to the value of the variable x_1, the values of α_2, β_2 to x_2, etc. Let the cardinalities of the sets M_{F_1} and M_{F_0} be equal to q_1 and q_2, respectively.

We take the first vector in order, $\tilde{\alpha}_1$ in M_{F_1}, and add its components (i.e., the values of the variables x_s, $s = 1, 2, \cdots, n$) mod 2 to the components of the first vector $\tilde{\beta}_1$. In exactly the same way $\tilde{\alpha}_1$ is added to $\tilde{\beta}_2$, etc. We denote the vectors obtained as a result of this operation (that contain only a single component) by $\alpha_i = 1$ $(i = 1, 2, \cdots, n)$, and all the others $\alpha_l = 0$,

$l \neq i$, $l = 1, 2, \ldots, n$ by $\tilde{\alpha}_j'$. We form from all such vectors the vector $\tilde{\alpha}_1'$

$$\tilde{\alpha}_1' = \bigvee_{j=1}^{q_3} \tilde{\alpha}_j', \tag{1}$$

where $q_3 \leq n$.

We put the α_i'-th component of the vector $\tilde{\alpha}_1'$ into a unique correspondence with the component α_i of the vector $\tilde{\alpha}_1'$. Then the vector $\tilde{\alpha}_1'$ will encode those components of the vector $\tilde{\alpha}_1$ that are the values of the obligatory variables and that correspond to the values equal to 1 in the vector $\tilde{\alpha}_1'$.

Putting into correspondence with each obligatory component of the vector $\tilde{\alpha}_1$ the variable whose value it takes on and connecting all such variables by the sign of logical multiplication, we obtain the elementary conjunction u_i, whose factors enter into all minimal terms covering the vector $\tilde{\alpha}_1$. This follows from the definition of obligatory literal x_i (a variable or its inversion), put into correspondence with the obligatory value of the vector $\tilde{\alpha}_1'$.

Those values of these variables in the vector

$$\tilde{\alpha}_1'' = \tilde{\alpha}_1 \wedge \tilde{\alpha}_1', \tag{2}$$

where $\wedge$ is the sign of logical multiplication, whose components are extracted by the unit values of the vector α_1' will correspond to an obligatory variable of the elementary conjunction u_i.

By definition of obligatory values of the variables, if for any vector $\tilde{\beta}_j$ $(j = 1, 2, \ldots, q_2)$ the relation

$$\tilde{\alpha}_1' \wedge \tilde{\beta}_j \oplus \tilde{\alpha}_1'' \not\equiv 0 \tag{3}$$

holds, then the elementary conjunction u_1, composed of the variables corresponding to their obligatory values in the vector $\tilde{\alpha}_1''$ and taking on the values 1 in the vector $\tilde{\alpha}_1'$, is a minimal term of the kernel.

Theorem 1. In order that the elementary conjunction $u_i \in F_1$, composed of obligatory literals (variables and their inversions), be a minimal term of the kernel, it is necessary and sufficient that the relation $U_iF_0 \equiv 0$ hold, where F_0 is the logical sum of the constituents on which the function takes on the value 0.

NECESSITY. Let $U_iF_0 \equiv 0$. Let the factors of U_i be the obligatory variables, and let the conjunction U_i be computed for the vector $\tilde{\alpha}_i$, on which $F_1(\tilde{\alpha}_i) = 1$. We denote by the symbol U_i' the conjunction obtained by removal from U_i of at least one arbitrary literal x_s contained in it. We assume that $U_i'F_0 \equiv 0$. This last contradicts the assumption that x_s is obligatory, since removal of any factor from U_i signifies complete coincidence of the values of the variables in U_i' in the vectors $\tilde{\alpha}_i$ and $\tilde{\beta}_j$ ($j = 1, 2, \ldots, q_2$) on which $F(\tilde{\alpha}_i) = 1$ and $F(\tilde{\beta}_j) = 0$, respectively. By definition it follows that the conjunction U_i is a minimal term. This proves necessity.

SUFFICIENCY. Let the minimal term U_i contain only obligatory variables, corresponding to their obligatory values in the vector $\tilde{\alpha}_i$ on which $F_1(\tilde{\alpha}_i) = 1$, and let there exist, aside from U_i, a minimal term U_i' also covering this vector $\tilde{\alpha}_i$. By virtue of the definition of obligatory literals the minimal term U_i' must then contain (at least) those literals that are factors of U_i. This means that the minimal terms U_i' and U_i coincide completely. In fact, let us assume the contrary; then U_i' absorbs U_i, and this contradicts the assumption that the conjunction U_i' is a minimal term. Consequently, the conjunction U_i is the unique minimal term covering the vector $\tilde{\alpha}_i$. Hence U_i enters into every particular mdnf covering the vectors of the set M_{F_1}. Consequently, U_i is a minimal term of the kernel.

Let the minimal term of the kernel U_1 be computed from $\tilde{\alpha}_1 \in M_{F_1}$ and $\tilde{\beta} \in M_{F_0}$. We tag in M_{F_1} those vectors $\tilde{\alpha}_m$ ($m = 1, 2, \ldots, q_1$) for which the relation $\tilde{\alpha}_1' \wedge \tilde{\alpha}_m \oplus \tilde{\alpha}_1'' \equiv 0$ holds. From the remaining untagged vectors $\tilde{\alpha} \in M_{F_1}$ we form the set M'_{F_1} and repeat the above procedure.

If a minimal term is not found from $\tilde{\alpha}_1$ and M_{F_0}, we take the vector $\tilde{\alpha}_2$ from $M_{F_1}(M'_{F_1})$ and repeat the above process. It is obvious that the entire kernel is found by repetition of the process as long as there remains in $M_{F_1}(M'_{F_1}(M''_{F_1}(, \ldots, (M^q_{F_1})\cdots)$ at least one vector $\tilde{\alpha}_t$ ($t \leq m$) for which a minimal term of the kernel has not yet been computed, and in M_{F_1} there will be tagged all the vectors $\tilde{\alpha}$ covered by it. If all the vectors $\tilde{\alpha} \in M_{F_1}$ are tagged, then the process of formation of the particular mdnf has terminated.

If untagged vectors $\tilde{\alpha}$ exist in M_{F_1} they are segregated in the set $M^0_{F_1}$

To compute the particular mdnf from the vectors $\tilde{\gamma} \in M^0_{F_1}$ and $\tilde{\beta} \in M_{F_0}$ we find the sets of all obligatory values of the variables encountered in the vectors $\tilde{\gamma}$.

The sets M_1 and $\bar{M}_1$ of all variables taking on the values 1 and 0 in the vectors $\tilde{\gamma}$ are defined as follows:

$$M_1 = \bigcup_{i=1}^{q_1'} \tilde{\gamma}_i, \qquad \bar{M}_1 = \bigcup_{i=1}^{q_1'} \tilde{\gamma}_i \urcorner,$$

respectively, where $\urcorner$ is the sign of inversion and $q_1' \leq q_1$. The sets M_2 and $\bar{M}_2$ are computed from the obligatory values of the variables of the vectors $\tilde{\gamma}$, defined by Eq. (1), i.e.,

$$M_2 = \bigcup_{i=1}^{q_1''} \tilde{\gamma}_i \wedge \tilde{\gamma}_i', \qquad \bar{M}_2 = \bigcup_{i=1}^{q_1''} \tilde{\gamma}_i \urcorner \wedge \tilde{\gamma}_i',$$

where $q_1'' \leq q_1'$ and $\tilde{\gamma}_i' = \bigvee_{j=1}^{q_3} \tilde{\gamma}_j'$. The set of nonobligatory variables taking on the values 1 and 0 are defined from the relations

$$M_3 = M_1 \backslash M_2, \qquad \bar{M}_3 = \bar{M}_1 \backslash \bar{M}_2,$$

respectively, where $\bar{M}_1$, $\bar{M}_2$, $\bar{M}_3$ are sets of variables taking on the value 0.

The following cases can arise naturally: (1) M_3 or $\bar{M}_3$ is nonempty, (2) $M_3 \equiv 0$ and $\bar{M}_3 \equiv 0$.

Let us consider the first case. It is obvious that the nonobligatory values of the variables can be successively removed (where the order determines the variant of the solution) from all vectors $\tilde{\gamma} \in M_{F1}^0$, if such values for any of the variables are encountered simultaneously for all vectors $\tilde{\gamma} \in M_{F1}^0$, since the condition $M_{F_1} \cap M_{F_0} \equiv 0$ continues to hold.

Let the values of the variables $x_{s_1}, \ldots, x_{s_i}, \ldots, x_{s_t}$, for example, be simultaneously nonobligatory over all vectors $\tilde{\gamma} \in M_{F1}^0$. Then, crossing off the values, for example, of the variable x_{s_1} in the vectors $\tilde{\gamma}$ and $\tilde{\beta} \in M_{F_0}$, it is easy to compute the values of the variables entering into the minimal terms. If after the values of the variable x_{s_1} have been crossed off no minimal terms are found, x_{s_2} is crossed off, etc. Assume that minimal terms are found after the values of x_{s_i} have been crossed off. We shall show that the minimal terms thus calculated belong in fact to a particular mdnf, and only to it.

In effect, at the first step after crossing off the values of the variable x_{s_i} in the vectors $\tilde{\gamma}$ and $\tilde{\beta}$, it is possible, from the vectors of the sets $M_{F1,x_{s_i}}^0$ and $M_{F_0,x_{s_i}}$ obtained by this, and using Eqs. (1)–(3), to calculate only such minimal terms as appear in the kernel of a function independent of the s_ith variable.

Since

$$M_{F_1}^0 = M_{F_1} \backslash \hat{M}_{F_1},$$

where $\hat{M}_{F_1}$ is the set of vectors $\tilde{\alpha}$ covered by the minimal terms of the kernel, then in $M_{F_1,x_{s_i}}^0$ it is possible to find at least one vector $\tilde{\gamma}$ that is characteristic.[5] We assume that such a vector exists and let the minimal term φ_1 be calculated from it and the set M_{F_0}. By virtue of Theorem 1, φ_1 cannot cover the singular vectors in $\hat{M}_{F_1}$. Similar considerations can be given for the second, third, etc., steps in the calculation of the particular mdnf.

Let us consider the kth step ($k < 2n$). Each step corresponds to the crossing off of the values of a nonobligatory variable, different from the previous crossed-off variables.

Let the nonobligatory variable x_k be selected at the kth step, and let the

[5] We call the vector $\tilde{\gamma}$ characteristic if a minimal term was calculated by Eqs. (1)–(3) from this vector and the vectors $\tilde{\beta} \in M_{F_0}$.

remaining nontagged vectors $\tilde{\gamma}$ and $\tilde{\beta}$ be united in the sets $M_{F_1}^{k-1}$ and M_{F_0}. If from the vectors of the sets M_{F_1,x_k}^{k-1} and M_{F_0,x_k} minimal terms are calculated that cover all the vectors $\tilde{\gamma} \in M_{F_1,x_k}^{k-1}$, the process of calculating the mdnf terminates. Otherwise we pass to the $k + 1$st step.

We shall show that the singular vectors, covered by the minimal terms calculated at the rth step, $|r < k|$, cannot cover the minimal terms calculated at the kth step. Let the minimal term φ_k, calculated at the kth step, cover the set $\tilde{\gamma}_r$ (the converse cannot hold by construction of $M_{F_1}^{k-1}$), covered by the minimal term φ_r, calculated at the rth step. This means that φ_k either contains variables whose values were crossed off in the vector $\tilde{\gamma}_r$, which contradicts the assumption about the simultaneous crossing off of the values of the same variables in the vectors $\tilde{\gamma}_r$ and $\tilde{\gamma}_k$ (the vector from which the minimal term φ_k was determined), or φ_k contains exactly the same variables as φ_r, which is impossible by construction of the set M_{F_1,x_k}^{k-1} and Theorem 1. This signifies that all the calculated minimal terms enter into the particular mdnf, Q.E.D.

Let us assume that in the vectors $\tilde{\gamma} \in M_{F_1}^0$ the values of a nonobligatory variable, equal only to one or only to zero, are crossed off, and let us show that in this case the minimal terms calculated from these vectors also enter into the particular mdnf.

Let the values of the variable x_s $(\bar{x}_s)$, equal only to one or to zero, be crossed off in the vectors $\tilde{\gamma} \in M_{F_1}^0$. We denote the set of vectors $\tilde{\gamma}$ in which the variable $\bar{x}_s$ (x_s) takes on the value zero (unity) by $M_{F_1,\bar{x}_s(x_s)}^0$. None of the minimal terms containing $\bar{x}_s$ (x_s) can cover vectors in $M_{F_1,x_s(\bar{x}_s)}^0$, since in the latter x_s $(\bar{x}_s)$ takes on a value equal to unity (zero). The case in which $\bar{x}_s$ (x_s) does not enter into minimal terms calculated from the vectors $M_{F_1,\bar{x}_s(x_s)}$ reduces to the assumption that all values of x_s $(\bar{x}_s)$ equal to both unity and to zero are crossed off simultaneously from the vectors $\tilde{\gamma} \in M_{F_1}^0$, i.e., to the case that has been considered above.

This case is easily generalized for several variables $x_{i_1}, x_{i_2}, \ldots, x_{i_s}$, since the values of these variables remain crossed off in the vectors $\tilde{\gamma} \in M_{F_1}^0$ during the time for calculating one particular mdnf.

Obviously, to obtain the various particular mdnf it is necessary to enumerate all possible variants of crossing off of the variables. Of the particular mdnf obtained, those are chosen that yield minimal cost of realization.

Practically, in the presence of complex Boolean functions (i.e., for which it is necessary to enumerate a large number of variants to find the smallest cost of realization) the enumeration can be limited, on the basis of the real expenditure of manual or machine time for examining several variants.

Let us consider the case in which, as mentioned, $M_3 \equiv 0$ and $\bar{M}_3 \equiv 0$. This case reduces to the case considered above. For this it is necessary to write out from $M^0_{F_1}$ those vectors $\tilde{\gamma}$ that contain obligatory values of, let us say, the variable x_s. Let the set of such vectors be M_{x_s}. If in this the nonobligatory values of any of the variables $x_{i_1}, \ldots, x_{i_{s-1}}, x_{i_{s+1}}, \ldots, x_{i_n}$, encountered simultaneously in all vectors $\tilde{\gamma} \in M_{x_s}$ are absent, then we find in M_{x_s} vectors containing obligatory values of the variable x_r and form the subset of vectors M_{x_s,x_r}, etc. In $M_{x_s,x_r,\ldots,x_t}$ the nonobligatory values of the variables $\{x_1, \ldots, x_n\}\backslash\{x_s, x_r, \ldots, x_t\}$ certainly appear, since otherwise all the values of the variables $x_1, \ldots, x_n$ will be obligatory in all vectors $\tilde{\gamma} \in M_{x_s,x_r,\ldots,x_t}$. This contradicts the hypothesis that in $M^0_{F_1}$ and, therefore, in $M_{x_s,x_r,\ldots,x_t}$ there appear residues obtained from M_{F_1} after tagging in it the vectors covered by the kernel. Determining the particular mdnf from the distinguished subset of vectors M_{x_s} and the vectors M_{F_0}, we tag in $M^0_{F_1}$ all the vectors covered by it.

All the vectors remaining untagged after this, marked by the symbol $\hat{M}^0_{F_1}$, will contain simultaneously nonobligatory values of the variable x_s.

Crossing off the nonobligatory values of this variable from all vectors $\tilde{\gamma} \in \hat{M}_{F_1}$ and $\tilde{\beta} \in M_{F_0}$, we continue the calculation of the particular mdnf up to complete coverage of the vectors of the initial set M_{F_1}, using the method described. The vectors in M_{x_s} cannot be covered by the minimal terms obtained from the vectors of the sets $\hat{M}^0_{F_1}$, since in the latter, first the nonobligatory values of the variable have been crossed off and, second, all the vectors of the subset M_{x_s} contain obligatory values of x_s. For proof it is sufficient to assume the contrary. Then the minimal terms calculated from the vectors $\hat{M}^0_{F_1}$ must also contain the variable x_s, since its values are obligatory in the vectors M_{x_s}, which contradicts the hypothesis about crossing off the values of this variable from the vectors $\hat{M}^0_{F_1}$.

Turning now to machine execution, we shall describe the method in LYaPAS. Here we first present a series of subroutines of the general L-program, realizing the method of calculating the particular mdnf. The L-program is represented in the form of an external program, but it is easily presented in the form of a subroutine.

Let the vectors of the sets M_{F_1} and M_{F_0} be represented by the elements of the sets $A = \{a_0, a_1, \ldots, a_{q_1-1}\}$ and $B = \{b_0, b_1, \ldots, b_{q_2-1}\}$, respectively, where there exist sets $D = \{d_0, d_1, \ldots, d_{n-1}\}$ and $\{x\} = \{x_1, x_2, \ldots, x_n\}$, where x_i is a Boolean variable.

To denote the sense of the operands we introduce a matrix of binary relations among the elements of the two sets A and B. We define a set C in the following manner: $r_i = (a_i \cup b_j)\backslash(a_i \cap b_j)$ for $j = 0, 1, \cdots, q_2 - 1$.

We define only those r_i for which $c_i = r_i \cap D = r_i$. The elements of the single-row matrix are defined from the relation

$$\mathbf{a}::\|\ \{C\} \ni X\ \|.$$

The elements in the complexes **A** and **B** are defined by the relations

$$\mathbf{A}::\| A \ni X \|,$$
$$\mathbf{B}::\| B \ni X \|.$$

In the LYaPAS subroutines and the external program for finding the particular mdnf, described below, the variable j corresponds to the obligatory values of the variables of the set $\{x\}$ in the vector $\tilde{\alpha}_j \in M_{F_1}$ or $\mathbf{a} \in \mathbf{A}$, and is defined by the relations

$$C^* = a_i \cap \mathbf{a},$$
$$\mathbf{j}::\| \{C^*\} \ni X \|.$$

Denoting the set of all zero and zero obligatory values of the variables $\{x\}$ by A^* and B^*, and the sets of all unit and all obligatory unit values of the variables $\{x\}$ by A^{**} and B^{**}, respectively, we define the variables **n**, **g**, and **k**, **p** in the following way:

$$\mathbf{n} \sim \mathbf{g}::\| \{C^{***}\} \ni x \|,$$
$$\mathbf{p} \sim \mathbf{k}::\| \{C^{**}\} \ni x \|,$$

where

$$C^{**} = A^* \backslash B^*;$$
$$C^{***} = A^{**} \backslash B^{**}.$$

The remaining operators will be introduced into the program by virtue of the need for storage or processing of information, but either their purpose or sense will easily be seen directly from the programs or they will be analogous to operands defined by matrices of binary operations.

In the subroutine represented by the operator *deker*, determination of the kernel, or the minimal terms entering into a particular mdnf, the sets A and B are represented by the complexes α and β. The set $\{A^*\}$ of all zero values of the variables $\{x\}$ is defined by the variable γ, the set $\{A^{**}\}$ of all unit values of the variable $\{x\}$ by the variable δ. The set $\{B^{**}\}$ of obligatory unit values of the variables $\{x\}$ is given by the variable ξ, and $\{B^*\}$ by ζ. The set of nonobligatory unit values of the variables $\{x\}$ is given by the variable η and zero by ϑ. The number of literals in the minimal term φ is defined by the variable κ. The set of obligatory values of the variables for which the subsets A and B are found is given by the variable λ. The number of cases of distinction of variables corresponds to the variable μ, and the number of cases of distinction of subsets A and B corresponds to the index ν. The presence of the variable x_i, b $\{x\}$, etc., corresponds to the unit value in the index л. The remaining variables in the operator *deker* denote exits from the subroutine to the external program.

The result of the operation of the subroutine is printed.

deker $\alpha\text{к}, \beta\text{к}, \gamma\text{п}, \delta\text{п}, \epsilon\text{п}, \zeta\text{п}, \eta\text{п}, \vartheta\text{п}, \kappa\text{п}, \lambda\text{п}, \mu\text{п}, \nu\text{п}, \xi\text{и}, \pi\text{ч}, \rho\text{п}, \sigma\text{п}\ (\mathbf{s}, \text{л})$
160 347 13 $(\mu\mathbf{a})$

§0 $\bar{\circ}\ c \circ \gamma \circ \delta \circ \epsilon \circ \zeta \circ \nu \circ \xi$
§1 $\triangle\, c \oplus b_\alpha \circ\!\!\rightarrow 13 \circ ab_\beta \circ\!\!\rightarrow \text{л}c_{37} \wedge \alpha_c$
$\quad \vert\!\!\rightarrow 1\ \bar{\circ}\ b$
§2 $\triangle\, b \oplus b_\beta \circ\!\!\rightarrow 3\rho \wedge \alpha_c \Rightarrow \mathbf{c}\sigma \urcorner \vee \alpha_c$
$\quad \wedge \beta_b \vee \mathbf{c} \Rightarrow \mathbf{c} \oplus \alpha_c \vert\!\!\leftarrow \oplus c_0 \vert\!\!\rightarrow 2\mathbf{c}$
$\quad \oplus \alpha_c \vee \mathbf{a} \Rightarrow \mathbf{a} \rightarrow 2$
§3 $\mu \circ\!\!\rightarrow 5\mathbf{a} \wedge \mu \oplus \mu \vert\!\!\rightarrow 1 \circ \mathbf{b}$

Detection of the obligatory values of the variables from complexes α and β

§4 $\triangle\, b \oplus b_\beta \circ\!\!\rightarrow 5\eta \wedge \alpha_c \Rightarrow \mathbf{c}\vartheta \urcorner \vee \alpha_c$
$\quad \wedge \beta_b \vee \mathbf{c} \Rightarrow \mathbf{c} \oplus \alpha_c \vert\!\!\leftarrow \oplus c_0 \rightarrow 4\mathbf{c}$
$\quad \oplus \alpha_c \vee \mathbf{a} \Rightarrow \mathbf{a} \rightarrow 4$

Detection of the obligatory values of the variables from α and β, taking into account η and ϑ

§5 $\triangle\, \xi\lambda \circ\!\!\rightarrow 6\alpha_c \Rightarrow S \vee \delta \Rightarrow \delta S \oplus \text{л} \vee \gamma$
$\quad \Rightarrow \gamma S \wedge \mathbf{a} \vee \epsilon \Rightarrow \epsilon S \urcorner \wedge \mathbf{a} \vee \zeta$
$\quad \Rightarrow \zeta \rightarrow 1$

Detection of the values of all the variables (including obligatory values) to determine the nonobligatory values of the variables

§6 $\mathbf{a} \circ\!\!\rightarrow 1\ \bar{\circ}\ b$
§7 $\triangle\, b \oplus b_\beta \circ\!\!\rightarrow 10\beta_b \oplus \alpha_c \wedge \mathbf{a} \vert\!\!\rightarrow 7 \rightarrow 1$
§10 $\mathbf{a} \wedge \alpha_c \Rightarrow \mathbf{ja} \nabla + \kappa \circ\!\!\rightarrow \pi \Rightarrow \kappa\ \bar{\circ}\ \mathbf{a}$

Test whether **a** among the obligatory values of the variables $\{x\}$ is contained in β or not

§11 $\triangle\, \mathbf{a} \oplus b_\alpha \circ\!\!\rightarrow 12\alpha_\alpha \wedge \mathbf{a} \oplus \mathbf{j} \vert\!\!\rightarrow 11c_{37}$
$\quad \vee \alpha_a \Rightarrow \alpha_a \rightarrow 11$

Tagging covered vectors

§12 $\mathbf{j} * \mathbf{a} * \triangle\, \nu \rightarrow 1$
§13 .

Printing and counting the number of such cases

retab. Reduce the table after removal from the complex α of its elements covered by the minimal terms of the kernel.

retab $\alpha\text{к}, \beta\text{п}, \gamma\text{и}, \delta\ (-, b)$
161 54 2

$\bar{\circ}\ a \circ b$
§1 $\triangle\, a \oplus b_\alpha \circ\!\!\rightarrow 2c_{37} \wedge \alpha_a \vert\!\!\rightarrow 1\alpha_a \Rightarrow \alpha_b \triangle\, b \rightarrow 1$
§2 $\beta * \gamma * b \circ\!\!\rightarrow \delta \Rightarrow b_\alpha.$

comtab. Compress the table after removal from the complex α of the elements covered by the minimal terms in the partial mdnf.

comtab αк $(-, b)$
162 42 2

$\bar{\circ}\ a \circ b$
§1 $\triangle\ a \oplus b_\alpha \circ \rightarrow 2c_{37} \wedge \alpha_a \rightarrow 1\alpha_a \Leftrightarrow \alpha_b \triangle\ b \rightarrow 1$
§2 .

eratag. Erase tags y of the elements in α covered by the minimal terms appearing in the particular mdnf.

eratag αк $(-, a)$
163 33 2

$\bar{\circ}\ a$
§1 $\triangle\ a \oplus b_\alpha \circ \rightarrow 2c_{37}\ \urcorner \wedge \alpha_a \Rightarrow \alpha_a \rightarrow 1$
§2 .

denonvar. Detect the nonobligatory values of the variables $\{x\}$ for the condition that the cardinality of the sets of all values of the variables is greater than the cardinality of the set of obligatory values of the variables x.

denonvar αп, γп, δп, ϵп, ζп, ηп, ϑп, κч (**b**, b)
164 102 4 (γab)

$\circ\ \alpha \circ \epsilon$
§1 $\alpha\ \urcorner \wedge \gamma \oplus \delta \Rightarrow$ **a** $\ddot{\mathbf{X}}\ 2a\alpha \vee C_a \Rightarrow \alpha \circ \vartheta \rightarrow \kappa$
§2 $\epsilon\ \urcorner \wedge \gamma \oplus \eta \wedge \alpha \Rightarrow$ **b** $\ddot{\mathbf{X}}\ 3b\epsilon \vee C_b \Rightarrow \epsilon \rightarrow \kappa$
§3 $\epsilon\ \urcorner \wedge \gamma \oplus \eta \Rightarrow$ **b** $\ddot{\mathbf{X}}\ 4b\epsilon \vee C_b \Rightarrow \epsilon \rightarrow \kappa$
§4 .

denonel. Determine the subset $M^0_{F_1}$ of nontagged elements of the complex α when the subset of nonobligatory variables is empty.

denonel αп, βп, γп, δп, ϵп, ζи, ηв, ϑч, ξп/**a**, a/
165 52 2

$\bar{\circ}\xi \circ \zeta \circ \gamma$
§1 $\alpha \vee \beta \wedge \xi \Rightarrow \xi\ \ddot{\mathbf{X}}\ 2aC_a \Rightarrow \eta \triangle\ \zeta C_a \vee \gamma \Rightarrow \gamma \Rightarrow \delta \Rightarrow \epsilon \rightarrow \vartheta$
§2 .

Below we give the external program of the method for calculating the particular mdnf from prescribed sets of elements of the complexes A and B.

External Program.

§0 $\bar{\circ}\ d\ \bar{\circ}\ \mathbf{a}$

§1 $\triangle\ d \oplus b_2 \circ\!\!\rightarrow 23 \circ \mathbf{m} \circ \mathbf{q} \circ \mathbf{s} \circ \mathbf{h} \circ \mathbf{j} \circ \mathbf{q} \circ \mathbf{k} \circ \mathbf{u} \circ \mathbf{l} \circ \mathbf{k}$
$\circ\ \mathbf{n}a \mid\!\rightarrow 7.$

§2 *deker* **ABbdefgkmqs***eg***np** $25//g \circ\!\!\rightarrow 4\mathbf{q} \rightarrow 13\ \bar{\circ}\ a \circ cd \mid\!\rightarrow 5e \circ\!\!\rightarrow 3$
retab **Am***d*$24//$

§3 $\mathbf{b}_0 \Rightarrow \mathbf{c} \circ e \rightarrow 1$

§4 $\bar{\triangle}\ k\mathbf{s} - \mathbf{i}_k \circ\!\!\rightarrow 15 \Rightarrow \mathbf{s} \Rightarrow \mathbf{u} \Rightarrow \mathbf{qz} + \mathbf{i}_k \Rightarrow \mathbf{zp} \Rightarrow \mathbf{kn} \Rightarrow \mathbf{g} \rightarrow 2$

§5 $e \mid\!\rightarrow 6c_1 \Rightarrow \mathbf{q} \rightarrow 2$

§6 *comtab* **A***b*$//b \circ\!\!\rightarrow 10 \Rightarrow b_0\,\mathbf{s} \mid\!\rightarrow 7c_1 \Rightarrow \mathbf{q} \rightarrow 2$

§7 $\mathbf{g} \Rightarrow \mathbf{hk} \Rightarrow \mathbf{jp} \Rightarrow \mathbf{kn} \Rightarrow \mathbf{g} \circ \mathbf{sc}_1 \Rightarrow \mathbf{q} \rightarrow 2$

§10 $\circ\ e\mathbf{m} * d * \mathbf{m} - \mathbf{a} \circ\!\!\rightarrow 11\mathbf{c} \Rightarrow b_0 \rightarrow 12$

§11 $\mathbf{m} \Rightarrow \mathbf{a}d \Rightarrow \mathbf{ic} \Rightarrow b_0$

§12 *eratag* **A**$// \rightarrow 1$

§13 *denonvar* **gdekbfq** $14//$

§14 $\mathbf{s} \mid\!\rightarrow 15\mathbf{g} \Rightarrow \mathbf{nk} \Rightarrow \mathbf{p}$

§15 $\circ\ \mathbf{q} \rightarrow 2$

§16 $\mathbf{s} \rightarrow 22\mathbf{u} \Rightarrow \mathbf{q} \Rightarrow \mathbf{s}k \circ\!\!\rightarrow 22$

§17 $\mathbf{k} \wedge \mathbf{q} \vee \mathbf{g} \wedge \mathbf{q} \rightarrow 20\mathbf{h} \Rightarrow \mathbf{gj} \Rightarrow \mathbf{k} \rightarrow 2$

§20 $\mathbf{g} \wedge \mathbf{q} \oplus \mathbf{q} \wedge \mathbf{d} \Rightarrow \mathbf{pk} \wedge \mathbf{q} \oplus \mathbf{q} \wedge \mathbf{b} \vee \mathbf{p} \Rightarrow \mathbf{q} \circ\!\!\rightarrow 21\mathbf{q}\ \urcorner \Rightarrow \mathbf{z} \Rightarrow \mathbf{j}$
$\Rightarrow \mathbf{sp} \Rightarrow \mathbf{kn} \Rightarrow \mathbf{g} \rightarrow 2$

§21 $\bar{\circ}\ \mathbf{z} \circ \mathbf{s} \Rightarrow \mathbf{q} \Rightarrow \mathbf{up} \Rightarrow \mathbf{kn} \Rightarrow \mathbf{g} \circ k \rightarrow 2$

§22 *denonel* **efqsu***k* $(\mathbf{i}_k)\ 24// \rightarrow 2$

§23 $\mathbf{a} * \mathbf{i} *$

§24 .

The complete LYaPAS program contains about 36 sentences. The method makes it possible to calculate one particular mdnf for 20–30 variables in several minutes for weakly defined functions with number of vectors for which $F(\tilde{\alpha}) = 1$ and $F(\bar{\alpha}) = 0$ up to 1000.

REFERENCES

1. Gavrilov, M. A., Minimization of Boolean Functions Characterizing Switching Networks. *Avtomat. i telemek.* No. 9, 1217–1238 (1959).
2. Zakrevskii, A. D., Description of LYaPAS. This volume.
3. Didenko, V. P., On Methods for Minimization and Construction of Bridge Structures of Switching Devices, *in* "Automatic Control," pp. 444–460 (in Russian). Moscow, Izd. AN SSSR, 1961.
4. Akers, S. B., A Truth Table Method for the Synthesis of Combinational Logic, *IRE Trans. Electron. Computers* **10**, 604–615 (1961).

AN ALGORITHM FOR OBTAINING FACTORED FORMS OF BOOLEAN FUNCTIONS

V. P. Didenko and V. Sh. Okudzhava

The method of detecting obligatory literals (variables or their negations) was widely used in [1] to find particular minimal disjunctive (conjunctive) normal forms (md(c)nf). We shall introduce the following definitions. A conjunction (disjunction) is elementary if it does not contain repeated literals and also literals corresponding simultaneously to a variable and its inversion.

The literal $x_i(\bar{x}_i)$, $(i = 1, 2, \ldots, n)$ of the elementary conjunction (disjunction) U_l, corresponding to the values of the variables in the vector $\tilde{\alpha}_l$, is called obligatory if the values of only this variable distinguish the vectors $\tilde{\alpha}_l$ $(l = 1, \ldots, q_1)$ and $\tilde{\beta}_j$ $(1 \leq j \leq q_2)$ on which $F(\tilde{\alpha}_l) = 1$ and $F(\tilde{\beta}_j) = 0$, respectively.

An elementary conjunction U_l $(l = 1, \ldots, q_1)$, composed of obligatory literals, is called insufficient[1] if and only if $U_l \cdot F_0 \not\equiv 0$, where F_0 is the logical sum of the constituents (the elementary conjunction (disjunctions) containing n literals each) on which the initial function takes on the value zero.

The logical sum of insufficient conjunctions U_l such that $\bigvee_{l=1}^{p} U_l F_l \equiv F_1$ and for which the relation

$$\bigvee_{\substack{l=1\\ l\neq j}}^{p} U_l F_1 \not\equiv F_1$$

[1] For conciseness we shall call it insufficient conjunction.

holds in the elimination of an arbitrary U_j $(j, l = 1, \ldots, p)$ is called a contradictory particular mdnf, where the notation

$$\bigvee_{\substack{l=1\\l\neq j}}^{p} U_l$$

signifies that the logical sum is taken over all l except $l = j$, and F_1 is the logical sum of the constituents on which the initial function takes on the value unity.

The logical sum of insufficient conjunctions U_m such that

$$\bigvee_{m=1}^{p_1} U_m F_1 \not\equiv F_1$$

is called an insufficiently contradictory particular mdnf. For $\bigvee_{m=1}^{p_1} U_m$ the relations

$$\bigvee_{m=1}^{p_1} U_m F_1 = F_1', \qquad F_1 \not\equiv F_1' \qquad \text{and} \qquad \bigvee_{\substack{m=1\\m\neq j}}^{p_1} U_m F_1' \not\equiv F_1'$$

hold. We denote by the symbols M_1 and M_0 the set of constituents on which $f(x_1, \ldots, x_n)$ takes on the values one and zero, respectively.

If the elementary conjunction U_l implies a logical sum of constituents of which not one does not belong to the set M_0, and if after crossing off from U_l of any single literal the U_l' obtained implies at least one constituent entering into the set M_0, then U_l is a minimal term. If U_l consists of obligatory literals and $U_l F_0 \equiv 0$, then, as shown in [1], the elementary conjunction U_l is a minimal term of the kernel, i.e., enters into every particular mdnf of the initial function $f(x_1, \ldots, x_n)$.

The method described in [1] for determining the minimal terms of the kernel and the obligatory literals makes it possible to formulate an algorithm for obtaining an irredundant factored form of a Boolean function.

This problem has also been considered by Gavrilov in [2]. However, to form the factored forms in [2] a method of partial inversion of constituents on which the function F takes on the value zero was used.

The literals factored outside the parentheses were not taken as obligatory, but were the literals having the largest number of occurrences in the constituents corresponding to the working states of the output of a switching circuit. This sometimes leads to the possible appearance in factored forms of redundant literals or conjunctions

Factored forms obtained by means of the method described below are free of this defect. They may be represented in the form of the first two types of contact trees considered in [3]. However, in contradistinction to these, the condition is not required that each network between the input nodes of the ith level of the tree and the corresponding output nodes of the $i + 1$st level must include only a single contact. These types of tree [3] have the most extensive practical application.

Let us assume that the function F is given by the vectors $\tilde{\alpha}_l$ ($l = 1, 2, \ldots, q_1$) and $\tilde{\beta}_j$ ($j = 1, \ldots, q_2$) on which they take the values $F(\tilde{\alpha}_l) = 1$ and $F(\tilde{\beta}_j) = 0$. We denote the set of vectors $\tilde{\alpha}_l$ and $\tilde{\beta}_j$ by M_1 and M_0, respectively.

The algorithm for finding factored forms of a Boolean function contains the following basic parts:

(1) finding from M_1 and M_0 the kernel and all the inessential conjunctions,

(2) absorption of the minimal terms of the kernel by the insufficient conjunctions,

(3) finding the contradictory or insufficiently contradictory particular mdnf, and

(4) formation of a factored form of the Boolean function.

Let us consider the first part of the algorithm. From the set M_1 we take the first vector in order. Let this be the vector $\tilde{\alpha}_1$.

We first add modulo 2 the components of the vector $\tilde{\alpha}_1$ with the components of the vector $\tilde{\beta}_1 \in M_0$, then the components of $\tilde{\alpha}_1$ with the components of the vector $\tilde{\beta}_2 \in M_0$, etc.; i.e.,

$$\tilde{\alpha}_1 \oplus \tilde{\beta}_j = \tilde{\alpha}_j' \qquad (j = 1, \ldots, q_2).$$

We denote the vectors of the form $\tilde{\alpha}_j'$, containing one and only one component of unity, by $\tilde{\alpha}_s'$.

We determine the vector $\tilde{\alpha}_1'$ from the relation $\tilde{\alpha}_1' = \bigvee_{s=1}^{q} \tilde{\alpha}_s'$, where $q \leq n$. We write down $\tilde{\alpha}_1$ and $\tilde{\alpha}_1'$. We tag the unit values of the components of vector $\tilde{\alpha}_1'$ with those corresponding components of the vector $\tilde{\alpha}_1$ that are obligatory; in other words, α_1 is the code of the obligatory values of the variables of the vector $\tilde{\alpha}_1$. Putting an obligatory literal into correspondence with each obligatory value of the variable ($0 - \bar{x}_i$ and $1 - x_i$) in the vector $\tilde{\alpha}_1$, and forming from them the product, we determine the elementary conjunction U_1. If $U_1F_1 \equiv 0$, then U_1 is a minimal term of the kernel. If $U_1F_0 \not\equiv 0$, then U_1 is an insufficient conjunction. We write down $\tilde{\alpha}_1'$ as an element of the set M_2 and associate it with the element $\tilde{\alpha}_1$ of the set M_1. If the vector $\tilde{\alpha}_1'$ is such that a minimal term U_1 is determined from its components and from the vector $\tilde{\alpha}_1$, then in M_1 we tag $\tilde{\alpha}_1'$ in some manner. Otherwise (when an insufficient conjunction U_j is determined from $\tilde{\alpha}_1$ and $\tilde{\alpha}_1'$) we leave $\tilde{\alpha}_1'$ untagged.

In the set M_1 we take the second vector. Let it be the vector $\tilde{\alpha}_2$. We repeat the above procedure for the vector $\tilde{\alpha}_2$. Then we take a third vector in the set M_1, etc., until all vectors in M_1 have been taken.

Now let us pass to a description of the second part of the algorithm. We small call a vector $\tilde{\alpha}_l$ on which there is at least one value of a variable that is obligatory for the given vector a characteristic vector of the insufficient conjunction U_j or a minimal term of the kernel φ_i.

We write down the minimal terms of the kernel and the insufficient conjunctions in the form $f_1 = \bigvee_{i=1}^{k} \varphi_i$ and $f_2 = \bigvee_{j=1}^{t} U_j$, where φ_i is a minimal term of the kernel and U_j is an insufficient conjunction. Obviously, $k + t \leq q_1$. We form from f_1 and f_2 the logical sum:

$$f_1 \vee f_2 = \bigvee_{i=1}^{k} \varphi_i \vee \bigvee_{j=1}^{t} U_j.$$

Let us introduce a definition. We say that the vector $\tilde{\alpha}_l$ is covered by the minimal term of the kernel φ_i or by insufficient conjunction U_j if the values of the variables in the vector $\tilde{\alpha}_l$ coincide with the values of the variables occurring in φ_i or U_j.

After all U_j have been found, a test is performed to see whether all except the characteristic vectors in M_1 are covered by insufficient conjunctions. At this stage two cases are possible.

(1) $\bigvee_{j=1}^{t} U_j$ covers all vectors $M_1' \subseteq M_1$, where M' is the set of vectors M_1 except for the characteristic vectors of the kernel.

(2) $\bigvee_{j=1}^{t} U_j$ does not cover all vectors in M_1'.

In the first case the minimal terms of the kernel absorbed by the insufficient conjunctions in $\bigvee_{j=1}^{t} U_j$ are removed from $\bigvee_{i=1}^{k} \varphi_i$. The remaining nonabsorbed minimal terms of the kernel $\bigvee_{i=1}^{k_1} \varphi_i$ cover in the general case a portion of the vectors in M_1. The noncovered vectors in M_1 are denoted by the symbol M_1''. Passage to the third part of the algorithm then takes place.

In the second case the set M_1^2 is formed from the noncovered vectors in M_1'. The values of an arbitrary variable[2] x_i ($i = 1, 2, \ldots, n$) are crossed

[2] By crossing off of the values of a nonobligatory variable x_i in the vectors $\tilde{\alpha}$ and $\tilde{\beta}$ we shall understand the reduction of these vectors to a form in which the ith components α_i and β_i in the vectors $\tilde{\alpha}_i$ qnd $\tilde{\beta}_i$ are identically equal, i.e., $\alpha_i \equiv \beta_i$.

off from the vectors $\tilde{\alpha} \in M_1^2$ and $\tilde{\beta} \in M_0$. Further we apply to the vectors of the sets M_1^2 and M_0 the first and second parts of the algorithm until passage is realized to the above first case of coverage of the vectors in M_1. Since crossing off is carried out fairly arbitrarily, the result of crossing off the values of the nonobligatory variables in all vectors $\tilde{\alpha} \in M_1^2$, $\tilde{\beta} \in M_0$ and applying the first part of the algorithm to these vectors will not be the obtaining of the minimal terms of the kernel, but the minimal terms of a pseudokernel, i.e., a kernel depending on a smaller number of variables than n. The obligatory literals obtained in this will be called pseudo-obligatory. Since the minimal terms of the kernel and the pseudokernel, as well as the insufficient conjunctions and the pseudoinsufficient conjunctions are united into a single set, they become indistinguishable with respect to the qualification "pseudo." This means that all the insufficient conjunctions and minimal terms obtained in the second case enter into $\bigvee_{i=1}^{k} \varphi_i$ and $\bigvee_{j=1}^{t} U_j$. Therefore the second case reduces completely to the first. In the third part of the algorithm a contradictory mdnf is determined from $\bigvee_{j=1}^{t} U_j$. For this the following method[3] is applied. In $\bigvee_{j=1}^{t} U_j$ we take U_1, and in M_1'' those $\tilde{\alpha}$ that are found to be covered by U_1 are tagged. If $\bigvee_{j=1}^{t-1} U_j$ covers U_1, then U_1 is excluded from $\bigvee_{j=1}^{t} U_j$. If it does not cover, then U_1 is written down in the contradictory particular mdnf f_1.

This procedure is continued until f_1 is obtained. Let

$$f_1 = \bigvee_{j=1}^{t_1} U_j, \qquad t_1 \leq t.$$

We now pass to a description of the fourth and last part of the algorithm.

Consider the minimal terms of the kernel. If there exists a kernel $\bigvee_{i=1}^{k_1} \varphi_i$, then it is printed out. We take in f_1 the first insufficient conjunction U_1. We take in M_1'' and M_0 those vectors that are covered by the insufficient conjunction U_1. We denote these vectors by the symbols $M_1^{1,1}$ and $M_0^{1,1}$, i.e., we tag these vectors in some way in the sets M_1 and M_0.

At this time the insufficient conjunction U_1 is listed, and then an opening parenthesis. Further, the four parts of the algorithm are again applied to the vectors of the sets $M_1^{1,1}$ and $M_0^{1,1}$, as a result of which the sets $M_1^{1,2}$ and $M_0^{1,2}$ are obtained, and so on until from the sets $M_1^{1,r}$, $M_0^{1,r}$ $(r < n)$ only minimal terms of the kernel or pseudokernel will be obtained. Then the latter are printed out, and directly behind them a closing parenthesis. Since in $M_1^{1,r}$, $M_0^{1,r}$ all vectors $\tilde{\alpha}$ are covered by the minimal terms that

[3] This method corresponds to a large number of operations, but the working storage is substantially reduced in computer automation of the method.

have been found, passage is effected to the vectors of the sets $M_1^{1,r-1}$, $M_0^{1,r-1}$, for which the above procedure is carried out, and so on until all vectors in the set $M_1^{1,1}$ are covered. Further, from $\bigvee_{j=0}^{t_1} U_j$ the insufficient conjunction U_2 is taken. From this conjunction in $M_1'\backslash M_1^{1,1}$ are extracted the sets $M_1^{2,1}$ and $M_0^{2,1}$ of vectors $\tilde{\alpha}$ and $\tilde{\beta}$, respectively. To the vectors $M_1^{2,1}$ and $M_0^{2,1}$ is applied exactly the same process that was used to determine the irredundant factored form for the vectors $M_1^{1,1}$ and $M_0^{1,1}$. This process for obtaining factored forms is continued until the subsets $M_1^{t_1,1}$ and $M_0^{t_1,1}$ are obtained from the insufficient conjunction U_t and the untagged vectors in M_1 and M_2. With the factored form of the function taking on the value 1 on the vectors of the set $M_1^{t_1,1}$, the process of formation of the factored representation of the initial function F, taking on the value 1 on the vectors $\tilde{\alpha} \in M_1$, terminates. As a result the factored expression of the initial function F_1 will have the form

$$\begin{aligned} F_g = F_{g0} \vee U_1{}^0\{\vee\ \varphi_1{}^0 \vee U_1{}'\{\vee\ \varphi_1{}' \vee U_1''[\vee\ \varphi_1'' \vee \cdots \\ \vee\ U_1^{m_1}(\vee\ \varphi_1^{m_1})\cdots]\}\} \vee \cdots \vee U_{t_1}^0\{\vee\ \varphi_{t_1}^0 \vee U_{t_1}'\{\vee\ \varphi_{t_1}' \\ \vee\ U_{t_1}''[\vee\ \varphi_{t_1}'' \vee \cdots \vee U_{t_1}^{m_v}(\vee\ \varphi_{t_1}^{m_v})\cdots]\}\}, \end{aligned}$$

where F_{g0} is the sum of minimal terms of the kernel; $\vee\ \varphi_i{}^j$ $(i = 1, 2, \ldots, t_1)$, $j = 0, 1, \ldots, m_l$ $(l = 1, \ldots, v)$, $m_v < n$ is the logical sum of minimal terms of the kernel or pseudokernel; and $U_i{}^j$ $(i = 1, 2, \ldots, t_1)$, $(j = 0, 1, \ldots, m_l)$, $l = 1, 2, \ldots, v$ are insufficient conjunctions.

The proof of irredundancy of the obtained factored form of representation of the initial function follows from the rules of construction of the factored form themselves, namely, the utilization of the pseudokernel in each stage of the progress of determination of the kernel, and the search for insufficient conjunctions with formation from them of insufficient contradictory particular mdnf. Thus, as can be seen from the procedure for construction of the factored form, on the one hand, it is impossible to eliminate a single term from F_g, and on the other, it is impossible to eliminate a single literal from $\varphi_i{}^j$ or $U_i{}^j$ without the expression F_g becoming contradictory. This proves the irredundancy of the expression F_g obtained.

We shall present the algorithm in LYaPAS [4]. One operator corresponds to each part of the algorithm: *deob* (determination of obligatory literals), *abker* (removal of absorbed minimal terms of the kernel), *ircov* (obtaining an irredundant coverage), *synthesis*.

The result of application of the operator *deob* to a Boolean function is the vector of all conjunctions consisting of obligatory literals (insufficient conjunctions and minimal terms of the kernel). The initial function is given by the list of vectors $\tilde{\alpha} \in M_1$ and $\tilde{\beta} \in M_0$, represented by the 15 leftmost components (from 0–14, inclusive) of complexes α and β. The operator is applied successively to the subsets $M_0^{i,r}$ and $M_1^{i,r}$ $(1 \leq i \leq t_1)$, which are defined by columns of tags (for which 15 components of complexes α and β, from 15–29, inclusive, are reserved) and the variable δ. $M_1^{i,r}$ and $M_0^{i,r}$ consist of those elements of the complexes α and β for which $\alpha^{15+r} = 1$, $\beta^{15+r} = 1$, $\delta^{15+r} = 1$, and the remaining components are equal to zero.

The variable ϵ determines the set of variables on which the function depends (the function depends on the variable x_i, $0 \leq i \leq n-1$ if $\epsilon^i = 1$).

The variable ζ (1's in the components from 0–14, inclusive) serves to extract the sets of values of the variables and the conjunctions of obligatory literals. The conjunctions of obligatory literals are defined by the pair of elements α_i, γ_i with the same value of the index, where α_i is a characteristic vector and γ_i the code of a conjunction of obligatory literals. For example, the conjunction $\bar{x}_1 x_3$ corresponds to the pair of elements $\alpha_i =$ 00 11 01 10 and $\gamma_i =$ 10 10 00 00.

deob $\alpha\kappa +, \beta\kappa +, \gamma\kappa +, \delta\pi +, \epsilon\pi +, \zeta\pi +$ $(\mathbf{b}, b)$
75 242 7

$\bar{\circ}\ b$

§1 $\triangle\, b \oplus b_\alpha \circ\!\!\rightarrow 7\alpha_b \wedge \delta \circ\!\!\rightarrow 1 c_{36} \vee c_{37}$ — Finding obligatory literals
$\Rightarrow a\alpha_b \wedge a \mid\!\rightarrow 1\ \bar{\circ}\ a$

§2 $\triangle\, a \oplus b_\beta \circ\!\!\rightarrow 3\beta_a \wedge \delta \circ\!\!\rightarrow 2\alpha_b \oplus \beta_a \wedge \epsilon$
$\Rightarrow \mathbf{a} \mid\!\leftarrow \oplus\ c_0 \mid\!\rightarrow 2\mathbf{a} \vee \gamma_b \Rightarrow \gamma_b \rightarrow 2$

§3 $\gamma_b \wedge \zeta \circ\!\!\rightarrow 1\ \bar{\circ}\ a\gamma_b \wedge \zeta \Rightarrow \mathbf{b}$ — Determining minimal terms of the kernel

§4 $\triangle\, a \oplus b_\beta \circ\!\!\rightarrow 6\beta_a \wedge \delta \circ\!\!\rightarrow 4\alpha_b \oplus \beta_a \wedge \mathbf{b}$
$\mid\!\rightarrow 4\ \bar{\circ}\ a$

§5 $\triangle\, a \oplus b_\alpha \circ\!\!\rightarrow 1\alpha_a \wedge \delta \circ\!\!\rightarrow 5\alpha_a \oplus \alpha_b \wedge \mathbf{b}$ — Cover M_1 by insufficient conjunctions
$\mid\!\rightarrow 5\gamma_a \vee c_{37} \Rightarrow \gamma_a \rightarrow 5$

§6 $\gamma_b \vee c_{36} \Rightarrow \gamma_b \rightarrow 1$
§7 .

The operator *abker* removes from the set represented by the complexes α and β those minimal terms of the kernel and the pseudokernel that are absorbed by insufficient conjunctions. The initial data and the result are written similarly to the operator *deob*. The complex α represents the set of vectors of the values of the variables M_1, and the complex β, the set of codes of conjunctions of obligatory literals. On removal of the absorbed minimal terms of the kernel, the corresponding codes of the conjunctions are erased. The variable γ corresponds to the variable ξ in the operator *deob*.

abker $\alpha\text{к}, \beta\text{к}, \gamma\text{п} + (\mathbf{b}, b)$
76 131 3

$\bar{\circ}\ b$

§1 $\triangle\ b \oplus b_\alpha \circ\!\!\rightarrow 3\alpha_b \wedge c_{36} \circ\!\!\rightarrow 1\beta_b \wedge \gamma \urcorner \Rightarrow \mathbf{a}\ \bar{\circ}\ a$ — Test minimal terms of the kernel for absorption
§2 $\triangle\ a \oplus b_\alpha \circ\!\!\rightarrow 1\alpha_a \wedge c_{37} \circ\!\!\rightarrow 2\alpha_a \wedge c_{36} \mid\!\!\rightarrow 2\beta_a$
$\wedge\ \gamma \Rightarrow \mathbf{b}\alpha_a \oplus \alpha_b \vee \mathbf{a} \wedge \mathbf{b} \mid\!\!\rightarrow 2\gamma \urcorner \wedge \beta_b$
$\Rightarrow \beta_b c_{36} \urcorner \wedge \alpha_b \Rightarrow \alpha_b \rightarrow 1$

§3 .

The result of applying the operator *ircov* to the set of insufficient conjunctions is a contradictory particular md(c)nf. The initial function and the set of insufficient conjunctions are given in the same way as in the operator *deob*. The complex α represents the set M_1, the complex β, the set of codes of conjunctions of obligatory literals. The codes of the conjunctions that do not enter into the contradictory particular md(c)nf are erased in the complex β. The variable γ (1's in the 15 leftmost components) serves to extract vectors of values of the variables α and the codes of conjunctions from the complexes α and β. The variable γ extracts the subset $M_1^{j,r}$ from $M_1^{j,r-1}$ $(1 \leq j \leq t_1{}^j)$, to which is applied in the given stage the operator *ircov*. The result is a conjunction of obligatory literals entering into the contradictory particular md(c)nf, represented as before by the complexes α and β.

ircov $\alpha\text{к}, \beta\text{к}, \gamma\text{п} +, \delta\text{п} + (\mathbf{c}, c)$
77 345 13

$\bar{\circ}\ c$

§1 $\triangle\ c \oplus b_\alpha \circ\!\!\rightarrow 13\beta_c \wedge \gamma \circ\!\!\rightarrow 1\beta_c \wedge c_{36} \rightarrow 1\beta_c$
$\wedge\ \gamma \Rightarrow \mathbf{a}\gamma \urcorner \wedge \beta_c \Rightarrow c_c\ \bar{\circ}\ a$ — Determination of the elements of the complex α covered by an insufficient conjunction

§2 $\triangle\ a \oplus b_\alpha \circ\!\!\rightarrow 13\alpha_a \wedge \delta \circ\!\!\rightarrow 2\alpha_a \wedge c_{36} \mid\!\!\rightarrow 2\alpha_a$
$\oplus\ \alpha_c \wedge \mathbf{a} \mid\!\!\rightarrow 2c_{37} \urcorner \wedge \alpha_a \Rightarrow \alpha_a \rightarrow 2$
§3 $\bar{\circ}\ b$
§4 $\triangle\ b \oplus b_\alpha \circ\!\!\rightarrow 12\beta_b \wedge \gamma \circ\!\!\rightarrow 4\beta_b \wedge c_{36}$
$\mid\!\!\rightarrow 4\beta_b \wedge \gamma \Rightarrow \mathbf{b}\ \bar{\circ}\ a \odot c$

§5 $\triangle\ a \oplus b_\alpha \circ\!\!\rightarrow 7\alpha_a \wedge \delta \circ\!\!\rightarrow 5\alpha_a \wedge c_{36} \mid\!\!\rightarrow 5\alpha_b$
$\oplus\ \alpha_a \wedge \mathbf{b} \mid\!\!\rightarrow 6c_{37} \vee \alpha_a \Rightarrow \alpha_a \rightarrow 4$ — Coverage of M_1 by all insufficient conjunctions

§6 $\alpha_a \wedge c_{37} \mid\!\rightarrow 5c_{37} \Rightarrow \mathbf{c} \rightarrow 5$
§7 $\mathbf{c} \mid\!\rightarrow 4$
§10 $\bar{\circ}\ a$
§11 $\triangle\ a \oplus b_\alpha \circ\!\!\rightarrow 1\alpha_a \wedge \delta \circ\!\!\rightarrow 11\alpha_a \wedge c_{36}$
$\mid\!\rightarrow 11\alpha_a \vee c_{37} \Rightarrow \alpha_a \rightarrow 11$
§12 $\beta_c \vee \mathbf{a} \Rightarrow \beta_c \rightarrow 10$
§13 .

The operator *synthesis* lists the insufficient conjunctions, the minimal terms of the kernel, and the parentheses (opening and closing), and forms the new sets $M_1^{i,r+1}$ and $M_0^{i,r+1}$. The complexes α and β and the variables ϵ and ζ (corresponding to the variables ζ and δ of the operator *deob*) represent the sets $M_1^{i,r}$ and $M_0^{i,r}$. The variable η takes on the value of the constant c_{37} if the factored form has been completely obtained. The value of the element δ_ϑ of complex δ represents the number of insufficient conjunctions that have already been printed from the given contradictory particular md(c)nf. The value of the index ϑ represents the number $r + 1$ (the $r + 1$st level of the Boolean tree, corresponding to the factored form obtained).

In printing conjunctions the code γ_i is first printed and then the corresponding characteristic vector α_i. The constant c_{37} is printed for the opening parenthesis, the constant c_{36} for the closing.

synthesis αк, βк, γк, δк, ϵп +, ζп, ηп, ϑи(**c**, c)
100 433 16

$\circ\ \eta\zeta > 1 \Rightarrow \mathbf{c}$

§1 $\delta_\vartheta \mid\!\rightarrow 4\ \bar{\circ}\ a$

§2 $\triangle\ a \oplus b\alpha \circ\!\!\rightarrow 3\gamma_a \wedge c_{36} \circ\!\!\rightarrow 2\gamma_a \wedge \epsilon * \alpha_a$ Print kernel
$\wedge\ \epsilon * \rightarrow 2$

§3 $\delta_\vartheta + 1 \Rightarrow \delta_\vartheta \rightarrow 1$ Pass to new level
§4 $\bar{\circ}\ a \circ \mathbf{a}$

§5 $\triangle\ a \oplus b_\alpha \circ\!\!\rightarrow 12\gamma_a \wedge \epsilon \circ\!\!\rightarrow 5\gamma_a \wedge c_{36}$ Print insufficient conjunction and opening parenthesis
$\mid\!\rightarrow 5\mathbf{a} + 1 \Rightarrow \mathbf{a} \oplus \delta_\vartheta \mid\!\rightarrow 5\gamma_a \wedge \epsilon$
$\Rightarrow \mathbf{a} * \alpha_a \wedge \epsilon \Rightarrow \mathbf{b} * c_{37} * \delta_\vartheta + 1$
$\Rightarrow \delta_\vartheta\ \triangle\ \vartheta\ \bar{\circ}\ a$

§6 $\triangle\ a \oplus b_\alpha \circ\!\!\rightarrow 7\alpha_a \wedge c_{37} \circ\!\!\rightarrow 6\gamma_a \wedge \zeta$ Obtain $M_1^{i,r+1}$
$\mid\!\rightarrow 6\mathbf{b} \oplus \alpha_a \wedge \mathbf{a} \mid\!\rightarrow 6\alpha_a \vee \mathbf{c} \Rightarrow \alpha_a\gamma_a$
$\vee\ \zeta \Rightarrow \gamma_a \rightarrow 6$
§7 $\bar{\circ}\ a$

§10 $\triangle\, a \oplus b_\beta \circ\!\!\rightarrow 11\beta_a \wedge \zeta \circ\!\!\rightarrow 10\beta_a \oplus \mathbf{b} \wedge \mathbf{a}$ Obtain $M_0^{i,r+1}$
$\mapsto 10\beta_a \vee \mathbf{c} \Rightarrow \beta_a \rightarrow 10$

§11 $\mathbf{c} \Rightarrow \zeta \rightarrow 16$ Print closing parenthesis

§12 $\circ\, \delta_\vartheta\, \bar{\circ}\, a\, \bar{\circ}\, b\, \bar{\circ}\, c\zeta \Rightarrow \mathbf{c} < 1$
$\Rightarrow \zeta c_{36} * \vartheta \circ\!\!\rightarrow 15\, \bar{\wedge}\, \vartheta$
§13 $\triangle\, a \oplus b_a \circ\!\!\rightarrow 14\zeta \urcorner \bar{\wedge}\, \alpha_a \Rightarrow \alpha_a \rightarrow 13$
§14 $\triangle\, b \oplus b_\beta \circ\!\!\rightarrow 16\zeta \urcorner \wedge \beta_b \Rightarrow \beta_b \rightarrow 14$

§15 $c_{37} \Rightarrow \eta$ Test end
§16 .

According to the algorithm the operators are connected by an external program. The 15 left components of the complexes A and B represent the sets M_1 and M_0, respectively. The 16th component of complexes A and B has the value 1 for all elements. The values of all components of the complexes C and D are equal to zero. The cardinality of the complex $D = 15$ (functions of not more than 15 variables are minimized). The results of the external program are listed.

§0 $\circ\, \mathbf{f}\, c_{17} \Rightarrow e$
§1 $d \Rightarrow \mathbf{g} \circ \mathbf{f} \rightarrow 5$
§2 $\mathbf{f} \mapsto 3c_0 \Rightarrow \mathbf{f} \rightarrow 4$
§3 $\mathbf{f} > 1 \Rightarrow \mathbf{f}$
§4 $\mathbf{g} - \mathbf{f} \Rightarrow \mathbf{g}$

§5 *deob* $ABC\mathbf{egd}//\,\bar{\circ}\, dc_{37} \vee c_{36} \Rightarrow \mathbf{i}$ Find insufficient conjunctions
§6 $\triangle\, d \oplus b_0 \circ\!\!\rightarrow 7c_d \wedge \mathbf{i} \vee a_d \Rightarrow a_d\mathbf{i} \urcorner \wedge c_d$
$\Rightarrow c_d \rightarrow 6$

§7 $\bar{\circ}\, d$ Test function for coverage

§10 $\triangle\, d \oplus b_0 \circ\!\!\rightarrow 11a_d \wedge \mathbf{e} \circ\!\!\rightarrow 10a_d \wedge \mathbf{i}$
$\mapsto 10 \rightarrow 2$

§11 *abker* $AC\mathbf{d}//\,\bar{\circ}\, d$ Absorption of kernel
§12 $\triangle\, d \oplus b_0 \circ\!\!\rightarrow 13a_d \wedge c_{36} \circ\!\!\rightarrow 12c_d \vee c_{36}$
$\Rightarrow c_d \rightarrow 12$

§13 $\bar{\circ}\, e$ Coverage of the set M_1^{ik}
§14 $\triangle\, e \oplus b_0 \circ\!\!\rightarrow 16c_e \wedge c_{36} \circ\!\!\rightarrow 14c_e \wedge \mathbf{d}$
$\Rightarrow \mathbf{h}\, \bar{\circ}\, d$
§15 $\triangle\, d \oplus b_0 \circ\!\!\rightarrow 14a_d \wedge \mathbf{e} \circ\!\!\rightarrow 15a_d \oplus a_e$
$\wedge\, \mathbf{h} \mapsto 15c_{37} \urcorner \wedge a_d \vee c_{36} \Rightarrow a_d \rightarrow 15$

§16 *ircov AC***de**//*synthesis ABCD***deif**//**i**
$\oplus\ c_{37}\ \circ\!\rightarrow 22\ \bar{\circ}\ ec_{37} \vee c_{36}\ \urcorner \Rightarrow \mathbf{gd}\ \urcorner$
$\wedge\ \mathbf{g} \Rightarrow \mathbf{h}$

§17 $\triangle\ e \oplus b_0\ \circ\!\rightarrow 20a_e \wedge \mathbf{g} \Rightarrow a_e \rightarrow 17$
§20 $\bar{\circ}\ e$
§21 $\triangle\ e \oplus b_0 \rightarrow 1c_e \wedge \mathbf{h} \Rightarrow c_e \rightarrow 21$
§22 .

Initialization of the complex C and the components of complex A reserved for tags

REFERENCES

1. Didenko, V., The Construction of Particular Minimal Normal Forms of Boolean Functions. This volume.
2. Gavrilov, M. A., The Minimization of Boolean Functions Characteristic of Switching Networks, *Avtomat. i Telemek.* No. 9, 1217–1238 (1959).
3. Caldwell, S., "Switching Circuits and Logical Design" [Russian translation]. IL, Moscow, 1962.
4. Zakrevskii, A. D., Description of LYaPAS. This volume.

APPROXIMATE METHOD FOR OBTAINING MINIMAL FACTORED FORMS OF A CERTAIN CLASS

V. I. Ostrovskii

In problems connected with the synthesis of logical circuits it is generally useful to obtain a factored form of representation of the Boolean function to be realized, optimized according to given criteria. The circuit corresponding to the factored form of representation of the Boolean function can be substantially simpler than one corresponding to the function initially given (for example, in minimal disjunctive form). In particular, in the realization of a Boolean function by Zakrevskii's method [1] it is necessary to find a factored form of the function in the form of the superposition of operators

$$\varphi = (\alpha_i \vee \cdots \vee \alpha_k)\beta_l \cdots \beta_m$$

in the variable α_j (variables of type β can be only the arguments of functions).[1] To each such operator there corresponds a diode quasielement in the circuit. The method of connecting quasielements is uniquely given by the concrete form of the superposition of operators, i.e., the factored form. Multistage diode circuits contain substantially fewer diodes than diode networks that perform the same function. To illustrate this, we present the example of two diode circuits that realize the same Boolean function.

$$f = x_1\bar{x}_2x_3x_5 \vee x_1\bar{x}_2x_4x_5 \vee x_3x_5x_6 \vee x_3\bar{x}_4 \vee \bar{x}_1\bar{x}_4 \vee \bar{x}_1x_2\bar{x}_3. \quad (1)$$

[1] In referring to factored forms we shall mean factored forms of this class only.

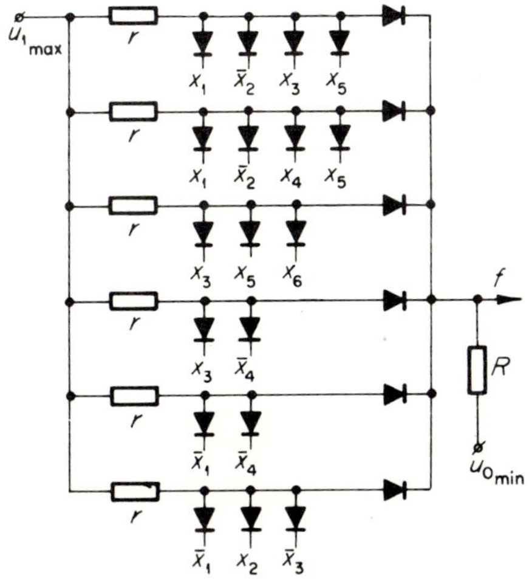

FIG. 1.

Figure 1 shows a diode circuit corresponding to the disjunctive form (1) of the function. It contains 24 diodes.[2] The diode circuit corresponding to the factored form of representation of the same function

$$f_f = x_5(x_1\bar{x}_2(x_3 \vee x_4) \vee x_3x_6) \vee \bar{x}_4(x_3 \vee \bar{x}_1) \vee \bar{x}_1x_2\bar{x}_3, \tag{2}$$

is given in Fig. 2. It contains 18 diodes, i.e., six fewer than the circuit in Fig. 1.[3] The number of diodes in the circuit depends on the corresponding scheme of the factored form representing the function. For example, if function (1) is represented in the form

$$f_f' = x_3(x_5(x_1\bar{x}_2 \vee x_6) \vee \bar{x}_4) \vee \bar{x}_1(x_2\bar{x}_3 \vee \bar{x}_4) \vee x_1\bar{x}_2x_4x_5,$$

the corresponding diode circuit will contain 20 diodes. In more complicated examples the difference can be more substantial.

Thus, obtaining an optimal factored form is an essential stage in the synthesis of diode logical circuits. A natural criterion according to which one factored form is preferable to another can be the number of diodes in

[2] Translator's note: Of course, the critical reader will have observed that (1) is not in minimal form, and that the first conjunction is redundant. The diode network corresponding to the true mdnf contains only 19 diodes.

[3] Translator's note: Correspondingly, the factored form corresponding to the true mdnf will have only 17 diodes, a difference of only two. Leaving aside questions of the possible technological interest of the circuit of Fig. 2, the factored form may be interesting in its own right.

the corresponding network. This criterion can be substituted by a simpler one, corresponding approximately to it. We shall say that one factored form is more nearly optimal than another if it contains fewer symbols (variables and their inversions). A factored form of representation of a given Boolean function, then, will be called minimal if it contains a minimal number of symbols.

In this chapter we describe an approximate method permitting a close-to-minimal factored form to be found for a Boolean function represented in disjunctive form. The possibility of minimization of the number of symbols is considered only through the application of the law $Ax \vee Bx = (A \vee B)x$.

The problem is formulated in the following manner. Let there be given a certain Boolean function $F(x_1, \ldots, x_n)$ in disjunctive form $F = \bigvee_{i=1}^{m} a_i$. It is required to find a minimal (or close-to-minimal) factored form of representation of the function F.

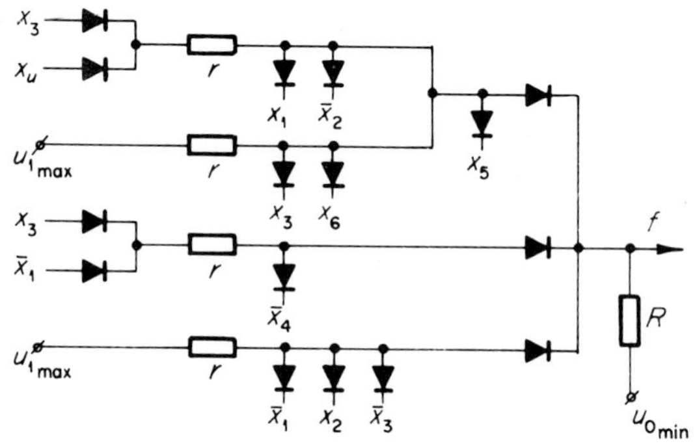

FIG. 2.

We denote the set of all conjunctions entering into the disjunctive form by A:

$$A = \{a_1, a_2, \ldots, a_m\}.$$

Two constraints are imposed on the set A without violation of the generality of the problem:

(1) No pair of conjunctions belonging to A is connected by the relation of inclusion. Otherwise the conjunction containing the larger number of symbols can be excluded from the initial disjunctive form. This transformation simply leads to a reduction in the number of symbols in the corresponding factored form.

(2) It is assumed that the conjunctions of the set A do not contain a common multiplier. If one exists, it must be factored out and the problem of obtaining the minimal form is considered for the expression inside the parentheses.

Each factored form of representation of the given function is put into correspondence with an ordered set W of subsets A_i of the set A and the ordered set W' of variables x_j ($j = 1, \ldots, n$) and their inversions, according to the following rule. A given subset A_i belongs to W if and only if the factored form represented by the set W contains the union of conjunctions listed in the subset, enclosed in parentheses, with factoring out of one or several symbols (variables of their inversions). If k symbols are factored out, the corresponding subset enters k times into the set W. The ordered list of variables x_j and their inversions, factored outside the parentheses, corresponding to the subsets A_i belonging to the set W, forms a set W'. For example, the factored form (2) corresponding to the disjunction (1) can be represented by the sets

$$W = \{\{a_1, a_2\}, \{a_1, a_2\}, \{a_1, a_2, a_3\}, \{a_4, a_5\}\},$$
$$W' = \{x_1, \bar{x}_2, x_5, \bar{x}_4\},$$

where

$$a_1 = x_1\bar{x}_2x_3x_5, \quad a_2 = x_1\bar{x}_2x_4x_5, \quad a_3 = x_3x_5x_6, \quad a_4 = x_3\bar{x}_4, \quad a_5 = \bar{x}_1\bar{x}_4.$$

Thus the sets W and W' uniquely define a certain factored form of Boolean function given in disjunctive form, and the problem of finding an optimal factored form can be reduced to the problem of finding the sets representing this form.

If subsets A_i and A_k enter into the set W, they must be related by one of the three following relations:

$$A_i \cap A_k = \begin{cases} \varnothing, \\ A_i, \\ A_k. \end{cases} \tag{3}$$

This property follows directly from the fact that each conjunction appears in the set A only once. Below we shall say that two subsets A_i and A_k are in contradiction if condition (3) is not satisfied.

Let us consider the set of subsets $\hat{Y}_i$ of the set A. Each $\hat{Y}_i$ contains two conjunctions of the set A, containing the symbol y_i:

$$a_k \in \hat{Y}_i \leftrightarrow (a_k \in A) \wedge (y_i \in a_k), \qquad y_i \in \{x_1, \ldots, x_n, \bar{x}_1, \ldots, \bar{x}_n\}.$$

We denote by W_0 the set of all $\hat{Y}_i$ whose cardinalities are not less than 2. The ordered set of symbols y_i (variables and their inversions) corresponding to the set W_0 is denoted by W_0'. For example, for the function (1) sets W_0 and W_0' have the following form:

$$\begin{aligned} W_0 &= \{\{a_1, a_2\}, \{a_1, a_3, a_4\}, \{a_1, a_2, a_3\}, \{a_5 a_6\}, \{a_1 a_2\}, \{a_4 a_5\}\}; \\ W_0' &= \{x_1, x_3, x_5, \bar{x}_1, \bar{x}_2, \bar{x}_4\}. \end{aligned} \tag{4}$$

If no two sets belonging to W_0 are in a state of contradiction, the initial function F can be represented in factored form containing $2n - k$ symbols (where k is the number of variables that enter into the initial disjunctive form either only with the sign of inversion or only without it), where W_0 and W_0' are the sets corresponding to this factored form. Since in the given statement of the problem it is impossible to obtain a factored form not containing some one symbol occurring in the disjunctive form, the factored form corresponding to the sets W_0, W_0' will be minimal.

If certain of the $\hat{Y}_i$ belonging to the set W_0 are in contradiction, there does not exist a factored form of the given class corresponding to the sets W_0 and W_0'.

Let us define the operation C_i^k of section of the sets W_0 and W_0' as the operation consisting of the partitioning of the set $\hat{Y}_i$, belonging to W_0, into two nonempty, nonoverlapping subsets $\hat{Y}_i^k$ and $\hat{Y}_i \backslash \hat{Y}_i^k$. At the same time the set $\hat{Y}_i$ in W_0 is replaced by those of the two newly obtained subsets $\hat{Y}_i^k$ and $\hat{Y}_i \backslash \hat{Y}_i^k$ whose cardinalities are not less than 2.

In the ordered set of symbols W_0', that symbol which previously corresponded to the set $\hat{Y}_i$ is associated with each of the newly introduced subsets in the set W_0. Let us take the example of the sets W_1, W_1'; W_2, W_2'; and W_3, W_3', which can be obtained from the sets W_0, W_0' (4) as a result of the section operations, which we denote by C_2^1, C_2^2, C_4^1:

$$\begin{aligned} W_1 &= \{\{a_1, a_2\}, \{a_1, a_3\}, \{a_1, a_2, a_3\}, \{a_5, a_6\}, \{a_1, a_2\}, \{a_4, a_5\}\}, \\ W_1' &= \{x_1, x_3, x_5, \bar{x}_1, \bar{x}_2, \bar{x}_4\}, \\ W_2 &= \{\{a_1, a_2\}, \{a_1, a_4\}, \{a_1, a_2, a_3\}, \{a_5, a_6\}, \{a_1, a_2\}, \{a_4, a_5\}\}, \\ W_2' &= \{x_1, x_3, x_5, \bar{x}_1, \bar{x}_2, \bar{x}_4\}, \\ W_3 &= \{\{a_1, a_2\}, \{a_1, a_3, a_4\}, \{a_1, a_2, a_3\}, \{a_1, a_2\}, \{a_4, a_5\}\}, \\ W_3' &= \{x_1, x_3, x_5, \bar{x}_2, \bar{x}_4\}. \end{aligned}$$

The superposition of several section operations will be described by the set of sections. For example, the set of sections $\{C_2^1, C_2^2, C_4^1\}$ transforms the sets W_0, W_0' into the sets W_4, W_4':

$$\begin{aligned} W_4 &= \{\{a_1, a_2\}, \{a_1, a_2, a_3\}, \{a_1, a_2\}, \{a_4, a_5\}\}, \\ W_4' &= \{x_1, x_5, \bar{x}_2, \bar{x}_4\}. \end{aligned}$$

It can be shown that any factored form of a given function F can be represented by the corresponding sets W and W', obtained from the sets W_0 and W_0' by means of a finite set of sections, where the number of symbols S in the factored form is equal to

$$S = 2n - k + c,$$

and c is the cardinality of the set of sections transforming W_0 and W_0' to W and W'.

Thus the problem of obtaining the minimal factored form can be reduced to the following set-theoretical problem: Find the set of sections of minimal cardinality that transforms the set W_0 to a certain set W free of contradictions.

Let us consider this problem in greater detail. We shall be interested only in the transformation of the set W_0, since the transformation of the set W_0' is trivial. Let two sets $\hat{Y}_i$ and $\hat{Y}_j$ ($\hat{Y}_i \in W_0$; $\hat{Y}_j \in W_0$) be in a state of contradiction, i.e.,

$$\hat{Y}_i \cap \hat{Y}_j = \gamma, \qquad (\gamma \neq \hat{Y}_i, \gamma \neq \hat{Y}_j, \gamma \neq \varnothing).$$

This contradiction can be liquidated by means of a single section of either the set $\hat{Y}_i$ or the set $\hat{Y}_j$ with respect to the boundary of the set γ. Aside from this, the sets $\hat{Y}_i$ and $\hat{Y}_j$ can be removed from the state of contradiction by sectioning, and the set $\hat{Y}_i$ and the set $\hat{Y}_j$, if these sections occur inside the set γ, where each of them partitions the set γ into two nonoverlapping subsets γ_1 and γ_2 ($\gamma_1 \cup \gamma_2 = \gamma$; $\gamma_1 \cap \gamma_2 = \varnothing$). In this case one contradiction is eliminated by two sections. Naturally, when only two sets in the set W_0 are in contradiction, the more nearly optimal (from the viewpoint of the chosen criterion) way to liquidate the contradiction is to section one of the sets with respect to the boundary of their common part. However, in the general case it may be found that to liquidate an arbitrary contradiction it will be more nearly optimal to section both the mutually contradictory sets.

The consideration of various methods of liquidating all contradictions among sets belonging to W_0 leads in certain conventional section coordinates to a system of logical equations whose solution is the set of irredundant sets of sections bringing the sets belonging to W_0 out of a state of contradiction. In the set of irredundant sets of sections it is possible by enumeration to obtain a set of section sets having minimal cardinality, and thereby to solve the problem defined. However, this method is connected with repeated computations, since the number of equations in the system and the mean number of symbols in each of the equations rapidly increases with the number of variables on which the initial equation depends and the number of conjunctions in the given disjunctive form.

The following method, intended for computer execution, is proposed for the approximate solution of the problem.

(1) For the set W_0 the set of distinct sections of the subsets $\hat{Y}_i$ in W_0 that take these subsets out of the state of contradiction is found. Only those sections are considered that pass over the boundary of the common part of two subsets in contradiction.

(2) To each section is assigned a definite weight, equal to the number of contradictions the given section can eliminate.

(3) A section having the maximum weight is obtained. In the result the set W_0 is transformed to the set W_1.

(4) The first three steps of the algorithm are repeated. The initial set is now W_1.

(5) This process is repeated until a certain contradiction-free set W_r is obtained. It is assumed that the set W_r corresponds to a minimal (or close-to-minimal) factored form.

A LYaPAS subroutine has been written for this algorithm, which performs the standard L-operator *minpar*. The initial Boolean function is given by the set $A = \{a_1, \ldots, a_m\}$ of conjunctions in its disjunctive form. Each conjunction a_i is represented by an element of the complex α: $\alpha{::}\| A \ni X \|$, where $X = \{x_1, \ldots, x_n, \bar{x}_1, \ldots, \bar{x}_n\}$ is the set of variables and their inversions appearing in the initial disjunctive form of the function. The operator *minpar* is designed for a maximal cardinality of the sets A and X equal to 32. By priming certain operands of *minpar* this number can be increased.

If the minimal factored form contains the union of certain conjunctions within parentheses, all symbols appearing in each of the component conjunctions are factored outside the parentheses. Otherwise the factored form will not be minimal. From this it follows that to represent the minimal factored form it is possible to use one set U, obtained from W by the reduction of similar terms, in place of the two sets W and W'. The set U corresponding to the factored form synthesized by the operator *minpar* is represented by the complex β: $\beta{::}\| U \ni A \|$.

The complex γ, equal in cardinality to the complex β, serves to represent intermediate results of the computation. The cardinalities of the complexes β and γ are not fixed, since they depend on the concrete problem, and can substantially exceed the cardinality of the set U in the intermediate stages of the computation.

The operator *minpar* uses the following L-operators as subroutines:

reduc. The operator for reduction of a complex.

The operator *reduc* forms a complex γ from those elements of the complex α in which the number of bits having the value 1 is not less than β (β is a natural constant).

$$reduc\ \alpha\text{к} +, \beta\text{ч}, \gamma\text{к}/-, b/\alpha\gamma/(\alpha\gamma)$$

71 45 2

$$\circ\ a \circ b$$

§1 $\alpha_a \nabla - \beta \circ\!\!\rightarrow 2\alpha_a \Rightarrow \gamma_b \triangle b$

§2 $\triangle a - b_\alpha \circ\!\!\rightarrow 1b \Rightarrow b_\gamma.$

carprel. The operator for finding the Cartesian product of a complex with its element, with elimination.

The operator *carprel* forms a complex δ whose element δ_i is equal to

$$\delta_i = \begin{cases} \beta_i \odot \beta_j, & \text{if } \beta_i \odot \beta_j \neq \beta_i \text{ and } \beta_i \odot \beta_j \neq \beta_j; \\ 0, & \text{if } \beta_i \odot \beta_j = \beta_i \text{ or } \beta_i \odot \beta_j = \beta_j; \end{cases}$$

where β_i and β_j are elements of the complex β, $i \in \{0, 1, \ldots, b_\beta - 1\}$, $j = \gamma$, $\odot = \alpha$.

$$carprel\ \alpha\text{о}, \beta\text{к} +, \gamma\text{и} +, \delta\text{к}??/(-, a)/(\beta\delta)$$

72 63 3

$\bar{\circ}\ ab_\beta \Rightarrow b_\delta??$

§1 $\triangle a \oplus b_\beta \circ\!\!\rightarrow 3\beta_a\alpha\beta_\gamma \Rightarrow \delta_a \oplus \beta_a \circ\!\!\rightarrow 2\delta_a \oplus \beta_\gamma \mid\!\!\rightarrow 1$

§2 $\circ\ \delta_a \rightarrow 1$

§3 .

parmaxwe. The operator for finding a partition with maximal weight.

The variable γ is a binary vector defining a subset A of a certain set M whose cardinality is not greater than 32: $\gamma{::}\|\ \{A\} \ni M\ \|$. Each nonzero element α_i ($i \in \{0, 1, \ldots, b_\alpha - 1\}$, $\theta(\alpha_i, M) \subset A$) of the complex α defines a partitioning of the subset A: $sep^2A = \{\theta(\alpha_i, M), A\backslash\theta(\alpha_i, M)\}$.[4] A given partition of the subset A can be defined by several different elements of the complex α. We shall call the number of such elements the weight s of this partition. The operator *parmaxwe* finds the partition whose weight is maximal and greater than a certain quantity prescribed by the value of the

[4] $\theta(a, M)$ is a subset of the set M where $a :: \|\{\theta(a, M)\} \ni M\ \|$.

variable δ. If such a partition $sep^2 A = \{A_1, A_2\}$ with weight $s > \delta$ is found, the variables δ, ϵ, ζ, and the index η take on the following values as a result of the functioning of the operator:

$$s = \delta, \quad \epsilon::\|\{A_1\} \ni M\|, \quad \zeta::\|\{A_2\} \ni M\|, \quad \eta = \beta,$$

where β is the index, whose value is given together with the vector of values of the elements of the complex α and corresponds to the number of this vector. If the maximum weight of the partition defined by the complex α does not exceed the value of the variable δ, the values of the variables δ, ϵ, ζ, and the index η remain unchanged after the operation.

parmaxwe αк $+$, βи $+$, γп $+$, δп $+$, ϵп, ζп, ηи$/(\mathbf{a}, b)/(\alpha\gamma\epsilon\zeta)$
73 126 5

$\circ\ a$
§1 $\alpha_a \circ\!\!\rightarrow 51 \Rightarrow \mathbf{a}a \Rightarrow b$
§2 $\triangle\, b - b_\alpha \mid\!\!\rightarrow 4\alpha_b \circ\!\!\rightarrow 2\alpha_b \oplus \alpha_a \circ\!\!\rightarrow 3\alpha_b \oplus \alpha_a \oplus \gamma \mid\!\!\rightarrow 2$
§3 $\circ\ \alpha_b \triangle\, \mathbf{a} \rightarrow 2$
§4 $\mathbf{a} - \delta \circ\!\!\rightarrow 5\mathbf{a} \Rightarrow \delta\alpha_a \Rightarrow \epsilon \oplus \gamma \Rightarrow \zeta\beta \Rightarrow \eta$
§5 $\triangle\, a - b_\alpha \circ\!\!\rightarrow 1.$

The operator *minpar* in second-level LYaPAS is represented as follows:

minpar $(332, 71, 72, 73, 40)/\alpha$к $+$, βк?, γк??$/(\mathbf{d}, d)/(\beta\gamma)$
74 121 3 $(\beta\mathbf{cd})$

transmat $\alpha\beta//$
§1 *reduc* $\beta 2\beta// \circ\ \mathbf{c} \circ \mathbf{b}$
§2 *carprel* $\wedge\ \beta c\gamma//$*parmaxwe* $\gamma c(\beta_c)\mathbf{bcddd}//\triangle\, c - b_\beta \circ\!\!\rightarrow 2\mathbf{b} \circ\!\!\rightarrow 3\mathbf{c}$
$\Rightarrow \beta_d(\mathbf{d}) \Rightarrow \beta? \rightarrow 1.$
§3 *redsim* $\beta\beta//\alpha * \beta *.$

Factored forms have been obtained for several Boolean functions by means of the operator *minpar*. Two examples follow.

Example 1. The given Boolean function has the following form:

$$f = \bar{x}_1x_2\bar{x}_3x_4\bar{x}_5\bar{x}_6 \vee x_1x_2\bar{x}_3x_5\bar{x}_6 \vee x_1x_2\bar{x}_3x_4x_6 \vee x_1\bar{x}_2x_3\bar{x}_4x_6 \vee \bar{x}_1x_2x_3x_4x_6 \vee \bar{x}_1\bar{x}_3\bar{x}_4x_5x_6 \vee \bar{x}_1\bar{x}_2\bar{x}_3x_6.$$

As a result of the operation a set U is obtained:

$$U = \{\{a_1, a_3, a_5\}, \{a_1, a_5\}, \{a_1a_2a_3a_5\}, \{a_4, a_6, a_7\}, \{a_6, a_7\}\},$$

corresponding to the factored form

$$f_f = x_6(x_1\bar{x}_2x_3\bar{x}_4 \vee \bar{x}_1\bar{x}_3(\bar{x}_2 \vee \bar{x}_4x_5)) \vee x_2(x_1\bar{x}_3x_5\bar{x}_6 \\ \vee x_4(x_1\bar{x}_3x_6 \vee \bar{x}_1(\bar{x}_3\bar{x}_5\bar{x}_6 \vee x_3x_6))).$$

The synthesized factored form contains 25 symbols. The minimal factored form of the given function contains 24 symbols.

Example 2. For the Boolean function

$$f = x_2\bar{x}_3x_5x_6 \vee \bar{x}_1x_2\bar{x}_4\bar{x}_5\bar{x}_6 \vee \bar{x}_1\bar{x}_2\bar{x}_5x_6 \vee x_1x_3x_4\bar{x}_5x_6 \vee x_1\bar{x}_2\bar{x}_4\bar{x}_5\bar{x}_6 \\ \vee x_1x_3\bar{x}_4x_5x_6 \vee \bar{x}_2\bar{x}_3\bar{x}_4\bar{x}_5\bar{x}_6 \vee x_1\bar{x}_2x_3x_6 \vee x_1x_2\bar{x}_3x_6$$

the factored form

$$f_f = x_6(x_2\bar{x}_3(x_5 \vee x_1) \vee x_1x_3(x_4\bar{x}_5 \vee \bar{x}_4x_5 \vee \bar{x}_2)) \\ \vee \bar{x}_5(\bar{x}_1\bar{x}_2x_6 \vee \bar{x}_4\bar{x}_6(\bar{x}_1x_2 \vee \bar{x}_2(x_1 \vee \bar{x}_3))),$$

is found, containing 23 symbols. The form found is minimal.

The time for solution of the problems given in the examples is estimated at approximately 10^5 machine operations. In all the examples considered the synthesized factored form differed from the minimal by not more than three symbols.

REFERENCES

1. Zakrevskii, A. D., A Method for the Synthesis of Diode Logical Circuits, *Tr. SFTI* **40**, 73–88 (1961).
2. Kazakov, V. D., The Minimization of Boolean Functions with Utilization of Factoring Outside of Parentheses, *in* "The Structural Theory of Switching Devices," pp. 163–169 [in Russian]. Izd. AN SSSR, Moscow, 1963.
3. Ostrovskii, V. I., Programmed Synthesis of Diode Logical Circuits, *Tr. SFTI* **42**, 93–101 (1963).

REALIZATION OF BOOLEAN FUNCTIONS BY THRESHOLD ELEMENTS

E. A. Butakov, S. V. Bykova, and V. A. Vorob'ev

1. THE PROPERTIES OF THRESHOLD FUNCTIONS

A threshold element is a device combining a linear summation network with n inputs and a discriminator. The role of the discriminator is to compare the weighted sum of input signals with a certain fixed level called the threshold. We shall assume that binary signals arrive at the inputs, described by the variables $x_0, x_1, \cdots, x_{n-1}$, taking on the values 0 and 1. The values of the binary output signal can be associated with the values of a certain Boolean function $f(X) = f(x_0, x_1, \ldots, x_{n-1})$ by the following rule [1, 2]:

$$f(x_0, x_1, \ldots, x_{n-1}) = 1 \leftrightarrow \sum_{i=0}^{n-1} a_i x_i - T \geq 0. \tag{1}$$

The number a_i $(i = 0, 1, \ldots, n-1)$ is the weight of the variable x_i; T is the threshold, the basic characteristic of the discriminator.

Definition. A Boolean function for which a representation in the form (1) exists with real weights and threshold is called a threshold function. We shall term the system of coefficients $a_0, a_1, \ldots, a_{n-1}; T$ the realization of the threshold function or, simply, the realization.

The fact that a Boolean function $f(x_0, x_1, \cdots, x_{n-1})$ has a realization $a_0, a_1, \ldots, a_{n-1}; T$ will be denoted in the following way:

$$f(x_0, x_1, \cdots, x_{n-1}) \sim [a_0, a_1, \ldots, a_{n-1}; T].$$

The class of all threshold functions depending essentially on n arguments will be denoted by R_n.

As is well known, a Boolean function depending on n arguments can be considered as a function whose value is defined on a set of M vertices of the n-dimensional unit cube. We shall denote by δ an arbitrary vertex of the n-dimensional cube independently of the value of $f(X)$ at this vertex; we shall denote by α and β elements of the sets M_1 and M_0, respectively.

We shall also term the vertices of the n-dimensional cube points. It is sometimes convenient to consider a point as the tip of a vector emerging from the origin of coordinates. It is obvious that the ith component of the vector is equal to 1 (0) if the coordinate x_i of the point is equal to 1 (0). The point, the vertex, and the vector will be denoted by one of the three symbols α, β, δ, since it is always clear from the context what sense is given to the symbol.

The expression (1), prescribing the operator of the threshold element, now takes on a clear geometrical interpretation. Namely, the equation

$$\sum_{i=0}^{n-1} a_i x_i - T = 0 \tag{2}$$

is nothing else but the equation of a hyperplane in n-dimensional space, separating the set of unit vertices of the n-dimensional cube M_1 from the set of zero vertices M_0. Since (2) is linear with respect to a_i, the Boolean functions represented in form (1) are also called linearly separable functions [3, 4, 5].

In the presence of certain natural constraints a fairly broad class of physical devices may be considered as threshold elements [6, 7]. We shall list only the most widespread of them: multiwinding relays, magnetic cores with rectangular hysteresis loops, parametrons, tunnel diodes or transistors, combined with a resistive summation circuit. These last two types of element are very promising, since they permit very high switching speeds to be attained and, equally important, substantially increase the number of inputs compared to other types of element. In [8], for example, it is reported that in the laboratories of the General Electric Company a "hybrid" threshold device, a tunnel-diode–transistor combination, has been obtained with 51 inputs.

Finally, as shown by the results of neurophysiological investigations [9], the functional properties of the neuron can be described by a threshold model.

Definition. A Boolean function is called homogeneous if it can be represented in normal form in such a way that each variable enters into this form either only with inversion or only without.

Definition. A Boolean function, depending essentially on all of its arguments, is positive (negative) in the variable x_i if it can be represented in normal form in such a way that the variable x_i does not enter this form with (without) inversion. A function is called homogeneous with respect to some argument x_j if it is either positive or negative with respect to this argument.

Definition. A Boolean function is called positive (negative) if it is positive (negative) with respect to all its arguments.

Theorem 1.1.[1] Every threshold function is homogeneous.

The proof of this important theorem, giving one of the necessary conditions for realizability of a Boolean function by one threshold element, is given in [2, 3, 10].

Below the terms *homogeneous*, *positive*, and *negative* will be applied both to Boolean functions and to their normal forms.

Lemma. The reduced disjunctive normal form of a homogeneous function is homogeneous.

Proof: By definition, the homogeneous function F can be represented in a certain disjunctive normal form F_1 so that no variable enters into this form both with and without inversion. Consequently, in F_1 the operation of merging is not possible, but the operation of absorption is, which yields the reduced form of the function F. This form is obviously homogeneous.

Theorem 1.2. The reduced disjunctive normal form of a homogeneous function is minimal.

In a somewhat different form, this theorem was given in [13].

Corollary. A homogeneous Boolean function has a unique minimal form.

These results permit the formulation of a criterion for testing a Boolean function for homogeneity: To establish the homogeneity of a Boolean function it is sufficient to find its reduced form. A function is positive in the variable x_i if x_i enters into the reduced form only without inversion, and negative if only with inversion.

This method of testing for homogeneity has the advantage that it uses the well-known algorithms for minimization of Boolean functions (more precisely, for obtaining the reduced disjunctive normal form), which have

[1] See references [2, 3, 10–12].

been programmed for computer. However, when it is necessary to conduct the calculations by hand, it is somewhat clumsy, even for a small number of variables.

Below we describe a simple criterion for testing a Boolean function for homogeneity, using the matrix representation of Boolean functions [14–16].

The following theorem uses operators on the matrix representation described in [15 and 16].

Theorem 1.3.[2] A function $F(X)$ does not depend essentially on the variable x_i if

$$S_i^{\oplus}F(X) = 0; \tag{3}$$

for $S_i^{\oplus}F(X) \neq 0$ the function is positive in x_i if

$$\begin{aligned} x_i \wedge [S_i^{\oplus}F(X)] \wedge F(X) &\neq 0, \\ \bar{x}_i \wedge [S_i^{\oplus}F(X)] \wedge F(X) &= 0; \end{aligned} \tag{4}$$

negative if

$$\begin{aligned} \bar{x}_i \wedge [S_i^{\oplus}F(X)] \wedge F(X) &\neq 0, \\ x_i \wedge [S_i^{\oplus}F(X)] \wedge F(X) &= 0; \end{aligned} \tag{5}$$

and not homogeneous if

$$\begin{aligned} x_i \wedge [S_i^{\oplus}F(X)] \wedge F(X) &\neq 0, \\ \bar{x}_i \wedge [S_i^{\oplus}F(X)] \wedge F(X) &\neq 0. \end{aligned} \tag{6}$$

To establish homogeneity of the function by this criterion it is sufficient in practice to construct its matrix representation. Conditions (3–6) are easily tested visually, since the matrix is symmetrical with respect to the x_i-axis [14] if x_i is an inessential variable, while positiveness or negativeness of the function in x_i leads to asymmetry of the matrix on one or the other side. The criterion is particularly effective in the manual solution of problems, and is easily generalized to the case of incompletely defined Boolean functions. In using the matrix representation of a Boolean function for its representation in a computer, it is also possible to use Theorem 1.3. The memory volume of modern computers permits functions to be represented in matrix form of up to 14–16 variables. Below is given a description of an algorithm for determining the inessential variables and testing for homogeneity of functions in LYaPAS.

The variable β represents the matrix form of representation of a Boolean

[2] See reference [16].

function; the number of variables is represented by the value of the index γ; the result of the operator is represented by the variables δ and ϵ, so that the ith position of the code for δ is equal to 1 if the function is independent of x_i, and the ith position of the code for ϵ is equal to 1 if the function is negative in x_i, where $i = 0, 1, \ldots, n-1$; α, an auxiliary exit terminal, corresponds to inhomogeneity of the function.

homogeneity αч, βп +, γи +, δп, ϵп/**a**, $a/(\delta\epsilon)$
230 107 4 (β**a**)

$\bar{\circ}\ a \circ \delta \circ \epsilon$
§1 $\triangle\ a \oplus \gamma \circ\!\!\rightarrow 4\beta \veebar a \oplus \beta \Rightarrow \mathbf{a} \circ\!\!\rightarrow 2 \wedge \beta \wedge e_a \circ\!\!\rightarrow 3\mathbf{a} \wedge \beta \wedge d_a$
$|\!\!\rightarrow \alpha \rightarrow 1$
§2 $c_a \vee \delta \Rightarrow \delta \rightarrow 1$
§3 $c_a \vee \epsilon \Rightarrow \epsilon \rightarrow 1$
§4 .

A different method for testing a Boolean function for homogeneity is described in [17].

In the synthesis of a threshold element it is more convenient to operate with positive functions. Therefore below, if not otherwise stated, we shall consider positive functions. Any homogeneous function can be reduced to positive fairly easily. If the homogeneous function is given in dnf, to bring it to positive it is sufficient simply to drop the inversion signs. If the function is given in matrix form, to bring it to a positive function it is necessary to subject its representation to the operator $S_K{}^{\varphi}$. The numbers of the unit components of the vector K coincide with the indexes of the variables in which the function is negative. In the operator *posifun* given below,

$$\alpha ::\| \{M_1\} \ni M \|, \qquad [\beta] = K.$$

posifun αп, βп/—, a
232 23 2

§1 $\beta\ \dot{\mathbf{X}}\ 2a\alpha \veebar a \Rightarrow \alpha \rightarrow 1$
§2 .

After the realization of a positive function has been found, it is possible to determine from it the realization of the initial function, using the following feature, which is verified by direct substitution: if

$$f(x_0, x_1, \ldots, x_i, \ldots, x_{n-1}) \sim [a_0, a_1, \ldots, a_i, \ldots, a_{n-1}; T],$$

then

$$f(x_0, x_1, \ldots, \bar{x}_i, \ldots, x_{n-1}) \sim [a_0, a_1, \ldots, -a_i, \ldots, a_{n-1}; T - a_i]. \qquad (7)$$

Let $f(X)$ be a certain Boolean function, and let $X = \{x_0, x_1, \ldots, x_{n-1}\}$ be the set of its arguments; let A and B be two subsets of X, ($A \cap B = \varnothing$, $A \cup B \subseteq X$).

We let $1 \to A$ ($0 \to A$) be the operation of assigning to all variables in the set A the value 1 (0). The function $f(1 \to A; 0 \to B)$ is called the reduced function [4, 16]. If $f(X)$ is a threshold function, then by S_A and S_B we denote the sums of weights of the variables occurring in the subsets A and B, respectively.

Let

$$f(X) \sim [a_0, a_1, \ldots, a_{n-1}; T];$$

then, obviously,

$$f(0 \to x_i) \sim [a_0, a_1, \cdots, a_{i-1}, a_{i+1}, \ldots, a_{n-1}; T], \tag{8}$$

$$f(1 \to x_i) \sim [a_0, a_1, \ldots, a_{i-1}, a_{i+1}, \ldots, a_{n-1}; T - a_i]. \tag{9}$$

Let us consider two positive threshold functions:

$$f(X) \sim [a_0, a_1, \ldots, a_{n-1}; T], \tag{10}$$

$$\varphi(X) \sim [a_0, a_1, \cdots, a_{n-1}; t]. \tag{11}$$

Let $T \geq t$; then, obviously, $\varphi(X) \geq f(X)$, i.e., there does not exist a set of values of the variables $x_0, x_1, \ldots, x_{n-1}$ for which the function $\varphi(X)$ would take on the value 0 while $f(X)$ is equal to 1 on this set. A more general proposition also holds: threshold functions whose realizations differ only in threshold are comparable; i.e., if we know nothing about the relations of the thresholds in (10) and (11), it can still be stated that either $f(X) > \varphi(X)$ or $f(X) < \varphi(X)$, or $f(X) = \varphi(X)$.

Thus, for the threshold function $f(X)$ one of the following relations always holds:

$$f(1 \to \{x_i\}) > f(0 \to \{x_i\}), \tag{12}$$

$$f(1 \to \{x_i\}) < f(0 \to \{x_i\}), \tag{13}$$

$$f(1 \to \{x_i\}) = f(0 \to \{x_i\}), \tag{14}$$

where in the last case $f(X)$ does not depend essentially on x_i, while cases (12) and (13) hold for functions positive and negative in x_i, respectively. As follows from the definition of homogeneous function, relations (12–14) are valid for any homogeneous function.

In a more general case the constraints may be applied, not to one variable, as has been done above, but to k variables, where $1 \leq k \leq n$. It is clear that relations of the form (12–14) must hold for any threshold function in this more general case as well, since two reduced functions obtained from $f(X)$ differ only in threshold and are therefore always comparable.

Theorem 1.4.[3] If $f(X)$ is a threshold function, any two reduced functions $f(1 \to A; 0 \to B)$ and $f(1 \to A'; 0 \to B')$ are comparable on condition that $A \cup B = A' \cup B' = C$. Furthermore, from the relation

$$f(1 \to A; 0 \to B) > f(1 \to A'; 0 \to B')$$

it follows that $S_A > S_{A'}$ and from $f(1 \to A; 0 \to B) < f(1 \to A'; 0 \to B')$ it follows that $S_A < S_{A'}$. If A and B each contain only one element x_i and x_j, and

$$f(1 \to \{x_i\}; 0 \to \{x_j\}) = f(1 \to \{x_j\}; 0 \to \{x_i\}),$$

then among the sets of hyperplanes separating the vertices of the n-dimensional cube is found one such that its equation $a_i = a_j$.

This result was first published by Paull and McCluskey [10], and therefore we shall refer to Theorem 1.4 as the Paull–McCluskey theorem, remarking that in [10] these results were formulated as two theorems (2 and 3) and a corollary (2).

Corollary.[4] If $f(X)$ is a threshold function, the set of weights $a_0, a_1, \cdots, a_{n-1}$ can be ordered linearly.

REMARK: It must be kept in mind that the "weights" can be "ordered" even when the function is not threshold. Below we shall often order weights without knowing whether the function investigated is threshold, recalling that this ordering may be purely formal. The expression $a_i > a_j$ must be considered as a convenient way to express the fact that if the analyzed function is threshold, then in every realization of it $a_i > a_j$, while the relation $a_i = a_j$ indicates that if $f(X) \in R_n$, a realization is found in which $a_i = a_j$. When not stated explicitly, we shall everywhere below have in mind a realization satisfying the condition

$$f(0 \to \{x_i\}; 1 \to \{x_j\}) = f(1 \to \{x_i\}; 0 \to \{x_j\}) \to a_i = a_j. \qquad (15)$$

When the set $C = A \cup B$ contains more than two elements and

$$f(1 \to A; 0 \to B) = f(0 \to A; 1 \to B),$$

the relation between S_A and S_B is undefined, i.e., it may be found that $S_A = S_B$ or $S_A > S_B$ ($S_A < S_B$).

[3] See references [4, 10, 16].

[4] See references [1, 16].

If the condition of Theorem 1.4 is satisfied when the set C contains exactly k elements, we shall call the function k-comparable. With satisfaction of the condition of the theorem when the set C contains successively $1, 2, \cdots, k$ elements, the function is called k-monotonic [11, 16]; an n-monotonic function is called completely monotonic. The concepts of 1-comparability, 1-monotonicity, and homogeneity are equivalent. The property of k-monotonicity includes 1-, 2-, ..., k-comparability. These concepts permit the necessary condition for realizability of a threshold function by a single threshold element, following from the Paull–McCluskey theorem, to be expressed more concisely:

Every threshold function is totally monotonic [11].

The property of total monotonicity is a very important property of threshold functions, giving the necessary condition for realizability which, as has been shown by the investigation of standard threshold functions [1, 18], is sufficient, at least for functions of six or less variables. In the general case this condition is not sufficient. E. F. Moore has found an example, a function of 12 variables, which is totally monotonic but not threshold [11].

From the definition it follows that, in order to recognize whether a k-monotonic function is $k+1$-monotonic, it is sufficient to test it simply for $k+1$-comparability. The comparison of the reduced functions is carried out by the operator *comparability* which, in the subsets of variables $A \subset X, B \subset X, A \cup B \subseteq X$, and $A \cap B = \varnothing$, given by the indexes β and γ, finds the reduced functions $f(1 \to A; 0 \to B)$ and $f(0 \to A; 1 \to B)$ and compares them. ϵ::|| Int $\ni W$ ||, the index δ, is given the value $00 \cdots 01$ if $f(1 \to A; 0 \to B) < f(0 \to A; 1 \to B)$, and the value $11 \cdots 10$ if $f(1 \to A; 0 \to B) > f(0 \to A; 1 \to B)$. In the presence of incomparability of the reduced functions, exit from the operator takes place over the auxiliary terminal α; ζ and η are auxiliary complexes.

comparability (40, 41, 60, 61)/αч, βи, γи +, δп, ϵк +, ζк?, ηк/**a**, f
233 222 10

$\circ\ \delta$
§1 $\bar{\circ}\ e\ \bar{\circ}\ f$
§2 $\triangle\ e \oplus b_\epsilon \circ\!\!\to 3\epsilon_e \wedge \gamma \mid\!\!\to 2\epsilon_e \wedge \beta \circ\!\!\to 2 \oplus \epsilon_e \Rightarrow \zeta_f?\ \triangle f$ *redsim*
$\zeta\zeta//$ *lowlim* $\zeta\zeta// \to 2$
§3 $!\beta \Leftrightarrow \gamma$ *transfer* $\zeta\eta//$*cleanup* $\zeta// \to 1$
§4 $\bar{\circ}\ f \triangle\ e \oplus b_\eta \circ\!\!\to 10$
§5 $\triangle f \oplus b_\zeta \circ\!\!\to 4\eta_e \oplus \zeta_f \circ\!\!\to 5\eta_e \wedge \zeta_f \Rightarrow \mathbf{a} \oplus \eta_e \circ\!\!\to 6\mathbf{a} \oplus \zeta_f \circ\!\!\to 7 \to 5$
§6 $\delta \oplus c_{37} \circ\!\!\to \alpha c_{37} \urcorner \Rightarrow \delta \to 5$
§7 $c_{37} \urcorner \oplus \delta \circ\!\!\to \alpha c_{37} \Rightarrow \delta \to 5$
§10 .

Let a positive function in minimal disjunctive form be given. Let r be the minimal rank of the conjunctions in this form, and R be the maximal.

Theorem 1.5.[5] Let $F(X)$ be a certain positive Boolean function, and let x_i and x_j be two of its variables. If

$$F(1 \to \{x_i\}; 0 \to \{x_j\}) > F(0 \to (x_i\}; 1 \to \{x_j\}),$$

then there exists an integer t ($R \geq t \geq r$) such that in conjunctions of rank t, x_i occurs more often than x_j. Furthermore, if $t > r$, then in conjunctions of rank less than t, x_i and x_j occur the same number of times.

The algorithm based on this theorem consists of the successive division of the set of weights $a_0, a_i, \cdots, a_{n-1}$ into classes such that the variables whose weights occur in one class in the kth rank are contained the same number of times in the conjunctions of the rth, $r + 1$st, ..., $r + k - 1$st ranks.

Example:

$$\begin{aligned} F(x_0, x_1, \ldots, x_5) = {} & x_0x_1x_2 \vee x_0x_1x_3 \vee x_0x_1x_4 \\ & \vee x_0x_2x_3x_4 \vee x_0x_2x_3x_5 \\ & \vee x_0x_2x_4x_5 \vee x_0x_3x_4x_5 \vee x_1x_2x_3x_4x_5. \end{aligned}$$

In this case $r = 3$, and $R = 5$. In the conjunctions of third rank x_0 and x_1 occur three times, x_2, x_3, and x_4 once each, so

$$a_0, a_1 > a_2, a_3, a_4 > a_5.$$

In the conjunctions of fourth rank x_0 occurs four times, x_1 zero times, x_2, x_3, and x_4 three times each. It is not necessary to count the occurrences of x_5 since a_5 is the sole element in its class.

Consequently,

$$a_0 > a_1 > a_2, a_3, a_4, > a_5.$$

There remains a single class containing more than one element: a_2, a_3, a_4. In the conjunction of fifth rank x_2, x_3 and x_4 enter once each, and since $t = R$, the ordering process terminates, and we put the weights in a given class equal to each other; i.e.,

$$a_0 > a_1 > a_2 = a_3 = a_4 > a_5.$$

Below we shall denote the classes of weights by $A_0, A_1, \cdots, A_{l-1}$ and put in class A_0 the maximal weight, in A_{l-1} the minimal.

[5] See reference [4].

This algorithm is performed by the operator for linear ordering of weights *lowal*.

$$\alpha::\|\,\mathrm{Int} \subseteq W\,\|, \qquad \beta::\|\,A \ni a\,\|, \qquad [\epsilon] = n,$$
$$\gamma,\ \delta\text{—working complexes}, \qquad \sigma(\gamma) = n, \qquad \sigma(\delta) = n.$$

lowal (61, 60, 61)/ακ +, βκ, γκ, δκ, εи/**e**, *c*
234 256 12

$cleanup\ \delta//0 - c_\epsilon \Rightarrow \delta_0 \circ \mathbf{a} \Rightarrow \mathbf{c}$
§1 $transfer\ \delta\beta//cleanup\ \gamma//\mathbf{a} \oplus \epsilon \circ\!\!\rightarrow 12\triangle\, \mathbf{a}\ \bar{\circ}\ a$
§2 $\triangle\, a \oplus b_\alpha \circ\!\!\rightarrow 4\alpha_a \nabla \oplus \mathbf{a} \,|\!\!\rightarrow 2\alpha_a \wedge \mathbf{c} \Rightarrow \mathbf{b}$
§3 $\mathbf{b}\ \dot{\mathbf{X}}\ 2b \triangle \gamma_b \rightarrow 3$
§4 $\circ\ a\ \bar{\circ}\ b$
§5 $\triangle\, b\, \beta_b \nabla \circ\!\!\rightarrow 1 \oplus 1 \,|\!\!\rightarrow 6\beta_b \Rightarrow \delta_a \vee \mathbf{c} \Rightarrow \mathbf{c} \triangle a \rightarrow 5$
§6 $\beta_b \Rightarrow \mathbf{b} \circ \mathbf{e} \Rightarrow \mathbf{d}$
§7 $\mathbf{b}\ \dot{\mathbf{X}}\ 11c\mathbf{e} \oplus \gamma_c \,|\!\!\rightarrow 10c_c \vee \mathbf{d} \Rightarrow \mathbf{d} \rightarrow 7$
§10 $\mathbf{e} - \gamma_c \,|\!\!\rightarrow 7c_c \Rightarrow \mathbf{d}\ \gamma_c \Rightarrow \mathbf{e} \rightarrow 7$
§11 $\mathbf{d} \Rightarrow \delta_a \triangle a\mathbf{d} \urcorner \wedge \beta_b \Rightarrow \beta_b \,|\!\!\rightarrow 6 \rightarrow 5$
§12 .

It is clear that this algorithm can be applied to any positive function, while it is possible to speak of ordered weights only in the case of a threshold function (see the remark on the corollary to Theorem 1.4). Therefore after ordering of the coefficients it is necessary to test $F(X)$ for 2-monotonicity. It is obvious that in this case it is sufficient to compare $n - 1$ pairs of reduced functions.

To organize the discussion and shorten the subroutines, we shall assume that the variables have been numbered in decreasing order of weights; i.e., the weight of variable x_0 is maximal, and the weight of x_{n-1} is minimal. The corresponding transformation is performed by the operator for renumbering of variables of a function given in disjunctive normal form:

renumbering

$$\alpha::\|\,T_0 \subseteq W\,\|, \qquad \beta::\|\,A \ni a\,\|, \qquad \gamma::\|\,T_0' \subseteq W\,\|, \qquad \delta::\|\,A' \ni a\,\|.$$

The elements of corresponding sets in the new numbering are primed.

renumbering (60, 60)/ακ +, βκ +, γκ, δκ/**d**, *f*/(αγ) (βδ)
235 162 5

$\circ\ f\ \bar{\circ}\ b\ transfer\ \alpha\gamma//transfer\ \beta\delta//b_\beta - 1 \Rightarrow e$
§1 $\bar{\triangle} f \triangle b \oplus e \circ\!\!\rightarrow 5\delta_b \Rightarrow d$
§2 $\triangle f\, \mathbf{d}\ \dot{\mathbf{X}}\ 1c \oplus f \circ\!\!\rightarrow 2f + 1 \Rightarrow d$
§3 $\triangle\, d\delta_d \wedge c_f \circ\!\!\rightarrow 3 \urcorner \wedge \delta_d \vee c_c \Rightarrow \delta_d\ \bar{\circ}\ a$
§4 $\triangle\, a \oplus b_\alpha \circ\!\!\rightarrow 2c_f \vee c_d \wedge \gamma_a \nabla \oplus 1 \,|\!\!\rightarrow 4c_f \vee c_d \oplus \gamma_a \Rightarrow \gamma_a \rightarrow 4$
§5 .

The return to the original numbering is accomplished by the permutation of the weights in the realization found; this transformation is performed by the operator *rereal*.

$$[\alpha_i] = a_i', \quad \beta::|| A \ni a ||, \quad [\gamma_i] = a_i.$$

rereal αк $+$, βк $+$, γк/**a**, c
236 46 3

$\bar{\text{o}}\ a \circ c$
§1 $\triangle a \oplus b_\beta \circ\!\!\rightarrow 3\beta_a \Rightarrow$ **a**
§2 **a** $\dot{\mathbf{X}}$ $1b\alpha_c \Rightarrow \gamma_b \triangle c \rightarrow 2$
§3 .

Taking into account the transitivity of the relation comparability for $\sigma(C) = \sigma(A \cup B) = 2$ and $\sigma(A) = 1$, it is fairly simple to construct an algorithm to test for 2-monotonicity. In the set X all $n - 1$ pairs of successive variables of the form x_i, x_{i+1} $(i = 0, 1, \cdots, n - 2)$ are tested; for each pair the reduced functions

$$f(0 \rightarrow \{x_i\}; 1 \rightarrow \{x_{i+1}\}), \qquad f(1 \rightarrow \{x_i\}; 0 \rightarrow \{x_{i+1}\})$$

are found and compared. α is an auxiliary exit terminal, corresponding to incomparability of the reduced functions β::|| Int $\subseteq W$ ||; γ and δ are working complexes; $[\epsilon] = n$.

twomontest αч, βк $+$, γк?, δк, ϵи/**d**, h
237 44 2

$\circ\ h$
§1 $c_h \Rightarrow$ **c** $\triangle h \oplus \epsilon \circ\!\!\rightarrow 2c_h \Rightarrow$ **b**
comparability α **cb**β**d**$\gamma\delta$// $\rightarrow 1$
§2 .

Theorem 1.6. To establish that the 2-monotonic function $f(X)$ is also 3-monotonic, it is sufficient to compare not more than C_n^4 pairs of reduced functions, specifically, for each ordered triplet of variables x_i, x_j, x_l such that $a_i > a_j > a_l$ it is necessary to test the pair

$$F(1 \rightarrow \{x_i\}; 0 \rightarrow \{x_j, x_l\}), \qquad F(0 \rightarrow \{x_i\}, 1 \rightarrow \{x_j, x_l\}).$$

Theorem 1.7. To establish that a 3-monotonic function is also 4-monotonic, it is sufficient to compare not more than $2 \cdot C_n^4$ pairs of reduced functions.

Chow's Lemma.[6] If for some $k > 0$ points $\alpha_1, \alpha_2, \ldots, \alpha_k \in M_1$ and $\beta_1, \beta_2, \ldots, \beta_k \in M_0$ are found such that

$$\sum_{i=1}^{k} \alpha_i = \sum_{i=1}^{k} \beta_i, \tag{16}$$

the prescribed partitioning (M_0, M_1) of the Boolean function $F(X)$ is unrealizable. Here $\sum$ is the ordinary (not Boolean) componentwise sum of vectors.

Example: Let $F(x_0, x_1, x_2, x_3)$ be given by two sets of vectors

$$\begin{aligned} M_1 &= \{(0, 0, 1, 1); (0, 1, 1, 1); (1, 1, 1, 1); (1, 0, 1, 1); (1, 1, 0, 0); \\ &\quad (1, 1, 0, 1); (1, 1, 1, 0)\}, \\ M_0 &= \{(0, 0, 0, 0); (0, 0, 0, 1); (0, 0, 1, 0); (0, 1, 0, 0); (0, 1, 0, 1); \\ &\quad (0, 1, 1, 0); (1, 0, 0, 0); (1, 0, 0, 1); (1, 0, 1, 0)\}. \end{aligned}$$

Is this function realizable? We put

$$\begin{aligned} \alpha_1 &= (0, 0, 1, 1); \qquad \alpha_2 = (1, 1, 0, 0); \\ \beta_1 &= (0, 1, 0, 1); \qquad \beta_2 = (1, 0, 1, 0); \\ \alpha_1 + \alpha_2 &= \beta_1 + \beta_2 = (1, 1, 1, 1). \end{aligned}$$

Consequently, the function $F(x_0, x_1, x_2, x_3)$ is unrealizable.

Definition.[7] A Boolean function $F(X)$ is k-summable if in each of the sets M_1 and M_0 are found j vectors (not necessarily distinct) $2 \leq j \leq k$ for which relation (16) is satisfied. Otherwise $F(X)$ is called k-asummable. $F(X)$ is called summable if it is k-summable for some k, otherwise $F(X)$ is asummable.

Elgot [19] and Chow [20] have shown that 2-asummability and total monotonicity are equivalent.

Theorem 1.8.[8] In order that the Boolean function $f(X)$ be realizable, it is necessary and sufficient that it be asummable.

The necessary condition given by Chow's lemma is obtained from the necessary and sufficient condition supplied by Theorem 1.8 by imposition of a constraint: The vectors α_i and β_j cannot enter into the sum more than once.

[6] See reference [19].
[7] See reference [12].
[8] See reference [19].

2. REALIZATION OF A BOOLEAN FUNCTION BY A SINGLE THRESHOLD ELEMENT

Here we shall consider questions connected with the analysis and synthesis of threshold elements.

By analysis of threshold elements we shall understand the process of determining the minimal disjunctive form of the Boolean function $f(X)$ from its realization.

We shall refer to the process of finding the realization of the Boolean function $f(X)$ as the synthesis of a threshold element. Since a realization does not exist for every Boolean function, but only for threshold functions, the synthesis of the threshold element also includes the analysis of the initial function for realizability, i.e., the test of certain necessary conditions that are satisfied by all $f(X) \in R_n$.

Definition. That integer realization of a threshold function in which the linear ordered weights satisfy condition (15) and the sum $\sum_{i=0}^{n-1} | a_i |$ is minimal is called the minimized integer realization.

In the general case the minimized integer realization—i.e., a realization satisfying the conditions that (a) the weights a_i are integers and (b) the sum $\sum_{i=0}^{n-1} a_i$ is minimal—is not minimal [12], although for the majority of functions both realizations coincide. Below, if not otherwise stated, we shall mean the minimized realization.

As follows from the definition of threshold function, the problem of analysis of a Boolean function for realizability and the process of finding the realization are equivalent to determining the compatibility and finding the solution to a system of the form

$$\begin{aligned} a \times \alpha_i - T \geq 0, \qquad & \alpha_i \in M_1, \\ a \times \beta_j - T < 0, \qquad & \beta_j \in M_0. \end{aligned} \tag{17}$$

Since the vectors α_i and β_j have binary components, the system (17), which we shall term basic, is a system of linear inequalities with respect to the weights and threshold. If the basic system is compatible, the function $f(X) \in R_n$ and the solution of the system will give the realization $f(X)$; if the system is incompatible, the function is unrealizable.

At first glance this method appears fairly simple and reliable. However, it is necessary to consider the following: The number of inequalities in a system of form (17) increases as 2^n. For sufficiently large n (in practice for $n > 5$) the systems of inequalities obtained are so large that there can be no question of their solution by a direct method. The use of the properties of threshold functions permits the number of inequalities in (17) to be reduced by the elimination of dependent inequalities. Using the property of homogeneity of functions [1–5] permits the number of inequalities to be reduced to $\sigma(F_0 \cup T_0)$, where F_0 is the set of "maximally false" vertices [3] and T_0 is the set of "minimally true" vertices which, following Varshavskii [2], we shall call the reference sets. The system of inequalities written for the reference sets will be called the N-system, and its subsystems, associated with T_0 and F_0, respectively, the N_T- and N_F-subsystems. In turn, the N-system can be reduced by taking into account the linear orderability of the coefficients (2-monotonicity) [21] and total monotonicity [11]. The system of inequalities obtained by the reduction of the N-system by taking into account 2-monotonicity of the function will be called the K-system, and the system obtained by taking into account 3-, 4-, . . ., $[n/2]$-monotonicity, the W-system.

A. ANALYSIS OF A THRESHOLD ELEMENT

The problem of threshold-element analysis can be solved fairly simply in principle. For each element δ in the Boolean space the scalar product of the vectors $a \times \delta$ can be calculated and compared to the threshold. If $a \times \delta \geq T$, then $\delta \in M_1$; otherwise, $\delta \in M_0$. Since the set M_1 in the case of a positive function is convex, we have

$$(\delta_i \geq \delta_j) \wedge (\delta_j \in M_1) \rightarrow \delta_i \in M_1. \tag{18}$$

Thus, it is sufficient to find only the boundary elements of the set M_1, i.e., the set T_0. We note that the codes of the vertices belonging to the set T_0 coincide with the codes of the conjunctions in the mdnf of the function. Using the convexity of the set M_1 and the possibility of arranging the weights in decreasing order, it is possible to obtain directly the mdf of the threshold function, thereby reducing the volume of computation required.

The algorithm for reduced search uses the operator of elementary incrementation in matched search $\hat{\Delta}_\sigma{}^s$ [22]. The operator $\hat{\Delta}_\sigma{}^s$ acts on the vector b of binary components, representing the subsets of $\{x_0, x_1, \ldots, x_{n-1}\}$, where $b_i = 1$ $(i = 0, 1, \ldots, n-1)$ if x_i enters into the subset represented by the vector b. The parameters σ and s take on values in the set $\{0, 1\}$, changing the action of the operator $\hat{\Delta}_\sigma{}^s$.

The effect of the operator $\hat{\Delta}_\sigma{}^s$ on the vector b can be described in the following way.

(1) From the code of b is removed the group of 1's occupying the extreme right position; if $b_{n-1} = 1$, the parameter σ is assigned the value 0, the parameter s the value 1. In other cases the values of the components of vector b and parameter σ are conserved, while the parameter s is given the

TABLE I

Index	a_i 10	8	5	4	2	1	$f(b)$
1	1	0	0	0	0	0	0
2	1	1	0	0	0	0	1
3	1	0	1	0	0	0	1
4	1	0	0	1	0	0	1
5	1	0	0	0	1	0	1
6	1	0	0	0	0	1	1
7	0	1	1	0	0	0	1
8	0	1	0	1	0	0	1
9	0	1	0	0	1	0	0
10	0	1	0	0	1	1	1
11	0	0	1	1	0	0	0
12	0	0	1	1	1	0	1
13	0	0	1	1	0	1	0
14	0	0	1	0	1	1	0
15	0	0	0	1	1	0	0
16	0	0	0	1	1	1	0

value 0. Let us assume that after this operation $b_j = 1$ and $b_{j+1} = b_{j+2} = \cdots = b_{n-1} = 0$.

(2) The component b_{j+1} is assigned the value 1.

(3) If $\sigma = 0$, component b_j is given the value 0; if $s = 1$, then the component b_{j+2} is given the value 1.

Examples:

$$\hat{\Delta}_1{}^1\, 010101 = 010011$$
$$\hat{\Delta}_0{}^1\, 010101 = 010011$$
$$\hat{\Delta}_0{}^0\, 110100 = 110010$$
$$\hat{\Delta}_1{}^0\, 110100 = 110110$$

To find the set T_0 we utilize the following algorithm:

(a) the vector b is assigned the value $100\cdots0$;

(b) if $b \times a < T$, the parameter σ is given the value 1, and if $b \times a \geq T$, the vector b represents one of the boundary points of the set M_1; in the latter case it is written into the solution, and the parameter σ is given the value 0;

(c) the operation $\hat{\Delta}_\sigma{}^s\, b$ is performed; if after this $b = 00\cdots0$, then the analysis of the element has been completed; otherwise, we pass to step (b).

Example: Let there be given the realization of a certain Boolean function (10, 8, 5, 4, 2, 1; 11). Let us find an mdf of this function. The result of applying the algorithm is given in Table I, which gives the values of the vector b for which the quantity $b \times a$ is calculated. The set T_0 in the table corresponds to the rows for which $f(b) = 1$, so

$$f(x_0, x_1, \ldots, x_5) = x_0x_1 \vee x_0x_2 \vee x_0x_3 \vee x_0x_4 \vee x_0x_5 \\ \vee x_1x_2 \vee x_1x_3 \vee x_1x_4x_5 \vee x_2x_4x_5.$$

Below we give a subroutine for the operator for analysis of a threshold element *threlan.* The auxiliary exit α corresponds to termination of the analysis. The basic exit terminal corresponds to the finding of the next element of T_0. To obtain each successive element of T_0 return is effected to the second sentence.

$$[\beta_i] = a_i, \qquad [\gamma] = T, [\delta] = n, \qquad \epsilon::\| b \ni X \|, \\ [\zeta] = 0 \leftrightarrow b \times a > T.$$

threlan αч $+$, βк $+$, γи $+$, δи $+$, ϵи, ζи, $/\mathbf{b}$, d
240 167 5

$\bar{\circ}\ \mathbf{a} \circ \mathbf{b}\ \bar{\circ}\ a \circ d\, c_\delta < 1 \Rightarrow b$
§1 $\mathbf{a}\ \dot{\mathbf{X}}\ 2c\beta_c + d \Rightarrow d \rightarrow 1$
§2 $\bar{\circ}\ \zeta\ \triangle\ a \oplus \delta \circ\!\!\rightarrow 4\beta_a + d \Rightarrow d - \gamma \circ\!\!\rightarrow 3 \circ \zeta d - \beta_a \Rightarrow dc_a \vee \mathbf{b}$
$\Rightarrow \epsilon \rightarrow 5$
§3 $c_a \vee b \Rightarrow b \rightarrow 2$
§4 $\mathbf{b} + b \wedge b \Rightarrow \mathbf{b} \circ\!\!\rightarrow \alpha\ \urcorner + b \wedge \mathbf{b} \vdash \Rightarrow ac_a \oplus \mathbf{b} \Rightarrow \mathbf{b} \Rightarrow \mathbf{a}\ \triangle\ ac_a \vee \mathbf{b}$
$\Rightarrow \mathbf{b} \circ c \rightarrow 1$
§5 .

The symmetry of the function with respect to certain subsets of variables can be taken into account to carry out the analysis of the threshold element with simultaneous "reduction of similar terms" in the set T_0.

Definition. Two elements of the set T_0 will be called similar if one of them is obtained from the other by renumbering the variables within the class.

The reference set obtained from the set T_0 after reduction of similar terms will be called the reduced reference set and denoted by T_0^1.

When it is necessary to obtain the reduced reference set the effect of $\hat{\triangle}_\sigma^s$ is somewhat changed, in that steps 2 and 3 of the algorithm are modified as follows:

TABLE II

	a_i						
Index	8	5	5	5	3	3	$f(b)$
1	1	0	0	0	0	0	0
2	1	1	0	0	0	0	0
3	1	1	1	0	0	0	1
4	1	1	0	0	1	0	1
5	1	0	0	0	1	0	0
6	1	0	0	0	1	1	0
7	0	1	1	0	0	0	0
8	0	1	1	1	0	0	1
9	0	1	1	0	1	0	0
10	0	1	1	0	1	1	1
11	0	1	0	0	1	1	0
12	0	0	0	0	1	1	0

(2a) If $\sigma = 1$, the component b_{j+1} is assigned the value 1; if $\sigma = 0$, the value 1 is assigned to the component b_k, where k is the closest index of a variable to j such that $a_k > a_j$.

(3a) If $\sigma = 0$, the component b_j is given the value 0; if at the same time $s = 1$, the component b_{k+1} is assigned the value 1.

Example: Find the reduced reference set T_0' of the function having the realization [8, 5, 5, 5, 3, 3; 15].

In Table II are given the values of the vector b for which it is necessary to calculate the product $b \times a$ and compare it with the threshold. The reduced reference set consists of the following four elements:

$$T_0' = \{x_0x_1x_2,\ x_0x_1x_4,\ x_1x_2x_3,\ x_1x_2x_4x_5\}.$$

The operator is given the code name *threlans*; it is performed by a (2, 2)-terminal subroutine; the auxiliary output α corresponds to termination of the analysis, the auxiliary input to the second sentence, to which return is effected to obtain the 2nd, 3rd, ..., elements of T_0',

$$[\beta_i] = a_i,\quad \gamma::\|\, A \ni a \,\|,\quad [\delta] = T,\quad [\epsilon] = n,\quad \zeta::\|\, b \ni X \,\|,$$
$$i \in \{0, 1, \ldots, n-1\}.$$

threlans $\alpha\text{ч}\,+,\ \beta\text{к}\,+,\ \gamma\text{к}\,+,\ \delta\text{и}\,+,\ \epsilon\text{и}\,+,\ \zeta\text{и}/\mathbf{c},\ e$
241 257 10

$\bar{\circ}\ \mathbf{a}\ \circ\ \mathbf{b}\ \bar{\circ}\ a\ \circ\ dc_\epsilon < 1 \Rightarrow b$
§1 $\mathbf{a}\ \ddot{\mathbf{X}}\ 2c\beta_c + d \Rightarrow d \rightarrow 1$
§2 $\triangle\ a \oplus \epsilon\ \circ\!\!\rightarrow 5\beta_a + d \Rightarrow d - \delta\ \circ\!\!\rightarrow 4d - \beta_a \Rightarrow d\ \bar{\circ}\ e$
§3 $\triangle\ e \oplus b_\gamma\ \circ\!\!\rightarrow 6c_a \wedge \gamma_e\ \circ\!\!\rightarrow 3c_a \vee \mathbf{b} \Rightarrow \mathbf{c}\ \triangle\ e \oplus b_\gamma\ \circ\!\!\rightarrow 6\gamma_e\ \vdash -1$
$\Rightarrow a \rightarrow 10$
§4 $c_a \vee \mathbf{b} \Rightarrow \mathbf{b} \rightarrow 2$
§5 $\mathbf{b} + b \wedge \mathbf{b} \Rightarrow \mathbf{b}\ \circ\!\!\rightarrow \alpha$
§6 $\urcorner + b \wedge \mathbf{b}\ \vdash \Rightarrow ac_a \oplus \mathbf{b} \Rightarrow \mathbf{b} \Rightarrow \mathbf{a}\ \bar{\circ}\ e$
§7 $\triangle\ e \oplus b_\gamma\ \circ\!\!\rightarrow 6c_a \wedge \gamma_e\ \circ\!\!\rightarrow 7\ \triangle\ e \oplus b_\gamma\ \circ\!\!\rightarrow 6\gamma_e\ \vdash \Rightarrow e - 1$
$\Rightarrow a\ \circ\ c \rightarrow 1$
§10 .

B. THRESHOLD-ELEMENT SYNTHESIS BY A LINEAR PROGRAMMING METHOD

We consider here the threshold-element synthesis method based on the solution of a system of linear inequalities by a linear programming method [23]. The method utilizes an existing, traditional approach [1–3, 10, 11], consisting of the following.

(1) The initial Boolean function $f(X)$ is analyzed for homogeneity and minimized; when necessary, transition to a positive function is effected.
(2) The weights a_i are linearly ordered and 2-monotonicity is tested.
(3) The inversion of the function $f(X)$ is found, and from it is obtained the set F_0; the N-system is constructed.
(4) Dependent inequalities (not necessarily all) are removed from the N-system, where the Paull–McCluskey theorem is utilized in one form or another.
(5) The system obtained (in the best case this is the W-system) is solved. Usually a minimized solution is sought.

The solution of a fairly large system of inequalities, particularly in integers, is laborious and complicated. The best-known method for the solution of systems of linear inequalities is the simplex method of linear programming. However, in our view, it has essential disadvantages, expressing themselves in the given concrete application. First, the simplex method is purely a machine method, unadapted to manual calculations. Second, it does not guarantee integer solutions; to obtain these, and to eliminate looping (the problem is often degenerate) additional measures are necessary, rendering the program more complicated.

The characteristic feature of the proposed method is the application of a new algorithm for the solution of linear programming problems, proposed by Chernikov [23]. This algorithm is conveniently (from the viewpoint of the synthesis of threshold elements) distinguished from the simplex method. It may be expected that in the majority of cases (1) it can be applied to the manual solution of systems of 10–15 inequalities; (2) being related to the method of elimination of unknowns used by Elgot [19] and Winder [11], it permits an integer solution to be obtained fairly quickly; (3) in contradistinction to the method of elimination of variables, it is more algorithmic, and is therefore more easily programmed.

Let us consider the realizations of the individual steps.

(1) As pointed out in Section 1, in the representation of a function in dnf the problem of testing $f(X)$ for homogeneity reduces mainly to finding the reduced normal form, which falls into the domain of minimization of Boolean functions. Therefore we shall not here consider the corresponding operators. It is assumed that the initial function is given in mdf and is positive.

(2) The operator for linear ordering of weights, *lowal*, and the operator for testing the function for 2-monotonicity have been considered in Section 1.

(3) Since the process of finding the set F_0 reduces to expanding parentheses in the minimal conjunctive form, the R-operator [22, 24] can be applied to solve this problem with high efficiency.

A different approach, using the property of 2-monotonicity, has been proposed by Winder [11]; the subroutine that executes Winder's algorithm is faster than that of the R-operator.

$$\alpha::\| T_0 \ni a \|, \qquad \beta::\| F_0 \ni a \|, \qquad [\gamma] = n,$$
$$[\alpha_i] > [\alpha_{i+1}], \qquad i = 0, 1, \cdots, \sigma(T_0) - 1.$$

expacon $\alpha\text{к} +, \beta\text{к}\ ?, \gamma\text{к} + /\mathbf{b}, c$
242 134 3

$0 - c_\gamma \oplus c_\gamma \Rightarrow \mathbf{b}\ \bar{\circ}\ a \circ c$

§1 $\triangle\ a \oplus b_\alpha \circ\!\!\rightarrow 3\alpha_a \Rightarrow \mathbf{a}a + 1 \Rightarrow b \oplus b_\alpha \circ\!\!\rightarrow 2\alpha_a \oplus \alpha_b \wedge \alpha_a \vdash \Rightarrow bc_b$
$-1 \wedge \alpha_a \Rightarrow \mathbf{a}$

§2 $\mathbf{a}\ \ddot{\mathbf{X}}\ 1bc_b - 1 \vee \alpha_a \oplus c_b \wedge \mathbf{b} \Rightarrow \beta_c?\ \triangle\ c \rightarrow 2$

§3 $c \Rightarrow b_\beta.$

Example[9]*:*

$$T_0 = \{x_0x_1,\ x_0x_2,\ x_0x_3,\ x_1x_2,\ x_1x_3x_4\}.$$

$$\| T_0 \ni \mathbf{a} \| = \begin{array}{c|ccccc} & x_0 & x_1 & x_2 & x_3 & x_4 \\ 0 & 1 & 1 & & & \\ 1 & 1 & & 1 & & \\ 2 & 1 & & & 1 & \\ 3 & & 1 & 1 & & \\ 4 & & 1 & & 1 & 1 \end{array}$$

F_0 is obtained directly from the second and fourth rows:

$$F_0 = \{x_0x_4,\ x_2x_3x_4,\ x_1x_4,\ x_1x_3\}.$$

(4) Let v be the set of variables common to the elements $t_i, t_j \in T_0$; $v = t_i \cap t_j$; $i, j = 0, 1, \ldots, \sigma(T_0) - 1$. We form the set $A = t_i \backslash v$, $B = t_j \backslash v$ and find the reduced functions $F(0 \to A; 1 \to B)$ and $F(1 \to A; 0 \to B)$. On the basis of the Paull–McCluskey theorem it can be concluded that the inequality corresponding to the element $t_i \in T_0$ is dependent [1] if

$$F(1 \to A; 0 \to B) > F(0 \to A; 1 \to B), \tag{19}$$

and for $\sigma(A) = \sigma(B) = 1$ this inequality is also dependent for

$$F(1 \to A; 0 \to B) \geq F(0 \to A; 1 \to B). \tag{20}$$

Considering similarly the pair of elements $f_k, f_l \in F_0 (k, l = 0, 1, \ldots, \sigma(F_0) - 1)$ and putting $v = f_k \cap f_l$, $A = f_k \backslash v$, and $B = f_l \backslash v$, we conclude that in satisfaction of condition (19) or (20) the inequality corresponding to f_e is dependent.

The operator *redu* presented below reduces sets T_0 and F_0 on the basis of relations (19) and (20). The subroutine for the operator *redu* is (1, 2)-terminal. The auxiliary exit terminal α corresponds to the case of incomparability of the reduced functions

$$\begin{array}{lll} & \beta{:}{:}\| T_0 \ni a \|, & \\ \gamma = \circ\!\to, & \delta{:}{:}\| T_0 \ni a \|, & \text{if } T_0 \text{ is reduced}, \\ \gamma = |\!\to, & \delta{:}{:}\| F_0 \ni a \|, & \text{if } F_0 \text{ is reduced}. \end{array}$$

[9] See reference [11].

redu (233)/αч, βк $+$, γ_0, δк, ϵк ?, ζк/**g**, d
252 154 5

$\circ\ d$
§1 $\triangle d \Rightarrow c \oplus b_\delta \ \circ\!\!\rightarrow 5\delta_d \ \circ\!\!\rightarrow 1$
§2 $\triangle c \oplus b_\delta \ \circ\!\!\rightarrow 1\delta_c \ \circ\!\!\rightarrow 2\delta_d \oplus \delta_c \Rightarrow \mathbf{e} \wedge \delta_c \Rightarrow \mathbf{fe} \wedge \delta_d \Rightarrow \mathbf{e}$
comparability $\alpha\mathbf{fe}\beta\epsilon?\ \zeta\mathbf{g}//\mathbf{g}\ \circ\!\!\rightarrow 3 \wedge c_0\gamma 4 \ \circ\ \delta_d \rightarrow 2$
§3 $\mathbf{f} \ \nabla \oplus 1 \mid\!\!\rightarrow 2\mathbf{e} \ \nabla \oplus 1 \mid\!\!\rightarrow 2\mathbf{f} - \mathbf{e}\gamma 4 \ \circ\ \delta_d \rightarrow 2$
§4 $\circ\ \delta_c \rightarrow 2$
§5 .

(5) Let us consider a system of m linear inequalities over an n-dimensional space E^n.

$$
\begin{aligned}
f_j(a) - x_j &= x_{j0}a_0 + x_{j1}a_1 + \ldots + x_{jn-1}a_{n-1} - x_j \leq 0 \\
&\qquad\qquad (j = 1, 2, \ldots, m'), \\
f_j(a) - x_j &= x_{j0}a_0 + x_{j1}a_1 + \ldots + x_{jn-1}a_{n-1} - x_j < 0 \\
&\qquad\qquad (j = m' + 1, \ldots, m).
\end{aligned}
\tag{21}
$$

Let x_{pi} be an arbitrary positive, and x_{qi} an arbitrary negative coefficient of a_i in system (21). The new system, obtained from (21) in the following way, is called the a_i-convolution [23] of (21):

(a) In the a_i-convolution we write the inequality

$$x_{pi}f_q(a) - x_{qi}f_p(a) - (x_{pi}x_q - x_{qi}x_p) \leq 0$$

if $p, q \in \{1, 2, \ldots, m'\}$, and the inequality

$$x_{pi}f_q(a) - x_{qi}f_p(a) - (x_{pi}x_q - x_{qi}x_p) < 0$$

if $p \in \{m' + 1, \ldots, m\}$ or $q \in \{m' + 1, \ldots, m\}$.

(b) We include in the a_i-convolution all inequalities from (21) with zero coefficient for a_i.

Let the system (21) be compatible and let $f^{(r)}(a)$ $(r = 1, 2, \ldots, s)$ be a system of linear functions defined in E^n. The upper bound of values of the parameter t for which the system

$$
\begin{aligned}
f_j(a) - x_j &\leq 0, & (j &= 1, 2, \ldots, m'), \\
f_j(a) - x_j &< 0, & (j &= m' + 1, \ldots, m), \\
-f^{(r)}(a) + t &\leq 0, & (r &= 1, 2, \ldots, s)
\end{aligned}
\tag{22}
$$

is compatible is called the upper minimal value of the function on the set of solutions of system (21). The necessary and sufficient condition for the existence of an upper minimal value is formulated in [23]. The algorithm for solution of the system (22) consists in the successive convolution of (22), first for some variable a_i, then the a_i-convolution for a second variable a_k, etc., until a certain total convolution S will be obtained, containing only the parameter t. According to Theorem 2 in [23], in this case the system of functions $f^{(r)}(a)$ $(r = 1, 2, \ldots, s)$ has an upper minimal value on the set of solutions of (21), which coincides with the upper bound on the value of the parameter t satisfying S.

When the minimal value of t has been found, the values of the variables a_i are determined from the convolutions in the order inverse to that of convolution. The choice of variable for which convolution is realized is of importance, since it determines the number of inequalities in the convolution. It is obvious that the best case occurs when the variable a_i for which the convolution is effected enters one inequality with the plus (minus) sign and the others containing a_i with the minus (plus) sign. Such a situation can be set up artificially by transformation of the system (21).

Reduction of (21) to the form [23]

$$x_{j0}a_0 + \cdots + (x_{jk} + zx_j)a_k + \ldots + x_{jl}u_l + \ldots + x_{jn-1}a_{n-1} - x_j \le 0 \quad (j = 1, 2, \ldots, m')$$
$$x_{j0}a_0 + \cdots + (x_{jk} + zx_j)a_k + \cdots + x_{jl}u_l + \cdots + x_{jn-1}a_{n-1} - x_j < 0 \quad (j = m' + 1, \ldots, m),$$

where z is an arbitrary real number and $u_l = a_l - za_k$, is called elementary transformation. In [23] it was shown that convolution of the system for an unknown can alternate with elementary transformations of the convolutions obtained; with suitable choice of transformations it is possible to reduce substantially the number of inequalities in the convolutions.

Applying this method to the synthesis of a threshold function, the system of functions $f^{(r)}(a)$ is replaced by a single linear form of the type

$$L(a) = a_0 + a_1 + \ldots + a_{n-1} + T. \tag{23}$$

Since the problem of minimization is being solved, an additional constraint is written in the form

$$L(a) + t \le 0.$$

Example: We shall examine the actions of the basic stages of the algorithm by the example of finding a realization of the function

$$f(x_0, x_1, \ldots, x_5) = x_0x_1 \vee x_0x_2x_3 \vee x_0x_2x_4 \vee x_0x_2x_5 \vee x_0x_3x_4.$$

(1) The function is given in mdf and is positive.
(2) The weights are ordered as follows:

$$a_0 > a_1 > a_2 > a_4 = a_3 > a_5.$$

It is easily confirmed that the function is 2-monotonic.

$(C)=$

	x_0x_1	$x_0x_2x_3$	$x_0x_2x_4$	$x_0x_2x_5$	$x_0x_3x_4$
x_0	1	1	1	1	1
x_1	1				
x_2		1	1	1	
x_3		1			1
x_4			1		1
x_5				1	

FIG. 1.

(3) To find the set F_0 we use the R-operator. For this we construct a matrix (C), whose rows are associated with the variables and columns with the elements of the set T_0 (conjunctions of the mdf); the element c_{ij} is given the value 1 if the variable x_i ($i = 0, 1, \ldots, n - 1$) enters into the conjunction with index j. Figure 1 gives the matrix (C) of the above function. Finding the minimal sets of rows covering all columns, we obtain the function $f^*(x_0, \ldots, x_5)$ dual to the initial one:

$$x_0 \vee x_1x_2x_3 \vee x_1x_2x_4 \vee x_1x_3x_4x_5,$$

whence

$$F_0 = \{x_1x_2x_3x_4x_5,\ x_0x_4x_5,\ x_0x_3x_5,\ x_0x_2\}.$$

(4) It can be verified that the inequalities corresponding to the elements $x_0x_2x_3$, $x_0x_2x_4$, and $x_0x_4x_5$ are dependent. We write the W-system

$$\begin{array}{llllll}
a_0 + a_1 & & & & & \geq T \\
a_0 & + a_2 & & & + a_5 & \geq T \\
a_0 & & + a_3 & + a_4 & & \geq T \\
\quad\; a_1 & + a_2 & + a_3 & + a_4 & + a_5 & \leq T - 1 \\
a_0 & & + a_3 & & + a_5 & \leq T - 1 \\
a_0 & + a_2 & & & & \leq T - 1
\end{array}$$

(5) System (22) takes the form

$$\begin{aligned}
-a_0 - a_1 \qquad\qquad\qquad\qquad + T \quad &\le 0\\
-a_0 \qquad - a_2 \qquad\quad - a_5 + T \quad &\le 0\\
-a_0 \qquad\qquad - 2a_3 \qquad + T \quad &\le 0\\
a_1 + a_2 + 2a_3 + a_5 - T + 1 &\le 0\\
a_0 \qquad\qquad + a_3 + a_5 - T + 1 &\le 0\\
a_0 \qquad + a_2 \qquad\qquad\quad - T + 1 &\le 0\\
a_0 + a_1 + a_2 + 2a_3 + a_5 + T + t &\le 0.
\end{aligned}$$

Convoluting this system for a_1, we obtain the system

$$\begin{aligned}
-a_0 + a_2 + 2a_3 + a_5 \qquad + 1 &\le 0\\
-a_0 - a_2 \qquad\quad - a_5 + T \quad &\le 0\\
-a_0 \qquad - 2a_3 \qquad + T \quad &\le 0\\
a_0 \qquad + a_3 + a_5 - T + 1 &\le 0\\
a_0 + a_2 \qquad\qquad\quad - T + 1 &\le 0\\
+ a_2 + 2a_3 + a_5 + 2T + t &\le 0.
\end{aligned}$$

Convoluting for a_2:

$$\begin{aligned}
-2a_0 + 2a_3 \qquad + T + 1 &\le 0\\
- a_0 - 2a_3 \qquad + T \quad &\le 0\\
a_0 + a_3 + a_5 - T + 1 &\le 0\\
- a_5 \qquad + 1 &\le 0\\
- a_0 + 2a_3 \qquad + 3T + t &\le 0.
\end{aligned}$$

Convoluting for a_5:

$$\begin{aligned}
-2a_0 + 2a_3 + T + 1 &\le 0\\
- a_0 - 2a_3 + T \quad &\le 0\\
a_0 + a_3 - T + 2 &\le 0\\
- a_0 + 2a_3 + 3T + t &\le 0.
\end{aligned}$$

Convoluting this system for a_0, we obtain the system

$$\begin{aligned}
4a_3 - T + 5 \quad &\le 0\\
-a_3 \qquad + 2 \quad &\le 0\\
3a_3 + 2T + 2 + t &\le 0.
\end{aligned}$$

Further,

$$\begin{aligned}
-T + 13 \quad &\le 0\\
2T + 8 + t &\le 0.
\end{aligned}$$

Finally,

$$t + 34 \le 0.$$

We put $t = -34$, $T = 13$. Substituting these values in the convolution for a_0 we find $a_4 = a_3 = 2$, then $a_0 = 9$, $a_5 = 1$, $a_2 = 3$, $a_1 = 4$.

In this example, as in many others, no difficulties arise in the choice of values for the variable a_i. However, in the general case, with fairly large n, the range of possible values for the variable a_i determined from the convolution may be quite large, and the values of the variables obtained by setting the inequalities of the convolution to equality may not be solutions. In this case it is obviously necessary to sift the possibilities that arise. Since the value of the objective function and the compatibility of the system have been established, the number of possible variants is not large. One method of sifting is considered in Section 2D.

Another possible difficulty is connected with the requirement of an integer solution. The method of convolution does not guarantee an integer solution when the optimal solution is not integer. In this case it is obviously necessary to increase the value of the objective function (i.e., to decrease the value of the parameter t obtained from the last convolution) in order to find a value for which the solution is integer. In this case we can also use the method described in Section 2D. Naturally, the method for introducing additional constraints, proposed by Gomory [25], can also be applied in this case.

These difficulties must be taken into account in programming the method. At the present time we do not have sufficient experience with the solution of problems of this type by the method of convolutions, since the corresponding information can be obtained by computer experiments. The program presented below must be considered as a research instrument rather than a finished working program.

The algorithm for solving the system of inequalities includes the following operators.

(1) An operator for choice of optimal elementary transformation—*optimel*. This operator selects an elementary transformation such that the number of inequalities in the convolution be less than the number of inequalities in the system to be convoluted; if this cannot be done, the chosen transformation must give a convolution with the minimal number of inequalities. Since the elementary transformation consists of replacing the variable a_1 by another variable $u_1 = a_1 - za_k$, it is fully defined by the indexes of the variables l, k, and the value of z.

The operator code is *optimel*. α is an auxiliary terminal corresponding to the case of empty convolution; β, γ, and ς are working complexes, $\sigma(\beta) = \sigma(\gamma) = \sigma(\varsigma) = m$, $[\delta] = n + 3$, $[\epsilon] = m$; $[\eta]$ is the number of inequalities in the convolution; $[\vartheta] = i$, where i is the index of the variable for which the convolution is constructed; $[\kappa] = z$; $[\lambda] = l$, $[\mu_{(j-1)[\delta]+k}] = x_{jk}$; $\sigma(\mu) = [\epsilon] \times [\delta]$, $j \in \{1, 2, \ldots, m\}$.

Since LYaPAS utilizes only natural numbers, to make possible operations on negative numbers it is necessary to give them in complementary code, assuming that the zero-order position is assigned to the sign of the number.

optimel αч, βк, γк, δи, ϵи, ζк, ηи, ϑи, κи, λи, μк/**q**, p
263 1047 34

$\bar{\circ}\ k\ \delta - 2 \Rightarrow h\ \bar{\circ}\ \eta$
§1 $\triangle k \oplus h \circ\!\rightarrow 22\ \bar{\circ}\ g$
§2 $\triangle g \oplus h \circ\!\rightarrow 1 g \oplus k \circ\!\rightarrow 1\ \bar{\circ}\ a c_0 \urcorner \Rightarrow j c_0 + 1 \Rightarrow i$
§3 $\triangle a \oplus \epsilon \circ\!\rightarrow 13 a \times \delta + k \Rightarrow b - k + g \Rightarrow c \vartheta_c \circ\!\rightarrow 11 \mu_b \Rightarrow \mathbf{a} \wedge c_0$
$\circ\!\rightarrow 4\mathbf{a} - 1 \urcorner \Rightarrow \mathbf{a}$
§4 $\mu_c \Rightarrow \mathbf{b} \wedge c_0 \circ\!\rightarrow 5\mathbf{b} - 1 \urcorner \Rightarrow \mathbf{b}$
§5 $\mathbf{a}:\mathbf{b} \Rightarrow \mathbf{c}\mu_b \oplus \mu_c \wedge c_0 \mapsto 6\text{я} \urcorner + 1 \Rightarrow \text{я}c \urcorner + 1 \Rightarrow c$
§6 $\mathbf{c} \mapsto 7 f_{10} + \text{я} \Rightarrow \gamma_a + 2 \Rightarrow \beta_a \rightarrow 10$
§7 $\text{я} \Rightarrow \gamma_a + 1 \Rightarrow \beta_a c \wedge c_0 \circ\!\rightarrow 3\text{я} \Rightarrow \beta_a + f_{10} \Rightarrow \gamma_a$
§10 $\beta_a \Rightarrow \mathbf{a} i \Rightarrow \mathbf{b} \nrightarrow 30\mathbf{b} \Rightarrow i\gamma_a \Rightarrow \mathbf{a} j \Rightarrow \mathbf{b} \nrightarrow 30\mathbf{a} \Rightarrow j\mu_c \Rightarrow \mathbf{a}\gamma_a \Rightarrow \mathbf{b}\mu_b \Rightarrow \mathbf{g}$
$\nrightarrow 24 \rightarrow 12$
§11 $c_0 \urcorner \Rightarrow \beta_a \Rightarrow \gamma_a\mu_b \Rightarrow \mathbf{c}$
§12 $d_4 \Rightarrow \zeta_a \mathbf{c} \wedge c_0 \circ\!\rightarrow 3\zeta_a \urcorner \Rightarrow \zeta_a \rightarrow 3$
§13 $\circ\ \mathbf{d} \nrightarrow 15 f_{10} + i \Rightarrow \mathbf{d} \triangle i \Rightarrow e$
§14 $\triangle \mathbf{d} \oplus e \circ\!\rightarrow 22\mathbf{d} \circ\!\rightarrow 14 \nrightarrow 15 \rightarrow 14$
§15 $\bar{\circ}\ a \circ d \Rightarrow e \Rightarrow f$
§16 $\triangle a \oplus \epsilon \circ\!\rightarrow 20 a \times \delta + k \Rightarrow b - k + g \Rightarrow c\beta_a \Rightarrow \mathbf{a} \nrightarrow 30\beta_a \oplus \mathbf{a}$
$\circ\!\rightarrow 17\ \gamma_a \Rightarrow \mathbf{a} \nrightarrow 30\gamma_a \oplus \mathbf{a} \circ\!\rightarrow 17 \triangle d \rightarrow 16$
§17 $\triangle e\zeta_a \wedge c_0 \circ\!\rightarrow 16 \overline{\triangle} e \triangle f \rightarrow 16$
§20 $e \times f + d \Rightarrow d \circ\!\rightarrow 21 - \eta \mapsto 12 k \Rightarrow \kappa\eta \Rightarrow p\mathbf{d} \Rightarrow \mathbf{f}d \Rightarrow \eta - \epsilon \circ\!\rightarrow 22$
§21 !
§22 $\eta + 1 \circ\!\rightarrow \alpha \circ l$
§23 $\triangle l \oplus \epsilon \circ\!\rightarrow 34 l \times \delta + \vartheta \Rightarrow m - \vartheta + \eta \Rightarrow n\mu_n \Rightarrow \mathbf{af} \Rightarrow \mathbf{b}\mu_n \Rightarrow \mathbf{g}$
$\nrightarrow 24\mathbf{c} \Rightarrow \mu_m \rightarrow 21$
§24 $\mathbf{a} \wedge c_0 \circ\!\rightarrow 25\mathbf{a} - 1 \urcorner \vee c_0 \Rightarrow \mathbf{a}$
§25 $\mathbf{b} \wedge c_0 \circ\!\rightarrow 26\mathbf{b} - 1 \urcorner \vee c_0 \Rightarrow \mathbf{b}$
§26 $\mathbf{a} \times \mathbf{b} \Rightarrow \mathbf{ca} \oplus \mathbf{b} \wedge c_0 \circ\!\rightarrow 27 c_0 \urcorner + 1 \Rightarrow \mathbf{c}$
§27 $\mathbf{c} + \mathbf{g} \Rightarrow \mathbf{c}!$
§30 $\mathbf{a} \Rightarrow \mathbf{g} \wedge c_0 \circ\!\rightarrow 31\mathbf{g} - 1 \urcorner \Rightarrow \mathbf{g} \rightarrow 32$
§31 $\mathbf{g} \urcorner + 1 \Rightarrow \mathbf{g}$
§32 $\mathbf{g} + \mathbf{b} \wedge c_0 \circ\!\rightarrow 33\mathbf{a} \Leftrightarrow \mathbf{b}$
§33 !
§34 .

(2) The operator for elimination of a variable is *elimvar*, $[\alpha_{(j-1)[\delta]+i}] = x_{ji}$, $[\beta] = i$, i is the index of the variable for which the convolution is constructed. The latter is represented by the complex γ, $[\epsilon] = m$.

elimvar αк +, βи +, γк ?, δи +, ϵи +/**d**, i
264 323 11

$\bar{\circ}\ e \circ g$
§1 $\triangle\ e \oplus \epsilon \circ\!\!\rightarrow 11e \times \delta + \beta \Rightarrow a\ \bar{\circ}\ ic_0\ \urcorner \wedge \alpha_a \circ\!\!\rightarrow 6c_0 \wedge \alpha_a \rightarrow 1\ \bar{\circ}\ f$
§2 $\triangle\ f \oplus \epsilon \circ\!\!\rightarrow 1f \times \delta + \beta \Rightarrow b\alpha_b \Rightarrow \mathbf{d} \nrightarrow 7\mathbf{d} \Rightarrow \mathbf{a} \Rightarrow \mathbf{c}\alpha_a \Rightarrow \mathbf{d} \nrightarrow 7\mathbf{d} \Rightarrow \mathbf{b}$
$- \mathbf{c} \mid\!\!\rightarrow 3\mathbf{d} \Leftrightarrow \mathbf{c}$
§3 $\mathbf{d}:\mathbf{c} \circ\!\!\rightarrow 4 \Rightarrow \mathbf{d} \Leftrightarrow \mathbf{c} \rightarrow 3$
§4 $\mathbf{b}:\mathbf{c} \Rightarrow \mathbf{ba}:\mathbf{c} \Rightarrow a\ \bar{\circ}\ i$
§5 $\triangle\ i \oplus \beta \circ\!\!\rightarrow 5i \oplus \dot{\delta} \circ\!\!\rightarrow 2a - \beta + i \Rightarrow cb - \beta + i \Rightarrow d\alpha_c \times a$
$\Rightarrow \mathbf{c}\alpha_d \times b + \mathbf{c} \Rightarrow \gamma_g?\ \triangle\ g \rightarrow 5$
§6 $\triangle\ i \oplus \beta \circ\!\!\rightarrow 6i \oplus \delta \circ\!\!\rightarrow 1a - \beta + i \Rightarrow f\alpha_f \Rightarrow \gamma_g?\ \triangle\ g \rightarrow 6$
§7 $\mathbf{d} \wedge c_0 \circ\!\!\rightarrow \dot{1}0\mathbf{d} - 1\ \urcorner \Rightarrow \mathbf{d}$
§10 !
§11 $g \Rightarrow b_\gamma$.

(3) The operator for finding the upper and lower bounds on the values of the variable x_i from the convolution, $i \in \{0, 1, \ldots, n-1\}$-*bounds*. α is an auxiliary terminal, providing an exit from the subroutine in the case of incompatibility of the system; $[\beta]$ is the lower bound of values of the variable; $[\gamma] = i$; $[\delta] = n + 3$; $[\epsilon] = m$; $[\zeta_{(j-1)[\delta]+k}] = x_{jk}$, $j \in \{1, 2, \ldots, m\}$; the complex η contains the already calculated values of the variables; $[\vartheta]$ is the upper bound on the values of the variable.

bounds αч, βи, γи +, δи +, ϵи +, ζк +, ηк, ϑи/**c**, d
265 242 10

$\bar{\circ}\ \vartheta\ 1 \Rightarrow \beta\ \bar{\circ}\ d$
§1 $\triangle\ d \oplus \epsilon \circ\!\!\rightarrow 10d \times \delta + \gamma \Rightarrow a\zeta_a \circ\!\!\rightarrow 1d \times \delta - 1 \Rightarrow c\ \bar{\circ}\ b \circ \mathbf{a}$
§2 $\triangle\ b \oplus b_\eta \circ\!\!\rightarrow 3\ \triangle\ c\zeta_c \times \eta_b + \mathbf{a} \Rightarrow \mathbf{a} \rightarrow 2$
§3 $\zeta_a \Rightarrow \mathbf{b} \nrightarrow 6\mathbf{b} \Rightarrow \mathbf{ca} \Rightarrow \mathbf{b} \nrightarrow 6\mathbf{b} \wedge c_0 \mid\!\!\rightarrow 4c_0 \oplus \mathbf{c} \Rightarrow \mathbf{c} \mid\!\!\rightarrow \alpha\mathbf{c}:\mathbf{b} - \vartheta \mid\!\!\rightarrow 1$
$+ \vartheta \Rightarrow \vartheta \rightarrow 1$
§4 $c_0 \oplus \mathbf{b} \Rightarrow \mathbf{b}c_0 \wedge \mathbf{c} \mid\!\!\rightarrow 1\mathbf{c}:\mathbf{b} \Rightarrow \mathbf{b} - \vartheta \circ\!\!\rightarrow \alpha$я $\circ\!\!\rightarrow 5\ \triangle\ \mathbf{b}$
§5 $\mathbf{b} - \beta \circ\!\!\rightarrow 1b \Rightarrow \beta \rightarrow 1$
§6 $\mathbf{b} \wedge c_0 \circ\!\!\rightarrow 7\mathbf{b} - 1\ \urcorner \vee c_0 \Rightarrow \mathbf{b}$
§7 !
§10 .

The algorithm for solving the system of linear inequalities by the method of eliminating variables is assigned the code *sosinel.*

$\alpha(\beta)$ is an auxiliary terminal, providing exit from the subroutine if the convolution is empty (if the system is incompatible).

$$[\gamma_{(j-1)[\delta]+i}] = x_{ji}; \qquad [\delta] = n + 3; [\epsilon] = m; \qquad [\zeta_i] = a_i;$$
$$i \in \{0, 1, \ldots, n-1\}; \qquad j \in \{0, 1, \ldots, m-1\};$$
$$\eta, \vartheta, \kappa, \lambda \text{ — working complexes;}$$
$$\sigma(\vartheta) = \sigma(\kappa) = \sigma(\lambda) = m; \qquad [\mu] \leq [\epsilon] \times [\delta].$$

sosinel (263, 264, 60, 265)/αч, βч, γк?, δи, ϵи, ζк, ηк?, ϑк, κк, λк, μи + /**h**, y

262 332 7

$\delta \Rightarrow qk_a \Rightarrow \mathbf{h}\ \mu - 1 \Rightarrow w - 1 \Rightarrow x - 1 \Rightarrow y$

§1 *optimel* $\alpha\vartheta\kappa q\epsilon\lambda stxy\gamma//$

elimvar $\gamma t\eta q\epsilon \,||\, t \Rightarrow \gamma_\mu\epsilon \Rightarrow \gamma_w$

$\delta - q \times \mu \Rightarrow r \nearrow \mathbf{h}\mu r$ *transfer* $\eta\gamma//$

$s \Rightarrow \epsilon\ \bar{\triangle}\ q \oplus 2 \,|\!\rightarrow 1 \circ \gamma_\mu$

§2 *bounds* $\beta(\zeta_t t)q\epsilon\gamma\zeta\vartheta//\gamma_y \Rightarrow v\gamma_x \Rightarrow \mathbf{a}$

$\gamma_t \Rightarrow \mathbf{ba} \wedge c_0 \circ\!\!\rightarrow 3\mathbf{a} - 1 \urcorner \vee c_0 \Rightarrow \mathbf{a}$

§3 $\mathbf{b} \wedge c_0 \circ\!\!\rightarrow 4\mathbf{b} - 1 \urcorner \vee c_0 \Rightarrow \mathbf{b}$

§4 $\mathbf{a} \times \mathbf{b} \Rightarrow \zeta_v\mathbf{a} + \mathbf{b} \wedge c_0 \circ\!\!\rightarrow 5\zeta_v \urcorner + 1 \Rightarrow \zeta_v$

§5 $q \oplus \delta \circ\!\!\rightarrow 7(\triangle\ q\delta - q \times \mu \Rightarrow r \swarrow \mathbf{h}\mu r\gamma_\mu \Rightarrow t\gamma_w \Rightarrow \epsilon\eta_t \Rightarrow u$

§6 $u \Rightarrow v \triangle u\zeta_u \Rightarrow \zeta_v u \oplus t \,|\!\rightarrow 6 \circ \zeta_u \rightarrow 2$

§7 .

C. ITERATIVE METHOD FOR THE SYNTHESIS OF THRESHOLD ELEMENTS

Here we consider an iterative method, permitting a realization to be found (or unrealizability to be established) manually for a Boolean function of 8–10 variables [16]. The method provides integer values for the weights and threshold, and has been programmed for computer, but convergence has not been proved in the general case; for $n > 6$ branching of variants may arise, so that a minimal solution can be obtained only after sifting of the possibilities arising, although their number is not great.

As shown by computer experiment, for the overwhelming majority of threshold functions the method gives a minimal realization.

The proposed algorithm for analysis of a Boolean function for realizability and the synthesis of the threshold element consists of the following stages.

(1) The inessential variables are eliminated, and it is verified that the given function is homogeneous; the function is transformed to positive.

(2) The weights of the variables are linearly ordered and renumbered, and the function is analyzed for 2-monotonicity

$$a_0 \geq a_1 \geq \cdots \geq a_{n-1} > 0. \tag{24}$$

(3) The difference

$$d_i = a_i - a_{i+1}$$

is partially ordered.

(4) The information obtained in steps (2) and (3) is used to find the first approximation:

$$[a_0^{\mathrm{I}}, a_1^{\mathrm{I}}, \cdots, a_{n-1}^{\mathrm{I}}; T^{\mathrm{I}}].$$

(5) If the first approximation satisfies condition (1), the problem is solved; otherwise, from a comparison of the initial function and the function obtained by the first approximation, inequalities are found which must be satisfied by the weights of the next approximation.

The process terminates when one of the following conditions is satisfied:

(a) the system of coefficients obtained at the kth step satisfies condition (1);

(b) at a certain step the conditions that must be satisfied by the coefficients are found to be contradictory—the function is not realized by a single threshold element; or

(c) the sum of weights $\sum_{i=0}^{n-1} a_i |$ exceeds the maximum permissible value N for the given type of physical element.

The first two steps are common to the majority of known methods of threshold-element synthesis [2–5, 11] and have been considered in Section 1. We therefore begin the exposition of the algorithm with the third step, assuming that the function $f(X)$ is 2-monotonic and the weights are ordered according to (24).

Let us consider the third step. Since the weights of the variables are linearly ordered, to find their numerical values it is sufficient to find the differences between the adjacent branches of the network (24) $d_i = a_i - a_{i+1}$. We note first of all that in the general case the set d_i $(i = 0, 1, \ldots, n-1)$ cannot be linearly ordered (this follows from the Paull–McCluskey theorem), but is a partially ordered set.

For partial ordering of the differences d_i it is necessary to consider all possible combinations of two ordered pairs of variables of the form

$$x_i > x_{i+1} \qquad \text{and} \qquad x_j > x_{j+1}, \qquad i \pm 1 \neq j.$$

For each such combination we find the reduced functions

$$\begin{aligned} f(1 \to \{x_i, x_{j+1}\}, \quad & 0 \to \{x_{i+1}, x_j\}) = f_1; \\ f(1 \to \{x_{i+1}, x_j\}, \quad & 0 \to \{x_i, x_{j+1}\}) = f_2. \end{aligned} \tag{25}$$

Then

$$\begin{aligned} d_i &> d_j, && \text{if } f_1 > f_2; \\ d_i &< d_j, && \text{if } f_1 < f_2; \\ d_i &* d_j, && \text{if } f_1 = f_2, \quad * \in \{>, =, <\}. \end{aligned} \tag{26}$$

In the remaining cases the function is unrealizable. We adopt the convention that $d_i > d_j$ always when $a_i > a_{i+1}$, $a_j = a_{j+1}$.

The results of testing (26) are registered in the matrix of partial ordering (B), whose element

$$b_{ij} = \begin{cases} 1 & \text{if } d_i > d_j; \\ 0 & \text{otherwise.} \end{cases}$$

The matrix (B) is constructed by the operator of partial ordering of differences *pordi*. α denotes an auxiliary terminal which permits exit from the subroutine in the case of incomparability of the reduced functions;

$$\beta::\|\, T_0 \ni a \,\|; \quad \gamma::\|\, d > d \,\|; \quad d = \{d_0, d_1, \ldots, d_{n-1}\};$$
$$[\epsilon] = n; \quad \zeta, \eta \text{ — working complexes}; \quad \delta::\|\, A \ni a \,\|.$$

pordi (61, 233) / αч, βк +, γк, δк +, ϵи, ζк?, ηк/$\mathbf{c}g$
243 232 5

cleanup $\gamma // \circ\ d \circ c\, 0 - c_\epsilon \oplus c_\epsilon \Rightarrow \mathbf{b}\epsilon - 2 \Rightarrow \mathbf{c}$
§1 $\delta_d - 1 \wedge \delta_d \vee c \Rightarrow c \,\triangle\, d \oplus b_\delta \mid\!\to 1 \circ d \circ a$
§2 $\triangle\, a \oplus \beta_\gamma \circ\!\!\to 3c_a \wedge c \mid\!\to 2\gamma_a \vee c \Rightarrow \gamma_\alpha \to 2$
§3 $\triangle\, d \oplus \mathbf{c} \circ\!\!\to 5c_d \wedge c \mid\!\to 3d + 1 \Rightarrow e \Rightarrow f$
§4 $\triangle\, e \oplus \epsilon \circ\!\!\to 3c_e \wedge c \mid\!\to 4e + 1 \Rightarrow gc_d \vee c_g \wedge \mathbf{b} \Rightarrow \mathbf{d}c_e \vee c_f \Rightarrow \mathbf{e}$
comparability $\alpha\mathbf{de}\beta g\zeta\eta // c_e \wedge g \vee \gamma_d \Rightarrow \gamma_d \to 4$
§5 .

Let us pass to the fourth step. The matrix (B) is nothing else than the dominance matrix [26]; as in every dominance matrix, it can be represented in the form of a graph. To find the weights a_i ($i = 0, 1, \ldots, n-1$) in the first approximation, it is necessary to find such values for $d_0, d_1, \ldots,$

d_{n-1} that the sum $d_0 + d_1 + \cdots + d_{n-1}$ be minimal, and the relations among them, considered as the dominance relation, be described by the matrix (B). In other words, it is necessary to order the differences linearly, using the information presented by the matrix (B), and taking into account the linear ordering of the coefficients (24). Figure 2 gives an example of matrix (B) for six variables. Figure 3 represents the graph G whose nodes correspond to the differences; the presence of a branch (d_i, d_j) corresponds to the relation $d_i > d_j$. It is obvious that in the case of a threshold function the graph must have the following properties:

(a) it must be free of loops;
(b) at least one node is found to which no other node leads—the input;
(c) at least one node is found from which no branch leads—the exit.

	d_0	d_1	d_2	d_3	d_4	d_5
d_0				1		
d_1	1		1			
d_2						
d_3			1			
d_4				1		
d_5	1	1	1		1	

FIG. 2.

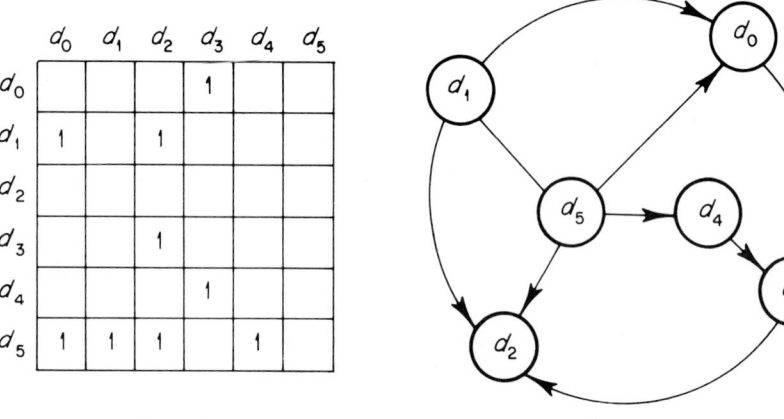

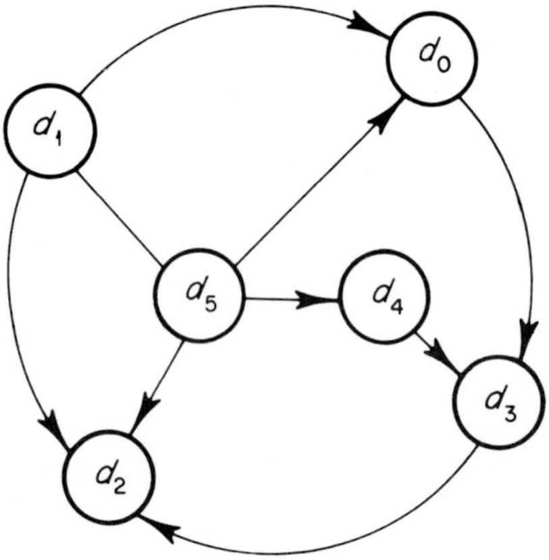

FIG. 3.

It can be shown that satisfaction of condition (a) implies the satisfaction of (b) and (c).

Let there exist a matrix (B) and a graph G, mapping the dominance relation. It is obvious that if conditions (b) and (c) are satisfied for the graph G, the matrix (B) contains a zero column and a zero row, corresponding, respectively, to the input and exit. Let the kth row consist completely of zeros. Removing the kth node from the graph G and the kth row and column from the matrix (B), we obtain a subgraph G_k and a submatrix (B_k); d_k is the smallest of the differences. In the matrix (B), in turn, a zero row is sought, corresponding in the subgraph G_k to the exit d_l, where $d_l > d_k$. The process of crossing out will continue either until all rows have been crossed off in (B) or until at a certain step a submatrix not having a zero row will be obtained. This last fact indicates the presence of loops in the graph G. Thus, if the initial Boolean function is realizable, the differences d_i will have been ordered in not more than n steps.

The application of this process to the matrix (B) shown in Fig. 2 gives the sequence

$$d_5 > d_1 > d_0 * d_4 > d_3 > d_2, \qquad * \in \{>, =, <\}. \tag{27}$$

It is obvious that by replacing all symbols $*$ by the sign "equals" and giving to the differences the minimal values satisfying the string (27), we find the minimal values of the differences satisfying (B). After replacement of all $*$ by the sign "equals" the differences are linearly ordered; in this case it is possible to speak of a partitioning of the differences into classes, defined by the string (27). We denote by D_0 the class of maximal differences, by D_{m-1} the class of minimal differences, $D = \{D_0, D_1, \ldots, D_{m-1}\}$. For the sequence (27), for example, $D_0 = \{d_5\}$, $D_2 = \{d_0, d_4\}$, etc.

We now present the operator *mast*, realizing this transformation of the matrix (B) to a string; α denotes an auxiliary exit terminal, corresponding to the presence of a loop in the graph,

$$\beta::\| d > d \|, \qquad \gamma::\| D \ni d \|.$$

mast $(43, 56, 43)/\alpha ч + \beta к +, \gamma к/\mathbf{b}, c$
244 135 4

$\mathbf{b}_\beta - 1 \Rightarrow c\, 0 - c_c \Rightarrow \mathbf{b}$
§1 $\bar{\text{o}}\ a\, \mathbf{b} \Rightarrow \mathbf{a}$
§2 $\triangle\ a \oplus b_\beta \circ\!\!\rightarrow 3\beta_a \mid\!\!\rightarrow 2c_a\ \urcorner\ \wedge\ \mathbf{a} \Rightarrow \mathbf{a} \rightarrow 2$
§3 $\mathbf{a} \oplus \mathbf{b} \circ\!\!\rightarrow \alpha\mathbf{a}\ \urcorner \Rightarrow \gamma_b\ \bar{\triangle}\ c$ *compress* $\beta \mathbf{a} \beta // \beta_\beta \circ\!\!\rightarrow 4$
carcomva $\beta \mathbf{a} \beta // \rightarrow 1$
§4 $0 - c_c \wedge \mathbf{b} \Rightarrow \mathbf{a}$ *compress* $\gamma \mathbf{a} \gamma //$.

Giving d_i the minimal values satisfying (27), we determine in the first approximation the weights $a_0{}^{\mathrm{I}}\ a_1{}^{\mathrm{I}}, \ldots, a_{n-1}^{\mathrm{I}}$:

$$a_{n-1}^{\mathrm{I}} = d_{n-1}, \qquad a_i{}^{\mathrm{I}} = a_{i+1} + d_i, \qquad i \neq n-1. \tag{28}$$

As the threshold we take $\min a^{\mathrm{I}} \times \alpha$. The threshold is found by the operator *comp*, in which

$$\alpha::\| T_0 \ni a \|, \qquad [\beta_i] = a_i, \qquad i \in \{0, 1, \ldots, n-1\}, \qquad [\gamma] = T.$$

comp $(50)/\alpha к +, \beta к +, \gamma и/\mathbf{a}, b$
245 45 2

$\bar{\text{o}}\ b \circ \gamma$
§1 $\triangle\ b \oplus b_\alpha \circ\!\!\rightarrow 2$ *concom* $+ \beta\alpha_\alpha\, \mathbf{a}//\mathbf{a} - \gamma \mid\!\!\rightarrow 1\mathbf{a} \Rightarrow \gamma \rightarrow 1$
§2 .

The weights are calculated by the operator *weights*: α::|| $D \ni d$ ||, $[\delta_i] = a_i^{\mathrm{I}}$, $i = 0, 1, \ldots, n - 1$; the jth position of the variable β is equal to 1 if $a_j = a_{j+1}$ in (24).

weights (47)/αк +, βи +, γи, δк/**b**, b
251 110 3

$\gamma \Rightarrow b \circ \mathbf{a}$
§1 $b - 1 \circ\!\!\rightarrow 3 \Rightarrow bb_\alpha \Rightarrow a \circ \mathbf{b}\beta \mid\!\rightarrow 2 \triangle \mathbf{b}$
§2 $a - 1 \circ\!\!\rightarrow 1 \Rightarrow a \triangle \mathbf{b}\alpha_a \wedge c_b \circ\!\!\rightarrow 2\mathbf{b} + \mathbf{a} \Rightarrow \mathbf{a} \Rightarrow \delta_b \rightarrow 1$
§3 *convol* + δb // $\triangle$ $b_\delta \mathbf{b} \Rightarrow b_\gamma$.

Theorem 2.1 follows from the method of finding the first approximation:

Theorem 2.1. The values of the coefficients a_i^{I}, obtained from (28), are the smallest of all values satisfying (24) and matrix (B).

From the weights a_i^{I} ($i = 0, 1, \ldots, n - 1$) and the threshold T^{I}, found in the first approximation, using the operator of threshold-element analysis we find the function $f^{\mathrm{I}}(X)$, which is the first approximation to the initial function $f(X)$.

We now pass to the fifth step in the algorithm. In the general case the function $f^{\mathrm{I}}(X)$, obtained by the first approximation, does not coincide completely with the initial function $f(X)$. As follows from the rule for determining the threshold, the difference between f^{I} and f can consist only in that $M_1^{\mathrm{I}} \supset M_1$, i.e., $f_1^{\mathrm{I}} > f$. This signifies that the mdf of $f^{\mathrm{I}}(X)$ contains conjunctions that do not exist in the mdf of $f(X)$.

Elements occurring in the set T_0^{I}, but not in the set T_0, will be called type-s elements, and will be denoted by $\alpha_{s0}, \alpha_{s1}, \ldots, \alpha_{sr}$.

In order that the functions $f(X)$ and $f^{\mathrm{I}}(X)$ coincide, it is necessary to so change the weights $a_0^{\mathrm{I}}, a_1^{\mathrm{I}}, \ldots, a_{n-1}^{\mathrm{I}}$, that the following inequality holds:

$$a \times \alpha_{sj} < T, \qquad i = 0, 1, \ldots, r. \tag{29}$$

The quantity $a \times \delta = I$, calculated for a certain element of Boolean space, will be called the input effect of this element.

Since $a \times \alpha_{sj}$ and T are expressed in the form of a sum of weights of variables, (29) is a system of inequalities with respect to the weights of the second approximation a_i^{II}. System (29) must be completed by n inequalities of the form $a_i^{\mathrm{II}} \geq a_i^{\mathrm{I}}$ (by virtue of Theorem 2-1) and by inequalities among the d_i, following from matrix (B).

Finally, we have

$$a^{\mathrm{II}} \times \alpha_{sj} < T^{\mathrm{II}}, \qquad j = 0, 1, \ldots, r - 1 \tag{30a}$$
$$a_i^{\mathrm{II}} \geq a_i^{\mathrm{I}}, \qquad i = 0, 1, \ldots, n - 1 \tag{30b}$$
$$d_k > d_t, \quad \text{if} \quad b_{kt} = 1. \tag{30c}$$

Solving the system (30) by the method of trial and error or the method of successive elimination of unknowns, we determine a_i^{II} ($i = 0, 1, \ldots, n-1$) and T^{II}. From the realization $a_0^{\text{II}}, a_1^{\text{II}}, \ldots, a_{n-1}^{\text{II}}, T^{\text{II}}$ we find $f^{\text{II}}(X)$. If $f^{\text{II}} \neq f$, a new system of inequalities of type (30) is found, from which are obtained a_i^{III} and T^{III}. If the system of type (30) is incompatible, $f(X)$ is not threshold. The number of inequalities in (30) depends on the degree of closeness of the current approximation to the exact solution, and has a tendency to decrease as the number of the approximation increases. Experience with computer solutions has shown that for $n \leq 10$ the number of elements of type s rarely exceeds n, and if the function is threshold, the solution is found in from one to three steps. The existence of linear ordering of the coefficients permits the number of inequalities in system (30) to be substantially reduced.

Before passing to the description of the operator for solving the system of inequalities by the method of trial and error, let us consider certain operators on parts of complexes.

Let there exist two complexes, α and β. The first of these operators, called *choice*, realizes the transfer of r elements of α, beginning with α_i, to complex β, where they are ascribed the indexes $s, s+1, \cdots, s+r$. It is clear that it is necessary to respect the condition $s + r \leq b_\beta$, $r \leq b_\alpha$.

$$[\gamma] = r, \qquad [\delta] = i, \qquad [\epsilon] = s.$$

choice αк $+$, βк, γи $+$, δи $+$, ϵи $+$ / $-$, c
246 35 1

$\delta \Rightarrow a + \gamma \Rightarrow c\epsilon \Rightarrow b$
§1 $\alpha_a \Rightarrow \beta_b \bigtriangleup b \bigtriangleup a \oplus c \mid\!\rightarrow 1.$

The code of the next operator is *tracom*. This operator shifts the elements of complex α, beginning with α_i, to the positions $\alpha_{i+r}, \alpha_{i+r+1}, \ldots, \alpha_{\sigma(\alpha)+r-1}$ and transfers to the free positions r elements of the complex β, beginning with β_s.

$$[\gamma] = r, \qquad [\delta] = i, \qquad [\epsilon] = s.$$

tracom (246)/αк, βк, γи $+$, δи $+$ ϵи $+$ / $-$, b
247 47 1

$b_\alpha \Rightarrow a + \gamma \Rightarrow b_\alpha \Rightarrow b$
§1 $\bar{\bigtriangleup}\, a\, \bar{\bigtriangleup}\, b\alpha_a \Rightarrow \alpha_b a \oplus \delta \mid\!\rightarrow 1$ *choice* $\beta\epsilon\gamma\alpha\delta//.$

The next operator compares r elements of complex β, beginning with β_i, with the elements $\epsilon_s, \epsilon_{s+1}, \ldots, \epsilon_{s+r-1}$ of the complex ϵ. The program performing it is (1, 2)-terminal; the auxiliary terminal α corresponds to the situation in which each element of the complex β, $\beta_i, \beta_{i+1}, \ldots, \beta_{i+r-1}$ is greater than or equal to the corresponding element of complex ϵ. The operator code is *exceed*: $[\delta] = r$, $[\gamma] = i$, $[\zeta] = s$.

exceed αч, βк +, γи +, δи +, ϵк +, ζи + / −, c
250 44 1

$\delta \Rightarrow \mathbf{c}\gamma + c \Rightarrow a\zeta + c \Rightarrow b$
§1 $\bar{\triangle} c \oplus f_{10} \circ\!\!\rightarrow \alpha \bar{\triangle} a \bar{\triangle} b\beta_a - \epsilon_a \mid\!\rightarrow 1.$

Now we present the operator for solving a system of inequalities by the method of trial and error—*solsin*. Each inequality is represented by an element of the complex γ, divided into halves, so that the expressions for the input signals of elements defining the threshold are written in the 16 lowest-order positions, and the expressions for the input signals of the type-s elements, in the 16 highest-order. The algorithm for the operator can be briefly described in the following way.

Of all the weight vectors a_i (for concreteness, system (30) is considered, written for the weights of the second approximation) satisfying the jth inequality of subsystem (30a) and subsystems (30b) and (30c), a set A_j of vectors a is chosen such that no two vectors in the set A_j are comparable. From the set of vectors $A = A_0 \cap A_1 \cdots \cap A_{r-1}$ is taken a vector a^{II} satisfying the condition

$$\sum_{i=0}^{n-1} a_i^{\mathrm{II}} = \min_{a \in A} \sum_{i=0}^{n-1} a_i.$$

Since the sets A_j may not have common elements, i.e., it can happen that

$$A_0 \cap A_1 \cap \cdots \cap A_{r-1} = \varnothing,$$

the possibility of "deformation" is provided, i.e., of changing the values of the components of elements of the sets $A_0, A_1, \ldots, A_{r-1}$ in order to obtain in the result a vector satisfying system (30). Of all deformations a vector a^{II} is chosen such that $\sum_{i=0}^{n-1} a_i^{\mathrm{II}} = \min_A$.

A function is considered unrealizable if in the calculation of the current approximation a contradictory system is obtained or if $\sum_{i=0}^{n-1} a_i^{\mathrm{II}} > N$, where N is the maximum permissible value of the sum for the chosen type of physical element or the value of the upper bound of the sum [1]. In this case exit from the subroutine takes place over the auxiliary terminal α.

$$[\beta_i] = a_i^{\mathrm{I}}, \qquad i \in \{0, 1, \ldots, n-1\}, \qquad [\beta_n] = \sum_{i=0}^{n-1} a_i, \qquad [\delta] = N,$$

$$[\epsilon_i] = a_i^{\mathrm{II}} \qquad [\epsilon_n] = \sum_{i=0}^{n-1} a_i^{\mathrm{II}}, \qquad \kappa{:}{:}\,\|\, d > d \,\|,$$

ζ, η, ϑ — working complexes.

solsin (61, 246, 250, 250, 247, 246, 246, 247, 50, 50)/$\alpha\text{ч}$, $\beta\text{к}$, $\gamma\text{к}\ +$, $\delta\text{и}\ +$, $\epsilon\text{к}$, $\zeta\text{к}$, $\eta\text{к}?$, $\vartheta\text{к}?$, $\kappa\text{к}\ +$ /**d**, q

255 1020 27

$b_\beta \Rightarrow b_\epsilon \Rightarrow e - 1 \Rightarrow d \triangle b_\beta\delta \Rightarrow \epsilon_d b_\gamma \Rightarrow b_\zeta$ *cleanup* $\zeta // \bar{\circ}\ f$

§1 $\triangle f \Rightarrow i \oplus b_\gamma \circ\!\!\rightarrow 24\gamma_f \wedge e_4 < 20 \Rightarrow \mathbf{b}\gamma_f \wedge d_4 \Rightarrow \mathbf{a} \nrightarrow 26 \circ\!\!\rightarrow 1$
$\Rightarrow \beta_e \circ h$

§2 $\zeta_i + h \Rightarrow h\ \bar{\triangle}\ i \oplus f_{10} \mapsto 2h - \zeta_f \Rightarrow g \odot i \circ j \nrightarrow 25i + b_\beta \Rightarrow i\ \bar{\circ}\ k$

§3 $\triangle k \oplus d \circ\!\!\rightarrow 14i \Rightarrow \mathbf{c}\, 0 - b_\beta \Rightarrow l$

§4 $l + b_\beta \Rightarrow l \oplus \mathbf{c} \circ\!\!\rightarrow 3$ *choice* $\nu l b_\beta j // \beta_e \Rightarrow \mathbf{d} \circ \beta_d \circ \beta_e$

§5 $\kappa\text{к} \mapsto 6 \triangle k \Rightarrow a \rightarrow 5$

§6 $\triangle \beta_a\ \bar{\triangle}\ a \oplus f_{10} \,|\!\dashrightarrow 6d \Rightarrow ak + 1 \Rightarrow b\beta_k - \beta_b \Rightarrow \mathbf{a}$

§7 $\bar{\triangle}\ a \oplus f_{10} \circ\!\!\rightarrow 10\kappa\text{к} \wedge c_a \mapsto 7a + 1 \Rightarrow b\beta_a - \beta_b \oplus \mathbf{a} \mapsto 7a \Rightarrow k \rightarrow 6$

§10 *convol* $\alpha\alpha_\delta // \epsilon_d - \beta_d \circ\!\!\rightarrow 4g \oplus h \circ\!\!\rightarrow 12g - d \Rightarrow m$

§11 $m + d \Rightarrow m$ *exceed* $4\beta je\eta m // m \oplus h \mapsto 11$

§12 $\nrightarrow 26 \circ\!\!\rightarrow 13 \Rightarrow \beta_e - \mathbf{d} \rightarrow 4 \nrightarrow 25i + b_\beta \Rightarrow i \rightarrow 4$

§13 $g - 1 \Rightarrow m \circ p \Rightarrow q$

§14 $m + e \Rightarrow m \Rightarrow n - h \mapsto 17\eta_m - \beta_d \mapsto 14m - d \Rightarrow n - e \Rightarrow m$

§15 $m \oplus h \circ\!\!\rightarrow 17$ *exceed* $16\eta me\beta j // m + e \Rightarrow m \rightarrow 15$

§16 $\circ\ \eta_m \triangle m \triangle p \oplus e \mapsto 16 \triangle q \circ p \rightarrow 15$

§17 *tracom* $\eta ne\beta j // \zeta_f - q + 1 \Rightarrow \zeta_f h - q + 1 \Rightarrow h\ \bar{\triangle}\ n \Rightarrow a$

§20 $\triangle a \triangle n \oplus b_\eta \circ\!\!\rightarrow 21\eta_n \mapsto 20\ \bar{\triangle}\ a \rightarrow 20$

§21 $a \Rightarrow b_\eta \rightarrow 4$

§22 $\zeta_f \circ\!\!\rightarrow \alpha b_\eta \circ\!\!\rightarrow 27 - e \Rightarrow i$ *choice* $\eta ie\beta j // i \Rightarrow b_\eta b_\zeta \Rightarrow a$

§23 $\bar{\triangle}\ a\zeta_a \circ\!\!\rightarrow 23\ \bar{\triangle}\ \zeta_a\ \bar{\circ}\ f \rightarrow 1$

§24 *choice* $\beta je\epsilon j // \bar{\circ}\ f \rightarrow 1$

§25 *tracom* $\vartheta ie\beta j //!$

§26 *concom* $+ \beta \mathbf{a} a //$ *concom* $+ \beta \mathbf{b} b // a - b!$

§27 .

The algorithm for synthesis of a threshold element by the method described in this section can be realized by the operator *synthrel*: α is an auxiliary terminal and corresponds to the case of unrealizability of the function.

$$\beta::\|\, T_0 \ni a \,\|, \qquad \gamma::\|\, d > d \,\|, \qquad \delta::\|\, A \ni a \,\|,$$

$$[\zeta_i] = a_i\, i \in \{0, 1, \ldots, n-1\}, \qquad \mu, \eta, \vartheta, \kappa, \pi, \rho, \sigma \text{— working complexes,}$$

$$\sigma(\beta) = \sigma(T_0), \qquad \sigma(\gamma) = n, \qquad \sigma(\delta) \le n, \qquad \sigma(\zeta) \le n+2,$$

$$\sigma(\mu) \le n+1, \qquad [\epsilon] = n, \qquad [\lambda] = N, \qquad [\nu_i] = a_i, \qquad [\xi] = T.$$

synthrel (234, 235, 237, 243, 244, 250, 245, 240, 252, 252, 255, 236)/ αч, βк, γк, δк, ϵи +, ζк, ηк?, ϑк?, κк? λк +, μк, νк, ξи, πк?, ρк, σк/**j**, i

256 455 13

lowal $\beta\sigma\zeta\mu\epsilon$ // *renumbering* $\beta\sigma\beta\delta$ // $\circ\ a \Rightarrow \mathbf{j}$

§1 $\delta_a - 1 \wedge \delta_a \vee j \Rightarrow j \triangle a \oplus b_\delta \mid\!\rightarrow 1$ *twomontest* $\alpha\beta\pi\rho$ // *pordi* $\alpha\beta\gamma\delta\epsilon$ // *mast* $\alpha\gamma\delta$ // *weights* $\delta\mathbf{j}\epsilon\zeta$ //

§2 *comp* $\beta\zeta\xi$ // $\circ\ i\ \circ\ g \Rightarrow h$ *threlan* $4\zeta\xi\epsilon e f$ // $f\ \circ\!\rightarrow 3e \Rightarrow \eta_h?\ \triangle\ h \rightarrow$ *threlan* 2

§3 $\triangle\ i \oplus b_\beta\ \circ\!\rightarrow 4b_\beta \oplus e \mid\!\rightarrow 3e \Rightarrow \vartheta_g?\ \triangle\ g\ \bar{\circ}\ i \rightarrow$ *threlan* 2

§4 $g\ \circ\!\rightarrow 13 \Rightarrow b_\vartheta h \Rightarrow b_\eta$ *redu* $\alpha\beta\ \circ\!\rightarrow \vartheta\pi\rho\ \|$ *redu* $\alpha\beta \mid\!\rightarrow \eta\pi\rho$ // $\bar{\circ}\ a \Rightarrow b\ \circ\ c$

§5 $\triangle\ b \oplus b_\vartheta \mid\!\rightarrow 6\ \bar{\circ}\ b\ \triangle\ a \oplus b_\eta\ \circ\!\rightarrow 12$

§6 $\eta_b \oplus \eta_a \Rightarrow \mathbf{a} \wedge \eta_a \Rightarrow \mathbf{b}\ \bar{\circ}\ d\mathbf{a} \wedge \vartheta_b > 20 \vee \mathbf{b} \Rightarrow \mathbf{b} \wedge d_4 \nabla \Rightarrow \mathbf{a}$

§7 $\triangle\ d \oplus c\ \circ\!\rightarrow 11\kappa_d \wedge d_4 \nabla - \mathbf{a} \mid\!\rightarrow 7c \Rightarrow f + 1 \Rightarrow c$

§10 $\bar{\triangle}\ e \oplus d\ \circ\!\rightarrow 11\ \bar{\triangle}\ f\kappa_f \Rightarrow \kappa_e \rightarrow 10$

§11 $\mathbf{b} \Rightarrow \gamma_d\ \triangle\ c \rightarrow 5$

§12 $c \Rightarrow b_\kappa$ *solsin* $\alpha\zeta\kappa\lambda\mu\eta\vartheta\gamma$ // $\rightarrow 1$

§13 *rereal* $\mu\sigma\nu$ //.

Example:

$$f(x_0, x_1, \ldots, x_5) = x_0x_1 \vee x_0x_2x_3x_4 \vee x_0x_2x_3x_5 \vee x_0x_2x_4x_5 \vee x_0x_3x_4x_5 \vee x_1x_2x_3x_4 \vee x_1x_2x_3x_5. \tag{31}$$

In the solution of this example, we shall follow the above algorithm.

(1) All variables are essential, and the function is homogeneous.
(2) The weights a_i ($i = 0, 1, \ldots, 5$) are ordered in the following way:

$$a_0 > a_1 > a_2 = a_3 > a_4 = a_5 > 0. \tag{32}$$

The comparison of the corresponding reduced functions shows that the function (31) is 2-monotonic.

(3) We calculate the values of the elements of matrix (B):

$$
(B) = \begin{array}{c|cccccc}
 & d_0 & d_1 & d_2 & d_3 & d_4 & d_5 \\
\hline
d_0 & & & 1 & & 1 & \\
d_1 & & & 1 & 1 & 1 & 1 \\
d_2 & & & & & & \\
d_3 & & & 1 & & 1 & \\
d_4 & & & & & & \\
d_5 & 1 & & 1 & 1 & 1 &
\end{array}
$$

(4) We expand the matrix (B):

$$d_1 > d_5 > d_3 * d_0 > d_2 * d_4 = 0. \tag{33}$$

Giving to the differences d_i the minimal values satisfying (33), we obtain

$$d_4^{\mathrm{I}} = d_2^{\mathrm{I}} = 0; \qquad d_0^{\mathrm{I}} = d_3^{\mathrm{I}} = 1; \qquad d_5^{\mathrm{I}} = 2; \qquad d_1^{\mathrm{I}} = 3.$$

By formula (28) we calculate the values of the weights in the first approximation:

$$a_4^{\mathrm{I}} = a_5^{\mathrm{I}} = d_5^{\mathrm{I}} = 2; \qquad a_2^{\mathrm{I}} = a_3^{\mathrm{I}} = a_4^{\mathrm{I}} + d_3^{\mathrm{I}} = 2 + 1 = 3;$$
$$a_1^{\mathrm{I}} = a_2^{\mathrm{I}} + d_1^{\mathrm{I}} = 3 + 3 = 6; \qquad a_0^{\mathrm{I}} = a_1^{\mathrm{I}} + d_0^{\mathrm{I}} = 6 + 1 = 7.$$

We put the threshold equal to $\min_{\alpha \epsilon M_1} a^{\mathrm{I}} \times \alpha$, i.e., it is necessary to calculate the sum of weights of the variables occurring in each of the conjunctions of the function f and take the minimal sum for the threshold. In this example the minimal sum is formed by the weights of the variables appearing in the conjunction x_0x_1: $a_0^{\mathrm{I}} + a_1^{\mathrm{I}} = 13$; $T^{\mathrm{I}} = 13$.

Applying the operator for threshold-element analysis, we find f^{I}:

$$f^{\mathrm{I}}(x_0, \ldots, x_5) \sim [7, 6, 3, 3, 2, 2; 13]$$
$$\sim x_0x_1 \vee x_0x_2x_3 \vee x_0x_2x_4x_5 \vee x_0x_3x_4x_5 \vee x_1x_2x_3x_4 \vee x_1x_2x_3x_5.$$

(5) From the comparison of f^{I} and f we find that the element of the set T_0 corresponding to the conjunction $x_0x_2x_3$ is a type-s element. This signifies that

$$a_0^{\mathrm{II}} + a_2^{\mathrm{II}} + a_3^{\mathrm{II}} < a_0^{\mathrm{II}} + a_1^{\mathrm{II}},$$

whence

$$a_2^{\mathrm{II}} + a_3^{\mathrm{II}} < a_1^{\mathrm{II}}.$$

Considering (32) and the condition $a_i^{\mathrm{II}} \geq a_i^{\mathrm{I}}$ for all $i \in \{0, 1, \ldots, 5\}$, we find

$$a_1^{\mathrm{II}} = 7; \qquad a_0^{\mathrm{II}} = a_1^{\mathrm{II}} + d_0 = 7 + 1 = 8; \qquad a_2^{\mathrm{II}} = a_2^{\mathrm{I}} = 3;$$

$$a_3^{\mathrm{II}} = a_3^{\mathrm{I}} = 3; \qquad a_4^{\mathrm{II}} = a_5^{\mathrm{II}} = 2; \qquad T^{\mathrm{II}} = 15.$$

Thus

$$f(x_0, x_1, \ldots, x_5) \sim [8, 7, 3, 3, 2, 2; 15].$$

We may omit the step of partial ordering of the differences d_i, at least when $n \leq 6$; i.e., we put $d_i^{\mathrm{I}} = 1$ if $a_i > a_{i+1}$ in the string (24), and $d_i = 0$ if $a_i = a_{i+1}$. In this case the number of approximations will increase negligibly.

For a large number of variables the ordering of the differences plays an essential role, since it makes it possible to test a large number of reduced functions and to detect earlier nonthreshold functions, and even in the first approximation to arrive fairly close to the solution. Furthermore, together with the ordering of the differences d_i it is useful to order the differences among the differences, which we call second-order differences. This process can be continued; i.e., the third-order differences can be brought into consideration, etc.

Example[10]:

$$\begin{aligned} f(x_0, x_1, \ldots, x_7) = {} & x_0x_1 \vee x_0x_2 \vee x_0x_3 \vee x_0x_4 \vee x_0x_5x_6 \vee x_0x_5x_7 \\ & \vee x_0x_6x_7 \vee x_1x_2 \vee x_1x_3x_4 \vee x_2x_3x_4 \vee x_1x_3x_5 \vee x_1x_3x_6 \vee x_1x_3x_7 \vee x_1x_4x_5 \\ & \vee x_1x_4x_6 \vee x_1x_4x_7 \vee x_1x_5x_6x_7 \vee x_2x_3x_5 \vee x_2x_3x_6 \vee x_2x_3x_7 \vee x_2x_4x_5 \\ & \vee x_2x_4x_6 \vee x_2x_4x_7 \vee x_2x_5x_6x_7 \vee x_3x_4x_5 \vee x_3x_4x_6x_7. \end{aligned}$$

The variables $x_0, x_1, \ldots, x_7$ are associated with the variables A, B, C, D, E, G, H, I in [12]. This example is presented in [12] to illustrate that it is not possible to guarantee minimality of the sum $\sum_{i=0}^{n-1} |a_i|$, by giving equal values to the weights connected in the string by the equal sign; i.e., the minimized realization is not minimal.

The author of [12] remarks that this example can serve as the "touchstone" for new methods of threshold-element synthesis.

The above method yields a minimal solution for the given example.

(1) The function is homogeneous; all variables are essential.
(2) The weights are ordered as follows:

$$a_0 > a_1 = a_2 > a_3 = a_4 > a_5 > a_6 = a_7 > 0.$$

[10] See reference [12].

(3) The matrix of partial ordering is

$(B) =$

	d_0	d_1	d_2	d_3	d_4	d_5	d_6	d_7
d_0		1		1		1	1	
d_1								
d_2		1		1			1	
d_3								
d_4		1		1			1	
d_5		1		1			1	
d_6								
d_7		1		1		1	1	

(4) The values $d_1^{\mathrm{I}} = d_3^{\mathrm{I}} = d_6^{\mathrm{I}} = 0$; $d_2^{\mathrm{I}} = d_4^{\mathrm{I}} = d_5^{\mathrm{I}} = 1$; $d_0^{\mathrm{I}} = d_7^{\mathrm{I}} = 2$ satisfy (B) and are minimal. Putting $d_1^{\mathrm{I}} = d_3^{\mathrm{I}} = d_6^{\mathrm{I}} = 0$, we automatically make two assumptions.

Assumption 1: If the optimal values of the weights connected by the equals sign do not coincide, they differ slightly.

Assumption 2: The optimal values of the weights joined by the equals sign are not obtained in the first approximation, but in one of the later ones.

The process of finding the optimal solution is constructed in the following way: The first approximation is found in the usual way; to find the values of the weights in the subsequent approximations the equal sign is not taken into account.

Thus $a_0^{\mathrm{I}} = 7$, $a_1^{\mathrm{I}} = a_2^{\mathrm{I}} = 5$, $a_3^{\mathrm{I}} = a_4^{\mathrm{I}} = 4$, $a_5^{\mathrm{I}} = 3$, $a_6^{\mathrm{I}} = a_7^{\mathrm{I}} = 2$, $T^{\mathrm{I}} = 10$.

(5) Omitting the intermediate calculations, we present the sequence of realizations converging to the realization of the initial function.

$$a_0^{\mathrm{II}} = 8; \quad a_1^{\mathrm{II}} = 5; \quad a_2^{\mathrm{II}} = 6; \quad a_3^{\mathrm{II}} = a_4^{\mathrm{II}} = 4; \quad a_5^{\mathrm{II}} = 3;$$
$$a_6^{\mathrm{II}} = a_7^{\mathrm{II}} = 2; \quad T^{\mathrm{II}} = 11.$$

$$a_0^{\mathrm{III}} = 8; \quad a_1^{\mathrm{III}} = a_2^{\mathrm{III}} = 6; \quad a_3^{\mathrm{III}} = 4; \quad a_4^{\mathrm{III}} = 5; \quad a_5^{\mathrm{III}} = 3;$$
$$a_6^{\mathrm{III}} = a_7^{\mathrm{III}} = 2; \quad T^{\mathrm{III}} = 12.$$

$$a_0^{\mathrm{IV}} = 9; \quad a_1^{\mathrm{IV}} = 6; \quad a_2^{\mathrm{IV}} = 7; \quad a_3^{\mathrm{IV}} = 4; \quad a_4^{\mathrm{IV}} = 5; \quad a_5^{\mathrm{IV}} = 3;$$
$$a_6^{\mathrm{IV}} = a_7^{\mathrm{IV}} = 2; \quad T^{\mathrm{IV}} = 12.$$

$$a_0^{\mathrm{V}} = 9; \quad a_1^{\mathrm{V}} = 6; \quad a_2^{\mathrm{V}} = 7; \quad a_3^{\mathrm{V}} = 5; \quad a_4^{\mathrm{V}} = 5; \quad a_5^{\mathrm{V}} = 3;$$
$$a_6^{\mathrm{V}} = a_7^{\mathrm{V}} = 2; \quad T^{\mathrm{V}} = 13.$$

The fifth approximation is the minimal realization of the function $f(x_0, x_1, \ldots, x_7)$.

This method is easily transferred to the case of the matrix representation of Boolean functions [16]. The set of operators remains essentially the same; the description of the algorithm, whose code is *synthrem*, is given in [27].

D. THE METHOD OF DENSE ENUMERATION

These synthesis algorithms are universal in the sense that they are equally suitable for manual and for computer execution. From the viewpoint of the computer these algorithms have both advantages and drawbacks. The method considered in Section 2B requires much space in memory to store the intermediate results, while the iterative method, more adapted to manual calculation for the number of variables $n < 10$, does not give the assurance that the solution obtained is minimal in more complicated cases.

These two disadvantages are absent from the algorithm we shall now describe. We begin the discussion with the statement of the basic steps.

(1) Inessential variables are eliminated, the initial function $f(X)$ is brought to positive [we shall also designate the reduced function by $f(X)$].

(2) The weights are linearly ordered without testing for 2-monotonicity. Let this be the string (24).

(3) The matrix (B) is calculated; only nonzero differences are considered.

(4) The first approximation is found:

$$f^{\mathrm{I}}(X) \sim [a_0^{\mathrm{I}}, a_1^{\mathrm{I}}, \ldots, a_{n-1}^{\mathrm{I}}; T^{\mathrm{I}}]. \tag{34}$$

If $f^{\mathrm{I}}(X) = f(X)$, the solution has been found.

Let $f^{\mathrm{I}}(X) \neq f(X)$, and $N^{\mathrm{I}} = \sum_{i=0}^{n-1} a_i^{\mathrm{I}}$; then:

(5) Find, successively, all partitions $S_1, S_2, \ldots, S_p$ of the number $P = N^{\mathrm{I}} + g$ (g is a natural number) into n components satisfying string (24) and matrix (B).

(6) For each partition $S_j = (a_0^j, a_1^j, \ldots, a_{n-1}^j)$ we determine the threshold T^j and analyze the result obtained.

(7) If one of the partitions realizes the initial function, the realization is minimal, and the process has terminated; otherwise, we increase g by 1 and pass to step (5).

(8) The function is unrealizable if none of the partitions of

$$P = N^{\mathrm{I}} + c_0, N^{\mathrm{I}} + c_0 + 1, \ldots, 4n[(n+1)/4]^{(n+1)/2}$$

gives a realization[11] (the sense of c_0 will be clarified below.

As can be seen from the description just given, all steps of the algorithm except step (5) are realized by previously considered operators. Let us consider the operator for finding the partition.

Let the weights a_i be ordered according to (24). We shall consider that it is necessary to increase the weight of variable x_j by one, conserving at the same time the given ordering. One method obviously consists of increasing by one all weights in the same class as a_j together with the weights that are greater then a_j. This method has the obvious property that

$$K = \sum_{i=0}^{n-1} a_i$$

is increased by a minimal amount.

We denote by l the number of different classes of weights (cf. Section 1) and eliminate from consideration the differences equal to zero, denoting the remaining differences (these, obviously, are the differences between the classes of weights) by $d_0, d_1, \ldots, d_{l-1}$.

Definition. The minimal quantity c_j by which it is necessary to increase the sum N^{I} when increasing the weights of the jth class by 1, conserving the given ordering of the weights is called the cost of the jth class.

Example: Let the weights of the variables of a certain threshold function be ordered as follows:

$$a_0 = a_1 > a_2 > a_3 = a_4 = a_5 > 0. \tag{35}$$

In this case $l = 3$ and $c_0 = 2$, $c_1 = 3$, and $c_2 = 6$.

A partition of P into n components, satisfying (24'), can be replaced by the expansion of the number in the costs of the classes; i.e., we represent g in the form

$$g = \sum_{j=0}^{l-1} \Delta_j c_j + \sigma; \qquad \sigma < c_j, \qquad (j = 0, 1, \cdots, l-1), \tag{36}$$

[11] We give the upper bound on the sum of weights obtained for noninteger values of the weights [1], since the analogous bound for integer weights is not known to us.

where Δ_j are the increments to the weights of the jth class.

We shall give the expansion s of the number g by the set of increments $s = (\Delta_0, \Delta_1, \ldots, \Delta_{l-1})$. The expansion of g is called regular if $\sigma = g - \sum_{j=0}^{l-1} \Delta_j c_j = 0$, and irregular if $\sigma \neq 0$.

Let there be given a certain value $g = g_0$. The algorithm for finding all expansions $s_1, s_2, \ldots, s_p$ of the number g_0 in the costs $c_0, c_1, \cdots, c_{l-1}$ consists of the following:

(a) As the first expansion $s_1 = (\Delta_0^1, \Delta_1^1, \ldots, \Delta_{l-1}^1)$ we take as the values of the increments:

$$\Delta_{l-1}^1 = \left[\frac{g_0}{c_{l-1}}\right]; \quad \Delta_{l-2}^1 = \left[\frac{g_0 - \Delta_{l-1}^1 c_{l-1}}{c_{l-2}}\right]; \cdots; \left[\Delta_0^1 = \frac{g_0 - \sum_{i=1}^{l-1} \Delta_i c_i}{c_0}\right].$$

(b) Let the kth expansion $s_k = (\Delta_0^k, \Delta_1^k, \ldots, \Delta_{l-1}^k)$ be found; in it the minimal index r is found such that $\Delta_r^k \neq 0$, $1 \leq r \leq l - 1$.

(c) The $k + 1$st expansion is obtained from the kth in the following way: We put $\Delta_r^{k+1} = \Delta_r^k - 1$, $\Delta_{l+1}^{k+1} = \Delta_{r+1}^k, \ldots, \Delta_{l-1}^{k+1} = \Delta_{l-1}^k$; the values of the remaining increments are calculated by the formula

$$\Delta_j^{k+1} = \left[\frac{g_0 - \sum_{i=j+1}^{l-1} \Delta_i c_i}{c_j}\right] \qquad (j = r - 1, r - 2, \ldots, 0).$$

(d) The criterion for termination of the process is the satisfaction of the condition

$$\Delta_1^q = \Delta_2^q = \cdots = \Delta_{l-1}^q = 0$$

for the expansion with number q.

Example: We obtain all expansions of the number 16 in costs 1, 5, 7, and 8:

(1) 0, 0, 0, 2;	(4) 8, 0, 0, 1;	(7) 9, 0, 1, 0;	(10) 11, 1, 0, 0;
(2) 1, 0, 1, 1;	(5) 2, 0, 2, 0;	(8) 1, 3, 0, 0;	(11) 16, 0, 0, 0;
(3) 3, 1, 0, 1;	(6) 4, 1, 1, 0;	(9) 6, 2, 0, 0 .	

The algorithm is carried out by the operator *expansion* whose program is (2, 2)-terminal. The basic terminals correspond to the start of generation of the expansion and the obtaining of the current expansion. The auxiliary exit terminal α corresponds to the end of the process of obtaining expansions. To obtain 2, 3, $\cdots$ expansions it is necessary to return to sentence 10;

$$\beta::||d > d||; \quad [\gamma] = g; \quad [\delta_i] = \Delta_i;$$
$$[\epsilon_i] = c_i; \quad i, j \in \{0, 1, \ldots, l-1\}.$$

expansion (61)/αч +, βк +, γи +, δк, ϵк +/**c**, e
240 261 12

$\gamma \Rightarrow$ **c** *cleanup* $\delta // \bar{\circ}$ **a** $\Rightarrow bb_\beta - 1 \Rightarrow d$
§1 **a** $\circ\!\!\rightarrow 12 \triangle b \oplus d \,|\!\!\rightarrow 2 \circ$ **a**
§2 **c** $\oplus f_{10} \,|\!\!\rightarrow 4 \circ$ **c** $\Rightarrow e$
§3 $\epsilon_e \times \delta_e +$ **c** $\Rightarrow$ **c** $\triangle e \oplus b \,|\!\!\rightarrow 3\gamma -$ **c** $\Rightarrow$ **c**
§4 **c**$:\epsilon_b \Rightarrow$ **c** я $\Rightarrow \delta_b$ **c** $\circ\!\!\rightarrow 5 - \delta_d \,|\!\!\rightarrow 5$**a** $|\!\!\rightarrow 11 \rightarrow 10$
§5 $b \Rightarrow c$
§6 $\bar{\triangle}\, c \oplus f_{10} \circ\!\!\rightarrow 1\beta_b \wedge c_c \Rightarrow b\beta_c \wedge c_b \vee b \circ\!\!\rightarrow 6b \circ\!\!\rightarrow 7\delta_c - \delta_b \,|\!\!\rightarrow 6\delta_c$
$\Rightarrow \delta_b \,\bar{\circ}$ **c a** $\circ\!\!\rightarrow 10 \rightarrow 5$
§7 $\delta_b - \delta_c \,|\!\!\rightarrow 6$
§10 $\bar{\triangle}\, b \oplus f_{10} \circ\!\!\rightarrow \alpha\delta_b \circ\!\!\rightarrow 10$
§11 $\bar{\triangle}\, \delta_b \,\bar{\circ}$ **a** $\Rightarrow$ **c** $\rightarrow 5$
§12 .

The process of finding the partitions [step (5)] of the number $P = N^{\mathrm{I}} + g$ can be described fairly simply:

1. Each of the expansions obtained, not contradicting the matrix (B), of the number g in costs is tested for regularity. The irregular expansions are dropped and the next expansion is generated. In this case *contradiction* is understood to mean a situation in which $\Delta_j > \Delta_i$, while in the matrix (B) the element $b_{ij} = 1$.

2. A regular expansion, not containing contradictions, is used to obtain the partition of P. In each class of weights, one representative $a_0^{\mathrm{I}}, a_1^{\mathrm{I}}, \ldots, a_{l-1}^{\mathrm{I}}$ is selected, and the values of the elements of the classes are found from the formula

$$a_j = a_j^{\mathrm{I}} + \sum_{i=j}^{l-1} \Delta_i \qquad (j = 0, 1, \ldots, l-1). \tag{37}$$

Since the value P increases monotonically, and for each fixed value of P all feasible combinations of weights are sifted, the first partition of P satisfying condition (1) will be a minimal realization.

Example: Find the minimal realization of the function

$$
\begin{aligned}
f(x_0, x_1, \ldots, x_5) = {} & x_0x_1x_2 \vee x_0x_1x_3x_4 \vee x_0x_2x_3x_4 \\
& \vee x_0x_1x_3x_5 \vee x_0x_2x_3x_5 \vee x_0x_1x_4x_5 \\
& \vee x_0x_2x_4x_5 \vee x_0x_3x_4x_5 \vee x_1x_2x_3x_4 \\
& \vee x_1x_2x_3x_5 \vee x_1x_2x_4x_5 \vee x_1x_3x_4x_5.
\end{aligned}
$$

(1) The function is positive; all variables are essential.

(2) $$a_0 = a_1 > a_2 > a_3 = a_4 = a_5 > 0.$$

(3) $$(B) = \begin{array}{c|ccc} & d_0 & d_1 & d_2 \\ d_0 & & & \\ d_1 & & & \\ d_2 & 1 & 1 & \end{array}$$

(4) The values of the weights in the first approximation are

$$a_3^{\mathrm{I}} = a_4^{\mathrm{I}} = a_5^{\mathrm{I}} = 2, \qquad a_2^{\mathrm{I}} = 3, \qquad a_0^{\mathrm{I}} = a_1^{\mathrm{I}} = 4,$$

$$N^{\mathrm{I}} = \sum_{i=0}^{n-1} a_i^{\mathrm{I}} = 17.$$

It can be verified that $f(X) \not\equiv f^{\mathrm{I}}(X)$.

(5) The cost of the classes have the following values: $c_0 = 2$, $c_1 = 3$, and $c_2 = 6$. We put $g = c_0 = 2$ and find the expansion of g in costs 2, 3, and 6:

$g = 2$

(i) 100. The expansion contradicts the matrix (B): $\Delta_0 > \Delta_2$, whereas $b_{20} = 1$. For $g = 2$ no other expansions exist.

(6) We put $g = 3$, and again find the expansion.

$g = 3$

(ii) 010 contradicts (B);
(iii) 100 contradicts (B);

$g = 4$

(iv) 010 contradicts (B);
(v) 200 contradicts (B);

$g = 5$

(vi) 010 contradicts (B);
(vii) 200 contradicts (B);

$g = 6$

(viii) 001 does not contradict (B), i.e., is regular.

The last expansion permits the partition $P = N^{\text{I}} + g = 17 + 6 = 23$ to be written according to (36): $a_0 = a_1 = 4 + 1 = 5$; $a_2 = 3 + 1 = 4$; $a_3 = a_4 = a_5 = 3$; $T = 14$. This system of coefficients is the realization of the initial function.

Now we present the subroutine that finds the realization (if it exists) of the Boolean function by dense enumeration. The code of the operator is *denum*. The program is (1, 2)-terminal; the auxiliary exit α corresponds to the case in which a realization does not exist;

$$[\beta] = n, \qquad \gamma :: \| T_0 \ni a \|,$$
$$\delta, \epsilon, \zeta, \eta, \vartheta, \kappa, \lambda \text{ — working complexes,}$$
$$\sigma(\delta) = \sigma(\epsilon) = \sigma(\kappa) = \sigma(\vartheta) \leq n;$$
$$\sigma(\zeta) = \sigma(\eta) = n, \qquad [\mu_p] = a_p,$$

$$[\nu] = 4n\left(\frac{n+1}{4}\right)^{(n+1)/2}, \qquad 0 \leq p < n - 1.$$

denum (234, 235, 243, 244, 250, 43, 60, 332, 254, 245, 240, 236)/$\alpha\text{ч}\,+$, $\beta\text{и}\,+$, $\gamma\text{к}\,+$, $\delta\text{к}$, $\epsilon\text{к}\,+$, $\zeta\text{к}\,+$, $\eta\text{к}\,+$, $\vartheta\text{к}\,+$, $\kappa\text{к}$, $\lambda\text{к}$, $\mu\text{к}$, $\nu\text{и}\,+$/$\mathbf{i}$, e/$\delta\mu$
253 337 7

lowal $\gamma\delta\lambda\mu\beta$//*renumbering* $\gamma\delta\gamma\epsilon$//
pordi $10\gamma\zeta\epsilon\beta\mathbf{f}$//*mast* $10\zeta\epsilon$//
weights $\epsilon\mathbf{f}\beta\eta$//$\nu - \eta_\beta \circ\!\!\rightarrow \alpha \circ\!\!\rightarrow \alpha \Rightarrow \nu$
$b_\epsilon - 1 \Rightarrow h\beta \Rightarrow \vartheta_h \Rightarrow \mathbf{g}$
§1 $\epsilon_h \nabla \Rightarrow \mathbf{h}\ \bar{\triangle}\ h \oplus \mathbf{f}_{10} \circ\!\!\rightarrow 2\mathbf{g} - \mathbf{h} \Rightarrow \mathbf{g} \Rightarrow \vartheta_h \rightarrow 1$
§2 $\mathbf{f}\ \urcorner \Rightarrow i$ *compress* $\epsilon i \epsilon$//*transmat* $\epsilon\lambda$//
compress $\lambda i \lambda$//*transmat* $\lambda\epsilon$//$\bar{\circ}\, j$
§3 $\triangle\, \mathbf{j} + \nu \circ\!\!\rightarrow \alpha$ *expansion* $3\vartheta\epsilon j\kappa$//
$\beta - 1 \Rightarrow h\eta_h \Rightarrow \mathbf{g} \circ k$
§4 $\mathbf{f} \wedge c_k \mid\!\!\rightarrow 5\ \bar{\triangle}\ k$
§5 $\triangle\, k\kappa_k + \mathbf{g} \Rightarrow \mathbf{g} \Rightarrow \lambda_h\ \bar{\triangle}\ h \oplus f_{10} \mid\!\!\rightarrow 4$ *comp* $\gamma\lambda\mathbf{i}$//
threlan $7\gamma\mathbf{i}\beta\mathbf{g}$//$\bar{\circ}\ l$
§6 $\mathbf{g} \oplus \gamma_l \circ\!\!\rightarrow$ *threlan* $2 \triangle\, l \oplus b_\beta \mid\!\!\rightarrow 6 \rightarrow$ *expansion* 10
§7 *rereal* $\lambda\delta\mu$//.

E. REALIZATION TABLES

Let $f(X)$ be a certain threshold function $X = x_0, x_1, \ldots, x_{n-1}$. On the basis of the properties of threshold functions (cf. Section 1) it can be stated that the absolute values of the weights in the realizations of monotypical threshold functions will coincide. We denote by $f^*(X)$ the function dual to $f(X)$. In [1] it was shown that if

$$f(x_0, x_1, \ldots, x_{n-1}) \sim [a_5, a_1, \ldots, a_{n-1}; T], \tag{38}$$

then

$$f^*(x_0, x_1, \ldots, x_{n-1}) \sim [a_0, a_1, \ldots, a_{n-1}; \sum_{i=0}^{n-1} a_i + 1 - T].$$

Thus the set R_n can be divided into equivalence classes, closed with respect to the following operations:

(a) substitution of variables or their negation;
(b) permutation of variables;
(c) substitution of a function by its negation.

In turn, in each class one function can be distinguished, which we call the representative function of the given class [28]. It is convenient to take as the representative of the class a positive function whose weights are ordered in decreasing order as in relation (24). These two constraints define a representative of the function to within duality, and we therefore agree to take as the representative in a dual pair the function with the larger threshold value.

A method for finding all typical and representative threshold functions is presented in [18 and 28]. Here we shall consider the question of finding a realization of a given function of a class from the realization of the representative function.

Let $f(X)$ be given by a partition (M_1, M_0) and $\sigma(M_1) = h$. We define an $n + 1$-dimensional vector $g(f) = [g_0(f), g_1(f), \ldots, g_n(f)]$ in the following manner:

$$\begin{aligned} g_k &= \sum_{i=0}^{h-1} (2\alpha_{ik} - 1) \qquad (k = 0, 1, \ldots, n-1) \\ g_n &= h. \end{aligned} \tag{39}$$

We term the vector $g(f)$ the characteristic vector of $f(X)$.

TABLE III

No.	T	$a_0\, a_1\, a_2\, a_3\, a_4$	g_5	$g_0\, g_1\, g_2\, g_3\, g_4$	T^*	No.	T	$a_0\, a_1\, a_2\, a_3\, a_4$	g_5	$g_0\, g_1\, g_2\, g_3\, g_4$	T^*
1	2	1 1 0 0 0	1	1 1 0 0 0	1	31	6	3 3 1 1 1	10	8 8 2 2 2	4
2	3	1 1 1 0 0	1	1 1 1 0 0	1	32	7	4 2 2 1 1	10	10 4 4 2 2	4
3	3	2 1 1 0 0	3	3 1 1 0 0	2	33	7	3 3 2 2 1	11	7 7 5 5 1	5
4	2	1 1 1 0 0	4	2 2 2 0 0	2	34	6	3 2 2 1 1	11	9 5 5 3 3	4
5	4	1 1 1 1 0	1	1 1 1 1 0	1	35	9	5 4 2 2 1	11	9 7 3 3 1	6
6	5	2 2 1 1 0	3	3 3 1 1 0	2	36	5	3 1 1 1 1	11	11 3 3 3 3	3
7	4	2 1 1 1 0	4	4 2 2 2 0	2	37	8	5 3 2 1 1	11	11 5 3 1 1	5
8	3	1 1 1 1 0	5	3 3 3 3 0	2	38	8	4 3 3 2 1	12	8 6 6 4 2	6
9	5	3 2 1 1 0	5	5 3 1 1 0	3	39	4	2 1 1 1 1	12	10 4 4 4 4	3
10	4	2 2 1 1 0	6	4 4 2 2 0	3	40	7	4 3 2 1 1	12	10 6 4 2 2	5
11	5	3 2 2 1 0	7	5 3 3 1 0	4	41	6	4 2 1 1 1	12	12 4 2 2 2	4
12	4	3 1 1 1 0	7	7 1 1 1 0	3	42	5	2 2 2 1 1	13	7 7 7 3 3	4
13	3	2 1 1 1 0	8	6 2 2 2 0	3	43	6	3 2 2 2 1	13	9 5 5 5 3	5
14	5	1 1 1 1 1	1	1 1 1 1 1	1	44	9	5 4 3 2 1	13	9 7 5 3 1	7
15	7	2 2 2 1 1	3	3 3 3 1 1	2	45	5	3 2 1 1 1	13	11 5 3 3 3	4
16	6	2 2 1 1 1	4	4 4 2 2 2	2	46	8	5 3 3 1 1	13	11 5 5 1 1	6
17	5	2 1 1 1 1	5	5 3 3 3 3	2	47	7	5 2 2 1 1	13	13 3 3 1 1	5
18	8	3 3 2 1 1	5	5 5 3 1 1	3	48	8	4 3 3 2 2	14	8 6 6 4 4	7
19	4	1 1 1 1 1	6	4 4 4 4 4	2	49	6	3 3 2 1 1	14	8 8 6 2 2	5
20	7	3 2 2 1 1	6	6 4 4 2 2	3	50	7	4 3 2 2 1	14	10 6 4 4 2	6
21	6	2 2 2 1 1	7	5 5 5 3 3	3	51	6	4 2 2 1 1	14	12 4 4 2 2	5
22	9	4 3 2 2 1	7	7 5 3 3 1	4	52	7	3 3 2 2 2	15	7 7 5 5 5	6
23	7	3 3 1 1 1	7	7 7 1 1 1	3	53	9	5 4 3 2 2	15	9 7 5 3 3	8
24	8	3 3 2 2 1	8	6 6 4 4 2	4	54	7	4 3 3 1 1	15	9 7 7 1 1	6
25	6	3 2 1 1 1	8	8 6 2 2 2	3	55	8	5 3 3 2 1	15	11 5 5 3 1	7
26	7	3 2 2 2 1	9	7 5 5 5 1	4	56	7	5 2 2 2 1	15	13 3 3 3 1	6
27	5	2 2 1 1 1	9	7 7 3 3 3	3	57	5	4 1 1 1 1	15	15 1 1 1 1	4
28	9	5 3 2 2 1	9	9 5 3 3 1	5	58	3	1 1 1 1 1	16	6 6 6 6 6	3
29	7	4 3 1 1 1	9	9 7 1 1 1	4	59	4	2 2 1 1 1	16	8 8 4 4 4	4
30	8	4 3 2 2 1	10	8 6 4 4 2	5	60	5	3 2 2 1 1	16	10 6 6 2 2	5
						61	4	3 1 1 1 1	16	14 2 2 2 2	4

Theorem 2.2.[12] Let $f(X)$ and $\varphi(X)$ be certain Boolean functions and $g(f)$ and $g(\varphi)$ their characteristic vectors. If $g(f) = g(\varphi)$, then either f and φ are realizable by the same threshold element and $f(X) = \varphi(X)$, or neither function is threshold.

By the definition of a characteristic vector it follows that $g(f)$ and $g(f^*)$ coincide in all components except the nth; it is also clear that $g_n(f) \leq$

[12] See reference [29].

$g_n(f^*)$, where the equality is satisfied for self-dual functions and $g_1(f) > g_j(f)$ if $a_i > a_j$.

The introduction of the characteristic vectors permits the representative functions to be ordered, for example, in order of increasing characteristic vectors. In Table III of representative functions up to five variables, inclusive,[13] all functions are ordered by increasing g_n; if $g_n(f) = g_n(\varphi)$, then the place of $f(X)$ and $\varphi(X)$ is given by the value of g_0; if $g_0(f) = g_0(\varphi)$, the the order is given by g_1, etc. Along with the characteristic vectors are given the realizations of the representative functions, where for convenient use the threshold realizing the dual function T^* is given together with the threshold T of the realization of the function.

The table permits the realization of a function to be found rapidly or its unrealizability to be established. For this it is sufficient to calculate the characteristic vector, constructing the matrix representation of the function, and to order its components in decreasing absolute values. If $h > 2^{n-1}$, then $2^n - h$ is taken for g_n. In this case the value of the threshold is defined by column T^*. According to the vector found the realization of the representative function is found in the table. There remains only to return to the initial designations of the variables and to add the "minus" sign to the weights of the variables in which the function is negative, subtracting their sum from the threshold.

Example:

$$f(x_0, x_1, \ldots, x_4) = x_0x_1\bar{x}_2 \vee x_0x_1x_3x_4;$$

the vector $g(f) = (5, 5, -5, 3, 1, 1)$, the corresponding characteristic vector in the table has the number 18. Thus

$$x_0x_1\bar{x}_2 \vee x_0x_1x_3x_4 \sim [3, 3, -2, 1, 1; 8].$$

3. REALIZATION OF AN ARBITRARY BOOLEAN FUNCTION BY A SINGLE-ROW THRESHOLD NETWORK

Since by far not every Boolean function is threshold, the problem arises of synthesizing networks of threshold elements for the realization of an arbitrary Boolean function. Threshold functions represent a functionally

[13] An analogous table for six variables is given in [18].

complete system; i.e., any Boolean function may be represented in the form of a superposition of threshold functions. In particular, any Boolean function may be represented as the disjunction or conjunction of threshold functions:

$$F(X) = \bigvee_{i=0}^{k-1} f_i(X); \tag{40}$$

$$F(X) = \bigwedge_{j=0}^{l-1} f_j(X). \tag{41}$$

We call a threshold network corresponding to (40) or (41) a single-row network. By virtue of De Morgan's theorem, only the case of disjunction of threshold functions need be considered. The general form of a single-row network is shown in Fig. 4.

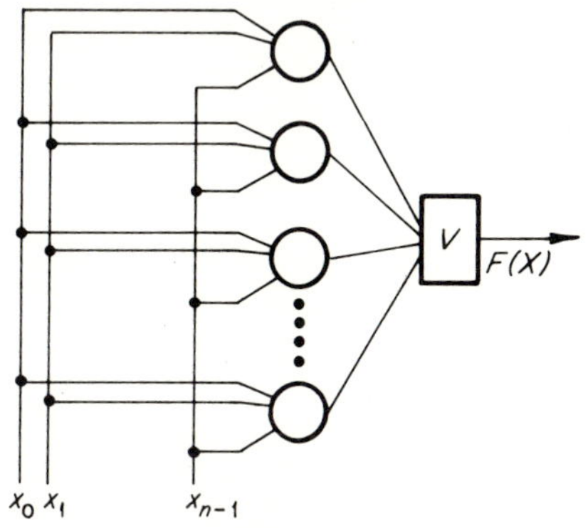

FIG. 4.

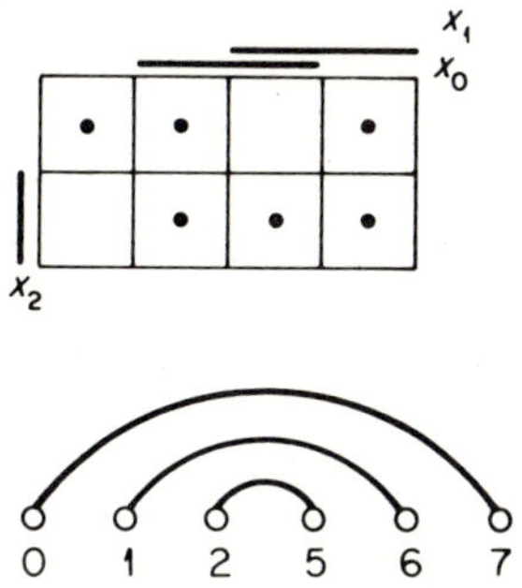

FIG. 5.

Let there be given an arbitrary Boolean function of n variables $F(X)$. It is required to represent it in the form of a disjunction of a minimal number of threshold functions. We shall term the corresponding threshold network the shortest network.

It is clear that only functions that do not absorb each other can enter into (40). We shall term such functions maximal.

As is well known [12], for a small number of variables (at least for $n \leq 6$) all totally monotonic functions are threshold. There is every basis for expecting that, for $6 < n < 10$, nonthreshold totally monotonic functions will be encountered extremely rarely.

With these considerations, it is possible to construct the following algorithm for the synthesis of the shortest single-row threshold network.

(1) In the initial Boolean function all maximal totally monotonic functions are distinguished.

(2) The realizations of these functions are found, using the operator *synthrem* or tables of characteristic vectors and realizations.

(3) The shortest coverages of the initial function by threshold functions are found, using the operators *oshco* or *fishco*, described by Zakrevskii [24].

Let us consider the first part of the algorithm.

We consider a Boolean function of n variables.

$$F(X) = sep^{(2)}M = \{M_1, M_0\}.$$

We construct for it the graph $G = (M_1, \Gamma)$. Each element $\alpha_i \in M_1$ is associated with the corresponding ith node of the graph and

$$j \in \Gamma_i \leftrightarrow \exists\ (\beta_k, \beta_l \in M_0) \qquad [\alpha_i + \alpha_j = \beta_k + \beta_l],$$

where the "plus" sign represents ordinary vector addition. Further, we shall call this the 2-summability graph.

Example:

$$F(X) = \bar{x}_0 x_1 \vee x_0 x_2 \vee \bar{x}_1 \bar{x}_2 \vee \bar{x}_0 \bar{x}_2.$$

This function corresponds to the following partition of the Boolean space:

$$M_1 = \{000, 001, 010, 101, 110, 111\},$$
$$M_0 = \{011, 100\}.$$

The matrix representation of the function and the corresponding 2-summability graph are given in Fig. 5.

From the properties of 2-asummability it follows that if in the 2-summability graph there are no pairs of adjacent nodes, i.e., the set of nodes is internally stable [26], the corresponding Boolean function is totally monotonic. If a pair of nodes $\{i, j\}$ is adjacent, the corresponding elements of Boolean space α_i and α_j cannot belong to the set M_1^i of a single threshold function $f_i(X)$, and must be separated.

This problem is solved by the operator for separating the nodes of the graph and its mapping. The code of the operator is *senoma*.

Let $N = \{n_0, n_1, \ldots, n_i, \ldots, n_{m-1}\}$ be the set of nodes of the loop-free graph $G = (N, \Gamma)$. A set $A \subseteq \tilde{N}$ is given with a distinguished subset $S \subseteq A$.

The operator *senoma* carries out the following transformation of the set A. For a certain node $n_i \in N$ and for all subsets of nodes of the graph $s_j \in S$, such that $n_i \in s_j$ and $s_j \cap \Gamma n_i \neq \varnothing$, there are constructed

$$s_j' = s_j \backslash \{n_i\},$$
$$s_j'' = s_j \backslash s_j \cap \Gamma n_i.$$

The transformed subsets of nodes s_j are dropped from the set S, while s_j' and s_j'' are joined to it if they do not contain any of the elements of A.

The set A is given by the complex δ:

$$\delta :: \| A \ni N \|,$$
$$[\epsilon] = \sigma(A) - 1.$$

The set S is always given by the last elements of the complex δ and is defined by the value of the index ζ of its first element.

The node of the graph n_i and its mapping Γn_i are represented by the variables β and γ, respectively:

$$\beta :: \| \{n_i\} \ni N \|,$$
$$\gamma :: \| \{\Gamma n_i\} \ni N \|.$$

The operator is (1, 2)-terminal.

The basic output corresponds to the case where all subsets of nodes $s_j \in S$ take part in the transformation and are dropped, but there is no s_j' or s_j'' connected to A, i.e., $[\epsilon] < [\zeta]$. The auxiliary exit α corresponds to all other cases.

senoma αч $+$, βп $+$, γп $+$, δк, ϵи, ζи $+/\mathbf{b}$, $c/(\beta\gamma\delta)$
257 220 6 (β**ab**)

$\zeta \Rightarrow a$
§1 $\delta_a \wedge \beta \circ\!\!\rightarrow 5\delta_a \wedge \gamma \Rightarrow \mathbf{a} \circ\!\!\rightarrow 5\delta_a \oplus \beta \Rightarrow \mathbf{b}$
$\delta_a \oplus \mathbf{a} \Rightarrow \mathbf{a} \circ b \circ c \circ \delta_a$
§2 $\delta_b \urcorner \wedge \mathbf{a} \circ\!\!\rightarrow 4 \triangle b \oplus b_\delta \mid\!\!\rightarrow \mathbf{2a} \Rightarrow \delta_a$
§3 $\delta_c \urcorner \wedge \mathbf{b} \circ\!\!\rightarrow 5 \triangle c \oplus b_\delta \mid\!\!\rightarrow 3 \triangle \epsilon \triangle b_\delta \mathbf{b} \Rightarrow \delta_\epsilon \rightarrow 5$
§4 $\delta_c \urcorner \wedge \mathbf{b} \circ\!\!\rightarrow 6 \triangle c \oplus b_\delta \mid\!\!\rightarrow 4b \Rightarrow \delta_a$
§5 $\triangle a \oplus b_\delta \mid\!\!\rightarrow 1 \rightarrow \alpha$
§6 $\delta_\epsilon \Leftrightarrow \delta_a \overline{\triangle} \epsilon \overline{\triangle} b_\delta \delta_a \mid\!\!\rightarrow 1\epsilon - \zeta \mid\!\!\rightarrow \alpha.$

In the particular case where $S = A = \{n_0, n_1, \ldots, n_{n-1}\}$ the application of *senoma* successively to all $n_i \in N$ leads to obtaining in the set A all maximal internally stable sets of the graph [26].

In the case of the 2-summability graph the set of nodes corresponds to the set M_1 prescribed by the Boolean function $F(X)$. If by means of the operator *senoma* all maximal internally stable sets $M_1{}^i$ are found, they will correspond to certain Boolean functions

$$f_i(X) = sep_i^{(2)} M = \{M_1^i, M \backslash M_1^i\}.$$

In general not all $f_i(X)$ will be totally monotonic, since the graph $G = (M_1, \Gamma)$ does not correspond to $f_i(X)$, but to $F(X)$. For example, the following maximal internally stable sets are contained in the graph in Fig. 5:

$$\{0, 1, 2\}, \quad \{0, 2, 6\}, \quad \{0, 1, 5\}, \quad \{0, 5, 6\},$$
$$\{1, 2, 7\}, \quad \{1, 5, 7\}, \quad \{2, 6, 7\}, \quad \{5, 6, 7\}.$$

Even homogeneous functions do not correspond to the sets $\{1, 2, 7\}$ and $\{0, 5, 6\}$.

Taking this circumstance into account, after all functions $f_i(X)$ have been obtained it is necessary to repeat analogous transformations for each of them. Then $f_i(X)$ is totally montonic if its 2-summability graph contains no branches; otherwise, $M_1{}^i$ is divided into maximal internally stable sets, etc.

In this case the set A of the operator *senoma* represents a set of Boolean functions given in matrix form. At the same time

$$A = P \cup Q \cup S,$$

where P is the set of maximal totally monotonic functions obtained in the preceding steps, Q is the set of Boolean functions remaining to be tested for total monotonicity, and S is the set directly transformed by the operator *senoma*.

In the execution of this algorithm it is not necessary to construct the adjacency matrix; i.e., after the mapping $\Gamma\alpha_i \subseteq M_1$ corresponding to each $\alpha_i \in M_1$ has been found, it is immediately possible to apply the operator *senoma*.

The adjacency of each pair $\{\alpha_i, \alpha_j\}$ is defined by means of a special operator *asummab*, which for each pair in M_1 tests each pair in M_0 for asummability. To reduce the enumeration, the following property [29] is utilized:

$$\alpha_i + \alpha_j = \beta_K + \beta_l \leftrightarrow (\alpha_i \cup \alpha_j = \beta_k \cup \beta_l) \wedge (\alpha_i \cap \alpha_j = \beta_k \cap \beta_l);$$
$$\alpha_i + \alpha_j = \beta_k + \beta_l \rightarrow \beta_k, \beta_l \in \text{int}_{\min} \{\alpha_i, \alpha_j\}.$$

The first property is tested directly; the second follows from the first and the fact that

$$\mathrm{int}_{\min}\{\alpha_i, \alpha_j\} = \mathrm{int}_{(\alpha_i \cap \alpha_j)}(\alpha_i \cup \alpha_j).$$

The operator *asummab* is (1, 2)-terminal, where the auxiliary exit α corresponds to the case of asummability of the pair $\{\alpha_i, \alpha_j\}$.

$$\begin{aligned} [\beta] &= \alpha_i, \\ [\gamma] &= \alpha_j, \qquad i, j \in \{0, 1, \ldots, \sigma(M_1) - 1\}, \\ [\delta] &= 2^n - 1, \qquad \text{where} \quad n = \sigma(X). \end{aligned}$$

The set M_0 is given by the variable ϵ:

$$\epsilon::\|\{M_0\} \ni M\|.$$

asummab αч $+$, βи $+$, γи $+$, δи $+$, ϵп $+$ /**a**, f
260 153 3 (ϵ**a**)

$\beta \oplus \gamma \nabla \oplus 1 \circ\!\!\rightarrow \alpha\beta \vee \gamma \urcorner \wedge \delta \Rightarrow a\beta \wedge \gamma \Rightarrow b \wedge a \Rightarrow c \urcorner \wedge \delta$
$\Rightarrow d\epsilon \Rightarrow$ **a**

§1 $a\ \dot{\mathbf{X}}\ 2e37 - e \Rightarrow ed_e \wedge$ **a** $\Rightarrow$ **a** $\circ\!\!\rightarrow \alpha \rightarrow 1$
§2 $b\ \dot{\mathbf{X}}\ 3e37 - e \Rightarrow ee_e \wedge$ **a** $\Rightarrow$ **a** $\circ\!\!\rightarrow \alpha \rightarrow 2$
§3 **a** $\dot{\mathbf{X}}\ \alpha e \wedge c \Rightarrow fe \urcorner \wedge d \vee f \Rightarrow ec_e \wedge$ **a** $\circ\!\!\rightarrow 3.$

Finding of all maximal totally monotonic functions in (40) now reduces to the successive application of the operator *asummab* to all pairs in $M_1{}^i$ for the function $f_i(X) \in Q$. If no summable pair exists in $M_1{}^i$, the function $f_i(X) \in P$; otherwise, it is assigned to A, and the operator *senoma* is applied to it, after which the entire set S is joined to Q. The initial form of the set A is

$$A = Q = \{F(X)\}.$$

The operator representing $F(X)$ in the form (40) has the code *tomof*. The initial function is given by variable α:

$$\alpha::\|\{M_1\} \ni M\|;$$

and the set A by the complex β:

$$\beta::\|A \ni M\|.$$

To define the positions assigned to the matrix representation $f_i(X)$, the variable γ is used, containing 1's in the 2^n lowest orders. The index δ has the same sense as in the operator *asummab*.

The complex β is divided into parts, representing the sets P, Q, and S, by means of the indexes g and h: $[g] = \sigma(P)$, and $[h] = \sigma(P) + \sigma(Q)$.

tomof (260, 257) $\alpha\text{п}\,+,\ \beta\text{к},\ \gamma\text{п}\,+,\ \delta\text{и}\,+/\mathbf{i},\ k/(\alpha\beta\gamma)$
261 215 7 (α **defgh**)

$1 \Rightarrow i \Rightarrow h2 \Rightarrow b_\beta\alpha \Rightarrow \beta_0\ \bar{\circ}\ g$
§1 $\triangle\, g \oplus h \cdot\!\circ\!\!\rightarrow 7\beta_g \Rightarrow \beta_h \Rightarrow \mathbf{d} \urcorner \wedge \gamma \Rightarrow \mathbf{e} \circ \mathbf{i} \circ \beta_g$
§2 $\mathbf{d}\ \dot{\mathbf{X}}\ 5j\mathbf{d} \Rightarrow \mathbf{f}c_i \Rightarrow \mathbf{g} \circ \mathbf{h}$
§3 $f\ \dot{\mathbf{X}}\ 4k$ *asummab* $3jk\delta\mathbf{e}//1 \Rightarrow \mathbf{i}c_k \vee \mathbf{h} \Rightarrow \mathbf{h} \rightarrow 3$
§4 $h \circ\!\!\rightarrow 2$ *senoma* $2\mathbf{gh}\beta ih//$
§5 $\mathbf{i} \circ\!\!\rightarrow 6i \Rightarrow h \oplus \mathbf{g} \circ\!\!\rightarrow 7\beta_i \Rightarrow \mathbf{d} \urcorner \wedge \gamma \Rightarrow \mathbf{e} \circ \mathbf{i} \rightarrow 2$
§6 $\beta_h \Leftrightarrow \beta_g \rightarrow 1$
§7 .

After *tomof* has terminated its action, the complex β contains all maximal totally monotonic functions $f_i(X)$ contained in the prescribed function $F(X)$. It remains to apply the operators listed in the second and third steps of the overall algorithm, and the problem will be solved. The L-program realizing this algorithm is called *shothnet*.

The initial Boolean function is given by the variable **a**: $\mathbf{a}::\|\ \{M_1\} \ni M\ \|$; $[a] = \sigma(X)$. The set of maximal totally monotonic functions is given by the complex **A**: $\mathbf{A}::\|\ A \ni M\ \|$. The shortest coverage is represented by the variable **e**: $\mathbf{e}::\|\ \{A_k\} \ni A\ \|$. The complex **I** contains the weights of the threshold-element realization, and the index e, the threshold T: $[e] = T$. The maximal total weight is given by the index d:

$$[d] = 4n\left(\frac{n+1}{4}\right)^{(n+1)/2} - 1.$$

B, **C**, **D**, **E**, **F**, **G**, **H**, and **I** are intermediate working complexes of the operators *oshco* and *synthrem*.

shothnet
§0 $37 - a \Rightarrow b\ \bar{\circ}\ \mathbf{b} > c_b \urcorner \Rightarrow \mathbf{b}\ \bar{\circ}\ b < a \urcorner \Rightarrow b\mathbf{a} \Rightarrow \mathbf{d}$
tomof $\mathbf{aAb}b//\bar{\circ}\ \mathbf{c} > b_a \urcorner \Rightarrow \mathbf{c}$ *oshco* $\mathbf{AcdIe}//\mathbf{A} * \mathbf{e} * \bar{\circ}\ c *$
§1 $\mathbf{e}\ \dot{\mathbf{X}}\ 3c$ *synthrem* $2\mathbf{a}_c\ \mathbf{BC}a\mathbf{DFGH}d\mathbf{EI}e//\mathbf{I} * \mathbf{e} * \bar{\circ}\ c * \rightarrow 1$
§2 $\mathbf{a}_c * \bar{\circ}\ c * \rightarrow 1$
§3 $\mathbf{a}\ *.$

TABLE IV

M	0	1	2	3	4	5	6	7	8	9	10	11	12	13	14	15	16	17	18	19	20	21	22	23	24	25	26	27	28	29	30	31
$F(x)$ \ i	1			1	1	1		1					1	1	1	1		1				1	1	1		1	1	1				1
1					1	1		1					1	1	1	1																
2																		1				1				1						
3																		1								1		1				
4				1		1		1						1		1						1		1								1
5					1	1		1						1		1						1		1								
6								1								1								1				1				1
7					1	1		1					1	1		1						1										
8						1		1										1				1		1								
9	1				1	1							1	1																		
10						1		1						1	1	1								1								1
11						1		1					1	1	1	1																1
12																										1	1	1				1
13						1		1								1						1	1	1								1

TABLE V

1	1		1	1	1	1	1	1					1	1	1	1	1	1	1	1	1	1	1	1	1	1	1	1	1	1	1	1
2	1	1		1	1	1		1	1	1		1	1	1	1	1	1	1		1	1	1	1	1	1	1	1	1	1	1	1	1
3	1	1	1	1	1	1	1	1	1	1	1	1	1	1	1	1		1	1	1		1	1	1		1	1	1				1
4	1	1		1	1	1	1	1	1	1	1	1	1	1	1	1		1			1	1	1	1	1	1	1	1	1	1	1	1

To illustrate the application of this algorithm, let us consider an example taken from [30]. In Table IV is presented the initial function $F(X)$ and the maximal totally monotonic functions obtained by means of *tomof* in URAL-1. The Boolean functions $f_i(X)$ are represented by the row matrices $\| \{M_1\} \ni M \|$.

The corresponding realizations of the threshold elements and certain shortest coverages are presented in Table V.

An appreciable economy in the number of elements as well as the sum of weights and thresholds in this case can be achieved by applying the operator *tomof* to the inversion of $F(X)$, and representing it in the form

$$\overline{F(X)} = \bigvee_{i=0}^{k-1} f_i(X).$$

In this case the given function is expressed in terms of the conjunction of threshold functions

$$F(X) = \bigwedge_{i=0}^{k-1} \overline{f_i(X)}.$$

In Tables VI and VII, analogously to Tables IV and V, one possible realization of this kind is presented. In the present case it is more economical than the first by two threshold elements.

This algorithm requires for its execution a large number of searches over the sets M_1 and M_0. Their number rapidly increases with increase in the number of variables. Let us show a way to accelerate substantially the process of finding all maximal totally monotonic functions contained in $F(X)$ in certain cases.

Let the Boolean function $F(X)$ be given by the set $\mathbf{I}$ of its maximal intervals, corresponding to the reduced dnf. As the first step, it is possible to represent it in the form of the disjunction of homogeneous functions.

We construct the graph $G = (\mathbf{I}, \Gamma)$. To each $\text{int}_{\alpha_k} \beta_k \in \mathbf{I}$ there corresponds a kth node of the graph, and

$$l \in \Gamma_k \leftrightarrow \text{int}_{\alpha_k} \beta_k \cap \text{int}_{\alpha_l} \beta_l = \phi.$$

To any internally stable set of this graph there corresponds a star in the set $\mathbf{I}$. If $\mathbf{I}$ is the set of maximal intervals, then the following proposition holds:

$$\bigcap [\mathbf{I}] \neq \phi \leftrightarrow (\exists \ \text{int}_{\alpha_k} \beta_k, \text{int}_{\alpha_l} \beta_l \in \mathbf{I}) \cdot [\text{int}_{\alpha_k} \beta_k \cap \text{int}_{\alpha_l} \beta_l = \phi].$$

TABLE VI

No.	Realization T	a_0	a_1	a_2	a_3	a_4	$\sum_{i=0}^{k-1} \| a_i \| + \| T \|$	Shortest coverage	$\sum$—min coverage
1	3	1	−1	3	1	−3	12		*
2	3	2	−2	−1	−1	2	11		*
3	4	2	−1	−2	1	2	12	*	
4	4	3	1	2	−1	−1	12	*	*
5	4	3	−1	4	−2	−2	16		
6	8	3	3	2	1	1	18		
7	3	2	−2	4	−1	−3	15		
8	4	3	−1	2	−3	1	14		
9	0	−1	−3	2	−1	−3	10	*	*
10	7	3	2	4	1	−2	19		
11	6	2	1	4	2	−3	18	*	
12	5	1	1	−1	2	2	12.	*	*
13	7	3	2	4	−2	1	19	*	*

TABLE VII

No.	T	a_0	a_1	a_2	a_3	a_4	$\sum \| \bar{a}_i \| + \| T \|$
1	0	−1	1	3	−2	3	10
2	0	2	−2	1	1	1	7
3	−2	2	2	−1	−1	−3	11
4	0	1	−1	2	2	−1	7

The last follows by induction from the fact that if two conjunctions of a reduced dnf are nonorthogonal, a third, nonorthogonal to both, is nonorthogonal to the conjunction of the first two.

As has been said, the problem of finding all maximal internally stable sets can be solved by means of the operator *senoma* for $S = A$.

The set A is given by the complex

$$\gamma::\| A \ni \mathbf{I} \|.$$

The set **I** is given by two complexes, α and β. The complex α contains the set of lower limits of the intervals, the complex β the set of upper limits. The operator code is *maxhof*.

$maxhof\ (257)/\alpha\kappa +, \beta\kappa +, \gamma\kappa/\mathbf{h}, g/(\alpha\beta)$
$\mathbf{266\ 135\ 4}\ (\gamma\mathbf{cdef})\ (\alpha\mathbf{g})$

$\bar{\circ}\ c > b_\alpha \urcorner > \Rightarrow \mathbf{c} \Rightarrow \gamma_0 \circ f\,1 \Rightarrow b_\gamma$
§1 $\mathbf{c}\ \dot{\mathbf{X}}\ 4d\mathbf{c} \Rightarrow \mathbf{e}c_d \Rightarrow \mathbf{d} \odot \mathbf{f}$
§2 $\mathbf{e}\ \dot{\mathbf{X}}\ 3e\alpha_d \oplus \alpha_e \Rightarrow \mathbf{g}\beta_d \oplus \beta_e \wedge \mathbf{g} \circ\!\rightarrow 2$
$c_e \vee \mathbf{f} \Rightarrow \mathbf{f} \rightarrow 2$
§3 $\mathbf{f} \circ\!\rightarrow 1\ senoma\ 1\mathbf{df}\gamma_f \circ //$
§4 .

Disposing now of all maximal homogeneous functions contained in the prescribed Boolean function $F(X)$, it is possible to apply the search algorithm, which can be shortened, using the following property of positive functions.

Theorem 3.1. Let a positive Boolean function be given by the partition (M_1, M_0), let $\text{int}_{\alpha_1} \alpha_2 \subset M_1$ and $\alpha_i \in \text{int}_{\alpha_1} \alpha_2$, and let there exist β_1, $\beta_2 \in M_0$, $\alpha_j \in M_1$ such that $\alpha_i + \alpha_j = \beta_1 + \beta_2$. Then in the partition of M_1 into a set of pairwise asummable elements the removal of α_i from M_1 must be accompanied by the removal of all α_s such that $\alpha_i > \alpha_s \geq \alpha_1$.

Theorem 3.2. Let the positive function $F(X)$ be given by the partition (M_1, M_0) and $\text{int}_{\alpha_1} \alpha_2 \subset M_1$, $\text{int}_{\alpha_3} \alpha_4 \subset M_1$. Then from the asummability of the pair α_1, α_3 there follows the asummability of any $\alpha_i \in \text{int}_{\alpha_1} \alpha_2$ and $\alpha_j \in \text{int}_{\alpha_3} \alpha_4$.

REFERENCES

1. Muroga, S., Toda, I., and Takasu, S., Theory of Majority Decision Elements, *J. Franklin Inst.* **271,** 376–418 (1961).
2. Varshavskii, V. I., Certain Questions of the Theory of Logic Networks, Constructed of Threshold Elements, *in* "Questions of the Theory of Mathematical Machines," No. 2 [in Russian]. Moscow, 1962.
3. McNaughton, R., Unate Truth Functions, *IRE Trans. Electron. Computers* **10,** 1–6 (1961).
4. Coates, C. Z., and Lewis, P. M., Linearly Separable Switching Functions, *J. Franklin Inst.* **272,** 366–410 (1961).
5. Coates, C. L., Kirchner, R. B., and Lewis, P. M., A Simplified Procedure for the Realization of Linearly Separable Switching Functions, *IRE Trans. Electron. Computers* **11,** 447–458 (1962).
6. Minnick, R. C., Linear-Input Logic, *IRE Trans. Electron. Computers* 6–16 (1961).
7. Rozenblat, M. A., and Gendler, M. B., Logical Possibilities of Real Threshold Elements, *Izv. Akad. Nauk SSSR, Tekhn. Kibernet.* No. 1, 50–64 (1964).
8. Sauer, W. A., How to Achieve Majority and Threshold Logic with Semiconductors. *Electronics* **36,** 12–15 (1963).

9. Eccles, J., "Physiology of Nerve Cells" [Russian translation], IL, Moscow, 1959.
10. Paull, M. C., and McCluskey, E. J., Jr., Boolean Functions Realizable with Single Threshold Devices, *Proc. IRE* **48,** 1335–1337 (1960).
11. Winder, R. O., Single Stage Threshold Logic. Switching Circuit Theory and Logical Design, AIEE Publ. S-134, pp. 321–332, September 1961.
12. Winder, R. O., More about Threshold Logic. Switching Circuit Theory and Logical Design, AIEE Publ. S-134, pp. 55–64, September 1961.
13. Glushkov, V. M., "The Synthesis of Digital Automata." Fizmatgiz, Moscow, 1962.
14. Zakrevskii, A. D., A Visual Matrix Method of Minimization of Boolean Functions, *Avtomat. i Telemek.*, **21,** No. 53, 369–373 (1960).
15. Butakov, E. A., Utilization of a Computer for the Synthesis of Threshold Elements, *Proc. Intern. Symp. Theory Switching Circuits and Finite Automata* [in Russian]. Moscow, 1962.
16. Butakov, E. A., and Zakrevskii, A. D., Certain Questions of the Realization of Boolean Functions by Threshold Elements, *Izv. Akad. Nauk SSSR, Tekhn. kibernet.* No. 1, 39–49 (1964).
17. Liss, D., A Test for Unate Truth Functions, *IRE Trans. Electron. Computers* **12,** 405–406 (1963).
18. Butakov, E. A., On the Systematization of Threshold Functions, *in* "Materials of Scientific Seminars on Theoretical and Applied Problems of Cybernetics." Sci. Council on Cybernetics, Academy of Sciences, USSR.
19. Elgot, C. C., Truth Functions, Realizable by Single Threshold Organs. Switching Circuit Theory and Logical Design, AIEE Publ. S-134, pp. 225–245, September 1961.
20. Chow, C. K., Boolean Functions Realizable with Single Threshold Devices. *Proc. IRE* 370–371 (1961).
21. Gabelman, I. J., The Synthesis of Boolean Functions Using a Single Threshold Element, *IRE Trans. Electron. Computers* **11,** 639–642 (1962).
22. Zakrevskii, A. D., On Reduced Enumeration for the Solution of Certain Problems in the Synthesis of Discrete Automata, *Izv. Vysshikh Uchnebn. Zavedenii, Radiofiz.* **7,** No. 1, 166–174 (1964).
23. Chernikov, S. N., The Solution of Linear Programming Problems by the Method of Elimination of Variables, *Dokl. Akad. Nauk SSSR,* **139,** 1314–1317 (1961).
24. Zakrevskii, A. D., Optimal coverage of Sets. This volume.
25. Dantzig, G. B., "Linear Programming and Extentions," Princeton Univ. Press, Princeton, New Jersey, 1962.
26. Berge, C., "Graph Theory and Its Applications" [Russian translation]. IL, Moscow, 1962.
27. Bykova, S. V., and Butakov, E. A., Threshold Element Synthesis Algorithms, *Tr. SFTI* **48** (1965).
28. Muroga, S., Toda, J., and Kondo, M., Majority Decision Functions of up to Six Variables, *Math. Comp.* **16,** 459–472 (1962).
29. Chow, C. K., On the Characterization of Threshold Functions, *Proc. AIEE Annual Symp. Switching Circuits Theory, 2nd, Detroit, 1961,* pp. 34–38 (1961).
30. Vorob'ev, V. A., and Butakov, E. A., Finding All Maximal Totally Monotonic Functions Contained in a Given Boolean Function, *Tr. SFTI* **48** (1965).
31. Stram, O., Arbitrary Boolean Functions of *N*-Variables in Terms of Threshold Devices. *Proc. IRE* **49,** 210–220 (1964).

AN ALGORITHM FOR THE SYNTHESIS OF MAJORITY-ELEMENT LOGICAL CIRCUITS

V. L. Pavlov

1. GENERAL ANALYSIS OF THE PROBLEM

A.

The problem of the synthesis of logical circuits from majority elements, which is taking on ever greater interest, has been considered in several papers. In particular, in [1 and 2] synthesis methods based on the application of canonical expansions of Boolean functions were considered. In [3] the problem was considered from the position of the functional decomposition of Boolean functions. Certain results of this work were generalized in [4]. The intuitive method for the synthesis of majority circuits, described in [5] also contains the idea of applying functional-decomposition algorithms.

The present article describes a multistep algorithm for the synthesis of irredundant logical circuits of majority elements, based on the iterative application of a relatively simple algorithm for the functional decomposition of the Boolean function realized by the synthesized circuit. The distinguishing property of the algorithm is the explicitly expressed tendency to the construction of minimal circuits in the framework of a special configuration, avoiding the need for additional temporal matching of the various stages of the circuit by means of delay elements. In this case the typical configuration has the form shown in Fig. 1, realizing a function

$f(a, b, c, d, e)$, which takes on the value 1 only if exactly two of its variables have the value 1. In the figure the inverted inputs of the elements are marked by black dots (for example, the extreme right element of the first stage of the circuit realizes the function $\psi(\bar{c}, d, e) = \bar{c}d \vee \bar{c}e \vee de$). Majority elements operating as delay lines are represented in the circuit by elements with a single input. It is assumed that they realize the function $\psi(a, a, a) = a$, if a is the input value.

We shall use the term *minimal circuit* for circuits containing the smallest possible number of elements and not requiring additional temporal matching.

By means of the algorithm described only a certain, apparently good degree of approximation to minimal circuits is achieved, and wide possibilities exist for its further improvement. The machine execution of the algorithm makes it possible to process arbitrary Boolean functions dependent on not more than 10–12 variables, and weakly defined Boolean functions

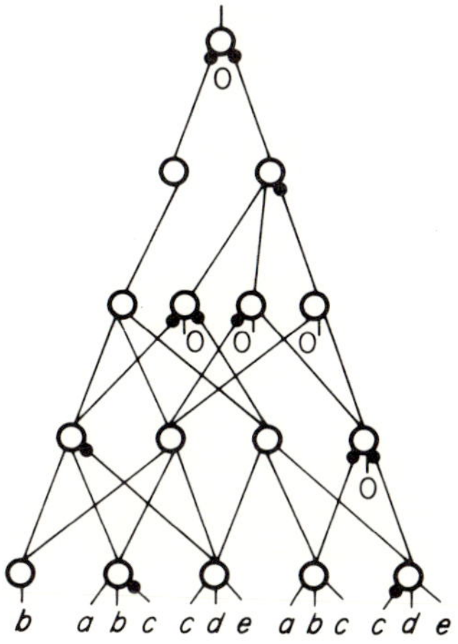

FIG. 1.

dependent on a greater number (up to 32) of variables. The algorithm can be applied to the manual realization of functions of a small (3–8) number of variables; to be sure, here intuition is often used, rather than the ability to organize sifting of variants.

The terminology used to describe the algorithm has been introduced in [6, 7, 8].

In the following sections the problem is first defined and certain possibilities for its exact solution are discussed. Then a multistep algorithm that finds an approximate solution is described, and then examples are presented along with a discussion of the results obtained.

B. STATEMENT OF THE PROBLEM

Let it be required to synthesize a circuit realizing a Boolean function $f(X)/X \equiv \{x_0, x_1, \ldots, x_{n-1}\}$, in the general case incompletely defined and prescribed by the partition $\{F_0, F_1, F_x\}$ of the Boolean space $X^{(2)}: f(m_i) = 0 \leftrightarrow m_i \in F_0$, $f(m_i) = 1 \leftrightarrow m_i \in F_1$. The values of the function in the set $F_x \subset X^{(2)}$ are not concretized, and can be defined as convenient during the synthesis process. With each Boolean function $f(X)$ we shall connect a set $\tilde{f}(X)$ of Boolean functions, up to which the function $f(X)$ can be completed with full definition of its value on the set F_x.

We shall synthesize the circuit realizing $f(X)$ by stages, beginning with the synthesis of the first stage of the circuit, at which arrive the input variables in X. We bring into consideration the set R_0 of Boolean functions realized by a single majority element and taking as their arguments elements in the set $X \cup 0$. The structure of the first stage is defined by the set $\xi_1 \equiv \{\psi_l, \psi_m, \ldots, \psi_n\} \subseteq R_0$, ensuring the existence of the Boolean function $g_1(\xi_1)$, satisfying the relation $\tilde{g}_1(\psi_l, \psi_m, \ldots, \psi_n) \subseteq \tilde{f}(X)$. The function $g_1(\xi_1)$, called transform of the function $f(X)$, must be realized in the synthesis of the next stages of the circuit. Therefore, as the final solution, defining the optimal structure of the first stage, the set $\xi_1 \subset R_0$ must be chosen, defining the transform with the lowest complexity, expressed in the approximate expenditure of majority elements necessary for the realization of the function g_1. We shall assume that there exists a numerical function $L(g)$ by means of which its complexity is estimated from the form of the Boolean function g. Generalizing the above, the synthesis of the ith stage of the circuit will be connected with finding a set $\xi_i \equiv \{\psi_l, \psi_m, \ldots, \psi_n\} \subseteq R_{i-1}$, defining the least complex Boolean function $g_i(\xi_i)$ satisfying the relation $\tilde{g}_i(\psi_l, \psi_m, \ldots, \psi_n) \subseteq \tilde{g}_{i-1}(\xi_{i-1})$. The form of Boolean function g_{i-1}, defined in the synthesis of the $i - 1$st stage, is known, where for $i = 1$ the function $f(X)$ appears as g_{i-1}. The set R_{i-1} is formed analogously to the set R_0 on the basis of the set $\xi_{i-1} \cup \{0\}$. The function $g_i(\xi_i)$ must be realized in the synthesis of the next stages of the circuit.

In summary, it is possible to note the possibility of stagewise synthesis of the circuit with optimization of the structure of each stage, carried out for the purpose of ensuring minimal cost of construction of the subsequent stages.

The problem of synthesis of each stage, which is nevertheless of substantial dimensions, can be solved in several steps with optimization carried out at each step. One possible multistep algorithm for finding an optimal solution for one stage is discussed in Section 2.

C. SET-THEORETICAL ANALYSIS

To analyze the situation encountered in the synthesis of each stage, we shall consider the problem of finding a set ρ of solutions $\xi \equiv \{\psi_l, \psi_m, \ldots, \psi_n\} \subseteq R_0$ for a superposition $g(\xi(X))$, satisfying the relation

$$\tilde{f}(X) \supseteq \tilde{g}(\psi_l, \psi_m, \ldots, \psi_n). \tag{1}$$

Let us analyze the relation between the sets associated with the Boolean functions in (1) for the purpose of establishing certain properties the solution $\xi \in \rho$ must have.

To the completely defined Boolean function $\psi_k \in R_0$ there corresponds a subset $\psi_k \subset X^{(2)}$ of elements of the Boolean space: $m_i \in \psi_k \leftrightarrow \psi_k(m_i) = 1$. The set ψ_k prescribes a partition $\{\nu_0, \nu_1\}$ of the space $X^{(2)}$ into two classes: $\nu_0 \equiv \bar{\psi}_k$; $\nu_1 \equiv \psi_k$. For any $m_i \in X^{(2)}$, $m_i \in \nu_0$, if and only if the proposition $m_i \in \psi_k$ is false, $(m_i \in \psi_k) = 0$. In this notation $m_i \in \nu_1 \leftrightarrow (m_i \in \psi_k) = 1$ and in the general case $m_i \in \nu_j \leftrightarrow (m_i \in \psi_k) = j/j \in (0, 1)$. The sets ψ_l, $\psi_m, \ldots, \psi_n \in \xi \subseteq R_0$ in their ensemble prescribe a partition $\{\nu_j/j \in J\}$: $(m_i \in X^{(2)}) \wedge (m_i \in \nu_j)$ if and only if the binary vector corresponding to the sequence of propositions $(m_i \in \psi_l)(m_i \in \psi_m), \ldots, (m_i \in \psi_n)$ is a binary representation of the value of the index j. The set of distinct values of the index j, defined in this way in sifting over the entire set $X^{(2)}$, forms a set J. Figure 2 represents schematically a partition of the Boolean space $X^{(2)}$ into the classes $\alpha_0, \alpha_1, \ldots, \alpha_6$, given by the sets $\psi_1, \psi_2, \psi_3 \subseteq X^{(2)}$.

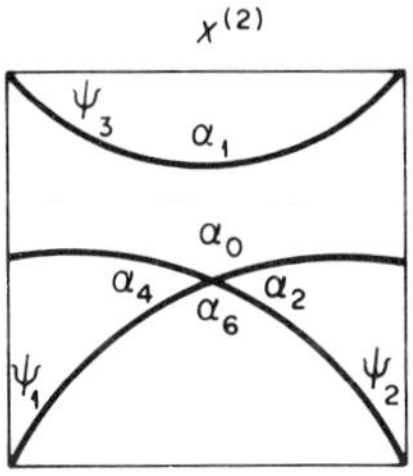

FIG. 2.

The partitioning into classes given by the set ξ of subsets in $X^{(2)}$ is denoted by $SEP^{\xi}X^{(2)}$.

It can also be stated that the set ξ defines a unique mapping of the space $X^{(2)}$ onto a certain subset γ of the space $\xi^{(2)}$:

$$X^{(2)} \xrightarrow{\xi} \gamma \subseteq \xi^{(2)},$$

by means of which each element $m_i \in X^{(2)}$ is associated with an element $\gamma_j \in \gamma$:

$$\gamma_j \ni \psi_k \leftrightarrow (\psi_k \in \xi) \wedge (\psi_k \ni m_i).$$

The proposition

$$(m_i \xrightarrow{\xi} \gamma_j) \wedge (m_i \in \nu_j \in SEP^{\xi}X^{(2)}) \leftrightarrow \nu_j \xrightarrow{\xi} \gamma_j$$

also holds, from which it follows that the sets $SEP^{\xi}X^{(2)}$ and $\gamma \subset \xi^{(2)}$ are isomorphic:

$$SEP^{\xi}X^{(2)} \overset{\xi}{\leftrightarrow} \gamma.$$

The Boolean function $f(X)$ is prescribed by the pair F_0, F_1 of nonintersecting subsets in $X^{(2)}$. The set $\xi \subset R_0$ ensures the existence of the Boolean function $g(\xi)$, satisfying relation (1) if and only if the sets $\gamma_0, \gamma_1 \subset \gamma$, defined by the mappings $F_0 \xrightarrow{\xi} \gamma_0$, $F_1 \xrightarrow{\xi} \gamma_1$, do not intersect: $\gamma_0 \cap \gamma_1 \equiv \varnothing$. This condition will be satisfied if and only if no class ν_j of partitions of the space $X^{(2)} \equiv F_0 \cup F_1 \cup F_x$, defined by the set ξ, contains simultaneously elements from F_0 and F_1.

Theorem 1.

$$\xi \in \rho \leftrightarrow \overline{\exists\,(\nu_j)}[(\nu_j \in SEP^{\xi}X^{(2)}) \wedge (\nu_j \cap F_0 \not\equiv \varnothing) \wedge (\nu_j \cap F_1 \not\equiv \varnothing)].$$

We shall say below that the solution $\xi \in \rho$ defines a partitioning of the space $X^{(2)}$ into internally compatible classes.

The set of solutions is convex with respect to the relation ($\subset$) and is therefore completely defined by its lower boundary *inf* ρ [8].

Theorem 2.

$$(\xi \in \rho) \wedge (\xi \subset \xi') \rightarrow \xi' \in \rho.$$

To each solution $\xi \in \rho$ there corresponds, in general, one incompletely defined Boolean function $g(\xi)$.

Theorem 3. If $\xi \in \rho$, the Boolean function $g(\xi)$ in (1) is unique and is given by the partition $\{\gamma_0, \gamma_1, \gamma_x\}$ of the space $\xi^{(2)}$, defined by the relations

$$F_0 \xrightarrow{\xi} \gamma_0; \qquad F_1 \xrightarrow{\xi} \gamma_1; \qquad \gamma_x \equiv \xi^{(2)} \backslash (\gamma_0 \cup \gamma_1).$$

The number of elements in the space $\xi^{(2)}$ for which the value of the function $g(\xi)$ is defined does not exceed $\sigma(F_0 \cup F_1)$.

Theorem 4.

$$\overline{\exists (\xi)}[(\xi \in \rho) \wedge (\sigma(F_0 \cup F_1) < \sigma(\gamma_0 \cup \gamma_1))].$$

It is natural to assume that with decrease of $\sigma(\gamma_0 \cup \gamma_1)$ the complexity of the function $g(\xi)$, in general, will also decrease. It can therefore be useful to analyze the set ρ for determining solutions of $\xi \in \rho$ yielding the smallest value of $\sigma(\gamma_0 \cup \gamma_1)$. However, it is also necessary to take into account the mutual positions of the elements in γ_0, γ_1, which may be characterized by the set of distances $d_{ij} = \sigma(\gamma_i + \gamma_j)$ between different elements $\gamma_i, \gamma_j \in \gamma$. The treatment of these two factors will be discussed below.

The set $\min_\sigma \rho$ can be of interest if it is required to find a function $g(\xi)$ in (1) dependent on the smallest possible number of variables.

Theorem 5.

$$\xi \in \min_\sigma \rho \rightarrow \sigma(\xi) \leq \sigma(X).$$

The minimization of the number of variables of the transform of the superposition (1), realized in the synthesis of each stage of the circuit, usually leads to obtaining a circuit containing an excessive number of stages, which, from the technical point of view, is undesirable.

The method of calculating the set inf_ρ and its various subsets is conveniently formulated using the terminology in [6 and 8].

D. GRAPHICAL INTERPRETATION AND REDUCTION TO THE PROBLEM OF COVERAGE OF SETS

We associate with the Boolean function $f(X)$ a symmetrical graph $G_f = (X^{(2)}, u_f)$, in which $X^{(2)} \equiv F_0 \cup F_1 \cup F_x$ is the set of nodes, and the nodes $m_i, m_j \in X^{(2)}$ are connected by the branch $u_{ij} \in u_f$ if and only if $(m_i \in F_0) \wedge (m_j \in F_1)$. The total number of branches of the graph G_f coincides with the cardinality of the set F_0 *cart* F_1.

To the partitioning of the space $X^{(2)} \equiv F_0 \cup F_1 \cup F_x$ into internally compatible classes there corresponds the partitioning of the graph into subgraphs not containing branches attainable when a series of cutsets are executed in the graph G_f. (We use the term *cutset* for a subset u_i of branches of the complete symmetrical graph $(X^{(2)}, u)$ emerging from the set of nodes $\psi_i \subset X^{(2)}/\psi_i \in R_0$). The realization of a cutset u_i in the graph G_f can be imagined as the removal of the branches of the graph occurring in the set $u_i \cap u_f$. The partitioning of the graph G_f into subgraphs not containing branches will be achieved by removal of all branches in u_f.

From this it follows that each solution $\xi \in \rho$ provides satisfaction of the condition $\bigcup[\{u_i/\psi_i \in \xi\}] \supseteq u_f$ and determines one possible coverage of the set u_f by elements in the set $U \equiv \{u_i/\psi_i \in R_0\}$. Using the notation introduced in [8], the above can be summarized as follows.

Theorem 6.

$$\xi \in \rho \leftrightarrow \{u_i/\psi_i \in \xi\} \in cov_U\, u_f;$$
$$\xi \in inf\, \rho \leftrightarrow \{u_i/\psi_i \in \xi\} \in inf\, cov_U\, u_f.$$

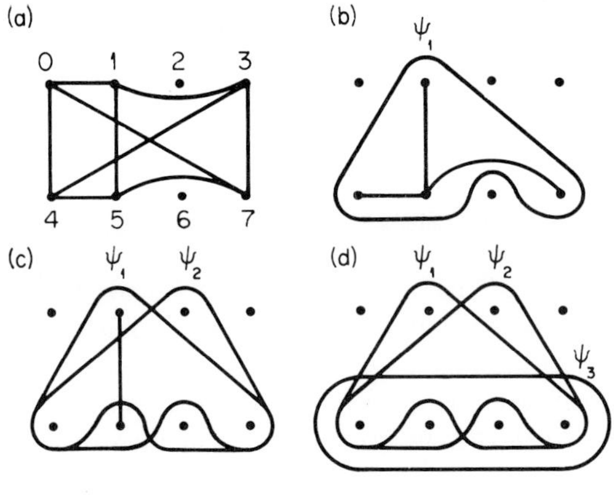

FIG. 3.

Disposing of the function $L(g)$, which estimates the complexity of the Boolean function g, we may attempt to find in the set ρ, defined by the bound $inf\, \rho$, a set of optimal solutions. Various algorithms for finding a coverage of sets, presented in [8], utilize a matrix interpretation of the initial information. In the present situation we use for this purpose the binary matrix $\| U \ni u_f \|$, which in a number of simple cases after traditional simplifications, can take on linear dimensions, which are convenient for the organization of the calculations. In complicated cases it is possible to use a multistep algorithm.

For illustration, Fig. 3a represents a graph G_f, corresponding to the Boolean function $f(a, b, c)$ presented in Table Ia. The realizations of three cutsets in succession are shown: $u_1/\psi_1 \sim \psi(a, \bar{b}, c)$, $u_2/\psi_2 \sim \psi(a, b, \bar{c})$, and $u_3/\psi_3 \sim a$, where the sets ψ_1, ψ_2, ψ_3 of nodes of the graph are enclosed by

TABLE I

$a\,b\,c$	f	$a\,b\,c$	$\psi_1\,\psi_2\,\psi_3$	$\psi_1\,\psi_2\,\psi_3$	g
000	0	000	000	000	0
001	1	001	100	001	X
010	X	010	010	010	X
011	0	011	000	011	X
100	1	100	111	100	1
101	0	101	101	101	0
110	X	110	011	110	X
111	1	111	111	111	1
(a)		(b)		(c)	

closed lines. After realization of the cutset u_1 there remain in the graph the branches u_{15}, u_{45}, u_{57} (Fig. 3b); the cutset u_2 contains the branches u_{45}, u_{57} (Fig. 3c), and the realization of cutset u_3 leads to removal of branch u_{15}.

The mapping $X^{(2)} \xrightarrow{(\xi)} \gamma \subset \xi^{(2)}/\xi \equiv \{\psi_1, \psi_2, \psi_3\}$, $X \equiv \{a, b, c\}$ is illustrated by Table Ib; the Boolean function $g(\xi)$ defined by this mapping is shown in Table Ic.

E. CHOICE OF OPTIMAL SOLUTIONS

The choice of optimal solutions in a set ρ of all solutions for relation (1) requires careful investigation, directed primarily to the construction of reliable and simple to realize methods for estimating the complexity of Boolean functions.

The algorithm formulated below for the selection of close-to-optimal solutions is based on the assumption that the complexity of the Boolean function $g(\xi)$, given by the sets γ_0, $\gamma_1 \subset \xi^{(2)}$, tends to be less as the mean distance between the elements within the sets γ_0, γ_1 decreases, and as the mean distance between the elements of γ_0 and γ_1 (i.e., the mean distance between the sets γ_0, γ_1 in the space $\xi^{(2)}$) increases.

The selection algorithm is realized in two stages. In the first stage the set $\min_\varphi \rho \subseteq \mathit{inf}\,\rho$ is calculated. The linear function φ is given on the set of solutions $\xi \in \rho$:

$$\varphi(\xi) = \sum_{\psi_i \in \xi} e(\psi_i),$$

where e is a positive function, given on the set R_0 and easily determined with the aid of the cutset concept: $e(\psi_i) = \sigma(u_i \backslash u_f)/\varphi_i \in R_0$. To calculate

φ-minimal coverages it is possible to utilize the algorithm *fiminco* from [8], first composing the table of values of the function $e(R_0)$. In the second stage from the set $\min_\varphi \rho$, whose cardinality in a number of cases is fairly great, a solution ξ is chosen, yielding the smallest value for the numerical function

$$\lambda(\xi) = \sigma(\xi) - \frac{1}{\sigma(\beta)} \sum_{\{\gamma i, \gamma j\} \epsilon \beta} d_{ij}/\beta \equiv \gamma_0 \; cart \; \gamma_1.$$

2. MULTISTEP ALGORITHM FOR FINDING AN APPROXIMATE SOLUTION

A.

The algorithm serves for finding a solution $\xi \in inf\ \rho$, close to optimal in the sense defined in Section 1E. The characteristic property of the algorithm consists in the sequential improvement of an already existing solution $\xi_0 \in \rho$, realized over several steps by the iteration of the same algorithm of functional decomposition.

Let us take as the initial solution the set $\xi_0 \equiv \{x_0, x_1, \ldots, x_n\}$, consisting of the elementary Boolean functions $x_0, x_1, \ldots, x_{n-1} \in R_0$, realizable by majority elements functioning as delay elements. It is understood that by this choice of solution, (1) is converted into the identity $\tilde{g}_0(\xi_0) \equiv \tilde{f}(X)$, since the function $g_0(\xi_0)$ coincides with the function $f(X)$.

In the first step according to the algorithm of selection in the current step, described in Section 2F, a set $\alpha_1 \subseteq \xi_0$ is chosen. In the solution ξ_0 the elementary terms from α_1 are replaced by a certain set $s_1 \equiv \{\varphi_{1l}, \psi_{1m}, \ldots, \psi_{1n}\}$ of Boolean functions, chosen from the set R_{α_1} of all Boolean functions realizable by a single majority element and having as their arguments the elements of the set $\alpha_1 \cup \{0\}$. The substitution must lead to obtaining a new solution $\xi_1 \equiv (\xi_0 \backslash \alpha_1) \cup s_1$, defining the least-complex Boolean function $g_1(\xi)$ satisfying the relation $\tilde{g}_1(s_1, \beta_1) \subseteq \tilde{f}(X)/\beta_1 \equiv \xi_0 \backslash \alpha_1$. It may be found that the finally chosen set s_1 contains some elements from α_1: $s_1 \cap \alpha_1 \neq \varnothing$. In such cases the elements of the sets s_1, β_1 are redistributed, as a result of which is formed a set $\beta_1^* \equiv \beta_1 \cup s_1 \cap \alpha_1$ of primary elements of the solution and a set $s_1^* \equiv s_1 \backslash s_1 \cap \alpha_1$ of secondary elements of the solution, after which a new solution $\xi_1 \in \rho$ takes the form $\xi_1 \equiv s_1^* \cup \beta_1^*$.

In the second step, $\alpha_2 \subseteq \beta_1^*$ is chosen in the set β_1^* and a substitution is sought in the form of the set $s_2 \equiv \{\psi_{2l}, \psi_{2m}, \ldots, \psi_{2n}\} \subseteq R_{\alpha_2}$, leading to obtaining the least-complex function g_2: $\tilde{g}_2(s_1^*, s_2, \beta_2) \subseteq \tilde{g}_1^*(s_1^*, \beta_1^*)/\beta_2 \equiv$

$\beta_1{}^*\backslash\alpha_2$. After formation of $\beta_2{}^* \equiv \beta_2 \cup s_2 \cap \alpha_2$; $s_2{}^* \equiv s_2\backslash(s_2 \cap \alpha_2)$ we obtain the second approximation to the solution $\xi_2 \equiv s_1{}^* \cup s_2{}^* \cup \beta_2{}^*$.

On the ith step $\alpha_i \subseteq \beta_{i-1}^*$ is chosen, and if $\sigma(\beta_{i-1}^*) < 2$, the process of forming the solution for relation (1) terminates, and the solution is given out in the form $\xi \equiv S_{i-1} \cup \beta_{i-1}^*/S_{i-1} \equiv s_1{}^* \cup s_2{}^* \cup \cdots \cup s_{i-1}^*$. If $\sigma(\beta_{i-1}^*) \geq 2$, then an optimal set $s_i \subseteq R_{\alpha_i}$ is sought, satisfying the relation $\tilde{g}_i(S_{i-1}; s_i, \beta_i) \subseteq \tilde{g}_{i-1}(S_{i-1}, \beta_{i-1}^*)$; $\beta_i \equiv \beta_{i-1}^*\backslash\alpha_i$. The approximation to the solution obtained on the ith step takes the form $\xi_i \equiv S_i \cup \beta_i{}^*/S_i \equiv S_{i-1} \cup s_i{}^*$; $\beta_i{}^* \equiv \beta_i \cup s_i \cap \alpha_i$; $s_i{}^* \equiv s_i\backslash s_i \cap \alpha_i$.

The cardinality of the set α_i, chosen on the ith step, defines in the last analysis the volume of computation and memory necessary for finding the optimal substitution set s_i and therefore must be reasonably limited. In programming the algorithm it was determined that the constraint $1 < \sigma(\alpha) \leq 3$ is perfectly acceptable.

The same algorithm is applied to realize each step; the different parts of the algorithm are described in the next paragraphs in an order that is convenient for the exposition.

B.

The problem solved at each step reduces, in general, to the following typical problem: when a Boolean function $f(X)$ is given, there exists a solution $\xi_0 \equiv \{x_0, x_1, \ldots, x_{n-1}\}$, satisfying the relation $\tilde{f}(X) \equiv \tilde{g}_0(\xi_0)$, a set $\alpha \subseteq \xi_0$ is given, and it is required to find the optimal set $s \equiv \{\psi_1, \psi_m, \ldots, \psi_n\} \subseteq R_\alpha$, satisfying the relation

$$\tilde{f}(X) \equiv \tilde{g}_0(\xi_0) \supseteq \tilde{g}_1(s, \beta)/\beta \equiv \xi_0\backslash\alpha. \tag{2}$$

Passing to the consideration of the sets connected with the Boolean functions in (2), we note that the set ξ_0, coinciding with the interval reference set W, defines a partitioning of the space $X^{(2)}$ into classes in $SEP^{\xi_0}X^{(2)}$, each of which contains one element from $X^{(2)}$, and is therefore internally compatible. For the same reason, the mapping

$$X^{(2)} \overset{\xi_0}{\leftrightarrow} \xi_0^{(2)}$$

is one-to-one, which explains the identity of the functions $f(X)$ and $g_0(\xi_0)$.

The set $\alpha \subseteq \xi_0$ may be replaced only by a set $s \in R \subseteq R_0$ such that it realizes all the necessary separations realized by the set α. The set $\xi_1 \equiv s \cup \beta$ obtained by the substitution must define a partitioning of the space $X^{(2)}$

TABLE II

		abc 0	1	2	3	4	5	6	7
d	0	0	1	×	0	1	1	0	1
	1	0	0	0	0	0	×	0	×

into internally compatible classes. All sets $s \subseteq R$ replacing the set α form a set $\rho_\alpha \subseteq R_\alpha^{(2)}$, convex with respect to the relation $(\subset)$: $(s \in \rho_\alpha) \wedge (s \subset s') \rightarrow s' \in \rho_\alpha$ and defined by the lower bound $inf\ \rho_\alpha$.

The set $\alpha \subseteq \xi_0 \equiv W$ defines a partitioning of the space into intervals $J_0, J_1, \ldots, J_{N-1} \in J \equiv SEP^\alpha X^{(2)}/N = 2^{\sigma(\alpha)}$, while the segmentation of each interval into compatible classes is realized by means of the elements in $\xi_0 \backslash \alpha$. On the other hand, each set $\psi_i \in R_\alpha$ may be composed from intervals in J: a subset $J_{\psi_i} \subset J$ always exists such that $\psi_i \equiv \cup [J_{\psi_i}]$. Hence it follows that any $s \subseteq R_\alpha$ defines a partitioning of the space $X^{(2)}$ into classes, consisting of intervals in J, where it is always possible to indicate an $s \subset R_\alpha$ such that $SEP^s X^{(2)} \equiv SEP^\alpha X^{(2)}$. However, such a detailed segmentation of the space $X^{(2)}$ is not always necessary. The set $s \subset R_\alpha$ may replace the set α if and only if each class of the partition $SEP^s X^{(2)}$ consists of mutually compatible intervals in J.

The relation of compatibility between elements of the set J can be defined in set-theoretical language, but it is much simpler to show it by an example.

Table II represents the Boolean function $f(a, b, c, d)$ in matrix form. Taking as the set α the set of elements w_a, w_b, w_c of the interval reference set W, corresponding to the variables $a, b, c \in X$, it can be stated that the columns of the matrix $0, 1, \ldots, 7$ correspond to the intervals $J_0, J_1, \ldots, J_7$. The intervals $J_i, J_j \in J$ are called compatible if and only if the contents of columns i and j are the same or may be rendered the same by choice of the undefined elements. For example, this is the case for the pairs of columns numbered 0, 2; 0, 3; 4, 5; etc. Correspondingly, the pairs of intervals J_0, J_2; J_0, J_3; J_4, J_5 are compatible. The relation of incompatibility between intervals holds in the absence of compatibility. The degree of incompatibility of intervals $J_i, J_j \in J$ is the number of elements in the intervals that do not coincide and cannot be made to coincide by completion of the function. For example, the intervals J_0, J_5 in the example are incompatible, and the degree of incompatibility is equal to 1.

It is convenient to interpret the relation of incompatibility between elements of the set J by a symmetrical graph $G_f^\alpha(J, v_\alpha)$, whose nodes

$J_i, J_j \in J$ are joined by a branch $v_{ij} \in v_\alpha$ if and only if the elements J_i, J_j are incompatible. Each branch v_{ij} is assigned a weight p_{ij} equal to the degree of incompatibility of the elements $J_i, J_j \in J$.

The graph G_f with nodes 0–7, corresponding to this example, is given in Fig. 4. The weight of each of its branches is equal to 1.

The partition $SEP^s X^{(2)}$, each class of which consists of mutually compatible intervals, corresponds to the separation of the graph G_f into subgraphs not containing branches, achieved by realization of cutsets.

According to Section 1D, we bring into consideration the cutset v_i, containing the branches of the complete symmetrical graph (J, v), emerging from the set of nodes of the graph $J_{\psi_i} \subset J/\psi_i \in R_\alpha$, the set of cutsets $V \equiv \{v_i/\psi_i \in R_\alpha\}$, and we reduce the problem of finding the set ρ_α to the problem of coverage of sets.

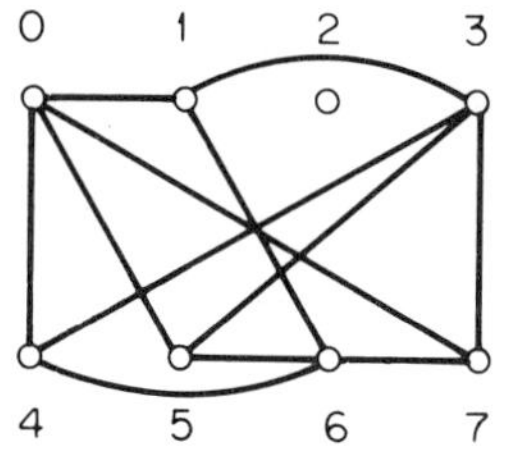

FIG. 4.

Theorem 6a.

$$s \in \rho_\alpha \leftrightarrow \{v_i/\psi_i \in s\} \in cov_V\, v_\alpha;$$
$$s \in inf\, \rho_\alpha \leftrightarrow \{v_i/\psi_i \in s\} \in inf\, cov_V\, v_\alpha.$$

For the graph G_f^α (Fig. 4) it is possible to indicate one cutset v_i, connected with the set $J_{\psi_i} \equiv \{J_1, J_4, J_5, J_7\}$, $\psi_i \sim \psi(a, \bar{b}, c)$ and containing all branches of the graph: $v_i \supset v_\alpha$. In this connection, the relation $\tilde{f}(a, b, c, d) \ni g(\psi(a, \bar{b}, c), d)$ holds for the function $f(a, b, c, d)$ (Table II). The matrix representation of the function g is given in Table III, whose left-hand column is obtained by merging the columns 0, 2, 3, 6 of Table II, and whose right-hand column is obtained by merging the columns 1, 4, 5, 7. This realizes the mapping

$$J \overset{\{\psi_i\}}{\longrightarrow} I/I \equiv SEP^{\{\psi_i\}}\, \xi^{(2)}, \qquad \xi \equiv \{\psi_i, w_d\}, \psi_i \sim \psi(a, \bar{b}, c),$$

in which the intervals $J_1, J_4, J_5, J_7 \subset \psi_i$ are mapped onto the interval $I_1 \in I$ of the space $\xi^{(2)}$, represented by the right-hand column of Table III, and the intervals $J_0, J_2, J_3, J_6 \subset \bar{\psi}_i$ are mapped onto the interval $I_0 \in I$, represented by the left-hand column.

TABLE III

$\psi(a, b, c)$			
d	0	0	1
	1	0	0

In the result we obtain the realization $\psi(\psi(a, \bar{b}, c), \bar{d}, 0)$ for the Boolean function $f(a, b, c, d)$.

C.

Following the algorithm for selection of optimal solutions, proposed in Section 1E, the set of solutions $s \in \rho_\alpha$ is tested for optimality; we retain the set $\min_\varphi \rho_\alpha$, defining the function φ by the formula

$$\varphi = \sum_{\psi_i \in s} e(v_i).$$

The function e is given on the set $V \equiv \{v_i/\psi_i \in R_\alpha\}$ by the relation

$$e(v_i) = \sum_{v_{ij} \in v_i} (P - p_{ij}), \tag{3}$$

where P is the maximum weight of a branch in the graph G_f. The value of the weight p_{ij} for all $v_{ij} \in v_i \backslash v_\alpha$ is taken equal to zero.

From the set $\min_\varphi \rho_\alpha$ we take elements defining the solution $\xi \equiv s \cup \beta$, to which corresponds the smallest value of the function $\lambda(\xi)$, introduced in Section 1E.

If the sets γ_0, γ_1, defining the transform $g_1(\xi_1)$ from (2), are given, then the value of the function λ is calculated by means of the operation *estimate* $\alpha\beta\gamma\delta\epsilon//$, having as its operands the complexes $\alpha::\|\, \gamma_0 \ni \xi_1 \,\|$; $\beta::\|\, \gamma_1 \ni \xi_1 \,\|$, and the variable $\gamma::\psi(\xi_1)$; the variable ϵ, represents the integer part of the value of λ and the variable δ represents the fractional part.

estimate $\alpha\text{к}\,+, \beta\text{к}\,+, \gamma\text{п}\,+, \delta\text{п}, \epsilon\text{п}, /\mathbf{c}, b/(\alpha\beta\gamma\delta\epsilon)$
100 103 1 $(\alpha\mathbf{abc})$

$\circ\ a \circ bb_\alpha \times b_\beta \Rightarrow \mathbf{a}\gamma \triangledown \times \mathbf{a} \Rightarrow \mathbf{b} \circ \mathbf{c}$
§1 $\alpha_b \oplus \beta_a \triangledown + \mathbf{c} \Rightarrow \mathbf{c} \vartriangle a \oplus b_\beta \mid\!\rightarrow 1 \circ a \vartriangle b \oplus b_\alpha \mid\!\rightarrow 1$
$\mathbf{b} - \mathbf{c}\!:\!\mathbf{a} \Rightarrow \delta\text{я} \Rightarrow \epsilon.$

D. PROGRAM FOR CALCULATING THE SET $\min_\varphi \rho_\alpha$

The algorithm for calculating the set $\min_\varphi \rho_\alpha$ must contain an operator by which the adjacency matrix and the table of branch weights of the graph G_f^α are calculated. The adjacency matrix is formed in the variable $\delta::\|\{v_\alpha\} \ni v\|$, where v is the set of branches of the complete symmetrical graph (J, v). The elements $v_{ij} \in v$ are ordered by increasing value of the smaller index i, while the elements having the same value of i are ordered by increasing values of the index j. In this sequence the element v_{ij} takes the ordinal number k, whose value is calculated by the formula

$$k(i,j) = \frac{i}{2}(2^{\sigma(\alpha)} + i - 1) + (j - i),$$

in which the left term of the sum serves to calculate the sum of the first i terms of the progression $N - 1, N - 2, \ldots, 1 (N = 2^{\sigma(\alpha)})$.

The elements $J_i, J_j \in J$ are incompatible if and only if the set F_0 *cart* F_1 contains the element $\{m_p, m_q\}$ such that

(1) $\mathbf{f} \oplus \mathbf{g} \wedge \bar{\mathbf{a}} = \circ/\mathbf{f}::\|\{m_p\} \ni X\|, \mathbf{g}::\|\{m_q\} \ni X\|,$
$\mathbf{a}::\|\{\alpha\} \ni \xi_0\|;$
(2) $(m_p \in J_i) \wedge (m_p \in J_j) = 1.$

During the process of calculation, elements $\{m_p, m_q\}$ satisfying condition (1) are successively selected from the set F_0 *cart* F_1; then on the basis of the values taken by the variables $\mathbf{b} = \mathbf{f} \wedge \mathbf{a}$, $\mathbf{c} = \mathbf{g} \wedge \mathbf{a}$, the values of the indexes i, j of the incompatible intervals J_i, J_j are determined, the value of $k(i,j)$ is calculated, a 1 is placed in the kth position of the variable $\delta::\|\{v_\alpha\} \ni v\|$, and the kth element of the complex ϵ, in which is formed the table of branch weights, receives a positive unit increment. The process terminates after the entire set F_0 *cart* F_1 has been examined, where, in passing, the value of P, the maximum branch weight is formed.

These transformations are realized by the (1, 1)-terminal *strugraph* $\alpha\beta\gamma\delta\epsilon\zeta//$ (determination of graph structure), whose operands have the following sense: $\alpha::\|F_0 \ni X\|; \beta::\|F_1 \ni X\|; \gamma::\|\{\alpha\} \ni \xi_0\|; \delta::\|\{v_\alpha\} \ni v\|;$ $[\zeta] = P$; ϵ is a complex, representing the table of weights: $[\epsilon_k] = p_{ij}$.

strugraph $\alpha\kappa +, \beta\kappa +, \gamma\pi +, \delta\pi, \epsilon\kappa, \zeta\pi/\mathbf{e}, d/(\alpha\beta\gamma\delta\epsilon\zeta)$
101 252 7 (α**abcde**)

$\circ\ a \circ b \circ \delta\gamma\ \urcorner \Rightarrow \mathbf{a} \circ \zeta 35 \Rightarrow b_\epsilon$ Initialization
§1 $\circ\ \epsilon_a \triangle a \oplus b_\epsilon \mid\!\rightarrow 1 \circ a$

TABLE IV

		01	02	03	04	05	06	07	12	13	14	15	16	17	23	24	25	26	27	34	35	36	37	45	46	47	56	57	67
1. $\psi(a, a, a)$	0	0	0	0	1	1	1	1	0	0	1	1	1	1	0	1	1	1	1	1	1	1	1	0	0	0	0	0	0
2. $\psi(b, b, b)$	0	0	1	1	0	0	1	1	1	1	0	0	1	1	0	1	1	0	0	1	1	0	0	0	1	1	1	1	0
3. $\psi(c, c, c)$	0	1	0	1	0	1	0	1	1	0	1	0	1	0	1	0	1	0	1	1	0	1	0	1	0	1	1	0	1
4. $\psi(b, 1, c)$	0	1	1	1	0	1	1	1	0	0	1	0	0	0	0	1	0	0	0	1	0	0	0	1	1	1	0	0	0
5. $\psi(\bar{b}, 0, c)$	0	1	0	0	0	1	0	0	1	1	1	0	1	1	0	0	1	0	0	0	1	0	0	1	0	0	1	1	0
6. $\psi(b, 0, \bar{c})$	0	0	1	0	0	0	1	0	1	0	0	0	1	0	1	1	1	0	1	0	0	1	0	0	1	0	1	0	1
7. $\psi(b, 0, c)$	0	0	0	1	0	0	0	1	0	1	0	0	0	1	1	0	0	0	1	1	1	1	0	0	0	1	0	1	1
8. $\psi(a, 1, c)$	0	1	0	1	1	1	1	1	1	0	0	0	0	0	1	1	1	1	1	0	0	0	0	0	0	0	0	0	0
9. $\psi(\bar{a}, 0, c)$	0	1	0	1	0	0	0	0	1	0	1	1	1	1	1	0	0	0	0	1	1	1	1	0	0	0	0	0	0
10. $\psi(a, 0, \bar{c})$	0	0	0	0	1	0	1	0	0	0	1	0	1	0	0	1	0	1	0	1	0	1	0	1	0	1	1	0	1
11. $\psi(a, 0, c)$	0	0	0	0	0	1	0	1	0	0	0	1	0	1	0	0	1	0	1	0	1	0	1	1	0	1	1	0	1
12. $\psi(a, 1, b)$	0	0	1	1	1	1	1	1	1	1	1	1	1	1	0	0	0	0	0	0	0	0	0	0	0	0	0	0	0
13. $\psi(\bar{a}, 0, b)$	0	0	1	1	0	0	0	0	1	1	0	0	0	0	0	1	1	1	1	1	1	1	1	0	0	0	0	0	0
14. $\psi(a, 0, \bar{b})$	0	0	0	0	1	1	0	0	0	0	1	1	0	0	0	1	1	0	0	1	1	0	0	0	1	1	1	1	0
15. $\psi(a, 0, b)$	0	0	0	0	0	0	1	1	0	0	0	0	1	1	0	0	0	1	1	0	0	1	1	0	1	1	1	1	0
16. $\psi(\bar{a}, b, c)$	0	1	1	1	0	0	0	1	0	0	1	1	1	0	0	1	1	1	0	1	1	1	0	0	0	1	0	1	1
17. $\psi(a, \bar{b}, c)$	0	1	0	0	1	1	0	1	1	1	0	0	1	0	0	1	1	0	1	1	1	0	1	0	1	0	1	0	1
18. $\psi(a, b, \bar{c})$	0	0	1	0	1	0	1	1	1	0	1	0	1	1	1	0	1	0	0	1	0	1	1	1	0	0	1	1	0
19. $\psi(a, b, c)$	0	0	0	1	0	1	1	1	0	1	0	1	1	1	1	0	1	1	1	1	0	0	0	1	1	1	0	0	0

§2 $\alpha_b \oplus \beta_a \wedge \mathbf{a} \mid\!\rightarrow 7\alpha_b \wedge \gamma \Rightarrow \mathbf{b}10 \Rightarrow \mathbf{e}$ $\beta_a \wedge \gamma \Rightarrow \mathbf{c} - \mathbf{b} \mid\!\rightarrow 3\mathbf{b} \Leftrightarrow \mathbf{c}$	Determination of the elements $\{m_p, m_q\} \in F_0\, cart\, F_1$, satisfying condition 1
§3 $\mathbf{c} - \mathbf{b} \wedge \gamma \Rightarrow \mathbf{c}\gamma \Rightarrow \mathbf{d} \circ c$ §4 $\mathbf{d}\, \dot{\mathbf{X}}\, 5dc + c \Rightarrow cc_d \wedge \mathbf{c} \circ\!\rightarrow 4\ 1 \vee c$ $\Rightarrow c \rightarrow 4$	Determination of the difference $j - i$
§5 $\mathbf{b} \circ\!\rightarrow 6\mathbf{b} - 1 \wedge \gamma \Rightarrow \mathbf{b}\ \bar{\triangle}\ \mathbf{e} + c$ $\Rightarrow c \rightarrow 5$	Calculation of the sum of the first i terms of the progression
§6 $c_c \vee \delta \Rightarrow \delta \triangle \epsilon_c - \zeta \circ\!\rightarrow 7\epsilon_c \Rightarrow \zeta$ §7 $\triangle a \oplus b_\beta \mid\!\rightarrow 2 \circ a \triangle b \oplus b_\alpha \mid\!\rightarrow 2.$	Formation of the values of the operands δ, ϵ, ζ

Disposing of the information on the structure of the graph G_f^α, we proceed to programming the operators directly connected with calculation of the set $\min_\varphi \rho_\alpha$, isomorphic to the set $\min_\varphi cov_V\, v_\alpha$ (Theorem 6a). The basic object of transformation will be the matrix $\| V \ni v_\alpha \|$, which for arbitrary α with cardinality $\sigma(\alpha) \leq 3$, serves as the minor of the matrix $\| V \ni v \|$ (Table IV) formed by the columns of the latter corresponding to the set $v_\alpha \subseteq v$. If $\sigma(\alpha) = 2$, the minor of the matrix $\| V \ni v \|$, formed by its first seven rows and columns, corresponding to the set v_α, is used for the calculation. The arrangement of the elements in the set V coincides with the order observed in Table IV, in which the rows corresponding to the cutset $v_i \in V$ contain the expressions for the Boolean functions $\psi_i \in R_\alpha$. The vector representation $\| \{\psi_i\} \supset J \|$ of the Boolean function ψ_i is contained in the first eight positions of the ith row of the table.

The first of the series of operations that prepares the information for the calculation of the φ-minimal coverage is the operation of compression of the matrix *com* $\alpha\beta\gamma//$, which transforms the matrix $\| V \ni v \|$ to the matrix $\| V_\alpha \ni v \|$, where $V_\alpha \equiv \{v_i/(v_i \in V) \wedge (v_i \cap v_\alpha \not\equiv \varnothing)\}$. The external operands of the (1, 1)-terminal *com* are the complexes $\alpha::\| V \ni v \|$; $\beta::\| V_\alpha \ni v \|$; and the variable $\gamma::\| \{v_\alpha\} \ni v \|$.

$$com\ \alpha\text{к} +, \beta\text{к} +, \gamma\text{п} +/-, b/(\alpha\beta\gamma)$$

102 46 2

$$\circ\ b\ \bar{\circ}\ a$$

§1 $\triangle a \oplus b_\alpha \circ\!\rightarrow 2\alpha_a \wedge \gamma \circ\!\rightarrow 1\alpha_a \Rightarrow \beta_b \triangle b \rightarrow 1$

§2 $b \Rightarrow b_\beta.$

In using the table of branch weights of the graph G_f^α and the values of P, a table of values of the function e [relation (3), Section 2C], taken on the set V_α, is calculated. The necessary calculations are organized by the (1, 1)-

terminal *taval* $\alpha\beta\gamma\delta//$: $\alpha::\|\ V_\alpha \ni v\ \|$; $[\beta] = P$; $[\gamma_k] = p_{ij}$; $[\delta_i] = e(v_i)/v_i \in V_\alpha$.

taval $\alpha\text{к} +, \beta\text{п} +, \gamma\text{к} +, \delta\text{к}/\mathbf{a}, b/(\alpha\beta\gamma\delta)$
103 62 3 $(\alpha\mathbf{a})$

$\bar{\circ}\ a$
§1 $\triangle\ a \oplus b_\alpha \circ\!\!\rightarrow 3\alpha_a \Rightarrow \mathbf{a} \circ \delta_a$
§2 $\mathbf{a}\ \dot{\mathbf{X}}\ 1b\beta - \gamma_b + \delta_a \Rightarrow \delta_a \rightarrow 2$
§3 $a \Rightarrow b_\delta$.

The operator *traw* $\alpha\beta//$ serves to transform the matrix $\|\ V_\alpha \ni v\ \|$ to the matrix $\|\ V_\alpha' \ni v\ \|$. The set V_α' may be distinguished from the set V only by the order of the elements; in the set V' the element v_j follows the element v_i if and only if $e(v_i) \leq e(v_j)$. In order not to disturb the correspondence established between the rows of the matrix $\|\ V_\alpha \ni v\ \|$ and the elements of the complex β, representing the table of values of the function e, the permutations of the elements of the complex $\alpha::\|\ V_\alpha \ni v\ \|$ are accompanied by corresponding permutations of the elements of complex β.

traw $\alpha\text{к} +, \beta\text{к}/\mathbf{a}, b/(\alpha\beta)$
104 75 3 $(\alpha\mathbf{a})$

§1 $\circ\ a \circ \mathbf{a}$
§2 $\beta_a - \mathbf{a} \circ\!\!\rightarrow 3\beta_a \Rightarrow \mathbf{a}a \Rightarrow b$
§3 $\triangle\ a \oplus b_\beta \mid\!\!\rightarrow 2\ \bar{\triangle}\ a \Rightarrow b_\beta\alpha_a \Leftrightarrow \alpha_b\beta_a \Leftrightarrow \beta_b a \mid\!\!\rightarrow 1b_\alpha \Rightarrow b_\beta$.

The closing operation of the algorithm for finding the set is the (1, 1)-terminal *fiminco'* $\alpha\beta\gamma\delta\epsilon//$, whose operands are interpreted in the following way: $\alpha::\|\ \{v_\alpha\} \ni v\ \|$; $\beta::\|\ V_\alpha' \ni v\ \|$; $\epsilon::\|\ \min_\varphi \rho_\alpha \ni R_\alpha'\ \|$, where the set $R_\alpha' \subseteq R_\alpha$ is isomorphic to the set V_α'. Aside from this, the complex γ serves to represent the auxiliary set S, the complex δ the table of values of the function e.

The operator *fiminco'* is written on the basis of *fiminco*, and so its functioning can be grasped from the description given in [8].

fiminco'[1] $\alpha\text{п} +, \beta\text{к} +, \gamma\text{к}, \delta\text{к} +, \epsilon\text{к}?/\mathbf{g}, c/(\alpha\beta\gamma\delta\epsilon)$
105 265 7

$\circ\ \mathbf{c}b_\beta \Rightarrow a \circ \mathbf{b}\alpha \urcorner \Rightarrow \mathbf{d} \Rightarrow \mathbf{f}$
§1 $\bar{\triangle}\ a\beta_a \vee \mathbf{d} \Rightarrow \mathbf{d} \Rightarrow \gamma_a a \mid\!\!\rightarrow 1\ \bar{\circ}\ g \Rightarrow a \circ \mathbf{a}$

[1] Translator's note: The prime on *fiminco'* does not appear in any other operator name in this volume, and for consistency could have been replaced by a syllable meaning the same thing, such as "bis." Then we would have written *fimincobis*.

§2 $\mathbf{a}\,\dot{\mathbf{X}}\,3b\beta_b \vee \mathbf{f} \Rightarrow \mathbf{f}\delta_b + \mathbf{b} \Rightarrow \mathbf{b} \rightarrow 2$

§3 $\triangle\, a \oplus b_\beta \circ\!\!\rightarrow 6\gamma_a \vee \mathbf{f} \oplus \mathbf{d} \mapsto 6\mathbf{f}\ \urcorner \wedge \beta_a \circ\!\!\rightarrow 3$

$\delta_a + \mathbf{b} \Rightarrow \mathbf{e} - 1 - \mathbf{g} \mapsto 3\beta_a \vee \mathbf{f} \oplus \mathbf{d} \mapsto 5\mathbf{e} \oplus \mathbf{g} \circ\!\!\rightarrow 4e \Rightarrow \mathbf{g} \circ c$

§4 $c_a \vee \mathbf{c} \Rightarrow \epsilon_c / \triangle\, c \Rightarrow b_\epsilon \rightarrow 3$

§5 $c_a \vee \mathbf{c} \Rightarrow \mathbf{ce} \Rightarrow \mathbf{b}\beta_a \vee \mathbf{f} \Rightarrow \mathbf{f} \rightarrow 3$

§6 $\mathbf{c} \circ\!\!\rightarrow 7\ \urcorner + 1 \wedge \underset{\smile}{\mathbf{c}} \vdash \Rightarrow a\alpha\ \urcorner \Rightarrow \mathbf{f} \circ \mathbf{b}c_a \oplus \mathbf{c} \Rightarrow \mathbf{c} \Rightarrow \mathbf{a} \rightarrow 2$

§7 .

In terms of these operators it is fairly simple to express the operator *calminco* $\alpha\beta\gamma\delta\epsilon\zeta\eta\vartheta//$, which calculates the set $\min_\varphi \rho_\alpha$: $\alpha::\| V \ni v \|$; $\beta::\| V_\alpha' \ni v \|$; the complex γ is the table of branch weights of the graph $G_f{}^\alpha: [\gamma_k] = p_{ij}$; it is then used to represent the auxiliary set S; the complex δ serves to represent the table of values of the function e; $\epsilon::\| \min_\varphi \rho_\alpha \ni R_\alpha' \|$; $\zeta::\| F_0 \ni X \|$; $\eta::\| F_1 \ni X \|$; $\vartheta::\| \{\alpha\} \ni \xi_0 \|$.

calminco (101 102 103 104 105)/αк +, βк, γк, δк, ϵк?, ζк, ηк, ϑп/**b**,
/($\alpha\beta\gamma\delta\epsilon\zeta\eta\vartheta$)
106 54 0 (α**ab**)

strugraph $\zeta\eta\vartheta$**a**γ**b**//*com* $\alpha\beta$**a**//*taval* β**b**$\gamma\delta$//*traw* $\beta\delta$//*fiminco'* **a**$\beta\gamma\delta\epsilon$ //.

E. PROGRAM FOR CALCULATING THE SUPERPOSITION TRANSFORM

To each element $s \in \min_\varphi \rho_\alpha$ there corresponds a new solution $\xi_1 \equiv s^* \cup \beta^* / s^* \equiv s \backslash \alpha$, $\beta^* \equiv \xi_0 \backslash \alpha \backslash s$ for the superposition (2). In order to determine the form of the Boolean function $g_1(\xi_1)$ in (2) it is necessary to calculate the sets $\gamma_0, \gamma_1 \subset \xi_1^{(2)}$, defined by the mappings

$$F_0 \xrightarrow{\xi_1} \gamma_0; \qquad F_1 \xrightarrow{\xi_1} \gamma_1.$$

The initial material for the calculation is the sets F_0, F_1, given by the matrices $\| F_0 \ni X \|$, $\| F_1 \ni X \|$, and the set $\xi_1 \equiv s^* \cup \beta^*$ about the dispositions of whose elements it is necessary to decide. We shall arrange the elements from β^* in the set ξ_1 at those positions which they occupied in the set ξ_0: $\| \{\beta^*\} \ni \xi_1 \| = \| \{\beta^*\} \ni \xi_0 \|$. In order to obtain the vector $\| \{\beta^*\} \ni \xi_0 \|$ it is sufficient to substitute zero for 1 in the positions of the vector $\Psi(\xi_0)$ corresponding to the elements in $\alpha \backslash s$. For this it is convenient to utilize the vector $\| \{\alpha \backslash s\} \ni \xi_0 \|$. To find this vector among the rows of the matrix $\| V_\alpha' \ni v \|$ corresponding to the set s and tagged by 1's in the code of vector $\| \{s\} \ni R_\alpha' \|$, it is necessary to recognize those rows that correspond to elements in α. In this recognition process information contained in the first eight positions of each row of the matrix is used. In passing, the vector $\| \{s\} \ni R' \|$ is transformed to the vector $\| \{s^*\} \ni R_\alpha' \|$. The disposition of the elements of the set $s^* \subset \xi_1$ can be prescribed by forming the vector $\| \{s^*\} \ni \xi_1 \|$, containing 1's only in the $k = \sigma(s^*)$ first positions not occupied by 1's in the vector $\| \{\beta^*\} \ni \xi_1 \|$, which corresponds to a dense "packing" of the elements of the set s^* among the elements of set β^*. The vector $\Psi(\xi_1)$ is easily obtained by elementwise addition of the vectors $\| \{s^*\} \ni \xi_1 \|$ and $\| \{\beta^*\} \ni \xi_1 \|$, either of which together with the vector $\Psi(\xi_1)$ defines the distribution of the elements in ξ_1.

These transformations are carried out by the operator *pack* $\alpha\beta\gamma\delta//$, using the initial information represented by the values of the complexes $\beta::\| V_\alpha' \ni v \|$ and γ: $\gamma_0::\Psi(\xi_0)$; $\gamma_1::\| \{\alpha\} \ni \xi_0 \|$; $\gamma_2::\| \{s\} \ni R_\alpha \|$. The complex δ represents the result $\delta_0::\Psi(\xi_1)$; $\delta_1::\| \{\alpha\} \ni \xi_0 \|$; $\delta_2::\| \{s^*\} \ni R_\alpha \|$; $\delta_3::\| \{s^*\} \ni \xi_1 \|$.

The operator *pack* is (1, 2)-terminal; its auxiliary exit terminal α is attained if $s^* \equiv \varnothing$, i.e., $s \equiv \alpha$.

pack αч, βк, γк, + δк/**g**, $c/(\alpha\beta\gamma\delta)$
107 223 6 (α**abcdefg**)

	a $\Leftarrow$ 037 700 000 000 γ $\Rightarrow$ (**bcd**)**b** $\oplus$ **c** $\Rightarrow$ **bd** $\Rightarrow$ **e**	Initialization
§1	**e** $\dot{\mathbf{X}}$ $4a \circ b$	Recognition of rows in $\| V_\alpha' \ni v \|$, corresponding to elements in $s \cap \alpha$
§2	$\triangle\, b \oplus 3 \circ\!\!\rightarrow 1\beta_a \wedge$ **a** $\Rightarrow$ **f**$e_b \wedge$ **a** $\oplus$ **f** $\mid\!\!\rightarrow 2c_a \oplus$ **d** $\Rightarrow d \circ\!\!\rightarrow \alpha$**c** $\Rightarrow$ **f** $\bar{\circ}$ $c2 - b \Rightarrow b$	Obtaining the vector $\| \{s^*\} \ni R_\alpha' \|$,
§3	**f** $\dot{\mathbf{X}}$ $1a \triangle c \oplus b \mid\!\!\rightarrow 3c_a \vee$ **b** $\Rightarrow$ **b** $\rightarrow 1$	Obtaining the vector
§4	**d** $\nabla \Rightarrow$ **ab** $\urcorner \Rightarrow$ **f** $\circ$ **g**	$\| \{\beta^*\} \ni \xi_1 \|$,
§5	**f** $\dot{\mathbf{X}}$ $6ac_a \vee$ **g** $\Rightarrow$ **g** $\triangle$ **e** $\oplus$ **a** $\mid\!\!\rightarrow 5$	Obtaining the vector
§6	**g** $\vee$ **b** $\Rightarrow$ **b** $\circ$ b_δ(**bcdg**) $\Rightarrow \delta$.	$\| \{s^*\} \ni \xi_1 \|$.

By way of illustrating the algorithm we present the vectors $\Psi(\xi_0)/\xi_0 \equiv \{w_a, w_b, w_c, w_d\} \equiv W$; $\| \{\alpha\} \ni \xi_0 \|/\alpha \equiv \{w_a, w_b, w_d\}$; $\Psi(\xi_1)$, $\| \{s^*\} \ni \xi_1 \|$, obtaining, if $s \equiv \{w_b, w_d, \psi_1, \psi_2\}$, $/\psi_1, \psi_2 \in s^*$,

$$\Psi(\xi_0) = (11011000)\Psi(\xi_1) = (11111000)$$
$$\| \{\alpha\} \ni \xi_0 \| = (11001000) \ \| \{s^*\} \ni \xi_1 \| = (10100000).$$

In order to obtain the sets $\gamma_0, \gamma_1 \subset \xi_1^{(2)}$ from the sets F_0, F_1 by means of the mappings

$$F_0 \overset{\xi_1}{\longrightarrow} \gamma_0, \qquad F_1 \overset{\xi_1}{\longrightarrow} \gamma_1,$$

it is sufficient to formulate an algorithm by means of which each element $m_i \in X^{(2)}$ has associated with it its transform $\gamma_j \in \xi_1^{(2)}$ with respect to the mapping

$$m_i \overset{\xi_1}{\longrightarrow} \gamma_j.$$

By the definition given in Section 1,C the transform of the element $m_i \in X^{(2)}$ is the subset $\gamma_j \subseteq \xi_1\colon \gamma_j \ni \psi_k \leftrightarrow (m_i \in \psi_k) \wedge (\psi_k \in \xi_1)$. The vector $\| \{m_i\} \in \xi_1 \|$, illustrating this binary relation, gives a complete representation of the set γ_j: $\| \{m_i\} \in \xi_1 \| = \| \{\gamma_j\} \ni \xi_1 \|$. The calculation of the transform γ_j is connected with the transformation of the vector $\| \{m_i\} \ni X \| = \| \{m_i\} \in \xi_0 \|$ to the vector $\| \{m_i\} \in \xi_1 \| = \| \{\gamma_j\} \ni \xi_1 \|$. The transform γ obtained for the set $F \subset X^{(2)}$ corresponds to the transformation of the matrix $\| F \ni X \|$ to the matrix $\| \gamma \ni \xi_1 \|$, where the ith row $\| \{m_i\} \ni X \| = \| \{m_i\} \in \xi_0 \|$ of the matrix $\| F \ni X \|$ is transformed to the form $\| \{m_i\} \in \xi_1 \| = \| \{\gamma_j\} \ni \xi_1 \|$. The set of distinct rows of the transformed matrix compose the matrix $\| \gamma \ni \xi_1 \|$. In the transformation of the vector $\| \{m_i\} \in \xi_0 \|$ to the vector $\| \{m_i\} \in \xi_1 \|$ it is necessary to clarify only the relation of inclusion of the element m_i in the elements of the set $s^* \subset \xi_1$, since the relations between the element m_i and the elements of the set $\beta^* \equiv \xi_1 \backslash s^*$ remain invariant.

It is known that each set $\psi_i \in s^*$ can be given by listing the elements of the partition $J = SEP^\alpha(X)^{(2)}$ contained in it and the fact that the vector $\| \{\psi_i\} \supset J \| = \| \{J_{\psi_i}\} \ni J \|$ for $\delta(\alpha) \leq 3$ is represented by the first eight positions of the row $\| \{v_i\} \ni v \|$ of the matrix $\| V_\alpha' \ni v \|$.

In effect, it is true that $m_i \in \psi_i \in s^* \leftrightarrow m_i \in J_j \in J_{\psi_i}$, and it is easy to determine the interval $J_j \in J$ containing the element m_i. To realize the transformation it is possible to utilize the matrix A, obtained as a result of the transposition of the minor of the matrix $\| V_\alpha' \ni v \|$ formed of the rows corresponding to the set s^* and the first eight columns. The columns of the transposed minor are arranged in positions marked by 1's in the code of the vector $\| \{s^*\} \ni \xi_1 \|$. The rows of the transposed minor are arranged in natural order. This transposition is performed by the (1, 1)-terminal *transmin* $\alpha\beta\gamma\delta//$, whose operands take on the sense $\alpha{::}\| V_\alpha' \ni v \|$; $\beta{::}A$; $\gamma{::}\| \{s^*\} \ni R_\alpha' \|$; $\delta{::}\| \{s^*\} \ni \xi_1 \|$.

$transmin\ \alpha\kappa\ +,\ \beta\kappa,\ \gamma\pi\ +,\ \delta\pi\ +/\mathbf{d},\ b/(\alpha\beta\gamma\delta)$
110 106 4 $(\alpha abcd)$

$\mathbf{a} \Leftarrow \mathbf{037\ 700\ 000\ 000} \circ a$
§1 $\circ\ \beta_a \triangle a \oplus b_\beta \mid\!\!\rightarrow 1\gamma \Rightarrow \mathbf{b}\delta \Rightarrow \mathbf{d}$
§2 $\mathbf{b}\ \dot{\mathbf{X}}\ 4a\alpha_a \wedge \mathbf{a} \Rightarrow \mathbf{cd}\ \dot{\mathbf{X}}\ 4b$
§3 $\mathbf{c}\ \dot{\mathbf{X}}\ 2ac_b \vee \beta_a \Rightarrow \beta_a \rightarrow 3$
§4 .

By way of illustrating the algorithm we present a matrix formed of the first eight columns of the matrix $\|\ V_\alpha' \ni v\ \|$. The rows of the matrix tagged by the asterisks form the minor defined by the vector $\|\ \{s^*\} \ni R_\alpha\ \|$. The vector $\|\ \{s^*\} \ni \xi_1\ \|$ indicates the position of the columns of the transposed minor. The matrix A is the result of operation *transmin.*

$$\|\ V_\alpha' \ni v\ \| = \begin{bmatrix} 0 & 0 & 0 & 0 & 1 & 1 & 1 & 1 \\ 0 & 0 & 1 & 1 & 0 & 0 & 1 & 1 \\ 0 & 1 & 0 & 1 & 0 & 1 & 0 & 1 \\ 0 & 0 & 0 & 0 & 0 & 0 & 1 & 1 \\ 0 & 0 & 0 & 1 & 0 & 1 & 1 & 1 \end{bmatrix} \begin{matrix} \\ \\ \\ * \\ * \end{matrix}$$

$$A = \begin{bmatrix} 0 & 0 & 0 & 0 & 0 \\ 0 & 0 & 0 & 0 & 0 \\ 0 & 0 & 0 & 0 & 0 \\ 0 & 0 & 0 & 1 & 0 \\ 0 & 0 & 0 & 0 & 0 \\ 0 & 0 & 0 & 1 & 0 \\ 0 & 1 & 0 & 1 & 0 \\ 0 & 1 & 0 & 1 & 0 \end{bmatrix}$$

$$\|\ \{s^*\} \ni R_\alpha\ \| = (0\quad 0\quad 0\quad 1\quad 1)$$
$$\|\ \{s^*\} \ni \xi_1\ \| = (0\quad 1\quad 0\quad 1\quad 0)$$

The transformation of the matrix $\|\ F \ni X\ \|$ to the matrix $\|\ F \in \xi_1\ \|$ is realized by the (1, 1)-terminal *transformat* $\alpha\beta\gamma\delta\epsilon\zeta//$, by means of which each element $\alpha_i::\|\ \{m_i\} \ni X\ \|$ of the complex $\alpha::\|\ F \ni X\ \|$ has associated with it an element β_j of the complex $\beta::A$. The index j belongs to the element $J_j \in J$ containing m_i: $m_i \in J_j$. The value of the index j is calculated by using the value of the variable $\mathbf{a} = \alpha_i \wedge \delta/\delta::\|\ \{\alpha\} \ni \xi_0\ \|$. Then the operation $\epsilon \oplus \zeta \wedge \alpha_i \vee \beta_j \Rightarrow \gamma_i/\epsilon::\Psi(\xi_1),\ \gamma::\|\ F \in \xi_1\ \|,\ \zeta::\|\ \{s^*\} \ni \xi_1\ \|$ is performed.

The matrix $\| \gamma \ni \xi_1 \|$ may be obtained from the matrix $\| F \in \xi_1 \|$ by means of the operator *redsim.*

transformat αк +, βк +, γк, δп, ϵп +, ζп +/**b**, $b/(\alpha\beta\gamma\delta\epsilon\zeta)$
111 111 3 (α**ab**)

$\circ\ ab_\alpha \Rightarrow b\gamma$
§1 $\alpha_a \wedge \delta \Rightarrow \mathbf{a}\delta \Rightarrow \mathbf{b} \circ b$
§2 $\mathbf{b}\,\dot{\mathbf{X}}\,3cb + b \Rightarrow bc_c \wedge \mathbf{a} \circ\!\rightarrow 21 \vee b \rightarrow 2$
§3 $\epsilon \oplus \zeta \wedge \alpha_a \vee \beta_b \Rightarrow \gamma_a \triangle a \oplus b_\alpha \rightarrow 1.$

By means of these operators it is possible to represent the (1, 2)-terminal *transform,* which calculates the sets $\gamma_0, \gamma_1 \subset \xi_1^{(2)}$, defining the form of the Boolean function $g_1(\xi_1)$ in (2). The operator *transform* $\alpha\beta\gamma\delta\epsilon\zeta\eta\vartheta\kappa//$ transforms the input information, represented by the complexes $\beta::\| F_0 \ni X \|$; $\gamma::\| F_1 \ni X \|$; $\delta::\| V_\alpha' \ni v \|$; ϵ: $\epsilon_0::\Psi(\xi_0)$; $\epsilon_1::\| \{\alpha\} \ni \xi_0 \|$; $\epsilon_2::\| \{s\} \ni R_\alpha' \|$. The result of the operator is represented in the complexes $\zeta::\| \gamma_0 \ni \xi_1 \|$; $\eta::\| \gamma_1 \ni \xi_1 \|$ ϑ: $\vartheta_0::\Psi(\xi_1)$; $\vartheta_1::\| \{\alpha\} \ni \xi_0 \|$; $\vartheta_2::\| \{s^*\} \ni R_\alpha' \|$; $\vartheta_3::\| \{s^*\} \ni \xi_1 \|$. The complex κ serves to represent the matrix A, and α is an auxiliary exit terminal of the operator, attained when $s \equiv \alpha$.

transform (107 110 111 40)/αч, βк +, γк +, δк + ϵк, ζк, ηк, ϑк
κк, /**d**, /$(\beta\gamma\delta\epsilon\zeta\eta\vartheta\kappa)$
112 75 0 (β**abcd**)

pack $\alpha\beta\epsilon\vartheta//\vartheta \Rightarrow$ (**abcd**) *transmin* $\delta\kappa$**cd**//
transformat $\beta\kappa\zeta$**bad**//*redsim* $\zeta\zeta$//*transformat* $\gamma\kappa\eta$**bad**//
redsim $\eta\eta$//.

F. ALGORITHM FOR CHOOSING NEXT STEP

We shall assume that the current approximation to the optimal solution for relation (1) has the form $\xi \equiv \xi'' \cup \xi'$, where ξ'' is the set of secondary elements of the solution (cf. Section 2A), and ξ' is the set of primary elements of the solution. Also, let the Boolean function $g(\xi)$ in (1) be given by the sets $\gamma_0, \gamma_1 \subset \xi^{(2)}$. For further calculations it is necessary to select the set $\alpha \subseteq \xi'$. In the present form of the program for the multistep algorithm the cardinality of the set α, taken at each step, is limited to $2 \leq \sigma(\alpha) \leq 3$, and therefore the problem of optimal choice of the set α arises only if $\sigma(\xi') > 3$. The quality of the current selection determines the quality of the sequence of selections and, in the final analysis, determines the degree of approximation of the final solution for relation (1) to the optimal solution.

In the practice of synthesis using the multistep algorithm, satisfactory results were obtained using the selection algorithm given below.

Following the terminology introduced in Section 1D, we associate with each element $\psi_i \in \xi$ a cutset u_i, realized in the graph $G_g = (\xi^{(2)}, u_g)$, corresponding to the function $g(\xi)$ in (1). The set of branches of the graph G_g contained only in the cutsets in $\{u_i/\psi_i \in \xi'\}$ is denoted by Z: $Z \equiv u_g \backslash \bigcup [\{u_i/\psi_i \in \xi''\}]$. The branches belonging to the set Z are removed from the graph G_g in the realization of all cutsets $u_i/\psi_i \in \xi'$, where the contribution of the individual cutset u_i is characterized by the set $z_i \equiv Z \cap u_i$.

We call the value taken by the numerical function

$$c(\psi_l) = \sum_{u_{ij} \epsilon z_l} [\sigma(\xi') - d_{ij} + 1] \tag{4}$$

the cost of the element $\psi_i \in \xi'$.

The formation of the set α, $\sigma(\alpha) = 3$ is carried out in three steps. In the ith step the element ψ_l having the greatest cost $c(\psi_l)$ is taken from the set $\xi' \backslash \alpha_i$ (α_i is the current value of the set α); cost is calculated by (4) in which the set z_l is defined in the following way:

$$z_l \equiv Z_l \cap u_l, \qquad Z_l \equiv u_g \backslash \bigcup [\{u_l/\psi_l \in \xi'' \cup \alpha_i\}].$$

If several elements in $\xi' \backslash \alpha_i$ have the greatest cost, one of them is taken at random and included in the set α_i. In a number of cases in the construction of the set α it is necessary to ensure satisfaction of the condition $b \cap \alpha \equiv \phi/b \subset \xi'$. For example, if the choice of set α, carried out in the preceding step, leads to no result in the sense that a set $\min_\varphi \rho_\alpha \equiv \{s\}/s \equiv \alpha$ was obtained in this step, then in the next step, in order to avoid repeating the choice already made, it is necessary to form the set b, the choice of whose elements is blocked. It is sufficient to include in the set b one of the elements of the set α, corresponding to the unsuccessful step. The choice of next step is obtained by the (1, 1)-terminal *sestep* $\alpha\beta\gamma\delta\epsilon\zeta\eta//$, using for the calculation the information represented in the complexes $\alpha::|| \gamma_0 \ni \xi ||$; $\beta::|| \gamma_1 \ni \xi ||$ and the variables $\delta::\Psi(\xi)$; $\epsilon::|| \{\xi''\} \ni \xi ||$; $\zeta::|| \{b\} \ni \xi ||$. The complex γ is used to represent the table of element costs in ξ'. The result of the calculation is represented by the variable $\eta::|| \{\alpha\} \ni \xi ||$.

sestep $\alpha\text{к} +, \beta\text{к} +, \gamma\text{к}, \delta\text{п}, \epsilon\text{п} +, \zeta\text{п}, \eta\text{п}/\mathbf{e}, c/(\alpha\beta\gamma\delta\epsilon\zeta\eta)$
113 257 007 (α**abcde**)

	$\circ\ b\ \circ\ \eta\delta \oplus \epsilon \nabla + 1 \Rightarrow \mathbf{a}\ \circ\ \mathbf{b},$	Initialization
§1	$\circ\ \gamma_b \triangle b - b_\gamma\ \circ\!\rightarrow 1\ \circ\ \mathbf{a}\ \circ\ b\epsilon \vee \zeta \vee \eta \Rightarrow \mathbf{c}$	
§2	$\alpha_b \oplus \beta_a \Rightarrow \mathbf{d} \wedge \mathbf{c} \mid\!\rightarrow 4\mathbf{d} \nabla \Rightarrow \mathbf{ea} - \mathbf{e} \Rightarrow \mathbf{e}$	Construction of table of costs of elements
§3	$\mathbf{d}\ \dot{\mathbf{X}}\ 4c\mathbf{e} + \gamma_c \Rightarrow \gamma_c \rightarrow 3$	
§4	$\triangle a - b_\beta\ \circ\!\rightarrow 2\ \circ\ a \triangle b - b_\alpha\ \circ\!\rightarrow 2\delta \oplus \mathbf{c} \Rightarrow \mathbf{c}\ \circ\ \mathbf{d}$	
§5	$\mathbf{c}\ \dot{\mathbf{X}}\ 7a\gamma_a - \mathbf{d}\ \circ\!\rightarrow 5\gamma_a \oplus \mathbf{d} \mid\!\rightarrow 6c_a \vee \mathbf{e} \Rightarrow \mathbf{e} \rightarrow 5$	Finding the set of elements with maximal cost
§6	$\gamma_a \Rightarrow \mathbf{d}c_a \Rightarrow \mathbf{e} \rightarrow 5$	
§7	$\mathbf{e}\ \ddot{\mathbf{X}}\ 1ac_a \vee \eta \Rightarrow \eta\ \circ\ b \triangle \mathbf{b} \oplus 3 \mid\!\rightarrow 1c_a \vee \zeta \Rightarrow \zeta.$	Obtaining sets α, b

G. OUTPUT OF INFORMATION ON THE CIRCUIT STRUCTURE

The information on the structure of the synthesized circuit is given out in parts after the completion of each calculation step. We shall assume that as a result of execution of the current step an optimal set $s \in \min_\varphi \rho_\alpha$ has been obtained, defining the structure of a certain part of the synthesized stage. The information about this subcircuit is packed in $k = \sigma(s^*)/s^* \equiv s\backslash\alpha$ simple variables, and printed or put out by punching. For each element $\psi_i \in s^*$ a variable is formed, in whose first eight bits is included information on the form of the Boolean function realized by the circuit element corresponding to ψ_i. Three portions of six bits in each are used to represent the numbers n_i, n_j, n_k of the terminals in the preceding stage to which the circuit element ψ_i is connected, and the last six bits of the variable are used to represent the number N_i, assigned to the described element in the synthesized stage. The information about the circuit stage is concluded by the list of numbers of the elements operating as delay lines, for which the number of the terminal in the preceding stage to which each element is connected is indicated.

For example, the resulting sequences corresponding to the circuit shown at the left in Fig. 5a have the form

0 2 7 0 1 0 2 0 3 0 1	3 0 0 0 1 0 2 0 0 0 1
0 0 3 0 4 0 5 0 0 0 2	0 5 3 0 3 0 4 0 5 0 2
0 5 3 0 1 0 2 0 3 0 4	(second stage)
0 6 0 0 4 0 5 0 0 0 5	
0 0 0 0 0 0 0 0 0 0 3	0 6 0 0 1 0 2 0 0 0 1
(first stage)	(third stage)

The terminals corresponding to the input variables *a*, *b*, *c*, *d*, *e* and the constant 0 are numbered 1, 2, 3, 4, 5, 0, respectively. The output of the information about the circuit structure is performed by the (2, 2)-terminal *linst* $\alpha\beta\gamma\delta//$, whose auxiliary input terminal corresponds to sentence 5. The initial information is given by the values of the complex $\beta::\|\ V_\alpha' \ni v\ \|$; the complex $\delta: \delta_0::\Psi(\xi_1); \delta \equiv \|\ \{\alpha\} \ni \xi_0 \|; \delta_2 \equiv \|\ \{s^*\} \ni R_\alpha' \|; \delta_3::\|\ \{s^*\} \ni \xi_1\ \|$, and the variable $\gamma::\|\ \{\xi''\} \ni \xi_1\ \|$.

linst αч, βк +, γп +, δк +/**f**, *b*/(βγδ)
114 165 7 (β**abcdef**)

$\delta \Rightarrow$ **(abcd)a** $\Leftarrow$ **037 700 000 000**
§1 ○ **e** ○ *b***b** ⇒ **f**
§2 **f** $\dot{\mathbf{X}}$ 3*aa* + 1 ∨ **e** ⇒ **ee** < 6 ⇒ **e** △ *b* → 2
§3 *b* ⊕ 3 ○→ 4**e** < 6 ⇒ **e**
§4 **d** $\dot{\mathbf{X}}$ *αaa* + 1 ∨ **e** ⇒ **ec** $\dot{\mathbf{X}}$ *αa***a** ∧ β_a ∨ **e**∗ → 1
§5 δ_0 ⇒ **aa** ⊕ γ ⇒ **a**
§6 **a** $\dot{\mathbf{X}}$ 7*aa*∗ → 6
§7 ○ **a**∗.

H. PROGRAM FOR THE MULTISTEP ALGORITHM

All the calculations connected with the realization of the Boolean function $f(X)$, given by the sets $F_0, F_1 \subset X^{(2)}$, by a logical circuit of majority elements are executed by the operator *synmac* $\alpha\beta\gamma\delta\epsilon\zeta\eta\vartheta\kappa\lambda\mu\nu\xi//$, which processes the initial information represented in the complexes $\alpha::\|\ F_0 \ni X\ \|$; $\beta::\|\ F_1 \ni X\ \|$; $\gamma::\|\ V \ni v\ \|$; $\delta:\delta_0::\Psi(X)$. The complexes ϵ, ζ, η, ϑ, κ, λ, μ, ν, ξ serve to represent the intermediate information. The result of the calculation, in the form of a sequence of codes bearing the information about the circuit structure, is printed during the operation of *linst*. During the *i*th step of the construction of a solution for superposition (1), the values of the internal variables of the operator *synmac* take on the following sense: **a**$::\|\ \{\xi''\} \ni \xi_{i-1}\ \|$; **b**$::\|\ \{b\} \ni \xi_{i-1}\ \|$; **c**$::\Psi(\xi_{i-1})$; **d**$::\|\ \{\alpha\} \ni \xi_{i-1}\ \|$; **e**$::\|\ \{s\} \ni R_\alpha'\ \|$; **f**$::\|\ \{s^*\} \ni \xi_i\ \|$.

synmac (113 106 112 100 60 114) ακ +, βκ +, γκ +, δκ, εκ, ζκ, ηκ, ϑκ, κκ, λκ, μκ?, νκ, ξκ/**m**, *a*/
115 360 10

§0 ○ **a** ○ **b**δ_0 ⇒ **c**
§1 **c** ⊕ **a** ⊕ **b** ⇒ **d** ∇ ⇒ **g** − 3 ○→ 2**g** ⊕ 3 ○→ 23 ⇒ **g** *sestep* αβε**cabd**//
§2 **g** − 2 ○→ *linst* 5 23 ⇒ b_γ**g** ⊕ 2 |→ 3 7 ⇒ b_γ
§3 *calminco* γηεζμαβ**d**// $\bar{\circ}$ *a* $\bar{\circ}$ **h** $\bar{\circ}$ **i**

§4 $\triangle\ a \oplus b_{\mu} \circ\!\!\rightarrow 7\,\mu_{a} \Rightarrow \mathbf{e}(\mathbf{cde}) \Rightarrow \delta\ \mathit{transform}\ 6\alpha\beta\eta\delta\epsilon\zeta\nu\lambda//\nu_{0} \Rightarrow \mathbf{m}$
$\mathit{estimate}\ \epsilon\zeta\mathbf{mkl}//\mathbf{h} - \mathbf{k} \circ\!\!\rightarrow 4\mathbf{h} \oplus \mathbf{k} \mapsto 5\mathbf{i} - \mathbf{1} \circ\!\!\rightarrow 4$
§5 $\mathit{transfer}\ \epsilon\vartheta//\mathit{transfer}\ \zeta\kappa//4 \Rightarrow b_{\nu}\ \mathit{transfer}\ \nu\xi//\mathbf{k} \Rightarrow \mathbf{hl} \Rightarrow \mathbf{i} \rightarrow 4$
§6 $b_{\mu} \oplus 1 \mapsto 4 \rightarrow 1$
§7 $\xi_{0} \Rightarrow \mathbf{c}\xi_{3} \vee \mathbf{a} \Rightarrow \mathbf{a} \circ \mathbf{b}$
$\mathit{transfer}\ \kappa\beta//\mathit{transfer}\ \vartheta\alpha//$
§10 $\mathit{linst}\ 1\eta\mathbf{a}\vartheta//b_{\alpha} \oplus 1 \mapsto 0b_{\beta} \oplus 1 \mapsto 0.$

Each repetition of the external cycle of the program, beginning with sentence 0 and terminating after execution of sentence 10, is connected with the synthesis of the next stage of the circuit. The synthesis of each stage is terminated after a certain number of repetitions of the inner loop, whose start is found in sentence 1 and the end in sentence 10. Each repetition of the inner loop is connected with finding the current approximation to the optimal solution for superposition (1) and terminates with the procedure for choice of next step (sentences 1, 2). Then the set $\min_{\varphi} \rho_{\alpha}$ is found, from which is selected the optimal solution (sentences 3–6). Sentences 7 and 10 organize the output of information about the structure of the next part of the synthesized stage and prepare the next repetition of the inner loop. To synthesize a circuit consisting of m elements, about $m/3$ repetitions of the inner loop are required.

It is fairly difficult to estimate the time needed for the realization of a Boolean function $f(X)$. For $\sigma(F_0)$, $\sigma(F_1) \leq 64$ and speed of the order of 20–30 thousand elementary operations per second, the execution time does not exceed two minutes.

3. CONCLUSION

The final judgment on the quality of the algorithm can be passed only after it has been tested in practical synthesis. However, the comparison with the results of known methods that has been carried out leads to a completely encouraging conclusion. For example, at the left in Fig. 5a is the logical circuit obtained by the multistep algorithm, modeling the operation of a five-input majority element. At the right is the equivalent circuit obtained in [5] and synthesized by the addition of delay elements. Figure 5b gives the analogous comparison for the second example solved in [5], and Fig. 5c gives a comparison with a result obtained by the combined method of [2].

The basic advantage of the algorithm is that by its use it is possible to synthesize circuits without applying other methods. In the process of

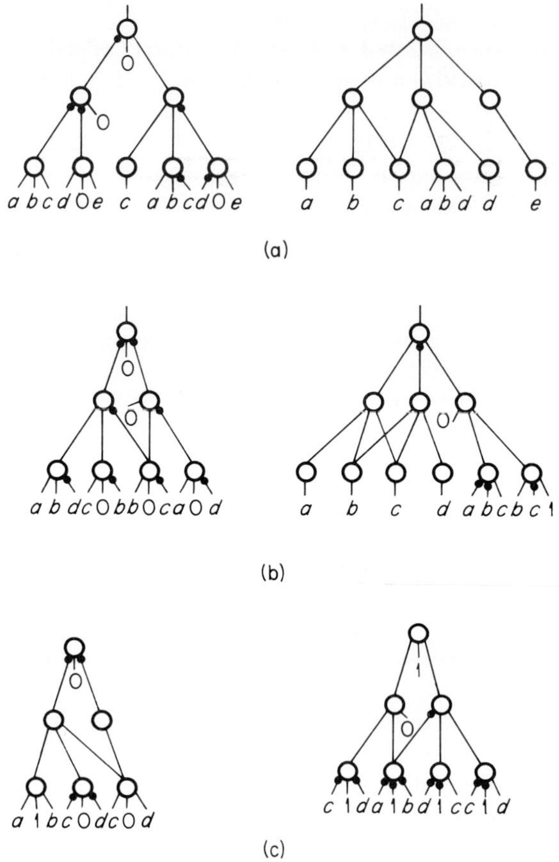

FIG. 5.

calculation it is possible to vary the operating regime of the individual operators of the algorithm according to the needs of the synthesis. In particular, it is possible to obtain a circuit of any desired configuration and to satisfy various requirements of a technical character, introducing definite constraints on the choice of next step. The algorithm can be modernized [sic!], perfecting the algorithms for choice of optimal solution and choice of next step.

The basic results obtained in the article can be utilized in the synthesis of circuits in any other basis of logical elements with small number (2–5) of inputs. For example, in the synthesis of circuits from elements that perform the operations "$\wedge$," "$\vee$," and inversion, it is possible to apply this algorithm by removing the last four rows in Table IV. To pass to a basis including the operation "⊕" it will be necessary to change the algorithms for choice of optimal solutions and choice of next step.

On the whole it can be assumed that the development of multistep algorithms for the synthesis of logical circuits will permit good solutions to be obtained with relatively low outlays of machine time.

REFERENCES

1. Cohn, M., and Lindaman, R., Axiomatic Majority-Decision Logic, *IRE Trans., Electron. Computers* **10**, 17–21 (1961).
2. Varshavskii, V. I., and Rozenblyum, L. Ya., The Minimization of Majority-Element Pyramids, *Izv. Akad. Nauk SSSR, Tekhn. Kibern.* No. 3, 24–29 (1964).
3. Zakrevskii, A. D., On the Synthesis of Majority-Element Circuits, *Tr. SFTI* **44,** 17–23 (1964).
4. Pavlov, V. L., Algorithms for the Functional Decomposition of Boolean Functions, *Tr. SFTI* **48,** 138–149 (1966).
5. Miller, H., and Winder, R., Majority-Logic Synthesis by Geometric Methods. *IRE Trans. Electron. Computers* **11,** 89 (1962).
6. Berge, C., "Theory of Graphs and Its Application" [Russian translation]. IL, Moscow, 1962.
7. Zakrevskii, A. D., Programming Boolean Computations. This volume.
8. Zakrevskii, A. D., Optimal Coverage of Sets. This volume.

PART II: APPLICATIONS

Section C: INVESTIGATION OF AUTOMATA

SIMULATION OF SWITCHING CIRCUITS

A. A. Utkin

1. INVESTIGATION OF AUTOMATA ON GENERAL-PURPOSE COMPUTERS

At the present time the theory of discrete automata synthesis has achieved great success. In particular, automated synthesis has appeared, reflecting a tendency to approach theory with practice.

The theory and practice of the analysis of discrete automata are in a different state. Although there can be no doubt of progress in this field, as a whole it develops more slowly than the theory and practice of synthesis. One possible explanation of this lag is that the analysis problem could not, apparently, serve as such a fruitful source of theoretically deep and practically important problems as synthesis. A less general consideration is the following. The classical form of presentation of the results of analysis—the flow table of a discrete automaton— is in essence useless (from the practical viewpoint) as soon as a complicated automaton having large internal memory capacity is involved. It is difficult to imagine how to use the flow table of an automaton having, say, 30 bits of memory, a table containing more than 10^9 columns! And such a memory is nothing compared to the memories of many modern automata.

At the same time, the investigation of existing computers and, in particular, those under development is an urgent problem, if only from the economic aspect. The development of a physical model of a complex device and the execution of experiments, having as their aim to establish the completeness with which intention has been embodied in reality, is expensive, requiring large outlays of means, time, and the work of highly qualified personnel. Even before the construction of such a model it is necessary to make sure of the absence of errors in the circuit of the device, to estimate its reliability, etc. Until now this laborious work has been, as a rule, carried out manually. The representation of the structure of the device in computer memory and the simulation of its operation by a suitable program not only permits the design time to be reduced, but makes it possible to obtain more complete test results, more accurate reliability estimates, etc. References [1–10] are devoted to the discussion of computer application.

The following remark must be made. Although it is obvious that in principle it is possible to represent the structure and simulate the operation of any automation by means of computer, if only the ratio of automaton memory capacity to computer memory capacity is not too great, the practical realization of this possibility encounters two great difficulties. The first is connected with the enormous volume of technical work on the development of computer programs, if simulation of a complicated automaton is concerned, the second with the outlay of a large quantity of machine time for simulation, since the processing of the input information in a computer is not developed in space-time, as in the physical model, but *almost exclusively in time.* This difficulty is aggravated further by the fact that with a large volume of information contained in the structure of the simulated automaton, frequent access to external computer storage is unavoidable, as a result of which the speed of the machine drops sharply.

A way to overcome the first of these difficulties exists at the present time. It consists in the utilization of already developed automatic programming systems, particular special-purpose systems, whose objects are the problems of discrete automata theory.

Concerning the second, more objective difficulty, it must be considered to be a certain quantitative constraint imposed on the complexity of the automata that can be investigated in a given stage of development of simulation theory and for a given state of the simulating means. As one and the other progresses, this constraint will weaken more and more.

An example of one of the most immediate practical problems in the simulation of discrete automata is the development of a system for representing the structure of a definite class of automata in computer memory and the choice of subroutines serving as models of its elements—the foundations on which the methodology and practice of simulation can be developed.

The present chapter is an attempt to solve this problem for a restricted class of automata—switching circuits consisting of elements with a single output terminal. The difference between the model of discrete automaton proposed here and the analogous models considered in [1, 4, 8, 9] consists, in particular, in the form of description of the structure of the simulated circuit, as well as in the orientation toward a broader program of utilization of models than the comparison of the output sequence obtained with that prescribed. The chapter must be considered as a first attempt at the application of LYaPAS to this problem.

2. REPRESENTATION OF SWITCHING-CIRCUIT STRUCTURE

A. CIRCUITS MODELED

We shall consider switching circuits satisfying the following conditions:

(1) All elements of the circuit are $(k, 1)$-terminal, the states of whose terminals are described by binary variables.

We shall also consider certain (k, p)-terminal circuits, introducing them, however, in the framework of condition (1) by a fictitious expansion of each (k, p)-terminal into p individual $(k, 1)$-terminals. For example, we shall represent the flipflop, having two output terminals—"one" and "zero"—as a set of two elements, a flipflop 1 and a flipflop 0, each of which has a single output terminal ("one" and "zero," respectively).

(2) The elements are divided into two types. The *first type* includes elements in which the time delay of the signal can be neglected. The elements of the first type can be both Boolean $(k, 1)$-terminals whose output state at a given instant is defined only by the states of its input terminals at the same moment, and $(k, 1)$-terminals not satisfying this condition (for example, the flipflop).

The sole function of the elements of the *second type* is to provide signal delay for a certain time interval, which is the same for all such elements, and which will be called the *clock cycle* [or, briefly, *cycle*].

B. BASIC AND AUXILIARY COMPLEXES

We now pass to the construction of the system representing the structure of the circuits in the described class in LYaPAS.

We include, by convention, among the elements of the circuit its input terminals, considering each of them as a $(0, 1)$-terminal, generating the input signal. We shall refer to the output terminals of its elements as circuit *nodes*, retaining the designation *terminal* for those that are input or output for the circuit as a whole.

We number the circuit nodes in arbitrary order, using the natural numbers from 1 to n. The circuit elements will also be numbered by this sequence of numbers. We introduce a fictitious node with number 0, which we shall consider to be the input terminal of the circuit, which is always found in the state 0.

Information on the concrete form of the ith circuit element will be represented by the element $\mathbf{a}_i$ of the *basic complex* **A**. We take the dimension of elements in complex **A** to be equal to 1. The majority of forms of circuit elements (delay element, inverter, conjunction, etc.) will be completely represented by a single element of the basic complex. To represent the others [threshold element and element realizing a Boolean function given in disjunctive normal form (dnf) or in perfect disjunctive normal form (pdnf)], a single element of the basic complex will not suffice. In this case the additional information will be constituted by one or several elements of the *auxiliary complexes* **B**, **C**, **D**. The concrete methods of coding the various forms of information will be considered below.

The element models will be subroutines that define the state of the output terminal of the element from the states of its input terminals. These subroutines will also be considered below.

C. CONNECTION COMPLEXES

We now pass to the question of representation of the connections among the circuit elements.

A convenient form of such representation would appear to be a square binary matrix of dimensions $(n + 1)^2$, whose columns correspond to the circuit nodes and rows to the circuit elements, and whose element at the intersection of the ith column and jth row takes the value 1 if and only if the node i is connected to one of the input terminals of the element j. We abandon this form for two reasons. First, in the presence of a large number of nodes n in the simulated circuit, and the small average number of element input terminals compared to it, such a representation is very sparse. Second, such a matrix completely represents the connection only for those circuit elements that realize symmetric functions; for elements that realize asymmetric functions it is still necessary to number their input terminals, which makes the system of representation still sparser.

We shall represent the connections of the input terminals of each element with the circuit nodes by enumerating the numbers of the nodes that are connected with these terminals. For elements that realize asymmetric functions, this enumeration will be realized in ascending order of the input terminals of the element, and this will be taken into account in the con-

TABLE I

Form of element	Code
Input terminal of circuit	00
Delay element	01
Inverter	02
Flipflop 1	05
Flipflop 0	06
Disjunction element	07
Conjunction element	10
Adder mod 2	11
Majority element	12
Threshold element	13
Element realizing a Boolean function given in dnf	14
Element realizing a Boolean function given in pdnf	15

struction of the models of the elements. We shall represent the information about the connections by the elements of the *connection complexes* **E**, **F**, **G** of dimensions 1, 2 and 3, respectively.

Let us introduce the upper limit N on the number of nodes n in the simulated circuit, and the upper limit K on the number of input terminals of an element, since it is just these limits that define the dimensions and number of complexes introduced. We shall take $N = 2^9 - 1$ and $K = 9$, so that the connections of an element having 2 or 3 input terminals will be represented by a single element of the complex **E**, and the connections of elements having 4–6 and 7–9 input terminals will be represented by elements of the complexes **F** and **G**, respectively. The number of the unique node connected with the input to a (1, 1)-terminal will be represented by the corresponding element of the basic complex **A**.

D. STRUCTURE OF THE ELEMENTS OF COMPLEXES

We shall consider the following twelve forms of circuit elements, using for the coding of each form a two-place octal number (Table I).

A certain informational redundancy of the code (6 bits in place of the necessary 4) is introduced both for considerations of the convenience of octal coding and the desire to reserve the possibility of extension of the set of elements considered.

We define the structure of the elements of the basic complex **A** in the following way. We assign the extreme right group of components 26–31 to the storage of the code K of the circuit element, while we put the value of

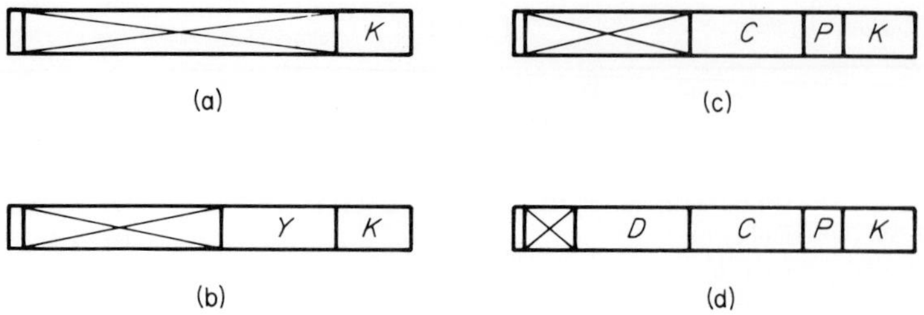

FIG. 1.

the zero component $\mathbf{a}_i^0$ equal to one if the node i is an *output terminal of the circuit,* and zero otherwise. The remaining components are utilized to represent various forms of elements in different ways.

To represent the input terminal of the circuit (the element with code 00) these components are not used at all, and the structure of the element of the basic complex has the form shown in Fig. 1a. Groups of unused components will be marked by the oblique cross; in real coding the unused components will be filled by zeros.

To represent the elements 01 and 02 (delay and inversion) the second group from the right, consisting of the nine components 17–25, will be used to store the number Y of the node connected to the input of the element. In this case the structure of the basic complex has the form shown in Fig. 1b.

To represent the elements 05, 06, ..., 12 the structure has the form shown in Fig. 1c: the value of the three components 23–25 is defined by the dimensions P of the coupling complex (and, consequently, prescribes a complex in the set $\{\mathbf{E}, \mathbf{F}, \mathbf{G}\}$), while the components 14–22 are used to store the number C of elements of this complex containing the information on the couplings of the element.

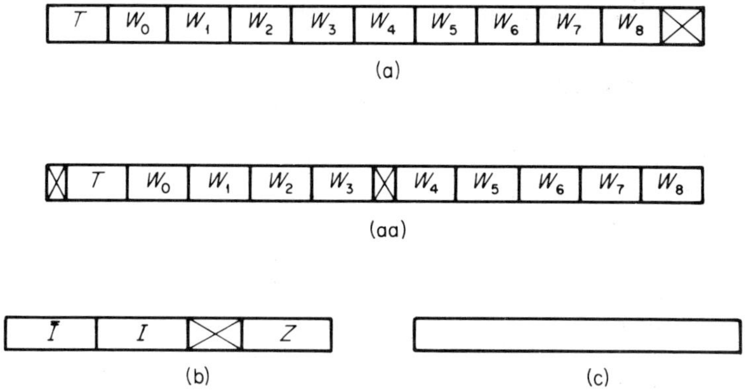

FIG. 2.

Finally, to represent the elements 13–15 the structure has the form shown in Fig. 1d. It is analogous to the form of Fig. 1c, except that components 5–13 are used to store the number D of elements in the auxiliary complex.

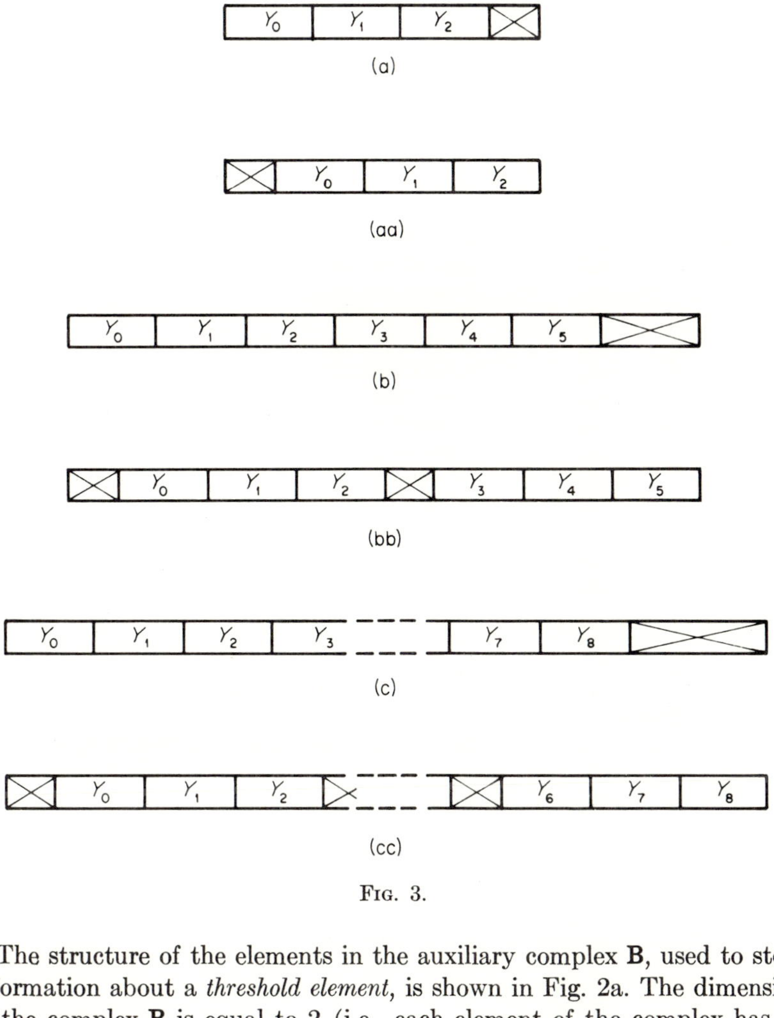

FIG. 3.

The structure of the elements in the auxiliary complex **B**, used to store information about a *threshold element,* is shown in Fig. 2a. The dimension of the complex **B** is equal to 2 (i.e., each element of the complex has 64 components); the six extreme left components represent the value T of the threshold, while the next nine groups of six components represent the values of the weights $W_0, W_1, \ldots, W_8$ of the threshold element inputs.

The complex **C** (dimension 1) is used to represent Boolean functions in *disjunctive normal form.* The structure of its elements is shown in Fig. 2b.

TABLE II

Complex	Elements of complexes	Value of elements of complexes
A	a_0	00···0
	a_1	00···0
	a_2	00···0
	a_3	00···0
	a_4	00···0
	a_5	00···0
	a_6	00···0
	a_7	00···0
	a_{10}	000000 016 01
	a_{11}	200000 022 01
	a_{12}	000000 015 02
	a_{13}	000000 017 02
	a_{14}	00000 000 1 05
	a_{15}	00000 000 1 07
	a_{16}	00000 002 1 10
	a_{17}	00 000 003 1 13
	a_{20}	20 001 004 1 13
	a_{21}	20 002 000 2 13
	a_{22}	20 000 005 1 15
	a_{23}	00 000 001 2 14
B	b_0	003 02 01 02 00 00···0
	b_1	005 04 01 01 00 00···0
	b_2	007 03 03 04 07 00···0
C	c_0	20010000023 (in binary code: 100000000010···)
	c_1	16000000023 (in binary code: 11100···000···)
	c_2	00···0
D	d_0	2010···0 (in binary code: 1000000100···)
E	e_0	00 011 000 013
	e_1	00 004 005 006
	e_2	00 001 002 012
	e_3	00 005 006 007
	e_4	00 001 010 023
	e_5	00 021 014 017
F	f_0	00 020 010 023 00 015 00···0
	f_1	00 010 002 003 00 004 00···0

The two left groups of nine components (0–8 and 9–17) represent an elementary conjunction; the first group represents the set I of arguments occurring in the conjunction without inversion, the second, the set $\bar{I}$ of arguments occurring with inversion. The extreme right group of components 23–31 represents the number Z of the simulated circuit element that realizes the Boolean function whose disjunctive term is the given conjunction. Thus a dnf is represented by several elements of the complex

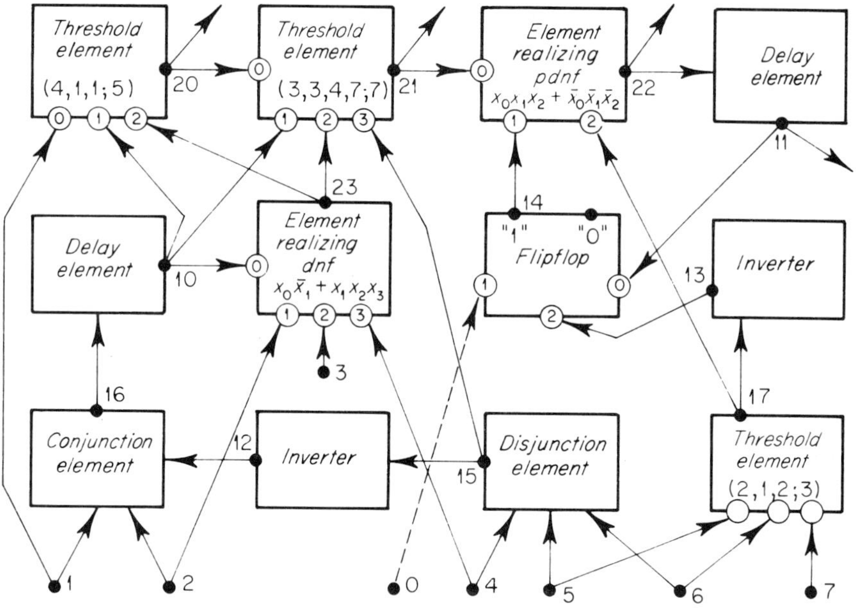

FIG. 4.

C in succession, containing the same code in their extreme right groups of components. The number of the first of these elements is written as D in the corresponding element of the basic complex (Fig. 1d).

We shall terminate the complex **C** by a zero element to eliminate the risk of erroneous inclusion of the next variable in memory in the given dnf.

To represent a Boolean function given in pdnf complex **D** of dimension 1 is used. The structure of its elements is homogeneous (Fig. 2c). The $2^{\pi(m-5)}$ elements in succession of this complex represent a Boolean function of m arguments ($\pi(x) = 0$ if $x \leq 0$, and $\pi(x) = x$ if $x \geq 0$). The number of the first of these elements is written as D in the corresponding element of the basic complex (Fig. 1d). The set of 2^m left components of these elements is put into one-to-one correspondence with the set of constituents of the expansion of unity. The component with number $i = \sum_{j=0}^{m} 2^j \xi_j$ corresponds to the constituent $x_0^{\xi_0} x_1^{\xi_1} \cdots x_m^{\xi_m}$, where $x_j^{\xi_j} = x_j$ if $\xi_j = 1$, and $x_j^{\xi_j} = \bar{x}_j$ if $\xi_j = 0$. The value of the components is equal to 1 if the corresponding constituent occurs in the pdnf and is equal to 0 otherwise.

Finally, let us consider the structure of the elements of the connection complexes **E**, **F**, **G** (Fig. 3a,b,c). Each element of these complexes contains several groups of nine components (3, 6, or 9, respectively), representing the numbers $Y_0, Y_1, \ldots$ of the nodes connected to the input terminals of the element. The groups are formed of successive components, beginning with the extreme left (0–8, 9–17, etc.).

The following remark must be made. The described structure of the connection-complex elements is used within the machine, since it permits a fairly simple programmed realization of the models of the elements. However, this form is not convenient for coding outside the machine, owing to the disagreement between the limits of the groups of components and the limits of binary triplets coded by octal digits. Therefore, the form shown in Fig. 3aa,bb,cc is used outside the machine. For the elements of the auxiliary complex **B** the structure shown in Fig. 2aa is used outside the machine. The adjustment of the data structures from aa, bb, and cc to the forms a, b, and c is carried out by the program.

Example: To conclude, we illustrate this system of switching-circuit representation by a simple example. Figure 4 gives a circuit, and Table II gives the values of the elements of the complexes **A**, **B**, ..., **F** representing this circuit.

3. MODELS OF THE ELEMENTS

A. PROCESSING OF THE INPUT SIGNALS

In the simulation of the operation of a switching circuit, the processing of the signals arriving at its inputs at the start of a cycle is developed in time. One elementary act of this process is the determination of the new state to which an output terminal of an individual element passes as a result of the change of state of the circuit nodes directly connected to the one under consideration. We shall carry out this process in two stages.

In the first stage we shall successively examine the elements of the first type (not having time delay), bringing their output terminals to the new states. It is natural that the sequence of examination must correspond to the propagation of the signals in the structure of the real circuit. We call such a sequence *permissible*. An exact formulation of the problem of finding a permissible sequence and its solution are considered in the next section.

In the second stage we consider the elements of the second type (delay elements), assuming that changes of state of their output terminals occur at the end of the cycle and that they therefore cause changes of state of the corresponding elements of the first type only at the start of the next cycle.

We shall represent the total state of the simulated circuit by the variable β, whose ith component β^i is equal to unity if the node i is in the state 1, and equal to zero if the node is in the state 0 ($i = 0, 1, \ldots, n$). The described elementary simulation acts are performed by the subroutines presented below, by means of which the values of the individual components of the variable β are changed, and which thus serve as the models of the circuit elements. The external operand of these subroutines is, aside from the variable β, the index α, prescribing the number $[\alpha]$ of that node of the circuit whose state is defined by access to the model of the element.

An inverter is modeled by the operator *inverter*, realized by the subroutine

$$\textit{inverter}\ \alpha\text{и}\ +,\ \beta\text{п}/-,\ a.$$

130 42 1

$$c_\alpha \vee \beta \Rightarrow \beta c_0 \urcorner \wedge \mathbf{a}_\alpha > 6 \Rightarrow a$$
$$\beta \vee c_a \circ\!\!\rightarrow 1 c_\alpha \oplus \beta \Rightarrow \beta$$

§1 .

This operator extracts from the element $\mathbf{a}_\alpha$ of the basic complex the number of the node connected to the input of the simulated element ($c_0 \urcorner \wedge \mathbf{a}_\alpha > 6 \Rightarrow a$), tests the state of this node ($\beta \wedge c_a \circ\!\!\rightarrow$), and, depending on the result of the test, assigns the value one or zero to the component $\beta^{[\alpha]}$ of the variable β ($c_\alpha \vee \beta \Rightarrow \beta$ or $c_\alpha \vee \beta \Rightarrow \beta \cdots c_\alpha \oplus \beta \Rightarrow \beta$).

A flipflop is modeled in both variants (flipflop 1 and flipflop 0) by a single operator *flipflop*. Flipflop 1 is "reduced" to flipflop 0 by interchange of the "one" and "zero" inputs. It is assumed that aside from these two inputs the flipflop can have one further counting input, to which is given the number 2. The "zero" and "one" inputs are numbered 0 and 1, respectively. If in the simulated circuit any of the flipflop inputs is not used, the corresponding group of components in the connection-complex element representing the connections of this flipflop is assigned the value zero. This input thereby is assumed to be connected normally to the fictitious input terminal of the circuit, which is always in state 0.

flipflop αи $+$, βп$/-$, d.
132 144 4

$c_0 \urcorner \wedge \mathbf{a}_\alpha > 11 \Rightarrow af_{10} + c_{26} \Rightarrow c \Rightarrow d\mathbf{e}_a > 27 \Rightarrow b$
$\mathbf{e}_a > 16 \wedge c \Rightarrow c\mathbf{e}_a > 5 \wedge d \Rightarrow d\mathbf{a}_\alpha \wedge c_{37} \circ\!\!\rightarrow 1b \Leftrightarrow c$
§1 $\beta \wedge c_b \circ\!\!\rightarrow 2c_\alpha \vee \beta \Rightarrow \beta \rightarrow 4$
§2 $\beta \wedge c_c \circ\!\!\rightarrow 3c_\alpha \urcorner \wedge \beta \Rightarrow \beta \rightarrow 4$
§3 $\beta \wedge c_d \circ\!\!\rightarrow 4c_\alpha \oplus \beta \Rightarrow \beta$
§4 .

B. ELEMENTS OF DISJUNCTION, CONJUNCTION, ADDITION MOD 2 AND MAJORITY DECISION

To simulate elements of disjunction, conjunction, addition mod 2, and majority decision an auxiliary operator *senode* is used, which transforms an element of the connection complex, representing the list of nodes connected to the inputs of the simulated element, to the variable β, representing the *set* of these *node*s in the space of all the circuit nodes. Here and below we will understand by the expression "the variable x represents the set M in the space P" that (1) the number of components of the variable x is not less than the cardinality $\sigma(P)$ of the space; (2) the value of the ith component x^i of the variable x is equal to 1 if the ith element p_i of the space belongs to M, and is equal to 0 otherwise (for all $p_i \in P$); (3) $[x^i] = 0$ if $i \geq \sigma(P)$.

The subroutine *senode* has the form

senode αи $+$, βп$/\mathbf{a}$, a
133 130 5

$c_0 \urcorner \wedge \mathbf{a}_\alpha > 11 \Rightarrow a \circ \beta\mathbf{a}_\alpha > 6 \wedge 7 \Rightarrow$ ж $- 2 \circ\!\!\rightarrow 3 - 1 \circ\!\!\rightarrow 2$
§1 $'\mathbf{g}_a \Rightarrow {}'\mathbf{a} \rightarrow 4$
§2 $'\mathbf{f}_a \Rightarrow {}'\mathbf{a} \rightarrow 4$
§3 $'\mathbf{e}_a \Rightarrow {}'\mathbf{a}$
§4 $'\mathbf{a} \Rightarrow a > 27 \Rightarrow a \circ\!\!\rightarrow 5c_a \vee \beta \Rightarrow \beta$
$'\mathbf{a} < 11 \Rightarrow {}'\mathbf{a} \rightarrow 4$
§5 .

Here $[\alpha]$ is the number of the circuit element; the dimensions of the operand following the symbol $'$ are defined by the value of the index ж. The contents of the connection-complex element found in the initial sentence are "rewritten" in the variable $'\mathbf{a}$ (sentence 1, 2 or 3) and then "unpacked" (sentence 4) by successive extraction of groups of components bearing the numbers of the nodes connected with the inputs to the element. The number zero is the end of the list.

The use of the operator *senode* enables the models of the elements of disjunction, conjunction, addition mod 2, and majority decision to be written in compact form:

disjun (133)/αи +, βп/**b**, *a*
134 37 1 (β**b**)

senode α**b**//$c_\alpha \vee \beta \Rightarrow \beta \wedge \mathbf{b} \mid\!\rightarrow 1 c_\alpha \oplus \beta \Rightarrow \beta$
§1 .

conjun (133)/αи +, βп/**b**, *a*
135 40 1 (β**b**)

senode α**b**//$c_\alpha \vee \beta \Rightarrow \beta \rceil \wedge \mathbf{b} \circ\!\rightarrow 1 c_\alpha \oplus \beta \Rightarrow \beta$
§1 .

motwo (133)/αи +, βп/**b**, *a*
136 42 1 (β**b**)

senode α**b**//$c_\alpha \vee \beta \Rightarrow \beta \wedge \mathbf{b} \nabla \wedge c_{37} \mid\!\rightarrow 1 c_\alpha \oplus \beta \Rightarrow \beta$
§1 .

major (133)/αи +, βп/**b**, *a*
137 52 1 (β**b**)

senode α**b**//$\mathbf{b} \nabla > 1 + 1 \Rightarrow a c_\alpha \vee \beta \Rightarrow \beta \wedge \mathbf{b} \nabla - a \mid\!\rightarrow 1$
$c_\alpha \oplus \beta \Rightarrow \beta$
§1 .

C. THRESHOLD ELEMENT AND ELEMENTS REALIZING dnf AND pdnf OF BOOLEAN FUNCTIONS

To model these elements an auxiliary operator *orsenode αβ*, analogous to the operation *senode αβ* is used; by means of this operator the list of nodes represented by one element of the connection complex is transformed to an *or*dered *set* of *node*s, represented by the *complex β*. The program *orsenode* is analogous to the subroutine *senode*.

orsenode αи +, βк/**a**, *b*/
140 134 5

$c_{15} - c_{26} \wedge \mathbf{a}_\alpha > 11 \Rightarrow a \circ b\mathbf{a}_\alpha > 6 \wedge 7 \Rightarrow \text{ж} - 2 \circ\!\rightarrow 3 - 1 \circ\!\rightarrow 2$
§1 $'\mathbf{g}_\alpha \Rightarrow '\mathbf{a} \rightarrow 4$
§2 $'\mathbf{f}_a \Rightarrow '\mathbf{a} \rightarrow 4$
§3 $'\mathbf{e}_a \Rightarrow '\mathbf{a}$
§4 $'\mathbf{a} \Rightarrow a > 27 \Rightarrow a \circ\!\rightarrow 5 \Rightarrow \beta_b \triangle b$
$'\mathbf{a} < 11 \Rightarrow '\mathbf{a} \rightarrow 4$
§5 $b \Rightarrow b_\beta$.

threshold (140)/αи +, βп/**a**, d
141 141 3

orsenode $\alpha\mathbf{H}//c_\alpha \vee \beta \Rightarrow \beta c_0 \urcorner \wedge \mathbf{a}_\alpha > 22 \Rightarrow b2 \Rightarrow$ ж
$'\mathbf{b}_b \Rightarrow 'a \Rightarrow b > 32 \Rightarrow b \;\bar{\circ}\; a \circ c$

§1 $\triangle\, a \oplus b_7 \circ\!\!\rightarrow 2\mathbf{h}_a \Rightarrow d\beta \wedge c_d \circ\!\!\rightarrow 1a + 1 \times 6 \Rightarrow d$
$'\mathbf{a} < 6 \Rightarrow d > 32 + c \Rightarrow c - b \mapsto 3 \rightarrow 1$

§2 $c_\alpha \oplus \beta \Rightarrow \beta$

§3 .

dnf (140)/αи +, βп/**b**, d
142 172 4 (β**ab**)

orsenode $\alpha\mathbf{H}//c_\alpha \vee \beta \Rightarrow \beta c_0 \urcorner \wedge \mathbf{a}_\alpha > 22 \Rightarrow b$

§1 $0 - c_{10} \Rightarrow c \Rightarrow d\mathbf{c}_b \wedge c \Rightarrow c\mathbf{c}_b < 11 \wedge d \Rightarrow d \circ \mathbf{a} \circ \mathbf{b} \not\leftrightarrow 2$
$\mathbf{a} \Leftrightarrow \mathbf{b}c \Rightarrow d \not\leftrightarrow 2\beta \urcorner \wedge \mathbf{a} \Rightarrow \mathbf{a}\beta \wedge \mathbf{b} \vee \mathbf{a} \circ\!\!\rightarrow 4 \triangle\, bf_{10} + c_{26} \wedge \mathbf{c}_b \oplus \alpha$
$\circ\!\!\rightarrow 1c_\alpha \oplus \beta \Rightarrow \beta \rightarrow 4$

§2 $d\, \dot{\mathbf{X}}\, 3a\mathbf{h}_a \Rightarrow ac_a \vee \mathbf{a} \Rightarrow \mathbf{a} \rightarrow 2$

§3 !

§4 .

pdnf (140)/αи +, βп/**a**, c
143 124 3

orsenode $\alpha\mathbf{H}//c_\alpha \vee \beta \Rightarrow \beta \;\bar{\circ}\; a \circ b$

§1 $\triangle\, a \oplus b_7 \circ\!\!\rightarrow 2\mathbf{h}_a \Rightarrow c\beta \wedge c_c \circ\!\!\rightarrow 1\ 37 - a \Rightarrow cc_c \vee b \Rightarrow b \rightarrow 1$

§2 $b{:}40 \Rightarrow bc_0 \urcorner \wedge \mathbf{a}_\alpha > 22 +$ я $\Rightarrow$ я**d**я $\wedge\, c_b \mapsto 3c_\alpha \oplus \beta \Rightarrow \beta$

§3 .

D. GENERALIZED ELEMENT

The foregoing subroutines are composed in a natural way into a single one, which can serve as the model of a certain *generalized element* of the first type. The operator *element* recognizes the code of a concrete element with number [α] and calls up the corresponding model.

element (130, 132, 134, 135, 136, 137, 141, 142, 143)/αи +, βп/**b**, d
144 177 12

$\mathbf{a}_\alpha \wedge 77 - 3 \circ\!\!\rightarrow 1 - 4 \circ\!\!\rightarrow 2 - 1 \circ\!\!\rightarrow 3 - 1 \circ\!\!\rightarrow 4 - 1 \circ\!\!\rightarrow 5$
$- 1 \circ\!\!\rightarrow 6 - 1 \circ\!\!\rightarrow 7 - 1 \circ\!\!\rightarrow 10 - 1 \circ\!\!\rightarrow 11 \rightarrow 12$

§1 *invertor* $\alpha\beta//\rightarrow 12$

§2 *flipflop* $\alpha\beta//\rightarrow 12$

§3 *disjun* $\alpha\beta//\rightarrow 12$

§4 *conjun* $\alpha\beta//\rightarrow 12$
§5 *motwo* $\alpha\beta//\rightarrow 12$
§6 *major* $\alpha\beta//\rightarrow 12$
§7 *threshold* $\alpha\beta//\rightarrow 12$
§10 *dnf* $\alpha\beta//\rightarrow 12$
§11 *pdnf* $\alpha\beta//$
§12 .

E. DELAY ELEMENTS

The change of state of the output terminals of the delay elements takes place at the end of the clock cycle, and as a result the changes are expressed only in the next cycle. Therefore the sequence of examination of the delay elements may be arbitrary, in contradistinction to the sequence of examination of the elements of the first type, which must satisfy definite conditions, as has already been mentioned. In the subroutine *delay* given here the elements are considered in order of increasing number. The external operands of the subroutine are the variables α and β, of the same dimensions. The first represents in the space of all circuit nodes those nodes that are the output terminals of the delay elements; the second variable is the total state of the circuit.

delay $\alpha\text{п} +, \beta\text{п}/\mathbf{b}, b/(\alpha\beta)$
145 67 2 $(\alpha\mathbf{ab})$

$\alpha \Rightarrow \mathbf{a} \vee \beta \Rightarrow \mathbf{b}$
§1 $\mathbf{a}\,\dot{\mathbf{X}}\,2ac_0 \urcorner \wedge \mathbf{a}_a > 6 \Rightarrow b$
$\beta \wedge c_b \mid\!\rightarrow 1c_a \oplus \mathbf{b} \Rightarrow \mathbf{b} \rightarrow 1$
§2 $\mathbf{b} \Rightarrow \beta.$

The simulation of a single clock cycle of operation of the circuit divides into two stages, mentioned in the beginning of this section.

Both stages are executed by a single subroutine, which performs the five-place operator *cycle*.

cycle $(146, 144, 145)/\alpha\text{п} +, \beta\text{п}, \gamma\text{п} +, \delta\text{п} +, \epsilon\text{к} +/\mathbf{b}, f/(\beta\gamma\delta)$
131 47 2

input $\alpha\beta\gamma//\bar{\text{o}}\ \jmath$
§1 $\triangle f \oplus b_\epsilon \circ\!\rightarrow 2\epsilon_f \Rightarrow e$ *element* $e\beta//\rightarrow 1$
§2 *delay* $\delta\beta//.$

The complex ϵ represents the permissible sequence of examination of the elements of the first type. The variables β, γ, and δ of the same dimensions represent the total state of the circuit, its set of input terminals, and the set of delay elements. The signals arriving at the input terminals of the circuit at the start of the cycle are represented by successive components of the variable α, beginning with the leftmost. The value of the leftmost component, representing the ficititious input node, is always equal to zero, since, by convention, this terminal is always found in the state 0. The effect of the input signals on the input terminals of the circuit is modeled by means of the operator *input.*

$$\textit{input}\ \alpha\text{п}\ +,\ \beta\text{п},\ \gamma\text{п}\ +/\mathbf{a},\ b/(\beta\gamma)$$
$$\mathbf{146\ 52\ 2}\ (\gamma\mathbf{a})$$

$$\gamma \Rightarrow \mathbf{a} \vee \beta \Rightarrow \beta\ \bar{\text{o}}\ a$$
§1 $\mathbf{a}\ \dot{\mathbf{X}}\ 2ba \triangle a\alpha \wedge c_a \mid\!\rightarrow 1c_b \oplus \beta \Rightarrow \beta \rightarrow 1$
§2 .

4. PRELIMINARY OPERATIONS

We have now constructed the apparatus for simulating switching circuits. It remains to consider certain operations of a preparatory nature, which have already been mentioned.

A. RESTRUCTURING THE ELEMENTS OF COMPLEXES

The rearrangement of the structures of the elements of complexes **B, E, F, G** is one of these operations. This restructuring consists of shifting the information components to the left so as to obtain the structures shown in Figs. 2a and 3a, b, and c. This operation is performed by the subroutine *restruc.*

$$\textit{restruc}\ \alpha\text{к},\ \beta\text{и}\ +/\mathbf{b},\ c$$
$$\mathbf{147\ 113\ 3}$$

$$\bar{\text{o}}\ b$$
§1 $\triangle b \oplus b_\alpha \text{ o}\!\rightarrow 3c_\beta + c_\beta - c_{37} \Rightarrow{}'\mathbf{a} \circ a\beta \Rightarrow c \circ{}'\mathbf{b}$
§2 $\alpha_b \wedge{}'\mathbf{a} < c \vee{}'\mathbf{b} \Rightarrow{}'\mathbf{b}c + \beta \Rightarrow c\,'\mathbf{a} > 40 \Rightarrow{}'\mathbf{a}$
$\triangle a \oplus \text{ж} \mid\!\rightarrow 2\,'\mathbf{b} \Rightarrow \alpha_b \rightarrow 1$
§3 .

The external operands of the subroutines are the complex α, the structure of whose operands is to be rearranged, and the index β, defining the value of the initial shift (for the complex **B** this magnitude is equal to 2, for **E**, **F**, and **G**, to 5).

B. VALUES OF THE REPRESENTATIVE VARIABLES

In the construction of certain subroutines we have utilized variables that represent in the space of all circuit nodes sets of nodes having a certain property (for example, the set of input terminals of the circuit). We construct a subroutine that will find the values of this type of variable. We construct it in the form of an L-operator for *par*titioning a *com*plex (*parcom*). This operator can find application in problems having no relation to the simulation of relay circuits.

parcom αп $+$, βп $+$, γп, δк $+/-$, $a/(\alpha\beta)$
150 43 2

$\circ\ \gamma\ \bar{\circ}\ a$
§1 $\triangle\ a \oplus b_\delta \circ\!\rightarrow 2\delta_a \oplus \alpha \wedge \beta \mid\!\rightarrow 1c_a \vee \gamma \Rightarrow \gamma \rightarrow 1$
§2 .

Here γ is a variable representing in the space of all elements of the complex δ the set of those elements for which the values of the components, the corresponding unit components of the variable β, coincide in value with the corresponding components of the variable α. In other words, the pair of variables (α, β) defines a particular attribute; the variable γ represents the set of those elements of the complex δ that have this attribute.

For our purposes it is necessary to represent the set of input terminals of the circuit, the set of delay elements, the set of flipflops, the set of output terminals of the circuit, and the set of elements of the first type. The values of the *va*riables *rep*resenting these sets will be found by means of the operator *varep*. The variables α, β, γ, δ, ϵ represent these sets, in order.

varep (150)/αп, βп, γп, δп, ϵп/**a**, $c/(\alpha\beta\gamma\delta\epsilon)$
151 142 3

b_0:40 $\circ\!\rightarrow 1 \triangle$ я
§1 я $\Rightarrow$ л $\circ$ $b77 \Rightarrow c \nleftrightarrow 2$
$''$**a** $\Rightarrow \alpha 1 \Rightarrow b \nleftrightarrow 2$ $''$**a** $\Rightarrow \beta 5 \Rightarrow b \nleftrightarrow 2$
$''$**a** $\Rightarrow \gamma 6 \Rightarrow b \nleftrightarrow 2$ $''$**a** $\vee\ \gamma \Rightarrow \gamma c_0 \Rightarrow b \Rightarrow c \nleftrightarrow 2$
$''$**a** $\Rightarrow \delta b_0 - 1 \Rightarrow a0 - c_a \Rightarrow$ $''$**a**$\alpha \vee \beta \oplus$ $''$**a** $\Rightarrow \epsilon \rightarrow 3$
§2 *parcom bc* $''$**aA**//!
§3 .

C. FINDING A PERMISSIBLE SEQUENCE

We map (partially) the circuit structure by an oriented graph [11], each of whose nodes corresponds one-to-one to a circuit node; each branch emerging from the node i and incident on the node j is a connection of the node i to the input terminal of the element j of the *first type*. A node on which no branch is incident is called a *root* node (others are called *nonroot* nodes).

We shall prescribe the graph by a square matrix **U** whose element u_j^i has the value 1 if from the node i emerges a branch incident on the node j (for all i, j constituting node numbers), and the value 0 otherwise.

In the model thereby constructed the problem of finding a permissible sequence may be formulated in the following way.

An oriented graph is given; it is required to find a sequence $v_0, v_1, \ldots$ of its nonroot nodes (if one exists) such that (1) each of the nonroot nodes occurs only once and (2) for any pair (v_l, v_m) of nodes no path exists from v_m to v_l if $l < m$.

This problem is solved by the operator of *enu*meration of *n*odes in an *or*iented *g*raph (*enunorg*). The algorithm for the solution examines in sequence the nonroot nodes, finding those at which only branches from root nodes arrive. These nodes are included in the sequence, and are conventionally transferred to the set of root nodes, and the procedure repeated for the set of remaining nonroot nodes. The procedure terminates when all nodes are transferred to the set of root nodes or it is found that not one of the remaining nonroot nodes can be transferred. This latter condition signifies that the required sequence does not exist, and the graph contains loops.

$$\textit{enunorg}\ \alpha\text{ч},\ \beta\text{п}\ +,\ \gamma\text{к},\ \delta\text{к}\ +/\mathbf{b},\ b/(\beta\delta)$$
$$\mathbf{152\ 60\ 1}\ (\beta\mathbf{ab})$$

$$\beta \Rightarrow \mathbf{a} \Rightarrow \mathbf{b} \circ b$$
$$\S 1 \quad \mathbf{b}\ \dot{\mathbf{X}}\ \alpha a \delta_a \wedge \mathbf{a} \mid\!\rightarrow 1a \Rightarrow \gamma_b \triangle b$$
$$c_a \oplus \mathbf{a} \Rightarrow \mathbf{a} \Rightarrow \mathbf{b} \mid\!\rightarrow 1b \Rightarrow b_\gamma.$$

Here the element δ_a of the complex δ represents the ath row of matrix **U**, the variable β the set of nonroot nodes, and the complex γ the required sequence. Exit from the operator over the auxiliary terminal $\dot{\mathbf{X}}\,\alpha$ signifies that the required sequence does not exist.

The operator for finding the *per*missible *seq*uence (*perseq*) differs from the one just considered in that it concerns, not the matrix prescription of connections among the circuit elements, but the coupling complexes **E**, **F**, **G**. The rows of the matrix **U** are found by means of the operator *senode*.

perseq (133)/αч, βп +, γк/**d**, *c*
153 65 1 (β**bcd**)

β ⇒ **c** ⇒ **d** ○ *c*
§1 **d** $\dot{\mathbf{X}}$ α *senode* *a***b**//**b** ∧ **c** |→ 1
$a \Rightarrow \gamma_c \triangle cc_a \oplus$ **c** ⇒ **d** ⇒ **c** |→ 1*c* ⇒ β_γ.

D. PREPARING THE MODEL FOR OPERATION

The foregoing subroutines are composed in the general program for preparation of the switching-circuit model for operation, which combines all the preparatory operations.

prepare (147, 151, 153)/αч, βп, γп, δп, εп, ζк/**e**, *c*/(βγδε)
154 155 2

1 ⇒ жк5 ⇒ *d* ⇸
$a_4 \Leftrightarrow a_1b_4 \Leftrightarrow b_1$ △ жк2 ⇒ *d* ⇸ 1
$a_4 \Leftrightarrow a_5b_4 \Leftrightarrow b_5 5 \Rightarrow d$ ⇸ 1
$a_4 \Leftrightarrow a_6b_4 \Leftrightarrow b_6$ △ жк ⇸ 1
$a_4 \Leftrightarrow a_6a_4 \Leftrightarrow a_5a_4 \Leftrightarrow a_1b_4 \Leftrightarrow b_6b_4 \Leftrightarrow b_5b_4 \Leftrightarrow b_1$
varep βγδε ″**e**//*perseq* α ″**e**ζ//→ 2
§1 *restruc* ′**E***d*//!
§2 .

5. MODEL OF THE SWITCHING CIRCUIT

To conclude, we construct the program (Fig. 5) that will serve as the *swi*tching *cir*cuit *m*odel of the class described.

swicirm (154, 146, 131, 156)/αч, βп +, γк +, δк/**k**, *g*
155 113 3

prepare α ″**f** ″**g** ″**h** ″**iI**//″**g** ∨ ″**h** ⇒ ″**j**
§1 *input* β ″**k** ″**j**//$\bar{\circ}$ *g*
§2 △ *g* ⊕ b_γ ○→ 3 *cycle* (γ_g) ″**k** ″**f** ″**gI**//
output (δ_g) ″**k** ″**i**//→ 2
§3 .

The program has two input and two output terminals. On entering the initial sentence all preparatory operations are executed; in particular, it is

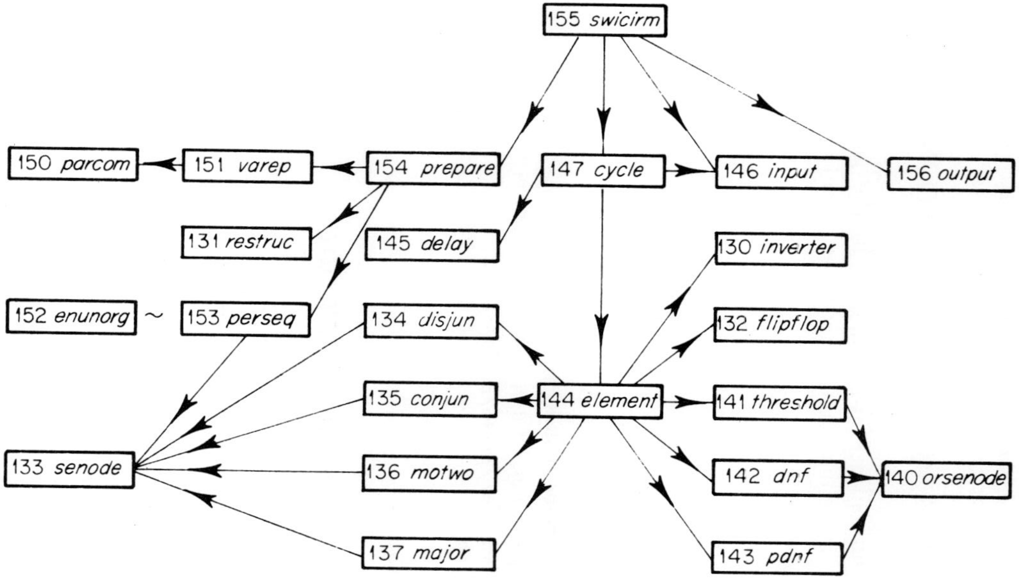

FIG. 5. The hierarchy of operators in the program for simulation of switching circuits.

established whether the structure of the simulated circuit is correct (i.e., whether it is free of loops without delay elements). If the structure is incorrect, exit takes place over the auxiliary terminal to sentence [α] of the external program; otherwise, the simulation of the circuit operation begins.

The operator *input*, considered previously, is used here to bring the output terminals of the delay and flipflop elements to a state defined by the values of the left components of the variable β, which thereby gives the initial *internal state* of the circuit. The sequence of input signals is represented by the complex γ; the values of the left components of the element γ_i of this complex define the states of the input terminals of the circuit in the ith cycle. The result of the operation of the circuit model is a sequence of output signals, represented by the complex δ (also by the left components of its elements).

The program uses the already-described operator *cycle* and the operator *output*, which forms the values of the elements of complex δ, considering the output terminals of the circuit in increasing order of their numbers:

$$\textit{output}\ \alpha\text{п},\ \beta\text{п}\ +,\ \gamma\text{п}\ +/\mathbf{a},\ b/(\beta\gamma)$$
$$\mathbf{156\ 47\ 2}\ (\gamma\mathbf{a})$$

$$\gamma \Rightarrow \mathbf{a} \circ \alpha\ \bar{\circ}\ a$$
§1 $\mathbf{a}\ \dot{\mathbf{X}}\ 2b \bigtriangleup a\beta \wedge c_b \circ\!\rightarrow 1\alpha \vee a_a \Rightarrow \alpha \rightarrow 1$
§2 .

When access to the operator *swicirm* is made by means of the auxiliary entry terminal (sentence 1) the preparatory operations are not executed.

A. SWITCHING-CIRCUIT ANALYSIS

The sequence of input signals can be considered to be a test program for the switching circuit. By prescribing various programs, it is possible to solve various problems. One of these is, for example, the classical problem of switching-circuit analysis, i.e., finding the flow table for a prescribed circuit structure.

We shall represent the next-state table by groups of 2^s elements of the complex β, and the rows of the output table by similar groups of elements of the complex γ. Here s is the number of circuit elements having one-bit memory (in our case the number of delay elements, flipflops 1 and flipflops

0[1]: Then the program for *sw*it*c*hing *cir*cuit *an*alysis can have the following form:

$$\textit{swiciran}\ (155, 156)/\alpha\text{ч},\ \beta\text{к}?,\ \gamma\text{к}?/\mathbf{k},\ m$$
$$\mathbf{157\ 135\ 2}$$

$$1 \Rightarrow b_{11}\ {}''\mathbf{f}\ \nabla\ -1 \Rightarrow j\ {}''\mathbf{j}\ \nabla\ -1 \Rightarrow k \circ h \circ i \circ m$$
$$\textit{swicirm}\ \alpha hiJ//$$
$$\S 1\quad \textit{output}\ l\ {}''\mathbf{k}\ {}''\mathbf{j}//l \Rightarrow \beta_m?\ \mathbf{j}_0 \Rightarrow \gamma_m?\ \triangle\ m$$
$$h + c_k \Rightarrow h \circ \rightarrow 2 \rightarrow \textit{swicirm}\ 1$$
$$\S 2\quad i + c_j \Rightarrow ic_0\ \urcorner\ \wedge\ i \mid\!\rightarrow \textit{swicirm}\ 1m \Rightarrow b_\beta \Rightarrow b_\gamma.$$

6. CONCLUSION

The solution of the analysis problem does not exhaust the range of application of the proposed simulation apparatus. It is assumed that by means of this apparatus it will be possible to carry out broader investigations of circuits of a fairly great complexity, developing the simulation methods—in particular, the methods of constructing test programs for circuit testing.

The practical possibilities of applying flow tables, as has been stated, are strongly limited. However, a program similar to the above may be applied to the analysis of subcircuits of a complex circuit having sufficiently small memory. The replacement of each such subcircuit by an equivalent element whose operator will be the pair of tables (next-state table and output table) permits a substantial reduction in the time necessary for simulating a single clock cycle of the circuit and consequently increases the speed of simulation of the system.

The effective application of this approach is hardly possible in the framework of the apparatus constructed here. It is necessary to extend the class of simulated circuits, developing the system of representation and the set of corresponding subroutines for circuits consisting of (k, p)-terminals. This conclusion also follows from the consideration of the structure of

[1] This will entail unreal internal states; in actual fact two output terminals of the same flipflop cannot occur simultaneously in the same state. This undesirable consequence of the artificial representation of the flipflop in the form of two independent elements could be eliminated by placing certain restrictions on the circuit-node numbering and correcting the corresponding subroutines. It would be more useful, however, to approach the problem from another position—in particular, to consider circuits consisting of (k, p)-terminal networks.

automata of the type of general-purpose computers, whose "elementary" components are decoders, registers, accumulators, memory registers, etc. A convenient and compact representation of such structures in the simulation system will be difficult to obtain if multioutput elements are not brought into consideration.

REFERENCES

1. Chang, Y. H., and Georg, O. M., Use of High-Speed Digital Computers to Study Performance of Complex Switching Networks Incorporating Time Delays, *Comm. & Electr.* **46,** 982–987 (1960).
2. Bennet, J. M., and Dakin, R. J., Computers as an Aid in Computer Design Assessment, *Computer J.* **4,** 253–255 (1961).
3. Goodman, H. P., The Simulation of the Orion Time-Sharing System on Sirius, *Comput. Bull.* **5,** No. 2, 51–55 (1961).
4. Stockwell, G. N., Computer Logic Testing by Simulation, *IRE Trans. Military Electron.* **6,** No. 3, 275–282 (1962).
5. Pecherskii, Yu. N., and Utkin, A. A., Programmed Analysis of Switching Circuits, *in* "Problems of the Methodology and Logic of Science" *Uch. Zap. Tomsk. Univ.* **41,** 180–193 (1962).
6. Schorr, H., A Program for the Analysis of Digital Systems. Princeton Univ., PB 157804-32, December 1962.
7. Linskii, V. S., Algorithmic Projecting of Digital Computer Devices, *in* "Commun. Comput. Engr." [in Russian], No. 2, Moscow, Izd. VTs AN SSSR, 1963.
8. Lehman, M., Raynor, E., and Netter, Z., The Checking of Computer Logic by Simulation on a Computer, *Computer J.* **6,** No. 2, 154–162 (1963).
9. Landau, I. Ya., Automated Computer Projecting, *Avtomat. i Telemek.* **25,** No. 11, 1581–1587 (1964).
10. Mikhnovskii, S. D., Determination of the Effective Speed of a Computer by the Method of Digital Simulation. *Voprosy Radiotekhni. Series* **7** No. 2, 3–10 (1964).
11. Berge, C., "Graph Theory and Its Application" [Russian translation]. IL, Moscow, 1962.

ALGORITHMS FOR STATE MINIMIZATION OF A DISCRETE AUTOMATON

Yu. V. Pottosin

1. SOME METHODS FOR MINIMIZATION OF THE NUMBER OF STATES OF A DISCRETE AUTOMATON

The problem of minimizing the number of states of a discrete automaton consists of finding an automaton, equivalent to the given one, that has the smallest number of states. The number of states of the equivalent automaton can be reduced because certain of its states may be merged without violating the law of mapping the sequences realized by the given automaton. Such states are usually termed compatible.

In the present chapter state-minimization algorithms are described in LYaPAS for two methods that are well adapted to machine execution.

One of these methods, proposed by Paull and Unger [1], involves finding in the set of states the smallest number of subsets such that merging of all the states in each of them yields an automaton equivalent to the given one. These subsets, in general, may overlap. Let us consider this method in greater detail.

We denote by X, Y, and Z the sets of inputs, outputs, and states, respectively. For the next-state function, given on the set $Z \times X$, we introduce the notation $\delta(z, x)$. The output function, which is also defined on the set $Z \times X$, is denoted by $\lambda(z, x)$. We shall consider that $\delta(z, x) = \Lambda$ (or $\lambda(z, x) = \Lambda$) if the next-state (output) function is undefined on a given pair (z, x). An automaton in which this case holds is called partially defined, in contradistinction to fully defined automata [2].

A set $P \subseteq Z$ implies a set $Q \subseteq Z$ if for some input $x \in X$, the set Q is the set of all defined values of $\delta(z, x)$ for all $z \in P$, i.e., Q satisfies the relation

$$Q = \{\delta(z, x)/z \in P\} \backslash \{\Lambda\}.$$

A certain set of states $B \subseteq Z$ is called a set of compatibles if any two states $z_i, z_j \in B$ are compatible.

A certain set S of sets of compatibles is called closed if

(1) for any set Q implied by the set $B \in S$, there is found a $P \in S$ such that $Q \subseteq P$;

(2) $\bigcup [S] = Z$.

The first step in the minimization process according to the method is to find all sets of compatibles that are maximal in the sense that none of them is contained in another such set. Then a minimal closed set L of sets of compatibles is found, and a new automaton is constructed in which each state is a set $Q \in L$.

The value of the new next-state function $\delta'(Q, x)$ can be any P satisfying the condition

$$\{\delta(z, x)/z \in Q\} \backslash \{\Lambda\} \subseteq P \in L.$$

The new output function will be defined as $\lambda'(Q, x) = \lambda(z, x)$, where z is any state in the set Q for which $\lambda(z, x) \neq \Lambda$ and $\lambda'(Q, x) = \Lambda$ if $\lambda(z, x) = \Lambda$ for all $z \in Q$.

Thus, the state-minimization problem is reduced by this method to finding the minimal closed set L of sets of compatibles. In the case of a completely defined automaton, the set L, as shown in [1], is the set of all maximal sets of compatibles. The problem is different if a partially defined automaton is given. Here, finding the set L involves substantial enumeration. It is therefore of interest to consider approximate methods of state minimization for partially defined automata. In general these methods may not give a minimal solution, but the time for their execution is enormously smaller than the execution of the exact methods.

In the present chapter we limit consideration to this problem for partially defined automata.

A second method considered in the present chapter for minimizing the number of states of an automaton is the method of successive reductions, developed by Zakrevskii [3–5]. It consists of the following.

A chain, generated by a pair of states $\langle z_i, z_j \rangle$, is a set C obtained by the following process: (1) a pair $\langle z_i, z_j \rangle$ is included in C as the first element of the chain; (2) if $\langle z_e, z_l \rangle \in C$, then all pairs of states implying the pair $\langle z_e, z_l \rangle \in C$ are included in C.

For a given automaton A let there exist some chain C, where the states occurring in a given pair of this chain are compatible. A single-step transformation of A is the passage to an automaton A' whose states are all pairs belonging to the chain C, with the states of the initial automaton A not entering into any of the pairs of the chain C, while the next state and output functions are defined as in the Paull and Unger method. The method of successive reductions consists in the repeated application of this type of transformation to A, leading to the reduction in the number of its states.

The number of distinct transformations of A is equal to the number of different chains in which each pair may contain only compatible states. In [3] it was proposed that a so-called informational criterion be applied to the choice of transformation. From the viewpoint of this criterion the best variant is that for which the mean decrease of the degree of indeterminacy of the functions δ and λ is minimal for unit reduction of the number of states. A second criterion of choice of transformation can be the number by which the states are reduced [5].

2. REPRESENTING THE INITIAL INFORMATION

Before passing to the description of the system of operators used in the programs for automaton minimization, it is useful to consider the method of representing the input information. This question is very important, since the concrete expression of each operator depends on it.

We shall utilize the specification of automata in the form of next-state and output matrices.

In the next-state matrix $\| Z \|$, as in the output matrix $\| Y \|$, the rows represent states, the columns inputs. It is convenient to number all inputs and states according to the columns and rows of $\| Z \|$ and $\| Y \|$ that represent them, and to consider Z and X to be the sets of corresponding numbers.

We shall represent the rows of the matrix $\| Z \|$ by the elements of the complex **Z**. The positions of the elements of this complex are divided into groups, representing the codes of the elements of the matrix. In each group there is an additional place, located to the left of all the other positions of the given group, representing the definition tag of the matrix element. This position has the value 1 if and only if the corresponding element of the matrix is defined. If the 32 positions of an element of the complex are insufficient to represent one row of $\| Z \|$, the entire set of columns of the matrix are divided into groups. The number of columns in each group is not greater than the number of elements of the matrix that can be placed in the 32 positions of an element of the complex. If the number of states is equal to n, the first group of columns is represented in the elements $z_0, z_1, \ldots, z_{n-1}$, the second in the elements $z_n, z_{n-1}, \ldots, z_{2n-1}$, etc.

We present an example of coding of the matrix $\| Z \|$:

$\| Z \| =$

$z \backslash x$	0	1	2
0	3	2	0
1	0	1	
2		0	2
3	3	1	

To encode one element of this matrix, taking into account the definition tag, three bits are required. Let us assume that the number of positions in the codes of the elements of complex **Z** is equal to 8. Then

$$
\mathbf{Z} = \begin{array}{l} 11111000 \\ 10010100 \\ 00010000 \\ 11110100 \\ 10000000 \\ 00000000 \\ 11000000 \\ 00000000 \end{array}
$$

The method for representing the matrix $\| Y \|$ is analogous.

Special constants are needed in the processing of the matrices $\| Z \|$ and $\| Y \|$. We shall call them ω-constants, and denote them by $\omega_1, \omega_2, \omega_3, \omega_4, \omega_5, \omega_6, \omega_7$. Each ω-constant is a certain p-dimensional binary vector

$$
\omega_j = \overset{0}{\omega_j} \overset{1}{\omega_j} \cdots \overset{p-1}{\omega_j},
$$

where p is the number of positions of the code of an element in the complex **Z**.

Let n be the number of states of the given automaton, m the number of columns in the matrix $\| Z \|$, and k the number of positions necessary to encode an element of $\| Z \|$. Then the values of the ω-constants will be defined by the following relations:

$$\begin{aligned}
\omega_1{}^i &= 1 \leftrightarrow p - k \leq i \leq p - 1; \\
[\omega_2] &= k; \\
[\omega_3] &= k \cdot m; \\
\omega_4{}^i &= 1 \leftrightarrow 0 \leq i < n; \\
\omega_5{}^i &= 1 \leftrightarrow i = p - k; \\
\omega_6{}^i &= 1 \leftrightarrow i \in \{0, k, 2k, \ldots, (m - 1)k\}; \\
\omega_7{}^i &= 1 \leftrightarrow 0 \leq i < k.
\end{aligned}$$

The symbol $[\omega]$ denotes the number whose positional code is ω.

For our example the ω-constants will have the form

$$\begin{aligned}
\omega_1 &= 00000111 & \omega_5 &= 00000100 \\
\omega_2 &= 00000011 & \omega_6 &= 10010000 \\
\omega_3 &= 00000110 & \omega_7 &= 11100000 \\
\omega_4 &= 11110000
\end{aligned}$$

These constants are calculated by means of the operator *comeco* from the number of states in the automaton and the number of its outputs. The former is represented by the variable α, the latter by β. The ω-constants obtained are represented by the following variables: $\gamma = \omega_1$; $\delta = \omega_2$; $\epsilon = \omega_3$; $\zeta = \omega_4$; $\eta = \omega_5$; $\vartheta = \omega_6$; $\kappa = \omega_7$.

comeco αп $+$, βп $+$, γп, δп, ϵп, ζп, ηп, ϑп, κп$/-$, a
271 116 1

$\alpha \vdash \Rightarrow a40 - a \Rightarrow a0 - c_a \Rightarrow \kappa \nabla \Rightarrow \delta\alpha - 1 \Rightarrow a0 - c_a \Rightarrow \zeta\kappa \leftleftarrows \delta$
$\Rightarrow \gamma c_0 \leftleftarrows \delta \Rightarrow \eta \circ a \circ \vartheta\delta \times \beta \Rightarrow \epsilon - 40 \circ\!\!\rightarrow 1\ 40{:}\delta$я $\times\ \delta \Rightarrow \epsilon$
§1 $c_a \vee \vartheta \Rightarrow \vartheta a + \delta \Rightarrow a - \epsilon \circ\!\!\rightarrow 1.$

For certain particular cases it is possible to propose one further compact method of representing the next-state function, which generates fairly simple expressions of the L-operators occurring in the minimization program. These cases include any automaton in which all the transitions to any state $z \in Z$ occur for the same input $x \in X$, unique for this state.

For such an automaton it is convenient to represent the next-state function by two binary matrices. One of them is the adjacency matrix of the graph $\| W \|$ of the automaton:

$$\| W \| = \| Z < Z \|,$$

where $z_i < z_j$ holds if and only if there exists an $x \in X$ such that $z_j = \delta(z_i, x)$.

The second matrix, denoted by $\| X \|$, represents the following:

$$\| X \| = \| \{\delta(Z, x)/x \in X\} \ni Z \|,$$

where $\delta(Z, x) = \{\delta(z, x)/z \in Z\} \backslash \{\Lambda\}$.

For an automaton with next-state matrix

$\| Z \| =$

$z \backslash x$	0	1	2
0	0	1	
1	3	1	2
2		1	2
3	3	1	

the matrices $\| W \|$ and $\| X \|$ will have the form

$$\| W \| = \begin{bmatrix} 1 & 1 & 0 & 0 \\ 0 & 1 & 1 & 1 \\ 0 & 1 & 1 & 0 \\ 0 & 1 & 0 & 1 \end{bmatrix} \qquad \| X \| = \begin{bmatrix} 1 & 0 & 0 & 1 \\ 0 & 1 & 0 & 0 \\ 0 & 0 & 1 & 0 \end{bmatrix}.$$

Among the automata whose function δ can be assigned in this way are, in particular, asynchronous automata, represented by a primary next-state table [6].

3. ALGORITHMS FOR STATE MINIMIZATION

Here we consider the L-programs that carry out certain algorithms for minimizing the number of states of an automaton, based on the above two methods.

The algorithm for minimization by the Paull–Unger method includes the following actions:

(1) Obtain the maximal sets of compatibles.
(2) Obtain the minimal closed set of sets of compatibles.
(3) Obtain the next-state and output matrices of the automaton obtained as a result of the minimization.

In the first stage, aside from obtaining the maximal sets of compatibles directly by means of the operator *maxcom*, two additional L-operators appear, determining the relation of compatibility in the set of states. One of them constructs the matrix of equivalence for outputs, the second transforms this matrix to the compatibility matrix, representing the relation of compatibility. The matrix of equivalence for outputs is constructed by the operator *equout*. To construct the compatibility matrix the operator *comat* or *comav* is used, the latter when the matrices $\| W \|$ and $\| X \|$ are given.

The second stage is carried out by the operator *closetcom* if the matrix $\| Z \|$ is given, or the operator *closetcov* if the matrices $\| W \|$ and $\| X \|$ are given. These operators utilize an approximate method for obtaining a closed set of sets of compatibles, close to minimal, presented in [7].

In the third stage the output matrix is constructed by means of the operator *tramout*, which is suitable for all cases considered here. The next-state matrix is transformed by means of the operator *tramast*. If the next-state matrix is constructed from the matrices $\| W \|$ and $\| X \|$, the operator *tramav* is used.

The number of states of the partially defined automaton can be minimized by the operator *minst*. The number of states and the number of inputs are represented by the variables μ and β, respectively; $\gamma::\| Y \|$; $\delta::\| Z \|$. ϵ, ζ, η, ϑ, κ, and λ are used as auxiliary complexes in the various L-operators entering into *minst*. The new next-state matrix obtained is represented by the complex κ, the new output matrix by γ. If it is impossible to reduce the number of states by the given operator, exit takes place over the auxiliary exit terminal, to which corresponds the operand α. The operator *comeco* calculates the ω-constants.

minst (270, 272, 274, 310, 277, 301, 302,)/αч, βп +, γк, δк, ϵк, ζк, ηк?,
ϑк?, κк, λк, μп +/**s**, d/($\epsilon\zeta\eta$)
305 125 0 (ϵ**p**)

comeco $\mu\beta$**lmnpqrs**//$\mu - 1 \Rightarrow b_\vartheta$ *equout* $\gamma\vartheta\mu$**rs**//
comat $\delta\vartheta\epsilon\eta\zeta$**msr**//*maxcom* $\vartheta\eta$?//
closetcom $\alpha\mu\eta$?ϑ?$\zeta\delta\epsilon$**nqm**//
tramout $\gamma\zeta\kappa\mu$//*tramast* $\delta\eta\kappa\lambda\mu$**lmnq**//.

If the function δ of the given automaton can be represented by the matrices $\| W \|$ and $\| X \|$, then to minimize the number of states of such an automaton it is possible to use the operator *minva*. Here the operands have the following sense: $\kappa::\| W \|$; $\beta::\| X \|$; $\gamma::\| Y \|$. δ, ϵ, ζ, η, ϑ are auxiliary complexes, and α corresponds to the auxiliary exit terminal. The results, i.e., the newly obtained matrices $\| Y \|$ and $\| Z \|$, are represented by the complexes γ and ϑ.

minva (272, 276, 310, 300, 301, 304)/αч, βк +, γк, δк, ϵк, ζк?, ηк, ϑк, κк +/**m**, $d(\beta\delta\epsilon\zeta\kappa)$
306 131 0

$b_\kappa - 1 \Rightarrow b_\eta \Rightarrow a0 - c_a \Rightarrow \mathbf{l}b_\kappa \vdash \Rightarrow a40 - a \Rightarrow a0 - c_a \Rightarrow \mathbf{m}$
equout $\gamma\eta(b_\kappa)$**ml**//*comav* $\kappa\eta\delta\zeta\epsilon\beta$//*maxcom* $\eta\zeta$?//
closetcov $\alpha\beta\zeta\eta$?$\epsilon\kappa$//*tramout* $\gamma\epsilon\vartheta(b_\kappa)$//*tramav* $\kappa\epsilon\eta\vartheta\beta$//.

A class of partially defined automata was studied in [8] for which the minimal closed set L of sets of compatibles may contain only maximal sets of compatibles. This class is composed of asynchronous automata whose functions δ can always be represented by a primary next-state table [6]. The undefined elements of this table must be due only to the constraints on the sequence of input symbols. In this case it is sufficient to find the shortest coverage of the set Z by the maximal sets of compatibles. This coverage will be the set L. For this purpose it is possible to apply the operator *minas*, whose operands have the same sense as for the operator *minva*. The set L is found by means of the operators *oshco* and *compress*.

minas (272, 276, 310, 4, 43, 301, 304)/αч, βк +, γк, δк, ϵк, ζк, ηк, ϑк, κк +/**j**, $d/(\beta\delta\epsilon\zeta\kappa)$
311 203 0 (β**efgh**)

$b_\kappa - 1 \Rightarrow b_\eta \Rightarrow a0 - c_a \Rightarrow \mathbf{i}b_\kappa \vdash \Rightarrow a40 - a \Rightarrow a0 - c_a \Rightarrow \mathbf{j}$
equout $\gamma\eta(b_\kappa)$ **ji**//*comav* $\kappa\eta\delta\zeta\epsilon\beta$//*maxcom* $\eta\zeta$?//
$b_\zeta - 1 \Rightarrow a0 - c_a \Rightarrow \mathbf{e}b_\kappa - 1 \Rightarrow a0 - c_a \Rightarrow \mathbf{f} \Rightarrow \mathbf{h}$ *oshco* ζ**ef**δ**g**//
g $\oplus$ **h** $\circ\!\!\rightarrow \alpha$ *compress* ζ**g**η//*tramout* $\gamma\eta\vartheta(b_\kappa)$//*tramav* $\kappa\eta\epsilon\vartheta\beta$//.

The algorithm for minimization by the method of successive reductions is presented in two variants, corresponding to different criteria for the choice of transformation. The operator *infstared* minimizes the number of states with the application of an information criterion for choice of single-step transformation of the matrices $\| Z \|$ and $\| Y \|$. The L-operators appearing in the L-program of *infstared* perform the following actions. The operator *comeco* calculates the ω-constants, the operator *equout* constructs the equivalence matrix for outputs, *inftran* carries out the choice of transformation, and the operators *tramout* and *sistramast* transform the matrices $\| Y \|$ and $\| Z \|$. The numbers of inputs, outputs, and states are given by the variables α, β, and γ, respectively. The complexes δ and ϵ represent the matrices $\| Y \|$ and $\| Z \|$. The complex ξ represents a table of logarithms, needed for the operator *inftran*. The remaining operands—ζ, η, ϑ, κ, λ, μ, and ν—are auxiliary complexes.

infstared (270, 272, 323, 301, 314)/αп +, βп +, γп, δк, ϵк, ζк, ηк, ϑк, κк, λк, μк, νк, ξк +/**w**, g/($\zeta\eta\vartheta\kappa$)
312 160 4 (ζ**s**)

comeco $\gamma\alpha$**pqrstuv**//$\beta \Rightarrow a\xi_a \Rightarrow$ **w** $\bar{\circ}$ g
§1 $\gamma - 1 \Rightarrow b_\kappa$ *equout* $\delta\kappa\alpha$**uv**//g $\circ\!\rightarrow 3$
§2 *inftran* 4ξ**quv**$\vartheta\delta\epsilon\kappa\eta\zeta$**sw**//*tramout* $\delta\vartheta\lambda\gamma$// $\circ$ $g \rightarrow 1$
§3 *sistramast* $\epsilon\vartheta\eta\kappa\zeta\nu$**rqpt**$\mu\lambda$//$b_\vartheta \Rightarrow \gamma - 1 \Rightarrow a0 - c_a \Rightarrow$ **s** $\rightarrow 2$
§4 .

The operator *maxstared* minimizes the number of states in an automaton by the method of successive reductions, using for choice of transformation the criterion of maximum reduction. Here for choice of single-step transformation the operator *transtamax* is used. The operands α and γ are the number of inputs and the number of states, the complexes δ and ϵ are the matrices $\| Y \|$ and $\| Z \|$, the complexes β, ζ, η, ϑ, κ, λ, and μ are auxiliary.

maxstared (270, 272, 324, 301, 314)/αп +, βк +, γп, δк, ϵк, ζк, ηк, ϑк, κк, λк, μк/**t**, g/($\zeta\eta\vartheta\kappa$)
313 202 4 (ζ**q**)

comeco $\gamma\alpha$**unpqrst**// $\bar{\circ}$ g
§1 $\gamma - 1 \Rightarrow b_\kappa$ *equout* $\delta\kappa\gamma$**ts**//g $\circ\!\rightarrow 3$
§2 $\gamma - 1 \Rightarrow b_x \Rightarrow b_v \Rightarrow b_\eta \Rightarrow b_\zeta$ *transtamax* 4ϵ**nst**$\kappa\vartheta\zeta\eta$**q**γ//*tramout* $\delta\zeta\lambda\gamma$// $\circ$ $g \rightarrow 1$
§3 $\gamma \Rightarrow b_\eta b_\xi \Rightarrow b_\mu \Rightarrow b_x$ *sistramast* $\epsilon\zeta\eta\kappa\vartheta\beta$**pnmr**$\mu\lambda$//$b_\vartheta \Rightarrow \gamma - 1 \Rightarrow a0 - c_a \Rightarrow$ **q** $\rightarrow 2$
§4 .

4. CONSTRUCTION OF THE EQUIVALENCE MATRIX FOR OUTPUTS

We call the states i and j equivalent for outputs if $(\lambda(i, x) = \lambda(j, x)) \vee (\lambda(i, x) = \Lambda) \vee (\lambda(j, x) = \Lambda)$ for all $x \in X$.

The element a_{ij} of the equivalence matrix for outputs $\| A \|$ takes the value 1 if the states i and j are equivalent for outputs and $i < j$; otherwise, $a_{ij} = 0$.

The operator *equout* obtains the matrix $\| A \|$ for a partially defined automaton. The initial information is given by the complex α: : $\| Y \|$; [γ] is the number of states; β: : $\| A \|$; $\delta = \omega_6$; $\epsilon = \omega_7$.

equout αк $+$, βк, γп $+$, δп $+$, ϵп $+/\mathbf{b}$, d
272 166 5

$\circ\ b\ \circ\ \mathbf{b}\gamma - 1 \Rightarrow a0 - c_a \Rightarrow \mathbf{a}\ \circ\ a$
§1 $c_a \oplus \mathbf{a} \Rightarrow \mathbf{a} \Rightarrow \beta_a \triangle a - b_\beta\ \circ\!\!\rightarrow 1$
§2 $b + 1 \Rightarrow a - \gamma \mid\!\!\rightarrow 5$
§3 $a + \mathbf{b} \Rightarrow cb + \mathbf{b} \Rightarrow d\alpha_c \wedge \alpha_d \wedge \delta \overline{\times}\ \epsilon < 1 \Rightarrow \mathbf{a}\alpha_c \oplus \alpha_d \wedge \mathbf{a}\ \circ\!\!\rightarrow 4c_a$
$\urcorner \wedge \beta_a \Rightarrow \beta_b$
§4 $\triangle a - \gamma\ \circ\!\!\rightarrow 3 \triangle b \rightarrow 2$
§5 $\circ\ b\gamma + \mathbf{b} \Rightarrow \mathbf{b} - b_\alpha\ \circ\!\!\rightarrow 2.$

5. OBTAINING A CHAIN GENERATED BY A GIVEN PAIR OF STATES

The chain C is represented by a binary matrix $\|C\|$, which we call the chain matrix, in which $c_{ij} = 1$ if and only if $\langle i, j\rangle \in C$ and $i < j$. To construct the matrix $\|C\|$ a certain auxiliary matrix $\|C'\|$ is constructed, for which $c'_{ij} = 1$ if and only if as yet the entire set of pairs implying the pair $\langle i, j\rangle \in C$ has not been found and $i < j$. The presence of at least one unit element in $\|C'\|$ indicates that the construction of the chain matrix has not been terminated. From the coordinates of such elements are found the corresponding rows of the matrix $\|Z\|$ or the matrix $\|W\|$, depending on the way in which the function δ is prescribed and $\|C\|$ is completed.

The search for unit elements of the matrix $\|C'\|$ is carried out by means of the operator *funel*. To find all pairs of states implied by any given pair, the operator *impast* or *impav* is used, depending on the method for defining δ. The corresponding variants of the operator for finding the chain will be *chest* or *chev*.

The operands of the operator *chest* have the following sense: the complex α represents the rows of the next-state matrix; β may be an expression included in the given operator during compilation; γ and δ are indexes corresponding to the numbers of the given states; $\epsilon{:}{:}\|C'\|$; $\kappa{:}{:}\|C\|$; $\zeta = \omega_2$; $\eta = \omega_7$; $\nu = \omega_6$.

chest (317, 325)/αк $+$, βв, γи, δи, ϵк, ζп $+$, ηп $+$, ϑп $+$, κк/$\mathbf{c}$, $b/(\epsilon$к$)$
321 104 2 $(\epsilon\mathbf{c})$

impast $1\alpha\gamma\delta(b_\kappa)\zeta\eta\vartheta ab$//*funel* $2\epsilon\gamma\delta$//$\rightarrow$ *impast* 1
§1 $c_a \Rightarrow \mathbf{c} \wedge \kappa_b \mid\!\!\rightarrow$ *impast* $3\beta\mathbf{c} \vee \epsilon_b \Rightarrow \epsilon_b \vee \kappa_b \Rightarrow \kappa_b \rightarrow$ *impast* 3
§2 .

For the operator *chev* we adopt the following: $\eta::\|C\|$; $\zeta::\|C'\|$; $\alpha::\|W\|$; $\epsilon::\|X\|$. γ and δ are indexes, corresponding to the numbers of the given states, β has the sense of an expression that is included in the operator during compilation.

chev $(316, 325)/\alpha к +, \beta в, \gamma и, \delta и, \epsilon к +, \zeta к, \eta к/\mathbf{b}, c/(\alpha\epsilon\zeta\eta)$
322 73 3 $(\alpha\mathbf{b})$

§1 *impav* $2\alpha\gamma\delta\epsilon\, c\mathbf{b}//\mathbf{b} \wedge \eta_c \mid\!\rightarrow$ *impav* $1\beta\mathbf{b} \vee \zeta_c \Rightarrow \zeta_c \vee \eta_c \Rightarrow \eta_c \rightarrow$ *impav* 1
§2 *funel* $3\zeta\gamma\delta//\rightarrow 1$
§3 .

The necessary condition for correct operation of these two operators is the zero initial value of all elements of the matrices $\|C\|$ and $\|C'\|$.

6. FINDING THE COORDINATES OF THE UNIT ELEMENTS OF A MATRIX

The operator *funel* finds the numbers of the rows and columns of the unit elements of a binary matrix, represented by the complex β. The row and column numbers are represented by the indexes γ and δ. If all the elements of the given matrix are zero, exit to the external program is effected over the auxiliary output terminal, connected with the operand α.

funel $\alpha ч, \beta к, \gamma и, \delta и/-, -$
325 26 2

$\bar{o}\ \gamma$
§1 $\triangle\, \gamma - b_\beta \mid\!\rightarrow \alpha$
§2 $\beta_\gamma\, \dot{\mathbf{X}}\, 1\delta.$

7. FINDING THE IMPLIED PAIRS OF STATES

In all the algorithms for minimization of automata considered here to solve the question of compatibility of two states, it is always necessary to find the set of all pairs of states implied by the given pair; i.e., for the given pair $\langle z_i, z_j\rangle$, it is necessary to find the set

$$\mathbf{C}_{z_i,z_j} = \{\langle\delta(z_i,x),\delta(z_j,x)\rangle/x\in X, \delta(z_i,x)\neq\Lambda, \delta(z_j,x)\neq\Lambda, \delta(z_i,x)\neq\delta(z_j,x)\}.$$

We present below two variants of such an operator, depending on the way in which the next-state function δ is prescribed. If the next-state function is given by the matrix $\| Z \|$, the operator *impast* is used. If the adjacency matrix of the graph of the automaton is used, then the operator *impav* is used.

In the case of the operator *impast* $\beta : : \| Z \|$; $[\alpha]$ is the number of the sentence in the external program to which the auxiliary exit terminal of the operator leads; γ and δ are indexes, corresponding to the state numbers of the given pair; $[\epsilon]$ is the number of states; $\zeta = \omega_2$; $\eta = \omega_7$; $\vartheta = \omega_6$; κ and λ are indexes corresponding to the state numbers of the pair implied by the given pair.

impast αч, βк +, γи, δи, ϵп +, ζп +, ηп +, ϑп +, κи, λи/**b**, –
317 147 4

§1 $\beta_\gamma \wedge \beta_\delta \wedge \vartheta \circ\!\!\rightarrow 4 \overline{\times}\, \eta < 1 \Rightarrow \mathbf{a} \wedge \beta_\gamma \Rightarrow \mathbf{b}\beta_\delta \wedge \mathbf{a} \Rightarrow \mathbf{a}$
§2 $|\!\leftarrow \Rightarrow \mathbf{a} \oplus c_0 \wedge \eta \Leftarrow \zeta \Rightarrow \kappa\mathbf{b}\, |\!\leftarrow \Rightarrow \mathbf{b} \oplus c_0 \wedge \eta$
$\Leftarrow \zeta \Rightarrow \lambda - \kappa \circ\!\!\rightarrow \alpha \circ\!\!\rightarrow 3\lambda \Leftrightarrow \kappa \rightarrow \alpha$
§3 $\eta \urcorner \wedge \mathbf{b} \circ\!\!\rightarrow 4 \Rightarrow \mathbf{b}\eta \urcorner \wedge \mathbf{a} \Rightarrow \mathbf{a} \rightarrow 2$
§4 $\gamma + \epsilon \Rightarrow \gamma\delta + \epsilon \Rightarrow \delta - b_\beta \circ\!\!\rightarrow 1.$

For the operator *impav*, $\beta : : \| W \|$; $[\alpha]$ is the number of the sentence in the external program to which the auxiliary exit leads; γ and δ are indexes, corresponding to the state numbers of the given pair; $\epsilon : : \| X \|$; ζ is an index corresponding to the number of one of the states in the implied pair. The number of the other state is the number of the unit position in η.

impav αч, βк +, γи, δи, ϵк +, ζи, ηп/**a**, $a/(\beta\epsilon\eta)$
316 54 1 (β**a**)

$\beta_\gamma \vee \beta_\delta \Rightarrow \mathbf{a}\ \bar{\circ}\ a$
§1 $\triangle a - b_\epsilon |\!\rightarrow \alpha\mathbf{a} \wedge \epsilon_a \Rightarrow \eta \circ\!\!\rightarrow 1\eta\, |\!\leftarrow \oplus c_0 \circ\!\!\rightarrow 1\eta\, \dot{\mathbf{X}}\, 1\zeta.$

8. CONSTRUCTION OF THE COMPATIBILITY MATRIX

The equivalence matrix for outputs $\| A \|$ is transformed to the compatibility matrix $\| B \|$ for which $b_{ij} = 1$ if and only if the states i and j are compatible and $i < j$.

In [1] it was shown that two states i and j are compatible if and only if the chain C generated by this pair of states does not contain a pair of states not equivalent for outputs.

Hence follows the algorithm for constructing the compatibility matrix. We introduce an auxiliary matrix $\| D \|$ such that $d_{ij} = 1$ if and only if $i < j$, and it has not yet been established whether the states i and j are compatible or not. The chains generated only by those pairs $\langle i, j \rangle$ for which $a_{ij} \wedge d_{ij} = 1$ are sought. If the chain C generated by the pair $\langle i, j \rangle$ contains a pair $\langle k, l \rangle$ such that $a_{kl} = 0$, then a_{ij} and d_{ij}, as well as the elements a_{pq} and d_{pq}, corresponding to that pair $\langle p, q \rangle$ which implies the pair $\langle k, l \rangle$ take the value 0. Changing in this way the values of the elements in matrix $\| A \|$, we arrive at the matrix $\| B \|$. For the chain C, generated by the pair $\langle i, j \rangle$, a matrix $\| F \|$ is constructed such that

$$f_{pq} = 1 \leftrightarrow (\langle p, q \rangle \in C \backslash \langle i, j \rangle) \wedge (a_{pq} \wedge d_{pq} = 1).$$

In the process of constructing $\| F \|$ an auxiliary matrix $\| F' \|$ is constructed that plays the same role as the matrix $\| C' \|$ in the construction of the chain matrix. If the chain C has no pair $\langle k, l \rangle$ such that $a_{kl} = 0$, then in the matrix $\| D \|$ all those elements d_{pq} take zero values to which correspond elements f_{pq} of $\| F \|$ with unit values. The process of transformation of matrix $\| A \|$ into the compatibility matrix $\| B \|$ terminates when all elements of matrix $\| D \|$ have taken on zero values.

Let us consider the two variants of the operator for constructing the compatibility matrix for the different ways of representing the initial information. These variants *comat* and *comav* differ only in that they contain different variants of the operator for obtaining chains. The operator *comat* contains the operator *chest*, and is used for a partially defined automaton with prescribed next-state matrix. If the adjacency matrix of the graph of the automaton is given, the operator *comav* is used, which contains the operator *chev*. However the operators *chest* and *chev* in the corresponding variants of the operator for constructing the compatibility matrix are written in expanded form, since they are used with certain modifications. These changes are due to the fact that in this case it is not necessary to construct the complete chain. In matrix $\| F \|$ are fixed only those elements (pairs) that contain states not yet tested for compatibility.

For the operator *comat* we adopt the following: $\eta = \omega_7$; $\vartheta = \omega_6$; $\zeta = \omega_2$; $\alpha :: \| Z \|$; $\beta :: \| A \|$; $\gamma :: \| F \|$; $\delta :: \| D \|$; $\epsilon :: \| F' \|$.

comat $(235, 317)/\alpha\text{к} +, \beta\text{к}, \gamma\text{к}, \delta\text{к}, \epsilon\text{к}, \zeta\text{п} +, \eta\text{п} +, \vartheta\text{п} +/\mathbf{d}, f/(\beta\gamma\delta\epsilon)$
274 326 11 $(\beta\mathbf{c})$

$\circ\ ab_\beta + 1 \Rightarrow \mathbf{d}$
§1 $\beta_a \Rightarrow \delta_a \circ \gamma_a \circ \epsilon_a \triangle a - b_\beta \circ\!\rightarrow$ *funel* $11\delta cd//c \Rightarrow ad \Rightarrow b$
impast $5\alpha ab\mathbf{d}\zeta\eta\vartheta fe//c \Rightarrow a$

§2 $\epsilon_a \dot{\mathbf{X}}\ 3b \rightarrow$ *impast* 1
§3 $\triangle\ a - b_\beta \circ\!\rightarrow 2c \Rightarrow a$
§4 $\gamma_a \oplus \delta_a \Rightarrow \delta_a \circ \gamma_a \triangle\ a - b_\beta \circ\!\rightarrow 4 \rightarrow$ *funel* 2
§5 $c_f \Rightarrow \mathbf{c} \wedge \gamma_e \mid\!\rightarrow$ *impast* $3\mathbf{c} \wedge \beta_e \rightarrow 7\mathbf{c} \wedge \delta_e \circ\!\rightarrow$ *impast* 3
§6 $\mathbf{c} \vee \epsilon_e \Rightarrow \epsilon_e \vee \gamma_e \Rightarrow \gamma_e \rightarrow$ *impast* 3
§7 $c_b \urcorner \wedge \beta_a \Rightarrow \beta_a \wedge \delta_a \Rightarrow \delta_a c_d \urcorner \wedge \beta_c \Rightarrow \beta_c \wedge \delta_c \Rightarrow \delta_c c \Rightarrow a$
§10 $\circ\ \gamma_a \circ \epsilon_a \triangle\ a - b_\beta \circ\!\rightarrow 10 \rightarrow$ *funel* 2
§11 .

The operator *comav* acts on the following operands: $\alpha{:}{:}\| W \|;\ \beta{:}{:}\| A \|{:}\ \gamma{:}{:}\| F \|;\ \delta{:}{:}\| D \|;\ \epsilon{:}{:}\| F' \|;\ \zeta{:}{:}\| X \|$.

comav $(325, 316)/\alpha\kappa +, \beta\kappa, \gamma\kappa, \delta\kappa, \epsilon\kappa, \zeta\kappa +/\mathbf{c}, f/(\alpha\beta\gamma\delta\epsilon\zeta)$
276 312 12 $(\beta\mathbf{c})$

$\circ\ a$
§1 $\beta_a \Rightarrow \delta_a \circ \gamma_a \circ \epsilon_a \triangle\ a - b_\beta \circ\!\rightarrow 1$ *funel* $12\delta cd//c \Rightarrow fd \Rightarrow b$
§2 *impav* $6\alpha f b\zeta e\mathbf{c}//\mathbf{c} \wedge \gamma_e \mid\!\rightarrow$ *impav* $1\mathbf{c} \wedge \beta_e \circ\!\rightarrow 4\mathbf{c} \wedge \delta_e \circ\!\rightarrow$ *impav* 1
§3 $\mathbf{c} \vee \epsilon_e \Rightarrow \epsilon_e \vee \gamma_e \Rightarrow \gamma_e \rightarrow$ *impav* 1
§4 $c_b \urcorner \wedge \beta_f \Rightarrow \beta_f \wedge \delta_f \Rightarrow \delta_f c_d \urcorner \wedge \beta_c \Rightarrow \beta_c \wedge \delta_c \Rightarrow \delta_c c \Rightarrow a$
§5 $\circ\ \gamma_a \circ \epsilon_a \triangle\ a - b_\beta \circ\!\rightarrow 5 \rightarrow$ *funel* 2
§6 $c \Rightarrow a$
§7 $\epsilon_\alpha \dot{\mathbf{X}}\ 10b \rightarrow 2$
§10 $\triangle\ a - b_\beta \circ\!\rightarrow 7c \Rightarrow a$
§11 $\gamma_a \oplus \delta_a \Rightarrow \delta_a \circ \gamma_a \triangle\ a - b_\beta \circ\!\rightarrow 11 \rightarrow$ *funel* 2
§12 .

Let us consider an example.

$\| Z \| =$

z \ x	0	1	2	3
0	4	2		
1	1	3		
2	0	2		
3		1		3
4	2		3	0

$$\| A \| = \begin{bmatrix} 0 & 1 & 1 & 0 & 0 \\ 0 & 0 & 1 & 1 & 1 \\ 0 & 0 & 0 & 1 & 0 \\ 0 & 0 & 0 & 0 & 0 \end{bmatrix}.$$

The initial value of $\| D \|$ coincides with the value of matrix $\| A \|$. In the investigation of the pair $\langle 0, 1 \rangle$ for compatibility, we obtain

$$\| F \| = \begin{bmatrix} 0 & 0 & 0 & 0 & 0 \\ 0 & 0 & 1 & 0 & 1 \\ 0 & 0 & 0 & 1 & 0 \\ 0 & 0 & 0 & 0 & 0 \end{bmatrix}.$$

Then the matrix $\| D \|$ takes the following value:

$$\| D \| = \begin{bmatrix} 0 & 0 & 1 & 0 & 0 \\ 0 & 0 & 0 & 1 & 0 \\ 0 & 0 & 0 & 0 & 0 \\ 0 & 0 & 0 & 0 & 0 \end{bmatrix}.$$

The final result, obtained in the investigation for compatibility of the pairs $\langle 0, 2\rangle$ and $\langle 1, 3\rangle$, has the following form:

$$\| B \| = \begin{bmatrix} 0 & 1 & 0 & 0 & 0 \\ 0 & 0 & 1 & 1 & 1 \\ 0 & 0 & 0 & 1 & 0 \\ 0 & 0 & 0 & 0 & 0 \end{bmatrix}.$$

9. FINDING THE MAXIMAL SETS OF COMPATIBLES

To find the maximal sets of compatibles, we use an algorithm described in [1]. It consists of the following.

Let there exist a set S of certain subsets S_i $(i = 0, 1, \ldots, \sigma(S) - 1)$ of a set Z of the states of the given automaton. A certain state $z \in Z$ is chosen, and for each of the sets $S_k \in S$ the condition

$$(z \in S_k) \wedge (\exists\, z')[(z' \in Z) \wedge (z' * z)] \tag{1}$$

is tested, where $z' * z$ signifies that the states z' and z are incompatible.

When this condition is satisfied, S_k is excluded from S and sets $S_k' = S_k \backslash \{z\}$ and $S_k'' = S_k \backslash \{z'/z' \in Z, z' * z\}$ are found. Each of these sets is included in S if and only if it is not contained in any $S_i \in S$. If condition (1) is not fulfilled, S_k remains in S without change.

If $S = \{Z\}$ is given as the initial value and the above transformation is applied to S $\sigma(Z)$ times, taking in order all states $z \in Z$, in the result S will represent a family of maximal sets of compatibles.

The initial information for this operator, to which we give the code *maxcom*, is the compatibility matrix $\| B \|$. First this matrix is transformed to the matrix $\| B' \|$, where $b_{ij}' = 0$ if and only if $i * j$ and $i < j$. This matrix is more convenient for calculations than $\| B \|$.

In sentences 2 and 3 condition (1) is tested and S_k' and S_k'' are found. Then S_k' and S_k'' are tested for maximality; the principle of the operator *uplim* is applied to this. The set S is represented by the complex β. The complex α represents the compatibility matrix $\| B \|$.

If S_k' and S_k'' are not maximal, i.e., S_l and $S_m \in S$ different from S are found such that $S_k' \subseteq S_l$ and $S_k'' \subseteq S_m$, the complex β is reduced by one element, where the value of the last element of complex β is assigned to that element β_k which represented S_k. Otherwise,

$$\beta_k :: \| \{S_k''\} \ni Z \|,$$

if S_k'' is maximal, or

$$\beta_k :: \| \{S_k'\} \ni Z \|.$$

maxcom $\alpha\kappa, \beta\kappa?/\mathbf{e}, c/(\alpha\beta)$
310 327 13 $(\alpha\mathbf{de})$

$b_\alpha \Rightarrow a \circ \mathbf{d} - c_a \Rightarrow \beta_0 \urcorner \Rightarrow \mathbf{d} \circ a$
§1 $c_a \vee \mathbf{d} \Rightarrow \mathbf{d} \oplus \alpha_a \Rightarrow \alpha_a \triangle a - b_\alpha \circ\!\!\rightarrow 1 \quad 1 \Rightarrow \mathbf{b} \Rightarrow b_\beta \circ b$
§2 $\alpha_b \urcorner \circ\!\!\rightarrow 10 \circ c$
§3 $c_b \urcorner \wedge \beta_c \Rightarrow \mathbf{e} \oplus \beta_c \circ\!\!\rightarrow 7\beta_c \wedge \alpha_b \Rightarrow \mathbf{d} \circ\!\!\rightarrow 7 \circ \beta_c \circ \mathbf{a} \circ a$
§4 $\beta_a \urcorner \wedge \mathbf{d} \circ\!\!\rightarrow 5 \triangle a - b_\beta \circ\!\!\rightarrow 4\mathbf{d} \Rightarrow \beta_c \triangle \mathbf{a}$
§5 $\circ\, a$
§6 $\beta_a \urcorner \wedge \mathbf{e} \circ\!\!\rightarrow 12 \triangle a - b_\beta \circ\!\!\rightarrow 6\mathbf{a} \mid\!\!\rightarrow 11\mathbf{e} \Rightarrow \beta_c$
§7 $\triangle c - b_\beta \circ\!\!\rightarrow 3$
§10 $\triangle b - b_\alpha \mid\!\!\rightarrow 13\mathbf{b} \Rightarrow b_\beta? \rightarrow 2$
§11 $\mathbf{b} \Rightarrow a\mathbf{e} \Rightarrow \beta_a \triangle \mathbf{b} \rightarrow 10$
§12 $\mathbf{a} \mid\!\!\rightarrow 7 \,\overline{\triangle}\, \mathbf{b} \oplus c \circ\!\!\rightarrow 10\mathbf{b} \Rightarrow a\beta_a \Rightarrow \beta_c\mathbf{b} - b_\beta \mid\!\!\rightarrow 7\mathbf{b} \Rightarrow b_\beta \rightarrow 3$
§13 .

Let us consider an example. Let there be given a set of states $Z = \{0, 1, 2, 3, 4, 5, 6\}$ with compatibility matrix

$$\| B \| = \begin{bmatrix} 0 & 1 & 1 & 0 & 0 & 0 & 0 \\ 0 & 0 & 1 & 0 & 0 & 0 & 1 \\ 0 & 0 & 0 & 0 & 1 & 0 & 1 \\ 0 & 0 & 0 & 0 & 0 & 0 & 0 \\ 0 & 0 & 0 & 0 & 0 & 1 & 1 \\ 0 & 0 & 0 & 0 & 0 & 0 & 0 \end{bmatrix}.$$

The result is represented by the following matrix:

$$\| S \ni Z \| = \begin{bmatrix} 1 & 1 & 1 & 0 & 0 & 0 & 0 \\ 0 & 1 & 1 & 0 & 0 & 0 & 1 \\ 0 & 0 & 1 & 0 & 1 & 0 & 1 \\ 0 & 0 & 0 & 1 & 0 & 0 & 0 \\ 0 & 0 & 0 & 0 & 1 & 1 & 0 \end{bmatrix}.$$

10. FINDING A CLOSED SET OF SETS OF COMPATIBLES

Let us consider an approximate method for finding a closed set L of sets of compatibles, which permits this set to be obtained fairly close to minimal, but does not use complete enumeration.

To consider this method we introduce the concept of the relation of partial implication, defined on the set of all sets of compatibles in the following way. We shall say that a set of compatibles $P \subset Z$ (maximal or nonmaximal) partially implies the set M if P implies the set Q while M is the only maximal set of compatibles that contains Q.

Let us construct a certain set S whose elements are the sets of compatibles selected through the process described below. Simultaneously with the construction of the set S we define on it a certain relation which we shall denote by the symbol $\circ$.

(1) Let $S_M \subseteq S$, where S_M is the set of all maximal sets of compatibles. We shall consider that $M \circ M'$ for $M \in S_M$ and $M' \in S_M$ if M partially implies M'.

(2) We include in S all those nonmaximal sets of compatibles P each of which: has cardinality greater than one, is implied by at least one M in S_M, is not contained in any M implying this P, but is contained in more than one of the remaining $M \in S_M$. For each P so defined and for all $M \in S_M$ implying the given P, we put $M \circ P$. If P partially implies M, then $P \circ M$.

(3) We include in S all Q each of which: is implied by some P already included in S in the preceding step, is not itself contained in it, and is contained in more than one $M \in S_M$. As in the preceding step, we define $P \circ Q$ and $Q \circ M$.

The third step is repeated as long as its execution changes the set S. We shall assume that $S_i \circ S_i$ holds for all $S_i \in S$.

We shall seek some $L \subseteq S$ such that it satisfy the conditions of closure and its number of elements be minimal.

For each $S_i \in S$ we find $K_i \subseteq S$ such that

$$S_i \bar{\circ} S_j \leftrightarrow S_j \in K_i,$$

where the symbol $\bar{\circ}$ is the transitive closure of the relation $\circ$. The first condition of closure is fulfilled for K_i, i.e., for any Q implied by some $S_j \in K_i$ there is always found such $S_k \in K_i$ that $Q \subseteq S_k$. This condition is preserved also for the set $N_i = \sup K_i$, which is the upper bound of the set K_i.

For the set N of the sets N_i thus defined, $B = \{\bigcup [N_i]/N_i \in N\}$ is found. A certain set $D \subseteq \tilde{N}$ is associated with the set $\inf \operatorname{cov}_B Z$ of irredundant coverages of the set Z by elements B_k of the set B in the following manner. With each $A_i \in \inf \operatorname{cov}_B Z$ is associated a $D_i \in D$ such that if $B_k \in A_i$, then $N_k \in D_i$.

The required set L will satisfy the condition

$$L \in \min_\sigma \{\sup \bigcup [D_i]/D_i \in D\}.$$

Let us consider the process of obtaining the set L by an example. Let there be given the next-state matrix and the set S_M:

$\| Z \| =$

z \ x	0	1	2	3	4
0			1	6	3
1		4	6		1
2			6	3	
3				2	1,
4	5	7			
5	0		6		
6		1		3	
7	6		3	0	

$S_M = \{S_0, S_1, S_2, S_3, S_4, S_5, S_6\}$,

where
$S_0 = \{2, 4, 7\}$;
$S_1 = \{2, 5, 6\}$;
$S_2 = \{2, 3, 4\}$;
$S_3 = \{2, 3, 6\}$;
$S_4 = \{1, 3, 4\}$;
$S_5 = \{1, 4, 7\}$;
$S_6 = \{0, 3, 4\}$.

The set S is obtained by addition to the set S_M of two nonmaximal sets of compatibles, $S_7 = \{4, 7\}$ and $S_8 = \{2, 6\}$. For clarity, we represent the relation $\circ$ by a graph (Fig. 1), where the set of nodes is the set S, and the S_i are connected to the S_j if and only if $S_i \circ S_j$. The elements of the set K will be

$$\{S_0, S_1, S_3, S_4, S_6, S_7, S_8\}; \quad \{S_1\}; \quad \{S_2\};$$
$$\{S_3\}; \quad \{S_1, S_4, S_7\}; \quad \{S_1, S_3, S_5\};$$
$$\{S_1, S_4, S_6, S_7, S_8\}; \quad \{S_1, S_7\}; \quad \{S_8\}.$$

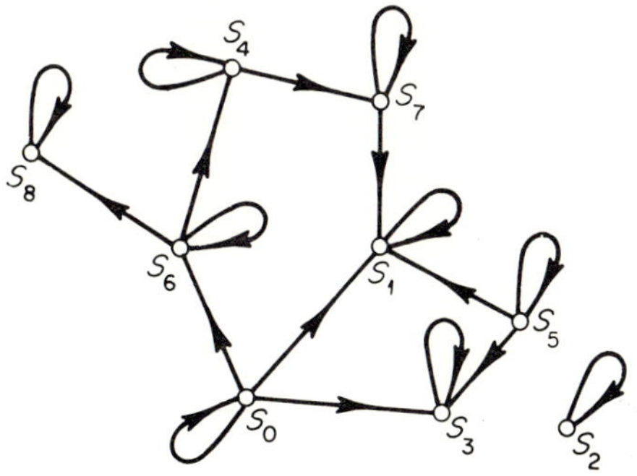

FIG. 1.

The elements of the sets N and B are conveniently arranged opposite each other:

$$\begin{aligned} N_0 &= \{S_0, S_1, S_3, S_4, S_6\} & B_0 &= \{0, 1, 2, 3, 4, 5, 6, 7\} \\ N_1 &= \{S_1\} & B_1 &= \{2, 5, 6\} \\ N_2 &= \{S_2\} & B_2 &= \{2, 3, 4\} \\ N_3 &= \{S_3\} & B_3 &= \{2, 3, 6\} \\ N_4 &= \{S_1, S_4, S_7\} & B_4 &= \{1, 2, 3, 4, 5, 6, 7\} \\ N_5 &= \{S_1, S_3, S_5\} & B_5 &= \{1, 2, 3, 4, 5, 6, 7\}. \\ N_6 &= \{S_1, S_4, S_6, S_7\} & B_6 &= \{0, 1, 2, 3, 4, 5, 6, 7\} \\ N_7 &= \{S_1, S_7\} & B_7 &= \{2, 4, 5, 6, 7\} \\ N_8 &= \{S_8\} & B_8 &= \{2, 6\} \end{aligned}$$

Hence we obtain $\inf \operatorname{cov}_B Z = \{\{B_0\}, \{B_6\}\}$ and $D = \{\{N_0\}, \{N_6\}\}$ and, finally, $\min_\sigma \{\sup \cup [D_i]/D_i \in D\} = \min_\sigma \{N_0, N_6\} = N_6$ or

$$L = \{S_1, S_4, S_6, S_7\}.$$

Thus the number of states of the prescribed automaton has been reduced by four.

This operator has two variants, one of which is intended for the case in which the matrix $\| Z \|$ is prescribed (*closetcom*), the other for the case in which the matrices $\| W \|$ and $\| X \|$ are prescribed (*closetcov*). In the first the operator *imset* is used to find the implied sets, in the second, *imsev*.

In sentences 3–5, $\sup \cup [D_i]$ and $\sup N_i$ are found. This work is carried out by the operator *uplim*, but in a different way. The difference is due to the fact that here not all the elements of the corresponding complex are tested for absorption, but only those that correspond to $S_i \in S \backslash S_M$.

The elements of the set $\inf \operatorname{cov}_B Z$ are found by means of the operator *fisirco*. The sets $\cup [N_i]$ and $\cup [D_i]$ are found by means of the operator *concom*. The operator *transclo* finds the set K_i.

The operands are the following: $\gamma :: \| S \ni Z \|$; β is the number of states, $[\alpha]$ is the number of the sentence to which the auxiliary exit terminal of the given operator leads. This exit is realized when no L has been found for which $\sigma(L) < \sigma(Z)$. $\zeta :: \| Z \|$; ϵ and η are used as auxiliary complexes for the operators *imset* and *fisirco*; ϑ, κ, and λ are ω-constants, used in the operator *imset*. The sense of δ changes in the process of passage from step to step, taking on the values $\| S \circ S \|$, $\| S \bar{\circ} S \|$, where $\bar{\circ}$ is the transitive closure of the relation $\circ$; $\| N \ni S \|$.

The result is represented by the complex $\epsilon :: \| L \ni Z \|$.

closetcom (330, 327, 40, 50, 5, 43)/αч, βп +, γк, δк?, ϵк, ζк +, ηк, κп +, ϑп +, λп +/**k**, $d/(\delta\epsilon)(\gamma\eta)$
277 556 20 (γ**e**) (δ**acfhj**)

○ $db_\gamma \Rightarrow b_\delta \Rightarrow a$ ○ δ_a
§1 $c_d \Rightarrow \delta_d$ *imset* $7\beta\lambda(\gamma_d)$**e**$\zeta\eta\vartheta\kappa\epsilon//\triangle\, d - b_\delta$ ○→ 1
transclo $\delta//b_\gamma - 1 \Rightarrow a0 - c_a \Rightarrow$ **h** ○ a
§2 $\delta_a \Rightarrow$ **j** $\nLeftrightarrow$ 3**j** $\Rightarrow \delta_a \triangle\, a - b_\delta$ ○→ 2 *redsim* $\delta\delta//$ → 15
§3 **h** ⌉ ∧ **j** ⇒ **c**
§4 **c** $\dot{\mathbf{X}}$ 6*b***h** ∧ **j** ⇒ **a**
§5 **a** $\dot{\mathbf{X}}$ $4c\gamma_c$ ⌉ ∧ γ_b |→ $5c_b$ ⊕ **j** ⇒ **j** → 4
§6 !
§7 **e** ○→ *imset* 5**e** |← ⊕ c_0 ○→ *imset* $5\gamma_d$ ⌉
∧ **e** ○→ *imset* 5 ○ **d** ○ a

§10 γ_a ⌉ ∧ **e** |→ 11**d** |→ 12 △ **d**$c_a \Rightarrow$ **f**
§11 △ $a - b_\gamma$ ○→ 10**f** ∨ $\delta_d \Rightarrow \delta_d$ → *imset* 5
§12 $b_\gamma \Rightarrow a$
§13 γ_a ⊕ **e** ○→ 14 △ $a - b_\delta$ ○→ 13**e** $\Rightarrow \gamma_a \triangle\, b_\delta$?
§14 $c_a \vee \delta_d \Rightarrow \delta_d$ → *imset* 5
§15 $b_\delta \Rightarrow b_\epsilon \Rightarrow b_\eta$ ○ a

Obtaining the set S and defining on it the relation $*$.

§16 $\delta_a \Rightarrow$ **a** *concom* ∨ γ**a**$(\epsilon_a)//\triangle\, a - b_\epsilon$ ○→ 16β
$\Rightarrow$ **g** $- 1 \Rightarrow a0 - c_a \Rightarrow$ **e** ○ **f**$b_\gamma - 1 \Rightarrow a0$
$- c_a \Rightarrow$ **h** *fisirco* 17ϵ**ei**$\eta//$**f** ○→ α → 20
§17 *concom* ∨ δ**ij**$//\nLeftrightarrow$ 3**j** ∇ ⇒ **k** − **g** |→ *fisirco* 4**k**
⇒ **gj** ⇒ **f** → *fisirco* 4
§20 *compress* γ**f**$\epsilon//$.

The operator *closetcov* differs from *closetcom* only in certain operations and in that the operator *imsev* is used in place of *imset* to find the implied sets. The operands α, γ, δ, η, and ϵ are defined as for *closetcom*. Further, $\beta::\|X\|$; $\rho::\|W\|$.

closetcov (331, 327, 40, 50, 5, 43)/αч, βк +, γк, δк?, ϵк, ζк +, ηк/**k**, $d(\gamma\beta\zeta)(\delta\epsilon)$

300 552 20 (γe) (δ**acfhj**)

○ $db_\gamma \Rightarrow b_\delta \Rightarrow a$ ○ δ_a
§1 $c_d \Rightarrow \delta_d$ *imsev* $7\beta\zeta(\gamma_d)$**e**$//\triangle\, d - b_\delta$ ○→ 1 *transclo* $\delta//b_\gamma - 1 \Rightarrow a$
$0 - c_a \Rightarrow$ **h** ○ a
§2 $\delta_a \Rightarrow$ **j** $\nLeftrightarrow$ 3**j** $\Rightarrow \delta_a \triangle\, a - b_\delta$ ○→ 2 *redsim* $\delta\delta//$→ 15

§3 $\mathbf{h} \urcorner \wedge \mathbf{j} \Rightarrow \mathbf{c}$

§4 $\mathbf{c}\, \dot{\mathbf{X}}\, 6b\mathbf{h} \wedge \mathbf{j} \Rightarrow \mathbf{a}$

§5 $\mathbf{a}\, \dot{\mathbf{X}}\, 4c\gamma_c \urcorner \wedge \gamma_b \mapsto 5c_b \oplus \mathbf{j} \Rightarrow \mathbf{j} \rightarrow 4$

§6 !

§7 $\mathbf{e} \circ\!\!\rightarrow \mathit{imsev}\ 2\mathbf{e} \mid\!\leftarrow \oplus c_0 \circ\!\!\rightarrow \mathit{imsev}\ 2\gamma_d \urcorner \wedge \mathbf{e} \circ\!\!\rightarrow \mathit{imsev}\ 2 \circ \mathbf{d} \circ a$

§10 $\gamma_a \urcorner \wedge \mathbf{e} \mapsto 11\mathbf{d} \mapsto 12 \triangle \mathbf{d}c_a \Rightarrow \mathbf{f}$

§11 $\triangle a - b_\gamma \circ\!\!\rightarrow 10\mathbf{f} \vee \delta_d \Rightarrow \delta_d \rightarrow \mathit{imsev}\ 2$

§12 $b_\gamma \Rightarrow a$

§13 $\gamma_a \oplus \mathbf{e} \circ\!\!\rightarrow 14 \triangle a - b_\delta \circ\!\!\rightarrow 13\mathbf{e} \Rightarrow \gamma_a \triangle b_\delta?$

§14 $c_a \vee \delta_d \Rightarrow \delta_d \rightarrow \mathit{imsev}\ 2$

§15 $b_\delta \Rightarrow b_\epsilon \Rightarrow b_\eta \circ a$

§16 $\delta_a \Rightarrow \mathbf{a}\ \mathit{concom} \vee \gamma\mathbf{a}(\epsilon_a)//\triangle a - b_\epsilon \circ\!\!\rightarrow 16b_\zeta \Rightarrow \mathbf{g} - 1 \Rightarrow a$
$0 - c_a \Rightarrow \mathbf{e} \circ \mathbf{f}b_\gamma - 1 \Rightarrow a\, 0 - c_a \Rightarrow \mathbf{h}\ \mathit{fisirco}\ 17\epsilon\mathbf{ei}\eta//\mathbf{f} \circ\!\!\rightarrow \alpha$
$\rightarrow 20$

§17 $\mathit{concom} \vee \delta\mathbf{ij}//\not\leftrightarrow 3\mathbf{j} \nabla \Rightarrow \mathbf{k} - \mathbf{g} \mapsto \mathit{fisirco}\ 4\mathbf{k} \Rightarrow \mathbf{g}$
$\mathbf{j} \Rightarrow \mathbf{f} \rightarrow \mathit{fisirco}\ 4$

§20 $\mathit{compress}\ \gamma\mathbf{f}\epsilon//.$

11. FINDING SETS OF STATES IMPLIED BY A GIVEN SET

For the given set $P \subseteq Z$ are found all $Q \subseteq Z$ that are implied by the set P. If the function δ is given by the next-state matrix $\| Z \|$, the operator *imset* is used, for the matrices $\| W \|$ and $\| X \|$, the operator *imsev*.

For the operator *imset* there must be given $\zeta :: \| Z \|$, $\delta :: \| \{P\} \ni Z \|$; $[\alpha]$ is the number of the sentence of the external program to which transfer is effected after the next implied set has been found; $\gamma = \omega_2$, $\vartheta = \omega_3$, $\kappa = \omega_7$; $[\beta]$ is the number of states. Two further working complexes, η and λ, are used. The elements of the complex λ are the rows of the matrix $\| Z \|$, corresponding to the states of the set P. The complex η represents the following. According to the adopted coding of the matrix $\| Z \|$, the value of the extreme left position in the group of positions coding an element of this matrix is the definition tag. We call these positions d-positions. The position of an element of the matrix $\| Z \|$ in the row (i.e., the number of its column) corresponds to some degree to the number of the d-position of the group coding this element. Elements of the complex η represent the numbers of the extreme left unit d-positions of the elements of complex ζ, distinguished in complex λ. To determine one of the required sets Q, which consists of elements of $\| Z \|$ belonging to the same column, it is necessary to find in λ all those elements whose numbers are the numbers of elements of the complex η having the same value.

The sets Q are defined successively, and are represented by the variable $\epsilon\colon :\| \{Q\} \ni Z \|$.

$$\textit{imset}\ \alpha\text{ч},\ \beta\text{п}\ +,\ \gamma\text{п}\ +,\ \delta\text{п}\ +,\ \epsilon\text{п},\ \zeta\text{к}\ +,\ \eta\text{к},\ \vartheta\text{п}\ +,\ \kappa\text{п}\ +,\ \lambda k/\mathbf{c},\ c$$
330 204 5

$$\circ\ \mathbf{b}\ \circ\ \epsilon(\delta\mathbf{c})$$

§1 $\circ\ c\ \circ\ \mathbf{a}\delta \Rightarrow \mathbf{c}\ \circ\ a$

§2 $\mathbf{c}\ \dot{\mathbf{X}}\ 3b + \mathbf{b} \Rightarrow b\zeta_b \vdash\ \Rightarrow \eta_c\zeta_b\ |\!\leftarrow\ \Rightarrow \lambda_c\ \triangle\ c \to 2$

§3 $\mathbf{a} \oplus \eta_a\ |\!\rightarrow 4\lambda_a\ \circ\!\!\to 4 \oplus c_0 \Lleftarrow \gamma \wedge f_{10} \Rightarrow b\kappa\ \urcorner\ \wedge\ \lambda_a \Rightarrow \lambda_a \vdash\ + \eta_a \Rightarrow \eta_a$
$\Rightarrow c_b \vee \epsilon \Rightarrow \epsilon\lambda_a\ |\!\leftarrow\ \Rightarrow \lambda_a$

§4 $\triangle\ a - c\ \circ\!\!\to 3 \to \alpha$

§5 $\circ\ \epsilon\ \circ\ a\gamma + \mathbf{a} \Rightarrow \mathbf{a} - \vartheta\ \circ\!\!\to 3\beta + \mathbf{b} \Rightarrow \mathbf{b} - b_\zeta\ \circ\!\!\to 1.$

To illustrate the functioning of the operator we present an example. Let the next-state matrix and the corresponding complex have the forms

$\|\mathbf{Z}\| =$

$z \backslash x$	0	1
0	4	
1		2
2	0	
3	2	1
4		3

$$\mathbf{Z} = \begin{array}{cccccccc} 1&1&0&0&0&0&0&0 \\ 0&0&0&0&1&0&1&0 \\ 1&0&0&0&0&0&0&0 \\ 1&0&1&0&1&0&0&1 \\ 0&0&0&0&1&0&1&1 \end{array}$$

We determine the sets of states implied by the set $\{0, 1, 3\}$. The complexes **C** and **D**, constituting the operands λ and η, respectively, take on the following values:

$$\mathbf{C} = \begin{array}{cccccccc} 1&1&0&0&0&0&0&0 \\ 1&0&1&0&0&0&0&0 \\ 1&0&1&0&1&0&0&1 \end{array} \qquad \mathbf{D} = \begin{array}{cccccccc} 0&0&0&0&0&0&0&0 \\ 0&0&0&0&0&1&0&0 \\ 0&0&0&0&0&0&0&0 \end{array} \qquad \begin{array}{c} (0) \\ (4) \\ (0). \end{array}$$

Opposite each element $\mathbf{d}_i$ in parentheses is shown the number whose positional code is the given element.

We first extract those $\mathbf{c}_i$ for which $[\mathbf{d}_i] = 0$. These are $\mathbf{c}_0$ and $\mathbf{c}_2$, whose highest-order positions give one of the required sets, $\{2, 4\}$.

After this we obtain new values of the complexes **C** and **D**:

$$\mathbf{C} = \begin{array}{cccccccc} 0&0&0&0&0&0&0&0 \\ 1&0&1&0&0&0&0&0 \\ 1&0&0&1&0&0&0&0 \end{array} \qquad \mathbf{D} = \begin{array}{cccccccc} 0&0&0&0&0&0&0&0 \\ 0&0&0&0&0&1&0&0 \\ 0&0&0&0&0&1&0&0 \end{array} \qquad \begin{array}{c} (0) \\ (4) \\ (4). \end{array}$$

On repeated access to *imset* those $\mathbf{c}_i$ are found for which $[\mathbf{d}_i] = 4$. The second required set will be $\{1, 2\}$.

The operator *imsev* is described by a simpler expression. Its operands are the complexes β and γ, representing the matrices $\| X \|$ and $\| W \|$, and the variable $\delta :: \| \{P\} \ni Z \|$. The result is represented by the variable $\epsilon :: \| \{Q\} \ni Z \|$; $[\alpha]$ is the sentence in the external program to which control is transferred after the next implied set is found.

imsev (50)/αч, βк +, γк +, δп, εп/**a**, *b*/(βγδε)
331 44 2 (β**a**)

concom $\vee$ γδ**a**// $\circ$ *b*
§1 $\mathbf{a} \wedge \beta_b \Rightarrow \epsilon \rightarrow \alpha$
§2 $\triangle\, b - b_\beta \circ\!\rightarrow 1.$

12. CONSTRUCTION OF THE TRANSITIVE CLOSURE MATRIX OF A BINARY RELATION

Let there be defined on the set $S = \{S_0, S_1, \ldots, S_{n-1}\}$ a certain binary relation $\circ$, represented by the matrix $\| S \circ S \|$. It is required to find a transitive closure of the relation $\circ$, i.e., to construct the matrix of a relation $\bar{\circ}$ on the set S such that if there exists a chain $S_i \circ S_{i_1}, S_{i_1} \circ S_{i_2}, \ldots, S_{ik} \circ S_j$, then $S_i \bar{\circ} S_j$.

The matrix $\| S \circ S \|$ is represented by the complex α. The result is obtained in the same complex.

transclo ακ/**b**, *b*
327 124 4 (α**ab**)

$\circ$ *a*
§1 $\alpha_a \Rightarrow \mathbf{a} \circ \mathbf{b}$
§2 $\mathbf{a}\, \dot{\mathbf{X}}\, 3b\alpha_b \vee \mathbf{b} \Rightarrow \mathbf{b}b - a \mid\!\rightarrow 2\alpha_b \vee \alpha_a \Rightarrow \alpha_a\alpha_b \urcorner \wedge \mathbf{a} \Rightarrow \mathbf{a} \rightarrow 2$
§3 $\alpha_a \urcorner \wedge \mathbf{b} \Rightarrow \mathbf{a} \circ\!\rightarrow 4\mathbf{a} \vee \alpha_a \Rightarrow \alpha_a \circ \mathbf{b} \rightarrow 2$
§4 $\triangle\, a - b_\alpha \circ\!\rightarrow 1.$

13. CHOICE OF SINGLE-STEP TRANSFORMATION OF THE AUTOMATON WITH RESPECT TO AN INFORMATION CRITERION

For each pair $\langle i, j \rangle$ for which the corresponding element a_{ij} of the equivalence matrix for outputs $\| A \|$ is equal to 1, a chain C is constructed, generated by this pair. If it is found that the chain C contains a pair $\langle k, l \rangle$

for which $a_{kl} = 0$, the construction of the chain is stopped and the element of matrix $\| A \|$ corresponding to the pair that implies $\langle k, l \rangle$ as well as the element a_{ij} takes the value 0. If $\sigma(C) \geq \sigma(\bigcup [C])$, the transformation on the basis of chain C does not lead to reduction in the number of states. We do not here consider such transformations. If $\sigma(C) < \sigma(\bigcup [C])$ and the chain C does not contain incompatible pairs, the quantity ΔH is calculated (the relative change of the degree of indeterminacy of the functions δ and λ, due to their possible further definition with merging of the states occurring in the pairs of the chain [3]). That chain is chosen for which ΔH is minimal. If there is more than one such chain, to reduce the time for transformation of the next state matrix that chain is taken for which the quantity $p = 2\,\sigma(C) - \sigma(\bigcup [C])$ is minimal [4].

The chain C is constructed by means of the operator *chest*. The unit elements a_{ij} are sought as in the case of the operator *funel*, but with conservation of their values. The quantity ΔH is calculated by the formula

$$\Delta H = [\log_2 \sigma(Z) \sum_{i=1}^{\sigma(C)} k_{Z}{}^{i} + \log_2 \sigma(Y) \sum_{i=1}^{\sigma(C)} k_{Y}{}^{i}]/r,$$

where $r = \sigma(\bigcup [C]) - \sigma(C)$; $k_Z{}^i$ is the number of elements of the next-state matrix that are additionally defined with merging of the states of the ith pair in the chain C; $k_Y{}^i$ is the same for the output matrix. The values of the logarithms are contained in a table that is represented by the complex β, where $[\beta_a] = k \log_2 [a]$, k being a scale factor similar to that used in machines with fixed-point representation of numbers.

The result is given in the form of a transformation matrix. We define the latter as a binary matrix $\| G \|$ whose rows correspond to the new states, columns to the old, and $g_{ij} = 1$ if and only if the old state j is transformed to the new state i. The matrix $\| G \|$ is represented by the complex ζ. The complex λ is used to represent the auxiliary matrix $\| C' \|$, necessary for the construction of the chain. The chain matrix is represented by the complex μ; $[\alpha]$ is the number of the sentence in the external program to which exit is realized when it has not been possible to find a single-step transformation of the automaton leading to a reduction in the number of states: $\gamma = \omega_2$; $\delta = \omega_7$; $\epsilon = \omega_6$; $[\xi] = k \log_2 \sigma(Y)$; $\nu = \omega_4$. The complexes κ, ϑ, and η represent the equivalence matrix for outputs, the next-state matrix, and the output matrix, respectively.

inftran (321, 325)/αч, βк +, γп +, δп +, ϵп +, ζк, ηк +, ϑк +, κк, λк, μк, νп +, ξп +/**p**, f/($\zeta\kappa\lambda\mu\nu$)
323 520 13 (ζ**cfim**)

$\circ$ e $\bar{\circ}$ **e** $\circ$ **f**$b_\kappa + 1 \Rightarrow a\beta_a \Rightarrow$ **g** $\bar{\circ}$ **k**

§1 $\kappa_e \Rightarrow$ **m** $\circ$ a

§2 $\circ\ \mu_a\ \circ\ \lambda_a \triangle a - b_\kappa \circ\!\rightarrow 2$

§3 $\mathbf{m}\, \dot{\mathbf{X}}\, 10f \Rightarrow de \Rightarrow cc_f \Rightarrow \mu_e \vee c_e \Rightarrow \mathbf{i1} \Rightarrow \mathbf{j}$
chest $\vartheta(\mathbf{c} \wedge \kappa_b \circ\!\rightarrow 12\mathbf{c} \vee c_b \vee \mathbf{i} \Rightarrow \mathbf{i} \triangle \mathbf{j})$
$dc\lambda\gamma\delta\epsilon\mu // \circ\ d$
$\mathbf{i} \nabla \Rightarrow \mathbf{pj} - \mathbf{p} \mid\!\rightarrow 2\mathbf{p} - \mathbf{j} \Rightarrow \mathbf{h} \circ \mathbf{a} \circ \mathbf{b}b_\kappa$
$+ 1 \Rightarrow \mathbf{d} \circ c$

funel $5\mu db // d \Rightarrow ac_a \vee c_b \Rightarrow \lambda_c \triangle c$ — Calculation of ΔH

§4 $\vartheta_a \oplus \vartheta_b \wedge \epsilon \nabla + \mathbf{a} \Rightarrow \mathbf{a}\eta_a \oplus \eta_b \wedge \epsilon \nabla$
$+ \mathbf{b} \Rightarrow \mathbf{bd} + a \Rightarrow ad + b \Rightarrow b - b_\vartheta$
$\circ\!\rightarrow 4 \rightarrow$ *funel* 2

§5 $\circ\ ag \times \mathbf{a} \Rightarrow \mathbf{a}\xi \times \mathbf{b} + \mathbf{a{:}h} \Rightarrow \mathbf{a2} \times \mathbf{j}$
$- \mathbf{p} \Rightarrow \mathbf{de} - я \circ\!\rightarrow 2 \mid\!\rightarrow 6\mathbf{h} - 1 \mid\!\rightarrow 2$
$\mathbf{d} - \mathbf{k} \mid\!\rightarrow 2$

§6 $\mathbf{d} \Rightarrow \mathbf{ki} \Rightarrow \mathbf{f}я \Rightarrow \mathbf{eh} \Rightarrow \mathbf{lj} \Rightarrow \mathbf{n}$ — Comparison of two transformations and retention of the better one
§7 $\lambda_a \Rightarrow \zeta_a \triangle a - c \circ\!\rightarrow 7 \circ a \rightarrow 2$

§10 $\triangle e - b_\kappa \circ\!\rightarrow 1\mathbf{f} \circ\!\rightarrow \alpha\mathbf{f} \oplus \nu \Rightarrow \mathbf{an} \Rightarrow c$ — Construction of the matrix $\| G \|$
§11 $a\, \dot{\mathbf{X}}\, 13ac_a \Rightarrow \zeta_c \triangle c \rightarrow 11$

§12 $c_c \urcorner \wedge \kappa_d \Rightarrow \kappa_d c_f \urcorner \wedge \kappa_e \Rightarrow \kappa_e \circ a \rightarrow 2$
§13 $c \Rightarrow b_\zeta$.

14. CHOICE OF SINGLE-STEP TRANSFORMATION OF THE AUTOMATON FOR THE CRITERION OF MAXIMUM REDUCTION

The operator *transtamax* operates basically in the same way as the operator *inftran*. For each chain C, generated by a pair of compatible states, the quantity r, by which the number of states is reduced for transformation on the basis of the given chain, is calculated. This quantity is calculated as in the case of the operator *inftran*. That chain is chosen for the transformation which gives maximal r.

The initial information is the value of the equivalence matrix for outputs and the next-state matrix, which are represented, respectively, by the complexes ζ and β; $\gamma = \omega_2$, $\delta = \omega_7$, $\epsilon = \omega_6$, $\lambda = \omega_4$, $\eta{::}\| C \|$, $\vartheta{::}\| G \|$, $\kappa{::}\| C' \|$; $[\alpha]$ is the number of the sentence in the external program to which control is transferred if a chain does not exist for which the corresponding transformation would reduce the number of states.

transtamax (321, 325)/αч, βк +, γп +, δп +, εп +, ζк +, ηк, ϑк, κк, λп +, /m, *f*/ζηϑκλ)
324 372 13 (ζcfil)

○ *e* ○ **e** ○ **f** $\bar{\circ}$ **k**

§1 $\zeta_e \Rightarrow$ **l** ○ *a*

§2 ○ ϑ_a ○ κ_a △ *a* − b_ζ ○→ 2

§3 **1** $\dot{\mathbf{X}}$ 7*f* ⇒ *de* ⇒ cc_f ⇒ ϑ ∨ c_e ⇒ **il** ⇒ **j** *chest* β(**c** ∧ ζ_b ○→ 6**c** ∨ c_b ∨ **i** ⇒ **i** △ **j**)*dc*κγδεϑ// ○ *a*
i ▽ ⇒ **mj** − **m** |→ 2 2 × **j** − **m** ⇒ **d**
m − **j** ⇒ **h** − **e** ○→ 2 |→ 4**d** − **k** |→ 2

§4 **d** ⇒ **ki** ⇒ **fh** ⇒ **e** ○ *a*

§5 $\vartheta_a \Rightarrow \eta_a$ ○ ϑ_a △ *a* − b_ζ ○→ 5 → 3

§6 c_c ⌉ ∧ $\zeta_d \Rightarrow \zeta_d c_f$ ⌉ ∧ $\zeta_e \Rightarrow \zeta_e$ ○ *a* → 2

§7 △ *e* − b_ζ ○→ 1**f** ○→ α ○ *c*

§10 *funel* 11η*ab*//c_a ∨ c_b ⇒ ϑ_c △ *c* → *funel* 2

§11 **f** ⊕ λ ⇒ **a**

§12 **a** $\dot{\mathbf{X}}$ 13ac_a ⇒ ϑ_c △ *c* → 12

§13 *c* ⇒ b_ϑ.

15. TRANSFORMATION OF THE OUTPUT MATRIX

Let $\{i_0, i_1, \ldots, i_{n-1}\}$ be the set of row numbers of the initial matrix $\| Y \|$ to be merged to obtain the *j*th row of the new output matrix. Then, for any $k \in X$ all defined elements $y_{i_pk} = \lambda(i_p, k)$ $(p \in \{0, 1, \ldots, n-1\})$ are equal to each other and to the elements of the new matrix y'_{jk} corresponding to them. If all y_{i_pk} are undefined, then y'_{jk} is also undefined. Thus, it is possible to obtain the *j*th row of the new matrix as the disjunction of all rows i_p $(p = 0, 1, \ldots, n-1)$. For this purpose the proposed operator uses the principle of the operator *concom*.

The complexes α and β represent the initial output and transformation matrices; [δ] is the number of states in the initial automaton. The result is represented by the complex γ. The operator can be applied both to partially and to completely defined automata.

tramout ακ, βκ +, γκ, δп +/**c**, *e*
301 206 6 (β**a**)

b_α:δ ○→ 1я + 1 × $b_\beta \Rightarrow b_\gamma$ ○ *a* → 2

§1 я × $b_\beta \Rightarrow b_\gamma$ ○ *a*

§2 ○ γ_a △ *a* − b_γ ○→ 2 $\bar{\circ}$ *b* ○ *a*

§3 △ *b* − b_β |→ 6β_b ⇒ **a**

§4 $\mathbf{a}\,\dot{\mathbf{X}}\,3c \circ \mathbf{c} \circ \mathbf{b}$
§5 $\mathbf{c} + c \Rightarrow e\mathbf{b} + b \Rightarrow d - b_\gamma \mid\!\rightarrow 4\alpha_e \vee \gamma_d \Rightarrow \gamma_d\mathbf{b} + b_\beta \Rightarrow \mathbf{bc} + \delta \Rightarrow \mathbf{c} \rightarrow 5$
§6 $\gamma_a \Rightarrow \alpha_a \triangle a - b_\gamma \circ\!\rightarrow 6b_\gamma \Rightarrow b_\alpha.$

16. TRANSFORMATION OF THE NEXT-STATE MATRIX

The single-step transformation used in minimization by the method of successive reductions is realized in the following way.

We denote by the first p natural numbers the states of the automaton A', representing the pairs $\langle i_0, j_0\rangle$, $\langle i_1, j_1\rangle$, $\ldots$, $\langle i_{p-1}, j_{p-1}\rangle$ of the transformation chain C. The remaining states of A' are denoted by $p, p+1, \ldots, p+q$, respectively. The elements of the new next-state matrix $\| Z' \|$ are primed ($'$), the symbols of the elements in the old matrix are left unchanged.

An element $z'_{lk} = \delta'(l, k)$ is given the value m if $l \in \{0, 1, \ldots, p-1\}$ and the pair of values of the elements z_{i_lk} and z_{j_lk} in $\| Z \|$ coincide with the mth pair in C.

If only one of the elements z_{i_lk} or z_{j_lk}—for example, z_{i_lk}—is defined, or $z_{i_lk} = z_{j_lk}$, then the element z'_{lk} may be any $m \in \{0, 1, \ldots, p-1\}$ corresponding to a pair that contains the state $j = z_{i_lk}$. (If there are several such pairs, we call z'_{lk} a nonuniquely defined element.) If the state $j = z_{i_lk}$ does not appear in any of the pairs $\langle i_m, j_m\rangle \in C$, the value of z'_{lk} is uniquely defined by one of the numbers $p, p+1, \ldots, p+q$. Similarly we define the value of z'_{lk} if $l \in \{p, p+1, \ldots, p+q\}$.

The element z'_{lk} remains undefined if $l \in \{0, 1, \ldots, p-1\}$, but both z_{i_lk} and z_{j_lk} are undefined or if l is a state i of the initial matrix not appearing in any of the pairs of the chain, for which z_{ik} is undefined.

For each column of the matrix $\| Z' \|$ we first find all defined and undefined elements. We then complete the definition of the nonuniquely defined elements.

As has been mentioned, to complete the definition of such elements one among several values is taken. This choice has a substantial influence on the possibilities of further reduction of the number of states. In the operator described here the choice of value of the nonuniquely defined elements is the one proposed in [4]. It consists in the following.

Let it be required to complete the definition of the nonuniquely defined element z'_{lk}. Let Z' be the set of states of the automaton A', E_i the set of states of A' equivalent to the states $i \in Z'$ for output, F is the set of possible values of the element z'_{lk}, Q the set of states that represent values of the defined elements of the set $\{z'_{jk}/j \in E_l\}$. Then the value of the element z'_{lk} is chosen in the set $P = \bigcup [\{E_i/i \in Q\}] \cap F$. If $P = \varnothing$, then the value of z'_{lk} is any $j \in F$.

The initial data, aside from the next-state matrix $\| Z \|$, subject to transformation, are the transformation matrix $\| G \|$ and the equivalence matrix for outputs $\| A \|$, constructed for the automaton A'. The auxiliary matrices are $T \| G \|$ (the transposed transformation matrix) and the matrix $\| Z' \circ Z' \|$, where the symbol "$\circ$" denotes a relation such that $i \circ j$ if the states i and j are equivalent for output and $i \neq j$. The last matrix is obtained by diagonal symmetrization of the matrix $\| A \|$; $\alpha\colon\colon \| Z \|$, $\beta\colon\colon \| G \|$, $\gamma\colon\colon T \| G \|$, $\delta\colon\colon \| A \|$, $\mu\colon\colon \| Z' \circ Z' \|$, $\eta = \omega_3$, $\vartheta = \omega_2$, $\kappa = \omega_1$, $\lambda = \omega_5$. The complex ϵ represents as many of the columns of the transformed next-state matrix as the elements of it that can be packed into the 32 bits of an element of the complex. In the course of execution of the operator a column of matrix $\| Z \|$ is extracted and from it is constructed a column of the matrix $\| Z' \|$. These columns are represented by the complexes ζ and ν, respectively. The result is represented by the complex α.

sistramast $(332, 326)/\alpha\text{к}, \beta\text{к}+, \gamma\text{к}, \delta\text{к}+, \epsilon\text{к}, \zeta\text{к}, \eta\text{п}+, \upsilon\text{п}+, \kappa\text{п}+,$
$\lambda\text{п}+, \mu\text{к}, \nu\text{к}/\mathbf{j}, d/(\beta\delta\mu)$
314 570 23 $(\beta\mathbf{cdef})$

transmat $\beta\gamma//$*diasym* $\vee\ \delta\mu//\circ\ a \circ \mathbf{i} \circ \mathbf{j}$

§1	$\circ\ \epsilon_a \triangle a - b_\beta \circ\!\!\rightarrow 1 \circ \mathbf{a}c_{31} \Rightarrow \mathbf{b}$	
§2	$\circ\ a\mathbf{i} \Rightarrow b\mathbf{a} - \eta \mid\!\!\rightarrow 22\vartheta + \mathbf{a} \Rightarrow \mathbf{ab} - \vartheta \Rightarrow \mathbf{b}$	
§3	$\alpha_b \Lleftarrow \mathbf{a} \wedge \kappa \Rightarrow \zeta_a \triangle b \triangle a - b_\gamma \circ\!\!\rightarrow 3 \circ c \circ \mathbf{e}$	
§4	$\beta_c \Rightarrow \mathbf{c}\ \bar{\circ}\ a$	Finding $z_{j_l k}$ and $z_{i_l k}$ for $l \in \{0, 1, \ldots, p-1\}$ and z_{ik} for $i \in \{p, p+1, \ldots, p+q\}$,
§5	$\mathbf{c}\,\dot{\mathbf{X}}\,11d\zeta_d \Rightarrow \mathbf{g} \triangle a \circ\!\!\rightarrow 7$	
	$\mathbf{g} \oplus \mathbf{h} \wedge \lambda \circ\!\!\rightarrow 11\mathbf{g} \wedge \lambda$	
	$\mid\!\!\rightarrow 6\mathbf{h} \Rightarrow \mathbf{g}$	
§6	$\mathbf{g} \oplus \lambda \Rightarrow b\gamma_b \mid\!\!\leftarrow \oplus c_0$	
	$\circ\!\!\rightarrow 10$	
	$c_c \vee \mathbf{e} \Rightarrow \mathbf{e}b \Rightarrow \nu_c \rightarrow 14$	
§7	$\mathbf{g} \Rightarrow \mathbf{h} \rightarrow 5$	
§10	$\gamma_b \vdash\ \vee\ \lambda \Rightarrow \nu_c \rightarrow 14$	
§11	$\circ\ \nu_c\mathbf{g} \wedge \bar{\lambda} \circ\!\!\rightarrow 14\mathbf{g} \oplus \mathbf{h}$	Defining z'_{lk}, if $z_{i_l k}$ and $z_{j_l k}$ are defined and $z_{i_l k} \neq z_{j_l k}$
	$\circ\!\!\rightarrow 6\mathbf{g} \oplus \lambda \Rightarrow a\mathbf{h} \oplus \lambda$	
	$\Rightarrow bc_a \vee c_b \Rightarrow \mathbf{c} \circ a$	
§12	$\beta_a \oplus \mathbf{c} \circ\!\!\rightarrow 13 \triangle a$	
	$- b_\beta \circ\!\!\rightarrow 12$	
§13	$a \vee \lambda \Rightarrow \nu_c$	
§14	$\triangle c - b_\beta \circ\!\!\rightarrow 4 \circ a\mathbf{j}$	
	$\Rightarrow b$	

(§4–§14: Detection of defined, undefined, and nonuniquely defined elements of the new matrix)

§15	$\mathbf{e}\ \dot{\mathbf{X}}\ 21c\mu_c \Rightarrow \mathbf{c}\nu_c \Rightarrow d\gamma_d \Rightarrow \mathbf{d}$	Finding E_l and finding F for z'_{lk}	Completion of definition of nonuniquely defined elements
§16	$\mathbf{c}\ \dot{\mathbf{X}}\ 17b\nu_b \wedge \lambda \circ\!\!\rightarrow 16\nu_b \oplus \lambda \Rightarrow dc_d \vee \mu_d \wedge \mathbf{d} \Rightarrow \mathbf{f} \circ\!\!\rightarrow 16 \rightarrow 20$	Finding P for z'_{lk}	
§17	$\mathbf{d} \Rightarrow \mathbf{f}$		
§20	$\mathbf{f} \vdash \vee \lambda \Rightarrow \nu_c \rightarrow 15$		

§21 $\nu_a \Leftarrow \mathbf{b} \wedge f_{10} \vee \epsilon_a \Rightarrow \epsilon_a \triangle a - b_\beta \circ\!\!\rightarrow 21 \rightarrow 2$
§22 $a + \mathbf{j} \Rightarrow b$
§23 $\epsilon_a \Rightarrow \alpha_b \triangle b \triangle a - b_\beta \circ\!\!\rightarrow 23b \Rightarrow \mathbf{j} \circ ab_\gamma + \mathbf{i} \Rightarrow \mathbf{i} - b_\alpha \circ\!\!\rightarrow 1b \Rightarrow b_\alpha.$

As an example, consider the single-step transformation of the matrix

$\| Z \| =$

z \ x	0	1	2	3	4
0			1		3
1		4	6		1
2				3	
3				2	5
4	5	7			
5	0		6		
6		1		3	
7	6		3	0	

according to the transformation matrix

$$\| G \| = \begin{bmatrix} 0 & 1 & 0 & 0 & 0 & 0 & 0 & 1 \\ 0 & 0 & 0 & 1 & 0 & 0 & 1 & 0 \\ 0 & 0 & 1 & 1 & 0 & 0 & 0 & 0 \\ 1 & 0 & 0 & 0 & 0 & 0 & 0 & 0 \\ 0 & 0 & 0 & 0 & 1 & 0 & 0 & 0 \\ 0 & 0 & 0 & 0 & 0 & 1 & 0 & 0 \end{bmatrix},$$

where the matrix $\| Z' \circ Z' \|$ has the form

$$\| Z' \circ Z' \| = \begin{bmatrix} 0 & 1 & 0 & 0 & 1 & 1 \\ 1 & 0 & 0 & 1 & 0 & 1 \\ 0 & 0 & 0 & 1 & 1 & 0 \\ 0 & 1 & 1 & 0 & 0 & 0 \\ 1 & 0 & 1 & 0 & 0 & 0 \\ 1 & 1 & 0 & 0 & 0 & 0 \end{bmatrix}.$$

In the new next-state matrix one nonuniquely defined element z'_{34} is obtained. The set of its possible values is $F = \{1, 2\}$. The value 1 is taken, and the new next-state matrix takes the form

$\| Z' \| =$

$z \backslash x$	0	1	2	3	4
0	1	4	1	3	0
1		0		2	5
2				2	5
3			0		1
4	5	0			
5	3		1		

If $z'_{34} = 2$ had been taken in the attempt at further state reduction, the states 2 and 3, which are equivalent for outputs, could not be merged, since this would require merging of states 2 and 5 which, as can be seen from the matrix $\| Z' \circ Z' \|$, are not equivalent for outputs. At the same time, if $z'_{34} = 1$, further state reduction is possible on the basis of the chain $\{\langle 2, 3\rangle, \langle 1, 5\rangle\}$.

In minimization by the Paull–Unger method, the next-state matrix is transformed once at the end of the entire minimization process. In this case no problems arise of completing the definitions of nonuniquely defined elements, optimal in the sense of further possibilities of state reduction. We therefore use here the operator *tramast*, which is simpler than the above.

Let there be found a certain closed family L of sets of compatibles S_l $(l = 0, 1, \ldots, \sigma(l) - 1)$. An element of the new matrix z'_{lk} remains undefined if all elements z_{lk} of the initial matrix $\| Z \|$ for all $i \in S_l$ are undefined. If the set of defined values of the elements z_{ik}^{-} for all $i \in S_l$ is Q, then by virtue of the closure of L at least one $S_j \in L$ such that $Q \subseteq S_j$ is always found. Any of the j for which $Q \subseteq S_j$ can be the value of the element z'_{lk}.

The operands have the following sense: $\alpha::\| Z \|$; $\beta::\| L \ni Z \|$; δ is a distinguished column of the matrix $\| Z \|$; ϵ is the number of states of the initial automaton; $\zeta = \omega_1$; $\eta = \omega_2$; $\vartheta = \omega_3$; $\kappa = \omega_5$. The result is represented by the complex γ.

tramast αк $+$, βк, γк, δк, ϵп $+$, ζп $+$, ηп $+$, νп $+$, κп $+$/**f**, c
302 317 13 (β**cd**)

$\circ\ a \circ \mathbf{e} \circ \mathbf{f} b\alpha{:}\epsilon \circ\!\rightarrow 1я + 1 \times b_\beta \Rightarrow b_\gamma \rightarrow 2$
§1 $я \times b_\beta \Rightarrow b_\gamma$
§2 $\circ\ \gamma_a \triangle a - b_\gamma \circ\!\rightarrow 2$
§3 $\circ\ \mathbf{a}c_{31} \Rightarrow \mathbf{b}$

§4	$\circ\ a\mathbf{f} \Rightarrow b\mathbf{a} - \nu \mid\!\!\rightarrow 13\mathbf{b} - \eta \Rightarrow \mathbf{ba} + \eta \Rightarrow \mathbf{a}$	
§5	$\alpha_b \Leftarrow \mathbf{a} \wedge \zeta \Rightarrow \delta_a \bigtriangleup b \bigtriangleup a - \epsilon \circ\!\!\rightarrow 5\ \bar{\circ}\ b$	Extraction of a column of the initial matrix $\|\| Z \|\|$
§6	$\bigtriangleup b - b_\beta \mid\!\!\rightarrow 4\beta_b \Rightarrow \mathbf{c} \circ \mathbf{d}$	Determination of the set Q
§7	$\mathbf{c}\ \dot{\mathbf{X}}\ 10a\delta_a \circ\!\!\rightarrow 7\delta_a \oplus \kappa \Rightarrow ac_a \vee \mathbf{d} \Rightarrow \mathbf{d} \rightarrow 7$	
§10	$\mathbf{d} \circ\!\!\rightarrow 6 \circ a$	Search for a set $s_j \supset Q$
§11	$\beta_a \urcorner \wedge \mathbf{d} \circ\!\!\rightarrow 12 \bigtriangleup a - b_\beta \circ\!\!\rightarrow 11$	
§12	$\mathbf{e} + b \Rightarrow ca \vee \kappa \Leftarrow \mathbf{b} \wedge f_{10} \vee \gamma_c \Rightarrow \gamma_c \rightarrow 6$	
§13	$\epsilon + \mathbf{f} \Rightarrow \mathbf{f}b_\beta + \mathbf{e} \Rightarrow \mathbf{e} - b_\gamma \circ\!\!\rightarrow 3.$	

To illustratc thc functioning of the operator, we present an example. The next-state matrix is

$\|\| Z \|\| =$

$z \backslash x$	0	1	2	3
0			1	0
1	2		3	1
2		0	3	
3	0			3

Let the matrix $\|\| L \ni Z \|\|$ be already found:

$$\|\| L \ni Z \|\| = \begin{bmatrix} 0 & 1 & 0 & 1 \\ & & & \\ 1 & 1 & 1 & 0 \end{bmatrix}$$

Then the new next-state matrix, the result of the transformation, will have the form

$\|\| Z' \|\| =$

$z \backslash x$	0	1	2	3
0	1		0	0
1	1	1	0	1

17. OBTAINING THE NEXT-STATE MATRIX

This operator is used if the function δ is given by the matrices $\|\| W \|\|$ and $\|\| X \|\|$. In merging states it may be found that the new automaton does not satisfy the conditions that permitted the function δ to be defined in this way. Therefore it is better to pass to a universal method of prescribing the δ-function—the next-state matrix.

The initial matrices $\| W \|$ and $\| X \|$ are represented by the complexes α and ϵ, respectively; $\beta::\| L \ni Z \|$. The set Q, implying some set in $S_i \ni L$ is found as in the operator *imsev*. A set $M = \{\bigcup [S_i]/S_i \in L\}$, represented by the complex $\gamma::\| M \ni Z \|$, is defined in sentence 3. The set Q is obtained in sentence 5 as the intersection of the sets M and $\delta(Z, x)$, where

$$\delta(Z, x) = \{\delta(z, x)/z \in Z\} \backslash \{\Lambda\}.$$

The result is represented by the complex δ.

tramav (50)/αк +, βк +, γк, δк, ϵк +/**e**, *e*/($\alpha\beta\gamma\delta\epsilon$)
304 275 12 (α**a**)

$\circ\ ab_\beta \vdash -1 \Rightarrow dc_d \Rightarrow$ **d**40 $- d \Rightarrow$ **e** $\times\ b_\epsilon$:40 $\circ\!\rightarrow$ 1я + 1 $\times\ b_\beta \Rightarrow b_\delta$
$\rightarrow 2$

§1 я $\times\ b_\beta \Rightarrow b_\delta$
§2 $\circ\ \delta_a\ \triangle\ a - b_\delta \rightarrow 2\ \circ\ a$
§3 $\beta_a \Rightarrow$ **a** *concom* $\vee\ \alpha$**a**$(\gamma_a)//\triangle\ a - b_\beta\ \circ\!\rightarrow 3\ \circ\ b\ \circ$ **c**
§4 $d \Rightarrow$ **b** $\circ\ a$
§5 $\gamma_a \wedge \epsilon_b \Rightarrow$ **a** $\circ\!\rightarrow 10\ \circ\ c$
§6 $\beta_c\ \urcorner\ \wedge$ **a** $|\ \circ\!\rightarrow 7\ \triangle\ c - b_\beta\ \circ\!\rightarrow 6$
§7 $a +$ **c** $\Rightarrow ec \vee d <$ **b** $\vee\ \delta_e \Rightarrow \delta_e$
§10 $\triangle\ a - b_\beta\ \circ\!\rightarrow 5\ \triangle\ b - b_\epsilon\ |\!\rightarrow$ 12**b** $-$ **e** $\circ\!\rightarrow 11 \Rightarrow$ **b** $\circ\ a \rightarrow 5$
§11 **c** $+ b_\beta \Rightarrow$ **c** $\rightarrow 4$
§12 .

18. DIAGONAL SYMMETRIZATION OF A BINARY MATRIX

For a given square binary matrix $\| A \|$ a matrix $\| B \|$ is found, each of whose elements is defined in the following way:

$$b_{ij} = a_{ij} \odot a_{ji},$$

where $\odot$ is one of the operators $\vee$, $\wedge$, or $\oplus$. The operands have the following sense: $\alpha = \odot$, $\beta::\| A \|$, $\gamma::\| B \|$.

diasym (332)/αo, βк +, γк/**a**, *b*
326 34 1

transmat $\beta\gamma//\circ\ a$
§1 $\beta_a\alpha\gamma_a \Rightarrow \gamma_a\ \triangle\ a - b_\beta\ \circ\!\rightarrow 1.$

19. TRANSPOSED BINARY MATRIX

For a given binary matrix $\| A \|$, represented by the complex α, a matrix $\| B \| = T \| A \|$ is found, represented by the complex β.

transmat αк +, βк/**a**, b
332 67 4 (α**a**)

$\circ\ a$
§1 $\circ\ \beta_a \triangle a - b_\beta \circ\!\rightarrow 1 \circ a$
§2 $\alpha_a \Rightarrow \mathbf{a}$
§3 $\mathbf{a}\,\dot{\mathbf{X}}\,4bc_a \vee \beta_b \Rightarrow \beta_b \rightarrow 3$
§4 $\triangle a - b_\alpha \circ\!\rightarrow 2.$

REFERENCES

1. Paull, M. C., and Unger, S. H., Minimizing the Number of States in Incompletely Specified Sequential Switching Functions, *IRE Trans. Electron. Computers* **8**, 356–367 (1959).
2. Glushkov, V. M., "The Synthesis of Digital Automata" [in Russian]. Fizmatgiz, Moscow, 1963.
3. Zakrevskii, A. D., The Synthesis of Sequential Automata and Its Programming, *in* "Application of Computers to Production Automation" [in Russian], pp. 525–534. Mashgiz, Moscow, 1961.
4. Zakrevskii, A. D., On the Synthesis of Sequential Automata, *Tr. SFTI* **40**, 89–94 (1961).
5. Butakov, E. A., and Zakrevskii, A. D., Minimization of the Number of States of a Switching Circuit Using the URAL Computer, *in* "Problems of Information Transmission" [in Russian], Vol. 11, pp. 66–76. Izd. Nauka, Moscow, 1962.
6. Caldwell, S., "The Logical Synthesis of Switching Circuits" [Russian translation], p. 515. IL, Moscow, 1963.
7. Pottosin, Yu. V., and Butakov, E. A., State Minimization of Sequential Automata, *Tr. SFTI* **48** (1965).
8. McCluskey, E. J., Jr., Minimum-State Sequential Circuits for a Restricted Class of Incompletely Specified Flow Tables, *BSTJ* **41**, 6 (1962).

SAK-LYaPAS—A SYSTEM OF CODING THEORY ALGORITHMS IN LYaPAS

G. P. Agibalov

1. PURPOSE OF THE SYSTEM SAK-LYaPAS

In the theory of optimal coding of information substantial results have been obtained in recent years in the development of concrete coding methods, suitable for the practical realization of information transmission with high reliability. The development of such methods is being carried on in two directions—probability-theoretical and algebraic. The book by Wozencraft and Reiffen [1] is an example of the former. A systematic exposition of the algebraic theory of error-correction codes was first given in the fundamental book by Peterson [2]. The direct introduction of these attainments into practice presupposes the application of computers able to carry out laborious computations and permitting the solution of coding-theory problems arising in engineering practice to be automated. In this connection it is necessary to develop an automatic programming system intended for the solution of these problems. At the present time the foundations for the development of such a system have been laid—a language for the representation of algorithms. Among the most highly developed languages of automatic programming systems are the universal language ALGOL-60 and the special-purpose logical language LYaPAS. The choice of the latter for the representation of coding-theory algorithms is preferable from the viewpoints of speed and compactness of translation, clarity of the written programs, etc.

We shall call the proposed system of algorithms for the solution of coding-theory problems, based on LYaPAS, and the corresponding programming system, SAK-LYaPAS. This system can be useful not only for the solution of practical problems of information coding and decoding, but for the development of coding theory itself as its experimental basis. For the construction of optimal coding and decoding switching devices this system is of interest because it is used simultaneously with a system of algorithms for the solution of problems in the theory of discrete-automata synthesis.

We shall present some of the simpler algorithms of the SAK-LYaPAS system, concerned essentially with the algebraic theory of error-correcting group codes [2]. They include programs for constructing linear-code decoding tables, Galois-field and polynomial-algebra calculations, solution of a system of linear equations over a finite field, decoding of Bose–Chaudhuri codes by Peterson's algorithm, etc. It is assumed that this system will be improved further, completed with new algorithms, and brought to a level answering the needs of practice.

2. SOME ALGORITHMS FOR THE SOLUTION OF PROBLEMS IN GROUP CODES[1]

(1) Reduction of a matrix with elements in $GF(2)$ to reduced-echelon form.

The initial matrix is represented by complex α, the result is assigned to the same complex.

redechmat αк/**d**, $d/(\alpha)/$
333 307 10 (α**bce**)

$\bar{\circ}\ b\ \circ\ \mathbf{c}b_\alpha \Rightarrow a > 5 + 1 \Rightarrow$ ш$c_a \urcorner + 1$
$\oplus\ c_a \Rightarrow {}''''\mathbf{a}$

§1 $\triangle\ b \oplus b_\alpha \circ\!\!\rightarrow 3\alpha_b \circ\!\!\rightarrow 1 \vdash \Rightarrow ac_a \vee \mathbf{c}$ $\Rightarrow \mathbf{c}\ c_b \oplus {}''''\mathbf{a} \Rightarrow {}''''\mathbf{d}$ — Determination of unit component in row and addition of this row to the other rows containing 1 in the given column

§2 $''''\mathbf{d}\ \dot{\mathbf{X}}\ 1cc_a \wedge \alpha_c \circ\!\!\rightarrow 2\alpha_b \oplus \alpha_c \Rightarrow \alpha_c$ $\mapsto 2c_c + {}''''\mathbf{a} \Rightarrow {}''''\mathbf{a} \rightarrow 2$

§3 $\mathbf{c} \Rightarrow \mathbf{b}\ \bar{\circ}\ c\ \bar{\circ}\ d$

§4 $\mathbf{c}\ \dot{\mathbf{X}}\ 6a\ {}''''\mathbf{a} \Rightarrow {}''''\mathbf{d}$ — Interchange rows

[1] These problems are discussed in detail in [2, Chapter 3].

§5 ''''d $\dot{\mathbf{X}}$ $4bc_a \wedge \alpha_b \circ\!\!\rightarrow 5 \bigtriangleup c\alpha_c \Leftrightarrow \alpha_b \rightarrow 4$

§6 $\bigtriangleup d$**b** $\dot{\mathbf{X}}$ $10a \oplus d \circ\!\!\rightarrow 6c_a \vee c_d \Rightarrow$ **e** $\bar{\circ}\ a$ Interchange columns

§7 $\bigtriangleup a \oplus c \circ\!\!\rightarrow 6\alpha_a \wedge$ **e** $\Rightarrow$ **c** $\circ\!\!\rightarrow 7$**e**

$\oplus$ **c** $\circ\!\!\rightarrow 7$

e $\oplus\ \alpha_a \Rightarrow \alpha_a \rightarrow 7$

§10 $c \Rightarrow b_\alpha$.

(2) Transformation of $k \times n$ matrix in reduced-echelon form $G = [I_k P]$ to the form

$$G' = \begin{bmatrix} P \\ \\ I_{n-k} \end{bmatrix}. \qquad \alpha::G, \qquad [\beta] = n;\ \gamma::G'.$$

shiftmat αк +, βи +, γк/ −, $b/\alpha\gamma/(\alpha)(\gamma)/$
334 53 2

$\circ\ a$

§1 $\alpha_a < b_\alpha \Rightarrow \gamma_a \bigtriangleup a \oplus b_\alpha \mid\!\rightarrow 1 \circ b$

§2 $c_b \Rightarrow \gamma_a \bigtriangleup b \bigtriangleup a \oplus \beta \mid\!\rightarrow 2 \Rightarrow b_\gamma$.

(3) The above algorithms together with *transmat* permit one of the matrices (parity check or generator) of an (n, k) group code to be calculated each time the other is known. This can be done by the following simple program:

redechmat $\alpha//$*shiftmat* $\alpha\beta\gamma//$ *transmat* $\gamma\alpha//$.

(4) Calculation of an (n, k) code from its generator matrix G.

The subroutine can function in three modes, depending on the values given the expression δ, according to the following table:

Value of δ	Characteristic of mode
/	Calculates code words
$\gamma\nabla\Rightarrow\epsilon$	Calculates code words and their weights
$\gamma\nabla\Rightarrow\epsilon - \varsigma \mid\!\rightarrow \alpha\epsilon\Rightarrow\varsigma$	Calculates code words, their weights, and minimum weight of code

To realize this last mode the index ς must be given initially the maximal value 11 $\cdots$1. After all code words have been calculated this index represents the minimal weight of the code.

The subroutine is (2, 2)-terminal, exit over the auxiliary terminal is realized each time the next code word (represented by the variable γ) and its weight are calculated if the latter is calculated in the given mode (represented by the index ϵ). To find the next code word the subroutine should be closed on itself over the auxiliary input terminal sentence 1.

The complex β represents the generator matrix of the code.

cogenmat αч, βк +, γп, δв/**a**, $c/(\beta\gamma)/$
335 56 4

$\circ\ b1 < b_\beta \Rightarrow c$
§1 $\triangle\ b \oplus c \circ\!\!\rightarrow 4b \Rightarrow \mathbf{a} \circ \gamma$
§2 $\mathbf{a}\ \dot{\mathbf{X}}\ 3a\beta_a , \oplus\ \gamma \Rightarrow \gamma \rightarrow 2$
§3 $\delta \rightarrow \alpha$
§4 .

(5) Determination of the vector N in the modular representation of an (n, k) code with generator matrix G.

$\alpha::G$; γ is the complex of values of the components of the vector N; $[\beta] = n$.

modvecode αк +, βи +, γк/**a**, $c/(\alpha)/$
336 107 4

$\bar{\circ}\ b\,1 < b_\alpha \Rightarrow b_\gamma\ \bar{\circ}\ \gamma 40 - b_\alpha \Rightarrow c$
§1 $\triangle\ b \oplus \beta \circ\!\!\rightarrow 4\ \bar{\circ}\ a \circ \mathbf{a}$
§2 $\triangle\ a \oplus b_\alpha \circ\!\!\rightarrow 3c_b \wedge \alpha_a \circ\!\!\rightarrow 2c_a \vee \mathbf{a} \Rightarrow \mathbf{a} \rightarrow 2$
§3 $\mathbf{a} > c \Rightarrow a\ \triangle\ \gamma_a \rightarrow 1$
§4 .

(6) Finding the generator matrix G from the vector N of the modular representation of a code.

Here α is the complex representing N; $\beta::G$, and $[\gamma]$ is the code length which is calculated in sentence 1 of the program.

genmatvec αк +, βк, γи/**b**, $d/$
337 154 5

$\circ\ b \circ \gamma$
§1 $\triangle\ b \oplus b_\alpha \circ\!\!\rightarrow 2\alpha_b + \gamma \Rightarrow \gamma \rightarrow 1$
§2 $\gamma > 5 + 1 \Rightarrow$ ш$b_\alpha \vdash\ \Rightarrow d\,40 - d \Rightarrow b_\beta \circ\ ''''\beta \circ c$
§3 $\bar{\triangle}\ b \circ\!\!\rightarrow 5\alpha_b \circ\!\!\rightarrow 3 - 1 \Rightarrow a$
$c_a \Rightarrow ''''\mathbf{a}\ \urcorner + 1 > c \Rightarrow ''''\mathbf{a}\alpha_b + c \Rightarrow cb < d \Rightarrow \mathbf{b}$
§4 $\mathbf{b}\ \dot{\mathbf{X}}\ 3a\ ''''\mathbf{a} \vee ''''\beta_a \Rightarrow ''''\beta_a \rightarrow 4$
§5 .

(7) Investigation of the vector v for its linear dependence on a given set A of vectors (all vectors have the same dimensions and binary components).

The set of vectors A is considered as a matrix that is first reduced to the reduced-echelon form B, after which it is realized by the following operator:

testlindep αч $+$, βк $+$, γп/**a**, $a/(\beta\gamma)/$
340 47 3 (γ**a**)

$\gamma \Rightarrow$ **a**
§1 **a** $\dot{\mathbf{X}}$ $2a - b_\beta \mid\!\rightarrow 2\beta_a \oplus \gamma \Rightarrow \gamma \rightarrow 1$
§2 $\gamma \mid\!\rightarrow 3 \rightarrow \alpha$
§3 .

Here $\beta::B$, and γ is a variable initially representing the tested vector v; after processing by the program γ is assigned the value of the vector that is the difference between the vector v and a certain vector linearly generated by A. If this value is zero (v depends on A), exit is realized over the auxiliary terminal to an address represented by the operand α.

(8) Calculation of the syndrome s of a vector v from the transposed parity-check matrix H^T of a group code.

Here $\beta::H^T$; the variables α and γ are the vectors v and s, respectively.

syndrome αп, βк $+$, γп/$-$, $a/(\alpha)(\beta\gamma)/$
341 26 2

$\circ\ \gamma$
§1 α $\dot{\mathbf{X}}$ $2a\beta_a \oplus \gamma \Rightarrow \gamma \rightarrow 1$
§2 .

(9) Generating combinations of n taken i at a time for all $i = 1, 2, \ldots, m$.
The program is (2, 2)-terminal, exit from which takes place over the auxiliary terminal to the address $[\alpha]$ when the current combination, represented by the variable δ, has been obtained. Return to the subroutine to calculate the next combination is carried out over the auxiliary input sentence 2. $[\beta] = n$; $[\gamma] = m$.

gencomb αч $+$, βи $+$, γи $+$, δп /**c**, $b/(\delta)/$
342 116 3 (δ**abc**)

$\circ\ a\ \circ\ \delta\ c_\beta < 1 \Rightarrow$ **c**
§1 $a \oplus \gamma \circ\!\!\rightarrow 3 \triangle ac_a \vee \delta \Rightarrow \delta \rightarrow \alpha$
§2 $\delta +$ **c** $\wedge\ \delta \Rightarrow$ **a** $\circ\!\!\rightarrow 1\delta +$ **c** $\oplus\ \delta < 1 +$ **c** $\Rightarrow$ **b**
a $-$ **c** $\oplus$ **a** $\vdash\ \Rightarrow b$**b** $< b \oplus \sigma \Rightarrow \delta \rightarrow \alpha$
§3 .

polynomial $p(x)$ of degree m. The operation of addition over the elements of this field, as in general, over the elements of any polynomial algebra, is equivalent to the operation of addition mod 2 over the LYaPAS operands representing these elements. The operations of multiplication and division have a more complicated character, and are carried out by program.

(5) Multiplication in a Galois field: $a(x) \cdot b(x) = c(x) \bmod p(x)$. For a description of the algorithm, see [2, p. 108].

$\rangle\alpha\langle = a(x)$; $\rangle\beta\langle = b(x)$; $\rangle\gamma\langle = p(x)$; $[\delta]$ is the degree of $p(x)$. The product is assigned to variable α.

multgal αп, βп, γп $+$, δи $+$ /
347 72 3

$$\alpha > 1 \Rightarrow \alpha$$
$$\S 1 \quad \beta \oplus c_0 \circ\!\!\rightarrow 3\beta > 1 \Rightarrow \beta \wedge c_\delta \circ\!\!\rightarrow 2\gamma \oplus \beta \Rightarrow \beta$$
$$\S 2 \quad \alpha < 1 \Rightarrow \alpha \wedge c_0 \circ\!\!\rightarrow 1\gamma \oplus \alpha \Rightarrow \alpha \rightarrow 1$$
$$\S 3 \quad \alpha < 1 \Rightarrow \alpha.$$

(6) An algorithm suitable for the multiplication in any algebra of polynomials modulo a polynomial $p(x)$ is considered in [2, p. 111]; the program realization has the form:

multmod αп $+$, βп $+$, γп $+$, δп, ϵи $+$ /$-$, a /
350 60 3

$$\circ\ \delta\epsilon \Rightarrow a$$
$$\S 1 \quad a \circ\!\!\rightarrow 3\ \overline{\triangle}\, a\, c_a \wedge \beta \circ\!\!\rightarrow 2\alpha \oplus \delta \Rightarrow \delta$$
$$\S 2 \quad \delta > 1 \Rightarrow \delta \wedge c_\epsilon \circ\!\!\rightarrow 1\gamma \oplus \delta \Rightarrow \delta \rightarrow 1$$
$$\S 3 \quad .$$

Here α and β are factors, δ is the result of multiplication, $\rangle\gamma\langle = p(x)$, and $[\epsilon]$ is the degree of $p(x)$.

(7) Division in Galois field: $a(x) : b(x) = c(x) \bmod p(x)$.

$\rangle\alpha\langle = a(x)$; $\rangle\beta\langle = b(x)$; $\rangle\gamma\langle = p(x)$; $[\delta]$ is the degree of $p(x)$. The result of the division is assigned to the variable α.

divgal αп, βп, γп $+$, δи $+$ /
351 60 3

$$\S 1 \quad \beta \oplus c_0 \circ\!\!\rightarrow 2\beta > 1 \Rightarrow \beta \wedge c_\delta \circ\!\!\rightarrow 2\gamma \oplus \beta \Rightarrow \beta$$
$$\S 2 \quad \alpha > 1 \Rightarrow \alpha \wedge c_\delta \circ\!\!\rightarrow 1\gamma \oplus \alpha \Rightarrow \alpha \rightarrow 2$$
$$\S 3 \quad .$$

(8) Frequently, in the solution of problems in the theory of cyclic codes it is necessary to execute the operations of multiplication and division over the elements of a Galois field (in particular, the operation of raising to a power, realized by successive multiplication, is widely used). The computation processes, saturated by these operations (for example, Peterson's algorithm [2] for the decoding of Bose–Chaudhuri codes), have the following essential defect: as a rule they require much machine time. The efficiency of such computational processes can be improved in two ways—minimization of the number of multiplications and divisions, and increase in the speed of these operations. The former reduces to finding a new method of solution of the problem as a whole. Thus, methods are proposed in [3 and 4] for decoding Bose–Chaudhuri double-error correcting codes that contain a smaller number of multiplication and division operations than Peterson's algorithm. Essentially the same goal—reduction of the number of such operations—is sought by Blokh [5], who proposes a method for decoding Bose–Chaudhuri triple-error correcting codes.

A substantial increase in the speed of computation by the second way is possible thanks to a certain well-known method of representing the Galois field [6]. This method is based on the presence in $GF(2^m)$ of polynomials called primitive elements of the field. The first $2^m - 1$ powers of such an element coincide exactly with all $2^m - 1$ nonzero elements of the field. Thus, if α is a certain primitive element of the field $GF(2^m)$, then for every nonzero element $\beta \in GF(2^m)$ there exists a unique power $j \in \{1, 2, \ldots, 2^m - 1\}$, also called the *index* of the element β, such that $\beta = \alpha^j$.[2] This representation of the elements of $GF(2^m)$ permits the substitution of the laborious operations of multiplication, division, and raising to a power of elements of the field by the almost elementary operations of addition, subtraction, and multiplication (respectively), mod $2^m - 1$, on the indexes of these elements. On the other hand, to perform addition over the elements of the field, it is more expedient to give the values of these elements, rather than their indexes. Therefore, to speed up the computation processes it is convenient to store in memory all nonzero elements of the field in order of increasing index. We shall call this the *standard* representation of the field. For small m (in the range of several tens) the standard representation of $GF(2^m)$ is completely feasible, and the advantages connected with it are practicable.

(9) Standard representation of the Galois field $GF(2^m)$.

α is a primitive element of the field; $\rangle\beta\langle = p(x)$ is an irreducible poly-

[2] We consider the index of the zero element of the field $GF(2^m)$ to be arbitrary.

nomial of degree $m = [\gamma]$; δ is the complex representing the field in standard form.

galfield (350) αп +, βп +, γи +, δк/**b**, b /
352 55 2

$1 < \gamma - 1 \Rightarrow b_\delta \circ b\, c_0 \Rightarrow \mathbf{a}$
§1 $\Rightarrow \delta_b \triangle b \oplus b_\delta \circ\!\!\rightarrow 2$ *multmod* $\alpha\mathbf{a}\beta\mathbf{b}\gamma// \mathbf{b} \Rightarrow \mathbf{a} \rightarrow 1$
§2 .

When the generating polynomial $p(x)$ is a primitive polynomial of degree m, and the primitive element of the field is the residue class containing x, the standard representation of the field $GF(2^m)$ can be obtained by the faster algorithm [2, p. 110], executed by the following program:

field βп +, γи +, δк/**a**, a /
353 62 2

$1 < \gamma - 1 \Rightarrow b_\delta \circ a\, c_0 \Rightarrow \mathbf{a}$
§1 $\mathbf{a} \Rightarrow \delta_a \triangle a \oplus b_\delta \circ\!\!\rightarrow 2\mathbf{a} > 1 \Rightarrow \mathbf{a} \wedge c_\gamma \circ\!\!\rightarrow 1\beta \oplus \mathbf{a} \Rightarrow \mathbf{a} \rightarrow 1$
§2 .

For the standard representation of the Galois field $GF(2^m)$ the complex **Z** is especially assigned, putting $\rangle \mathbf{z}_j \langle\; = \alpha^j$, $j = 0, 1, \ldots, 2^m - 2$. The attributes of this field are the irreducible polynomial $p(x)$ and the primitive element α, used for the construction of the field, the degree m and the quantity $2^m - 1$, which we shall always represent by the operands **y**, **z**, y, and z, respectively. This eliminates the need below to repeat the definitions of the operands representing the characteristics of the field. In this connection, in the previous two programs we adopted the notation that $\alpha = \mathbf{z}$, $\beta = \mathbf{y}$, $\delta = \mathbf{Z}$, $\gamma = y$, and $b_\delta = z$.

The following operators, although they are based on the representation of $GF(2^m)$ by means of the primitive polynomial, do not assume that the field is given in memory.

(a) Forward counter in the field $GF(2^m)$.

The sense of the operands is as follows: γ is a variable representing the sequence of states of the counter (the initial value of γ is given); $[\beta]$ is the volume of count; $[\alpha]$ is the address to which the subroutine exits over the auxiliary terminal after passage of the counter to the next state. To continue the count this terminal must be connected to the basic entry of the subroutine.

galcount αч +, βи, γп /
354 36 2

§1 $\beta \circ\!\!\rightarrow 2\overline{\triangle}\beta\gamma > 1 \Rightarrow \gamma \wedge c_y \circ\!\!\rightarrow \alpha\mathbf{y} \oplus \gamma \Rightarrow \gamma \rightarrow \alpha$
§2 .

(b) Reverse counter in the field $GF(2^m)$.

galrevcount αч +, βи, γп /
355 54 3

$\gamma > 1 \Rightarrow \gamma$
§1 $\beta \circ\!\!\rightarrow 3\ \overline{\triangle}\ \beta\gamma < 1 \Rightarrow \gamma \wedge c_0 \circ\!\!\rightarrow 2\mathbf{y} \oplus \gamma \Rightarrow \gamma$
§2 $\gamma < 1 \Rightarrow \gamma \rightarrow \alpha$
§3 .

(c) Calculation of the index of an element in $GF(2^m)$.
Here α is the variable representing the value of the element, $[\beta]$ its index.

index αп, βи /
356 50 2

$\bar{\text{o}}\ \beta$
§1 $\triangle\ \beta\alpha \oplus c_0 \circ\!\!\rightarrow 2\alpha > 1 \Rightarrow \alpha \wedge c_y \circ\!\!\rightarrow 1\mathbf{y} \oplus \alpha \Rightarrow \alpha \rightarrow 1$
§2 $z - \beta \Rightarrow \beta.$

(10) Calculation of the value of an element of Galois field from its index. $[\alpha]$ is the index of the element, β the value of the element.

galelval (354, 355) / αи, βп /
357 43 3

$c_0 \Rightarrow \beta z > 1 - \alpha \circ\!\!\rightarrow 2$
§1 *galcount* $1\alpha\beta$ // $\rightarrow 3$
§2 *galrevcount* $1\alpha\beta$ //
§3 .

(11) Calculation of the value of a polynomial with coefficients in $GF(2)$ on an element of $GF(2^m)$.

$[\alpha]$ is the index of the element; $\rangle\beta\langle$ is the polynomial; γ is the value of the polynomial on the given element.

valpol αи +, βп, γп / −, a /
360 34 2

$\circ\ \gamma$
§1 $\beta\ \dot{\mathbf{X}}\ 2a \times \alpha{:}z \Rightarrow a\mathbf{z}_a \oplus \gamma \Rightarrow \gamma \rightarrow 1$
§2 .

(12) Finding the set K of all roots of an irreducible polynomial of degree k with coefficients in $GF(2)$ from one known root $g \in GF(2^m)$ (cf. Theorem 6.26 in [2]).

$[\alpha]$ is the index of g; $[\beta] = k$; γ is a variable representing K.

fiseroo αи +, βи, γп / −, b / (γ) /
361 50 2

$1 \Rightarrow a\alpha \Rightarrow b\ \circ\ \gamma$
§1 $c_b \vee \gamma \Rightarrow \gamma\beta\ \circ\!\rightarrow 2\ \bar{\triangle}\ \beta a < 1 \Rightarrow a \times \alpha{:}z \Rightarrow b \rightarrow 1$
§2 .

4. THE SOLUTION OF SYSTEMS OF LINEAR EQUATIONS IN THE GALOIS FIELD $GF(2^m)$

The algorithm for the solution of linear equations in the Galois field $GF(2^m)$ is based on the method of successive elimination of variables, which consists in the following. We first consider the first equation of the system. It is multiplied by the field element inverse to the first nonzero coefficient of the equation. Let this coefficient be that for the variable x_i. Every other equation of the system, for example, the jth, is transformed to the sum of this equation with the first equation, multiplied by the coefficient of the variable x_i in the jth equation. The result is that the coefficient of x_i in all equations except the first, where it is equal to one, is reduced to zero. In the next step of the algorithm the calculation is carried out with respect to the second equation of the system and some other variable x_l is eliminated from the remaining equations, etc. At the end, if the solution is unique, a system is obtained in which each equation is trivial, i.e., contains not more than one variable whose coefficient is equal to one. The solution of such a system is found directly.

When during the calculation all variables are eliminated from some equation while its free term remains different from zero, it is concluded that the system is incompatible, and further calculation is broken off.

If r is the number of linearly independent equations in the system and n the number of variables, then by this algorithm the system is transformed to a set of r nondegenerate equations. In each of them there is one essential variable with unit coefficient, not appearing in any other equation, and

$n - r$ free variables, common to all the equations. Substituting arbitrary values from the field for the free variables, we reduce the system to a trivial one with obvious solution for the essential variables. Giving the free variables other values from the field, we obtain other solutions, $2^{m(n-r)}$ in all. Most often in practice it is only required to find any one solution. The program described below finds the solution for which all free variables have been given the value zero.

To increase its speed, the program uses three matrices of the same dimensions, representing the information about the system coefficients: the matrix A is the extended-system matrix [7], its components are the elements of the field $GF(2^m)$; the elements of matrix B are the indexes of the corresponding terms of matrix A; the elements of matrix C are 0 and 1, where $c_{ij} = 1 \leftrightarrow a_{ij} \neq 0$. The representation of these matrices by LYaPAS operands is realized in the following way. The complex α represents the matrix A, where initially the elements of the first row of A are enumerated, then the second, the third, etc. Analogously, the complex β represents the matrix B, $\gamma::C$. Also, $[\delta]$ is the number of equations, $[\epsilon]$ is the number of system variables, ξ is a variable representing the maximal subset of linearly independent equations, the complexes κ and λ represent the solution of the system, κ is the values of the variables (the elements of $GF(2^m)$), and λ the indexes of these elements.

If the rows of the matrix A are interpreted as certain vectors with components in $GF(2^m)$, the program can be used to find in the set of these vectors the maximal linearly independent subset. For this purpose the operation of testing the system for incompatibility must be blocked in the program, assigning the expression η the value of the empty symbol /, and the operation for forming the solution for trivial equations, putting $\vartheta = 10$. When the program is used for solving a system of equations, $\vartheta = 6$, $\eta = (\vdash \oplus \epsilon \circ\!\!\rightarrow \alpha\gamma_g)$, where $[\alpha]$ is the address of auxiliary exit from the subroutine when the system incompatibility has been established. If it is known that the system is compatible, it is possible to put $\eta = /$.

```
      resogal (357) ακ, βκ, γκ, δи +, εи +, ζп,
         ηв, ϑв, κκ, λκ/b, h /
      362 355 10 (ζa) (γb)

      ō bc_δ ⌉ + 1 ⊕ c_δ ⇒ ζε ⇒ b_λ ⇒ b_κ
§1    △ b ⊕ δ ○→ ϑγ_b ○→ 1 ⊢ ⇒ eb × ε
         ⇒ c + e ⇒ dc_b ⊕ ζ ⇒ a
      β_d ○→ 3z − β_d ⇒ dγ_b ⇒ b

§2    b Ẋ 3a + c ⇒ ad + β_a:z ⇒ β_a ⇒ f          Multiply equation by
      z_f ⇒ α_a → 2                               field element inverse to
                                                  coefficient of equation
```

§3 $\mathbf{a}\ \dot{\mathbf{X}}\ 1gc_e \wedge \gamma_g \circ\!\!\rightarrow 3$
$a \times \epsilon \Rightarrow c + e \Rightarrow d\gamma_g \Rightarrow \mathbf{b}$

§4 $\mathbf{b}\ \dot{\mathbf{X}}\ 3a + c \Rightarrow f\beta_d + \beta_a{:}z \Rightarrow h$ — Add equation multiplied by its coefficient to another equation
$\mathbf{z}_f \oplus \alpha_h \Rightarrow \alpha_h \circ\!\!\rightarrow 5 \Rightarrow h$
index $ha//a \Rightarrow \beta_f \rightarrow 4$

§5 $c_a \oplus \gamma_g \Rightarrow \gamma_g\eta \mid\!\!\rightarrow 4c_g \oplus \zeta \Rightarrow \zeta \rightarrow 4$
§6 $\zeta \Rightarrow \mathbf{a}$

§7 $\mathbf{a}\ \dot{\mathbf{X}}\ 10a\gamma_a \Rightarrow \mathbf{b} \vdash \Rightarrow bc_b \oplus \mathbf{b} \circ\!\!\rightarrow 7$ — Solution of trivial equations
$a \times \epsilon + \epsilon \Rightarrow d\alpha_d \Rightarrow \kappa_d\beta_d \Rightarrow \lambda_d \rightarrow 7$
§10 .

When a system of linear equations with coefficients in $GF(2)$ is considered, it is sufficient to define it by the single matrix C and utilize the following simple program.

reslineq γк, δи +, ϵи +, ζп, ηв, ϑв, κп/**a**, $c/(\gamma\eta)$ (ζ) /
363 154 5 $(\zeta\mathbf{a})$

$\bar{o}\ bc_\delta \urcorner + 1 \oplus c_\delta \Rightarrow \zeta$
§1 $\triangle b \oplus \delta \circ\!\!\rightarrow \vartheta\gamma_b \circ\!\!\rightarrow 1 \vdash \Rightarrow cc_b \oplus \zeta \Rightarrow \mathbf{a}$
§2 $\mathbf{a}\ \dot{\mathbf{X}}\ 1ac_c \wedge \gamma_a \circ\!\!\rightarrow 2\gamma_b \oplus \gamma_a \Rightarrow \gamma_a$
$\eta \mid\!\!\rightarrow 2c_a \oplus \zeta \Rightarrow \zeta \rightarrow 2$
§3 $\zeta \Rightarrow \mathbf{a} \circ \kappa$
§4 $\mathbf{a}\ \dot{\mathbf{X}}\ 5a\gamma_a \vdash \Rightarrow bc_b \oplus \gamma_a \circ\!\!\rightarrow 4c_b \vee \kappa \Rightarrow \kappa \rightarrow 4$
§5 .

This program can be used to extract from a set of binary vectors the maximal linearly independent subset. In this mode put $\eta = /$, $\vartheta = 5$. If a system of equations is being solved, then $\vartheta = 3$ and $\eta = (\vdash \oplus \epsilon \circ\!\!\rightarrow \alpha\gamma_a)$, where $[\alpha]$ is the address of the auxiliary exit used when the system is incompatible. If it is known that the system is compatible, it is possible to put $\eta = /$.

5. SOME ALGORITHMS OF THE THEORY OF CYCLIC GROUP CODES

(1) Finding the polynomial $h(x)$ generating a code dual to the code generated by the polynomial $g(x)$.

$\rangle\alpha\langle\ = g(x)$, $[\gamma]$ is the degree of $g(x)$;
$\rangle\beta\langle\ = h(x)$, $[\delta]$ is the degree of $h(x)$.

polyducode (345)/απ +, βπ, γи +, δи, εи/**c**, *b*/(αβ)
364 57 2 (α**bc**)

γ ⇒ ε
§1 > 5 + 1 ⇒ ж c_0 ∨ c_ϵ ⇒ **c** *divpol* **c**αβ**b**εγδ*b*//**b** ○→ 2 △ ε → 1
§2 .

By means of this program it is possible simultaneously to calculate the lengths of the codes generated by the polynomials $g(x)$ and $h(x)$; $n = [\epsilon]$.

(2) Construction of the generator matrix G of the code generated by the polynomial $g(x)$.
⟩α⟨ $= g(x)$; $[\beta]$ is the degree of $g(x)$; γ::G.

genmatpoly (364)/απ +, βи +, γк/**d**, *e*/
365 40 1

polyducode α**d**β*de*//*d* ⇒ b_γ ○ *a*
§1 α > *a* ⇒ γ_a △ *a* ⊕ *d* |→ 1.

(3) Construction of the parity-check matrix H of the code generated by the polynomial $g(x)$.
⟩α⟨ $= g(x)$; $[\beta]$ is the degree of $g(x)$, γ::H.

parmatpoly (364)/απ +, βи +, γк/**d**, *e*/
366 44 1

polyducode α**d**β*de*//**dI** ⇒ **d**β ⇒ b_γ ○ *a*
§1 **d** < *a* ⇒ γ_a △ *a* ⊕ β |→ 1.

(4) One of the methods for prescribing a cyclic group code consists in giving a certain subset $R \equiv \{\alpha_1, \alpha_2, \ldots, \alpha_r\}$ of nonzero elements of the Galois field $GF(2^m)$. The prescribed cyclic code is completely defined by the following proposition: The polynomial $f(x)$ belongs to the code space if and only if the elements of R are its roots. The generating polynomial of this code is

$$g(x) = \text{LCM } (m_1(x), m_2(x), \ldots, m_r(x)),$$

where $m_i(x)$ is the minimal function for α_i. The length of the code is defined as the LCM of the orders of the elements in the set R. The order of the element α^j, where α is a primitive element of the field, is

$$e = (2^m - 1)/\text{GCD } (2^m - 1, j).$$

The matrix H^T, transposed to the parity-check matrix of the code, is equal to

$$\begin{bmatrix} 1 & 1 & \cdots & 1 \\ \alpha_1 & \alpha_2 & \cdots & \alpha_r \\ \alpha_1^2 & \alpha_2^2 & \cdots & \alpha_r^2 \\ \cdot & \cdot & \cdots & \cdot \\ \cdot & \cdot & & \cdot \\ \cdot & \cdot & \cdots & \cdot \\ \alpha_1^{n-1} & \alpha_2^{n-1} & \cdots & \alpha_r^{n-1} \end{bmatrix}$$

Let us adopt the following method for representing the subset R of the set of nonzero elements of $GF(2^m)$ by means of the variable ξ in LYaPAS:

$$\xi^i = 1 \leftrightarrow \alpha^i \subset R,$$

where α is the primitive element used to construct the field. We shall state this fact by the notation

$$\xi::\| \{R\} \ni GF(2^m) \|.$$

(5) Testing the polynomial $f(x)$ for membership in the cyclic code generated by the set of roots R.

The operands are $\beta::\| \{R\} \ni GF(2^m) \|$; $\rangle\gamma\langle = f(x)$; $[\alpha]$ is the address of the auxiliary exit used when the test gives a positive reply.

testpolycode αч $+$, βп, γп/**b**, b /
367 54 3 (γ**a**)

§1 $\beta\ \dot{\mathbf{X}}\ \alpha b \gamma \Rightarrow \mathbf{a} \circ \mathbf{b}$
§2 $\mathbf{a}\ \dot{\mathbf{X}}\ 3a \times b{:}\mathbf{z} \Rightarrow a\mathbf{z}_a \oplus \mathbf{b} \Rightarrow \mathbf{b} \rightarrow 2$
§3 $\mathbf{b} \circ\!\!\rightarrow 1.$

(6) Calculation of the degree of the minimum function for a given field element.

$[\alpha]$ is the index of the element, $[\beta]$ is the required degree.

degminf αи $+$, βи $/-$, a
370 30 1

$\alpha \Rightarrow a \circ \beta$
§1 $\triangle\ \beta 2 \times a{:}z \Rightarrow a \oplus \alpha \mid\!\!\rightarrow 1.$

(7) Finding the minimum function $m(x)$ for a given field element in $GF(2^m)$ (cf. method 2 [2, p. 139]).

[α] is the index of the element; $\rangle\beta\langle = m(x)$; [$\gamma$] is the degree of $m(x)$; δ and ϵ are working complexes; $\sigma(\delta) = [\gamma] + 1$, $\sigma(\epsilon) = m$.

minfgalel (332, 363)/αи +, βп +, γи +, δк, ϵк/**b**, b /
371 72 1

$\gamma \Rightarrow b_\delta\ \bar{\circ}\ a \circ b$
§1 $\triangle\ a\mathbf{z}_b \Rightarrow \delta_a\alpha + b \Rightarrow b{:}z \Rightarrow b$
$a \oplus \gamma \mid\!\rightarrow 1$ *transmat* $\delta\epsilon//$
reslineq $\epsilon\gamma\gamma\mathbf{b}(/)$ (3) $\beta//c_\gamma \vee \beta \Rightarrow \beta$.

(8) Finding the generator polynomial $g(x)$ of the cyclic code, given by the set of roots R.

Here δ and ϵ are working complexes; $\sigma(\delta) = [\gamma] + 1$; $\sigma(\epsilon) = m$; $\rangle\beta\langle = g(x)$; [$\gamma$] is the degree of $g(x)$; $\alpha{::}\|\ \{R\} \ni GF\ (2^m)\ \|$.

genpolyroot (371, 344)/αп, βп, γи, δк, ϵк/**d**, e /
372 112 4

$c_0 \Rightarrow \beta \circ \gamma$
§1 $\alpha\ \dot{\mathbf{X}}\ 4b \Rightarrow c \circ d$
§2 $\triangle\ d2 \times b{:}z \Rightarrow b \oplus c \circ\!\!\rightarrow 3c_b \wedge \alpha \oplus \alpha \Rightarrow \alpha \rightarrow 2$
§3 *minfgalel* $\mathbf{c}cd\delta\epsilon//$*multpol* $\mathbf{c}\beta d\gamma\mathbf{d}e//\mathbf{d} \Rightarrow \beta e \Rightarrow \gamma \rightarrow 1$
§4 .

(9) The greatest common divisor of whole numbers (Euclid's algorithm). [α] and [β] are prescribed numbers, [γ] is their GCD.

gcd αи +, βи +, γи/−, b /
373 35 2

$\alpha \Rightarrow a\beta \Rightarrow \gamma$
§1 $a{:}\gamma \Rightarrow b \circ\!\!\rightarrow 2\gamma \Rightarrow ab \Rightarrow \gamma \rightarrow 1$
§2 .

(10) The lowest common multiple of whole numbers, not exceeding a certain n.

The sense of the operands is as follows: [α] and [β] are prescribed numbers; $\gamma{::}\|\ \{P\} \ni N\ \|$, where $N \equiv \{1, 2, \ldots, n\}$ and P is the subset of prime numbers in the set N; [δ] is the required LCM.

lcm αи, βи, γп +, δи/**a**, c /.
374 106 5 (γ**a**)

$1 \Rightarrow \delta\gamma \Rightarrow \mathbf{a}$
§1 $\mathbf{a}\ \dot{\mathbf{X}}\ 5a \circ b \circ c$

$$
\begin{aligned}
&\S 2 \quad b \mid\!\rightarrow 3\alpha{:}a \Rightarrow b \mid\!\rightarrow 3\text{я} \Rightarrow \alpha\\
&\S 3 \quad c \mid\!\rightarrow 4\beta{:}a \Rightarrow c \mid\!\rightarrow 4\text{я} \Rightarrow \beta\\
&\S 4 \quad b \times c \mid\!\rightarrow 1a \times \delta \Rightarrow \delta \rightarrow 2\\
&\S 5 \quad .
\end{aligned}
$$

(11) Length n of a cyclic code, given by the set of roots R.

Here $\alpha{::}\| \{R\} \ni GF(2^m) \|$; $[\beta] = n$; the variable γ is the subset of prime numbers smaller than 2^m.

codelen (373, 374)/αп +, βи, γп + / **b**, e /
375 70 3 (α**b**)

$$
\begin{aligned}
&\quad 1 \Rightarrow \beta\alpha \Rightarrow \mathbf{b}\\
&\S 1 \quad \mathbf{b}\ \dot{\mathbf{X}}\ 3c\ gcd\ zcd//z{:}d \rightarrow d \oplus z \mid\!\rightarrow 2z \Rightarrow \beta \rightarrow 3\\
&\S 2 \quad lcm\ d\beta\gamma e//e \Rightarrow \beta \rightarrow 1\\
&\S 3 \quad .
\end{aligned}
$$

(12) Reduction of similar elements in a set R in $GF(2^m)$. Here similar elements are elements with the same minimal function.

$\alpha{::}\| \{R\} \ni GF(2^m) \|$.

redsimel αп/**a**, b /
376 54 3 (α**a**)

$$
\begin{aligned}
&\quad \alpha \Rightarrow \mathbf{a}\\
&\S 1 \quad \mathbf{a}\ \dot{\mathbf{X}}\ 3a \Rightarrow b\\
&\S 2 \quad 2 \times b{:}z \Rightarrow b \oplus a \circ\!\!\rightarrow 1c_b \wedge \alpha \oplus \alpha \Rightarrow \alpha \rightarrow 2\\
&\S 3 \quad .
\end{aligned}
$$

(13) Construction of the transposed parity check matrix H^T of a cyclic code given by the set R of roots.

$\alpha{::}\| \{R\} \ni GF(2^m) \|$; $\beta{::}H^T$; the variable γ is the subset of prime numbers smaller than 2^m.

parmatroot (376, 375) αп, βк, γп + / **b**, f /
377 111 3

$$
\begin{aligned}
&\quad redsimel\ \alpha//\ codelen\ \alpha f\gamma//f \Rightarrow b_\beta \circ \beta\ \bar{\circ}\ d - y + 1 \Rightarrow d\\
&\S 1 \quad \alpha\ \mathbf{X}\ 3b \circ c\ \bar{\circ}\ ay + d \Rightarrow d\\
&\S 2 \quad \triangle\ d \oplus b_\beta \circ\!\!\rightarrow 1\mathbf{z}_c > d \vee \beta_a \Rightarrow \beta_a b + c{:}z \Rightarrow c \rightarrow 2\\
&\S 3 \quad .
\end{aligned}
$$

6. PETERSON'S ALGORITHM FOR BOSE–CHAUDHURI DECODING

Let α be a primitive element of the Galois field $GF(2^m)$. The code given by the set of roots $\alpha, \alpha^3, \ldots, \alpha^{2t-1}$ is a Bose–Chaudhuri code, correcting all combinations of errors $k \leq t$. Briefly, the general decoding procedure for such codes, developed by Peterson [2, 6], consists in the following.

(a) From the received polynomial $f(x)$ calculate the symmetric functions $S_j = f(\alpha^j)$ for $j = 1, 3, \ldots, 2t - 1$. If they are all equal to 0, $f(x)$ is a code polynomial (there are no errors). Otherwise, the symmetric functions S_j for even $j = 2, 4, \cdots, 2t$ are calculated constituting certain powers of the functions S_j with odd j: $S_2 = S_1^2$; $S_4 = S_1^4 = S_2^2$; $S_8 = S_1^8 = S_4^2$; $S_{10} = S_5^2$, etc.

(b) Compose the system of Newton's equations with respect to the elementary symmetric functions $\sigma_1, \sigma_2, \ldots, \sigma_k$, $k \leq t$. In matrix form this system has the following form:

$$\begin{bmatrix} 1 & 0 & 0 & 0 & \cdots & 0 \\ s_2 & s_1 & 1 & 0 & \cdots & 0 \\ s_4 & s_3 & s_2 & s_1 & \cdots & 0 \\ \cdot & \cdot & \cdot & \cdot & & \cdot \\ \cdot & \cdot & \cdot & \cdot & & \cdot \\ \cdot & \cdot & \cdot & \cdot & & \cdot \\ s_{2k-4} & s_{2k-5} & s_{2k-6} & s_{2k-7} & \cdots & s_{k-3} \\ s_{2k-2} & s_{2k-3} & s_{2k-4} & s_{2k-5} & \cdots & s_{k-1} \end{bmatrix} \begin{bmatrix} \sigma_1 \\ \sigma_2 \\ \sigma_3 \\ \cdot \\ \cdot \\ \cdot \\ \sigma_{k-1} \\ \sigma_k \end{bmatrix} = \begin{bmatrix} s_1 \\ s_3 \\ s_5 \\ \cdot \\ \cdot \\ \cdot \\ s_{2k-3} \\ s_{2k-1} \end{bmatrix} \qquad (1)$$

(c) It is first assumed that not more than two errors have occurred in transmission; solve the system (1) for $k = 2$.

(d) To test the correctness of the jth symbol in the received vector $f(x)$, calculate the value $\varphi(\alpha^{j-1})$ of the polynomial

$$\varphi(x) = x^k + \sigma_1 x^{k-1} + \sigma_2 x^{k-2} + \cdots + \sigma_k.$$

If $\varphi(\alpha^{j-1}) = 0$, the symbol is erroneous and must be inverted; otherwise this symbol has been correctly transmitted.

(e) The vector thereby corrected is tested for membership in the code, and in the case of a positive reply is given out as the decision of the decoding procedure. Otherwise, it is assumed that not more than four errors have occurred in transmission; solve system (1) for $k = 4$ and carry out the appropriate tests. If the test is unsuccessful, assume that not more than six errors have occurred, etc., until a vector is obtained that satisfies the tests. If the tests for $k = 2, 4, \cdots, t$, if t is even, or for $k = 2, 4, \ldots, t - 1, t$, if t is odd, are unsuccessful, it is concluded that an uncorrectable error has occurred.

This survey of the algorithm enables us to set up autonomous operators; some of them have already been presented in LYaPAS, the remainder are given below. Everywhere below it is assumed that the element α, used to prescribe the code, is also used to represent the Galois field $GF(2^m)$ in standard form.

(1) Calculation of the symmetric functions [see paragraph (a) of the algorithm].

The sense of the operands is as follows: $)\alpha(= f(x)$; β is the complex of symmetric functions; γ is the complex of indexes of these functions as field elements: $[\gamma_j] = i \leftrightarrow s_{j+1} = \alpha^i$, $[\gamma_j]$ (arbitrary) $\leftrightarrow s_{j+1} = 0, j = 0, 1, \ldots, 2t - 1, i = 0, 1, \ldots, 2^m - 2$; ϵ is a variable representing the nonzero symmetric functions: $\epsilon^j = 1 \leftrightarrow s_{j+1} \neq 0$; $[\delta] = 2t$.

syf (357)/αп +, βк, γк, ϵп, δп/**b**, c/
65 146 5 (α **a**)

$\bar{\circ}$ c $\circ$ ϵ
§1 $\triangle\ c \oplus \delta\ \circ\!\!\rightarrow 5\alpha \Rightarrow \mathbf{a} \circ \mathbf{b}$
§2 $\mathbf{a}\ \dot{\mathbf{X}}\ 3a\mathbf{z}_a \oplus \mathbf{b} \Rightarrow \mathbf{b} \rightarrow 2$
§3 $\mathbf{b} \Rightarrow \beta_c\ \circ\!\!\rightarrow 4$ *index* $\mathbf{b}b//b \Rightarrow \gamma_c c_c \vee \epsilon \Rightarrow \epsilon$
§4 $\triangle\ c > 1 \Rightarrow a\beta_a \Rightarrow \beta_c\ \circ\!\!\rightarrow 1\gamma_a \times 2 \Rightarrow \gamma_c \Rightarrow a\mathbf{z}_a \Rightarrow \beta_c \rightarrow 1$
§5 $\delta \Rightarrow b_\beta \Rightarrow b_\gamma$.

(2) Construction of the system of Newton's equations with respect to the elementary symmetric functions.

From known symmetric functions this operator constructs the matrices A, B, C representing for the operator *resogal* the input information about the coefficients of the matrix system (1) for given k.

The sense of the operands is as follows: α is the complex of symmetric functions; β is the complex of indexes of these functions as elements of $GF(2^m)$; $[\gamma] = k$; δ is the complex representing matrix A: initially the elements of the first row are calculated, then the second, etc.; ϵ is a complex that represents in analogous fashion the matrix B; $\zeta::C$.

$$\textit{newteq}\ \alpha\text{к} +, \beta\text{к} +, \gamma\text{и} +, \delta\text{к}, \epsilon\text{к}, \zeta\text{к}/-, e/(\zeta)$$

66 210 7

$\bar{\circ}\ b\ \bar{\circ}\ c\ \bar{\circ}\ d$

§1 $\triangle\ c\ \triangle\ b \oplus \gamma\ \circ\!\rightarrow 7\ \circ\ a2 + d \Rightarrow d \Rightarrow e$

§2 $\bar{\triangle}\ d\ \circ\!\rightarrow 4\alpha_d \Rightarrow \delta_c\ \circ\!\rightarrow 3\beta_d \Rightarrow \epsilon_c c_a \vee \zeta_b \Rightarrow \zeta_b$

§3 $\triangle\ c\ \triangle\ a \oplus \gamma\ |\!\rightarrow 2 \rightarrow 6$

§4 $c_0 \Rightarrow \delta_c\ \circ\ \epsilon_c c_a \vee \zeta_b \Rightarrow \zeta_b\ \triangle\ c$

§5 $\triangle\ \delta_c\ \triangle\ ca \oplus \gamma\ \vdots\!\rightarrow 5$

§6 $\alpha_e \Rightarrow \delta_c\ \circ\!\rightarrow 1\beta_e \Rightarrow \epsilon_c c_a \vee \zeta_b \Rightarrow \zeta_b \rightarrow 1$

§7 $c \Rightarrow b_\delta \Rightarrow b_\epsilon\gamma \Rightarrow b_\zeta.$

(3) Find the set R of all roots of the polynomial

$$f(x) = a_0 + a_1x + a_2x^2 + \cdots + a_{n-1}x^{n-1}$$

over the Galois field $GF(2^m)$.

The sense of the external operands is as follows: α is the complex representing the coefficients of the polynomial: $\alpha^i = a_i$, $i = 0, 1, \ldots, n - 1$; β is a complex that represents analogously the indexes of these coefficients as elements of the field $GF(2^m)$; γ is a variable indicating the nonzero coefficients of the polynomial: $\gamma^i = 1 \leftrightarrow a_i \neq 0$; $\delta::||\ \{R\} \ni GF(2^m)\ ||$.

$$\textit{roogalpol}\ \alpha\text{к} +, \beta\text{к} +, \gamma\text{п} +, \delta\text{п}/\mathbf{b}, c/$$

67 75 4 (γa)

$\bar{\circ}\ b\ \circ\ \delta$

§1 $\triangle\ b \oplus z\ \circ\!\rightarrow 4\gamma \Rightarrow \mathbf{a}\ \circ\ \mathbf{b}$

§2 $\mathbf{a}\ \dot{\mathbf{X}}\ 3a \times b + \beta_a \Rightarrow a\mathbf{z}_a \oplus \mathbf{b} \Rightarrow \mathbf{b} \rightarrow 2$

§3 $\mathbf{b}\ \circ\!\rightarrow 1c_b \vee \delta \Rightarrow \delta \rightarrow 1$

§4 .

(4) The program for Peterson's decoding algorithm for Bose-Chaudhuri codes.

The sense of the operands is as follows: $[\alpha]$ is the address to which exit is effected from the subroutine over the auxiliary terminal if an uncorrectable error is found; $\rangle\beta\langle = f(x)$; $[\gamma] = t$; δ is the variable representing the result of decoding; ϵ, ξ, η, ϑ, κ, λ, μ are working complexes whose maximal cardinalities are $2t$, $2t$, $(t + 1)t$, $(t + 1)t$, t, $t + 1$, $t + 1$, respectively

pedecal (65, 66, 362, 67, 360)/αч +,
βп +, γи +, δп, εк, ζк, ηк, ϑк, κк, λк,
μ_κ/**e**, *j*/(βδ) |
70 275 10 (β**d**)

$2 \times \gamma \Rightarrow i$ *syf* $\beta\epsilon\zeta\mathbf{c}i//\beta \Rightarrow \delta\mathbf{c} \circ\!\!\rightarrow 10 \circ j$

§1 $2 + j \Rightarrow j\gamma - j \mid\!\!\rightarrow 2 \oplus f_{10} \mid\!\!\rightarrow \alpha\ \bar{\triangle}\ j$

§2 *newteq* $\epsilon\zeta j\mathbf{c}\eta\vartheta\kappa//$
resogal $\eta\vartheta\kappa jj\mathbf{d}\,(/)\,(6)\,\lambda\mu//$
$\bar{\circ}\ aj \Rightarrow bc_j \Rightarrow \mathbf{d} \circ \mu_j\mathbf{c}_0 \Rightarrow \lambda_j\ 1 + b$
$\Rightarrow b_\lambda \Rightarrow b_\mu$

§3 $\triangle\ a \oplus b \circ\!\!\rightarrow 5\ \bar{\triangle}\ b\mu_a \Leftrightarrow \mu_b\lambda_a \Leftrightarrow \lambda_b$
$\circ\!\!\rightarrow 4\mathbf{c}_b \vee \mathbf{d} \Rightarrow \mathbf{d}$ — construction of polynomial $\varphi(x)$

§4 $\lambda_a \circ\!\!\rightarrow 3\mathbf{c}_b \vee \mathbf{d} \Rightarrow \mathbf{d} \rightarrow 3$

§5 $\lambda_a \circ\!\!\rightarrow 6\mathbf{c}_a \vee \mathbf{d} \Rightarrow \mathbf{d}$

§6 *roogalpol* $\lambda\mu\mathbf{d}\ \delta//\delta \oplus \beta \Rightarrow \delta \Rightarrow \mathbf{d}\ \bar{\circ}\ b$

§7 $2 + b \Rightarrow b - i \mid\!\!\rightarrow 10$ *valpol* $b\mathbf{de}//\mathbf{e}$
$\circ\!\!\rightarrow 7 \rightarrow 1$

§10 .

REFERENCES

1. Wozencraft, J., and Reiffen, B., "Sequential Decoding" [Russian translation]. IL, Moscow, 1963.
2. Peterson, W., "Error Correcting Codes" [Russian translation]. Izd. Mir, Moscow, 1964. [Translator's note: The page references in the text are to the original American edition, published by M.I.T. Press, Cambridge, Massachusetts, 1961.]
3. Banerji, R. B., A Decoding Procedure for Double-Error Correcting Bose–Ray-Chaudhuri Codes, *Proc. IRE* **49**, 10 (1961).
4. Kolesnik, V. D., The Analysis of Cyclic Group Codes and the Construction of Decoders. Dissertation [in Russian]. LIAP, Leningrad, 1964.
5. Blokh, E. L., A Decoding Method for Triple-Error Correcting Bose–Chaudhuri Codes, *Izv. Akad. Nauk SSSR, Tekhn. Kibern.* No. 3, 30–37 (1964).
6. Peterson, W., Coding and Error Correction for Bose–Chaudhuri Codes, *in* "Cybernetic Collection" [of Russian translations], Vol. 6, pp. 25–54. IL, Moscow, 1963.
7. Kurosh, A. G., "Course of Higher Algebra" [in Russian]. Fizmatgiz, Moscow, 1963.

PROGRAM INDEX

Certain programs do not have subroutine numbers. This primarily concerns parts of the LYaPAS operating system (SYS), the URAL-1 translator (UR1), and the external programs (EXT). However, programs given as examples and certain complete programs not registered as subroutines also have no numbers and are indicated by (—). In the case of *synthrem* it is cited as a subroutine in one program, but has neither number given nor text, as it has been previously published. For this case the corresponding reference has been given. Finally, in preparing the index it appeared that two subroutines had received two different English names, namely, *covsecom* for *closetsom* and *covsecov* for *closetcov*; this situation is also indicated in the index. In all there are 296 subroutines and system blocks listed here.

Name	Program No.	Text page	Brief description
abker	76	278	Absorbed kernels, removal of
abovin	211	237	Elements absorbed or overlapped by interval
absin	25	205	Absorption of intervals
adjin	205	235	Construct possible adjunctions in set of int.
anop	UR1	145	Analysis and processing of operands
ansigrac	15	218	Analyze simply connected graph
ansigrac two	16	219	Analyze simply connected graph, 2-connectivity
approshco	10	191	Approximately shortest coverage
assumab	260	345	Assumability, test for
assval	SYS	85	Assignment of value, construct addresses for
bounds	265	318	Bounds on variables, find
calltp	UR1	136	Call TP
calminco	106	369	Calculate minimal coverage
carcomva	56	45	Cartesian product of complex by variable
carlowlim	54	44	Cartesian product of complexes, lower limit of
carprel	72	289	Cartesian product of complex with its element
cartes	52	43	Cartesian product of complexes
cartred	53	44	Cartesian product of complexes with reduction
caruplim	55	45	Cartesian product of complexes, upper limit of
carvacom	57	45	Cartesian product of variable by complex
chest	321	415	Chain of states
chev	322	416	Chain of states (variant)
choice	246	325	Choice of elements to transfer
cleanup	61	46	Cleanup complex (assignment of zero value)
closetcom	277	425	Closed set of compatibles
closetcov	300	425	Closed set of compatibles (variant)
codelen	375	456	Code length, find
cogenmat	335	442	Code generator matrix
comat	274	418	Compatibility matrix
comav	276	419	Compatibility matrix (variant)

Name	Program No.	Text page	Brief description
comeco	271	410	Code omega constants
comparabilit	233	299	Comparability of Boolean functions
compiler	SYS	101	Compiler L-program
compress	43	41	Compression of a complex
com	102	367	Compress matrix
comp	245	323	Compare to threshold
comtab	162	269	Compress table
concom	50	43	Convolution of complexes with compression
condinf	35	206	Condense information
conjun	135	395	Conjunction element, model of
consinew	200	228	Construct set of intervals not intersecting
const	SYS	74	Constants, sets up table of
const	UR1	144	Constants, construct table of
convol	47	42	Convolution of complexes
coplu	UR1	162	Call OPLU
copoco	202	231	Construct possible coverages
corrector	SYS	62	Corrector of L-program
corrector	UR1	138	Corrector of L-program
cosmif	214	241	Construct set of maximal intervals
covsecom	—	—	Identical to *closetcom*
covsecov	—	—	Identical to *closetcov*
culp	UR1	137	Call LP
cycle	131	397	Clock cycle, model of
cycshift	SYS	82	Cyclic shift operator ←, macroinstruction for
decodeboom	77	246	Decomposability of Boolean function in matrix form
dectabgrouco	343	444	Decoding table for group code
degminf	370	454	Degree of minimal function
deker	160	268	Determine kernel
delay	145	397	Delay element, model of
denonel	165	269	Determine nontagged elements
denonvar	164	269	Detect nonobligatory variables
denum	253	337	Dense enumeration of Boolean functions
deob	75	277	Determine obligatory literals
depinc	201	229	Detect part of interval set not covered
desec	216	240	Delimit set of elements of complex
diasym	326	437	Diagonal symmetrization
differ	46	42	Difference of complexes
disjun	134	395	Disjunction element, model of
dissym	76	245	Disjunctive symmetrization of Boolean func.
divgal	351	446	Divide polynomials in GF
divpol	345	445	Divide polynomials
dnf	142	396	Disjunctive normal form, model of
element	144	396	Element, generalized logical, model of
elimvar	264	318	Eliminate variable

Name	Program No.	Text page	Brief description
enunorg	152	400	Enumerate nodes of graph
equout	272	415	Equivalence with respect to outputs
eretag	163	269	Erase tags
estimate	100	364	Estimate weight
exceed	250	326	Complex exceeds complex element by element
exch	SYS	86	Exchange of values between operands, analysis
expacon	242	310	Expand conjunctive form
expansion	240	335	Expansion of a number in costs
extin	210	236	Extension of an interval
extrank	177	230	Find interval of extremal rank
field	353	448	Galois field, representation of
fiminco	12	186	Find minimal coverage
fiminco′	105	368	The same
fiminel	22	199	Find minimal element
finexli	176	230	Find external set of literals (to set of intervals)
finin	206	235	Find next interval
fiseroo	361	450	Find set of roots
fishco	11	184	Find set of shortest coverages
fishort	14	184	Find set of shortest coverages
fisirco	5	181	Find set of irredundant coverages
flipflop	132	394	Flipflop, model of
funel	325	416	Find unit element
galcount	354	449	Galois field counter
galelval	357	449	Galois field element value
galfield	352	448	Galois field, standard representation of
Galrevcount	355	449	Galois field reverse counter
gcd	373	455	GCD of whole numbers
gcdpol	346	445	GCD of polynomial (Euclid's theorem)
genadj	170	227	General adjunction
gencomb	342	443	Generate combination
genco	212	237	Generate combinations, preassigned number of
gencombfw	7	202	Generate combinations, fixed weight
genmatpoly	365	453	Generator matrix of polynomial
genmatvec	337	442	Generator matrix from vector
genpolyroot	372	455	Generator polynomial from roots
homogeneity	230	296	Homogeneity of a Boolean function, test
impast	317	417	Implication by pair of states
impav	316	417	Implication by pair of states (variant)
imset	330	427	Implied sets of states
imsev	331	428	Implied sets of states (variant)
index	356	449	Index of element in GF
infstared	312	414	Information criterion for state reduction

Name	Program No.	Text page	Brief description
inftran	323	429	Information criterion for state transformation
init	UR1	141	Initialize translator
inmir	174	229	Interval, minimum rank covering set of intervals
input	146	398	Input node, model of
inters	45	42	Intersection of complexes
invert	SYS	82	Invert order operator I, macroinstruction for
inverter	130	393	Inverter, model of
ircov	77	278	Irredundant coverage, obtaining
irfaf	EXT	281	Irredundant factored forms, construction of
joinin	33	206	Join intervals
lcm	374	455	LCM of whole numbers
linst	114	376	List information about structure
lowal	234	301	Linear ordering of weights, construct
lowlim	41	40	Lower limit of a complex
major	137	395	Majority element, model of
mast	244	323	Matrix to string, transform
maxcom	310	421	Maximal compatible
maxhof	266	350	Maximal homogeneous functions
maxrow	—	33	Maximal row of matrix (example)
maxrow	6	191	Maximal row of matrix, find
maxstared	313	414	Maximal state reduction
mifboof	203	231	Minimize interval form of Boolean function
minas	311	413	Minimize asynchronous automaton
mincodefin	213	237	Minimize incomplete Boolean functions
mincol	2	188	Minimal column of matrix, find
mincovin	20	205	Minimal covering interval
minfgalel	371	455	Minimum function of GF element
minpar	74	290	Minimal parentheses
minst	305	412	Minimize number of states
minva	306	413	Minimize number of states, variant
modvecode	336	442	Modular vector of code
motwo	136	395	Sum mod 2 element, model of
multgal	347	446	Multiply polynomials in GF
multmod	350	446	Multiply polynomials mod *a* polynomial
multpol	344	445	Multiply polynomials
natass	SYS	91	Natural assignment operator $\Leftarrow$, macro for
newteq	66	459	Newton's equations, construct
nonsing	SYS	80	Nonsingular phrases, analysis of
op[1] §	UR1	149	Sentence operator
op $\downarrow$	UR1	149	Transfer-to-machine language operator
op •	UR1	150	End of program and stop operator

[1] The letters *op* followed by an operator symbol represent operator synthesis blocks.

Name	Program No.	Text page	Brief description
op $\Leftarrow, \Rightarrow$	UR1	150	Natural-assignment and assignment operators
op $\triangle$	UR1	150	Positive increment-by-one operator
op $\bar{\triangle}$	UR1	150	Negative increment-by-one operator
op $\circ, \bar{\circ}$	UR1	151	Null- and maximum-value assignment operators
op $\Leftrightarrow$	UR1	151	Interchange operator
op $\dot{\mathbf{X}}$	UR1	152	Enumeration-of-ones operator
op $\rightarrow$	UR1	152	Unconditional transfer operator
op $\nrightarrow$, !	UR1	153	Exit and return operators
op $\bigcirc\!\!\rightarrow, \mid\!\rightarrow$	UR1	154	Transfer-on-zero and -on-one operators
op ∇	UR1	155	Weighting operator
op $\mathbf{I}$	UR1	155	Order inversion operator
op $\mid\!\leftarrow$	UR1	155	Normalization operator
op $\vdash$	UR1	156	Leftmost 1 position-locating operator
op $\urcorner$	UR1	156	Componentwise inversion operator
op $\uparrow$	UR1	156	Input operator
op *	UR1	156	Print-one-element operator
op $\Leftarrow$	UR1	157	Unlimited shift operator
op $<$	UR1	157	Left shift operator
op $>$	UR1	157	Right shift operator
op $\leftarrow$	UR1	157	Cyclic shift operator
op $\underline{\vee}$	UR1	157	Reflection operator
op :	UR1	157	Division operator
op $\times$	UR1	158	Multiplication modulo operator
op $+$	UR1	158	Addition modulo operator
op $-$	UR1	158	Subtraction modulo operator
op $\vee$	UR1	159	Disjunction operator
op $\oplus$	UR1	159	Exclusive disjunction operator
optimel	263	317	Optimal elementary transformation
ord cdc	36	198	Order *cdc* (conjunction of disjunctions of elementary conjunctions
orsenode	140	395	Order set of nodes of simulated elements
oshco	4	190	One shortest coverage, find
oswep	215	239	Order set of weights of elements and partition
output	156	403	Output variable of simulated circuit
pack	107	370	Pack matrices
packva	30	200	Pack components of a variable
parcom	150	399	Partition complex
pardi	243	321	Partial ordering of differences
parmatpoly	366	453	Parity-check matrix of polynomial
parmatroot	377	456	Parity-check matrix from roots
parmaxwe	73	290	Partition of maximal weight, finding
parmdnf	EXT	270	Particular minimal dnf
passinvec	24	195	Pass from interval to vector form

Name	Program No.	Text page	Brief description
pdnf	143	396	Perfect dnf, model of
pedecal	70	460	Peterson's decoding algorithm
perseq	153	401	Permissible sequence
polyducode	364	453	Polynomial generating dual code
posifun	232	296	Positive function, reduce to
precheck	SYS	63	Prepares L-program for *checkinf*
prepare	154	401	Preparatory operations for simulation
prilp	UR1	140	Print LP
primp	UR1	161	Print MP
promp	UR1	160	Prepare output of MP
prop	SYS	77	Processor of operands
quine	—	29	Quine simplification (example)
quinesub	—	32	Quine with subroutines (example)
reco	SYS	78	Recognize code
reco	UR1	143	Recognize code
redcoin	171	227	Remove covered intervals
redechmat	333	440	Reduced-echelon form of matrix
redminor	1	188	Reduce minor
redsim	40	40	Reduce similar terms in a complex
redsimel	376	456	Reduce similar elements
redu	252	312	Reduce sets
reduc	71	289	Reduce complex
reflect	SYS	82	Reflection operator $\underline{\vee}$, macroinstruction for
reman	175	230	Remove antikernel from set of intervals
renumbering	235	301	Renumbering of variables
repeat	UR1	161	Repeat translation
rereal	236	302	Renumbernig of variables in realization
reslineq	363	452	Resolve system of linear equations
resogal	362	451	Resolve set of equations in GF
restruc	147	398	Restructure information about circuit models
retab	161	268	Reduction of table
roogalpol	67	459	Roots of Galois polynomial
roots cdc	17	198	Roots of Boolean system in form CDC
selco	UR1	144	Select next code
selco	SYS	73	Select next code
senode	133	394	Set of nodes corresponding to simulated element
senoma	257	343	Separate nodes of a graph and map
sepic	172	228	Separate set of intervals by classes
sequel	26	197	Seek equal element
sestep	113	375	Select next step
setin	21	205	Set of intervals, find
sevinex	207	236	Set of variables for interval extension
shiftmat	334	441	Shift matrix
shothnet	—	346	Shortest threshold element coverage network

Name	Program No.	Text page	Brief description
simpliseq	37	202	Simplify system of equations
sing	SYS	79	Singular phrases, analysis of
sinseq	24	204	Simplify interval form of system of equations
sisle	62	196	Sift systems of logical equations
sistramast	314	433	Single-step transformation of matrix, next state
solsin	255	327	Solve system of inequalities
solsysche	31	200	Solve Boolean system by Cherry's method
solsysgri	32	197	Solve Boolean system by Grigor'yan's method
sosinel	262	319	Solve systems of inequalities by elimination of variables
strugraph	101	365	Structure of graph, determine
swiciran	157	404	Switching circuit analysis
swicirm	155	401	Switching circuit model
syf	65	458	Symmetric functions, calculate
syndrome	341	443	Syndrome, calculate
synmac	115	376	Synthesize majority element circuit
synter	UR1	115	Syntactic error search algorithm
synthesis	100	279	Synthesis of factored form
synthrel	256	328	Synthesize threshold element
synthrem	—	—	Threshold element synthesis from matrix[2]
synth	SYS	84	Synthesize object program subsequence
synth	UR1	159	Synthesis of object program instructions
tape	UR1	162	Store MP on magnetic tape
taval	103	368	Table of values
tecosin	173	229	Test polynomial for membership in code
testidendisc	13	212	Test identity to unity of normal form
testlindep	340	443	Test for linear dependence
testpolycode	367	454	Test polynomial for membership in code
threlan	240	307	Threshold element analysis
threlans	241	309	Threshold element analysis, symmetrical
threshold	141	396	Threshold element, model of
tomof	261	346	Totally monotonic function in disjunctive form
tracom	247	325	Transfer elements of complex
tramast	302	435	Transform matrix, next state
tramav	304	437	Transform matrix, next state (variant)
tramout	301	431	Transform matrix, output
transclo	327	428	Transitive closure set
transfer	60	46	Transfer value of complex
transform	112	373	Transform input complex to matrix
transformat	111	373	Transform matrix
transmat	332	438	Transpose of matrix
transmin	110	372	Transpose minor

[2] From the work of Bykova and Butakov [27], p. 351, this volume.

Name	Program No.	Text page	Brief description
trans	SYS	89	Transfer instructions, construct
transtamax	324	431	Transformation of state matrix, maximal
traw	104	368	Transform by weights
tree	3	189	Tree search
twomontest	237	302	Two-monotonicity test
union	44	41	Union of complexes
unpack	27	200	Unpack components of a variable
uplim	42	41	Upper limit of a complex
valpol	360	450	Value of polynomial
varep	151	399	Variables, representative
vectin	23	203	Vector form of interval, find
weight	SYS	82	Weighting operator ∇, macroinstruction for
weights	251	324	Weights, calculate
write	UR1	159	Write machine instructions in core

GENERAL INDEX

D